Genetic Engineering & GMO's

Edited by Paul F. Kisak

Contents

Chapter 1

Genetic engineering

For a non-technical introduction to the topic, see Introduction to genetics. For the song by Orchestral Manoeuvres in the Dark, see Genetic Engineering (song).

Genetic engineering, also called **genetic modification**, is the direct manipulation of an organism's genome using biotechnology. It is a set of technologies used to change the genetic makeup of cells, including the transfer of genes within and across species boundaries to produce improved or novel organisms. New DNA may be inserted in the host genome by first isolating and copying the genetic material of interest using molecular cloning methods to generate a DNA sequence, or by synthesizing the DNA, and then inserting this construct into the host organism. Genes may be removed, or "knocked out", using a nuclease. Gene targeting is a different technique that uses homologous recombination to change an endogenous gene, and can be used to delete a gene, remove exons, add a gene, or introduce point mutations.

An organism that is generated through genetic engineering is considered to be a genetically modified organism (GMO). The first GMOs were bacteria generated in 1973 and GM mice in 1974. Insulin-producing bacteria were commercialized in 1982 and genetically modified food has been sold since 1994. GloFish, the first GMO designed as a pet, was first sold in the United States in December 2003.[1]

Genetic engineering techniques have been applied in numerous fields including research, agriculture, industrial biotechnology, and medicine. Enzymes used in laundry detergent and medicines such as insulin and human growth hormone are now manufactured in GM cells, experimental GM cell lines and GM animals such as mice or zebrafish are being used for research purposes, and genetically modified crops have been commercialized.

IUPAC definition

Process of inserting new genetic information into existing cells in order to

modify a specific organism for the purpose of changing its characteristics.

Note: Adapted from ref.[2][3]

1.1 Definition

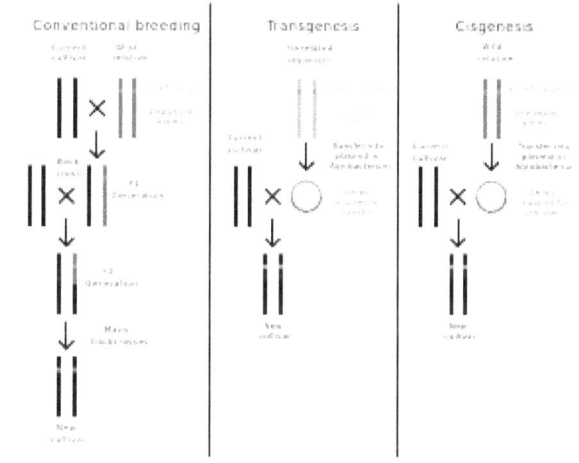

Comparison of conventional plant breeding with transgenic and cisgenic genetic modification.

Genetic engineering alters the genetic make-up of an organism using techniques that remove heritable material or that introduce DNA prepared outside the organism either directly into the host or into a cell that is then fused or hybridized with the host.[4] This involves using recombinant nucleic acid (DNA or RNA) techniques to form new combinations of heritable genetic material followed by the incorporation of that material either indirectly through a vector system or directly through micro-injection, macro-injection and micro-encapsulation techniques.

Genetic engineering does not normally include traditional animal and plant breeding, in vitro fertilisation, induction of polyploidy, mutagenesis and cell fusion techniques that do not use recombinant nucleic acids or a genetically modified organism in the process.[4] However the European Commission has also defined genetic engineering broadly as including selective breeding and other means of artificial selection.[5] Cloning and stem cell research, although not considered genetic engineering,[6] are closely related and genetic engineering can be used within them.[7] Synthetic biology is an emerging discipline that takes genetic engineering a step further by introducing artificially synthesized material from raw materials into an organism.[8]

If genetic material from another species is added to the host, the resulting organism is called transgenic. If genetic material from the same species or a species that can naturally breed with the host is used the resulting organism is called cisgenic.[9] Genetic engineering can also be used to remove genetic material from the target organism, creating a gene knockout organism.[10] In Europe genetic modification is synonymous with genetic engineering while within the United States of America it can also refer to conventional breeding methods.[11][12] The Canadian regulatory system is based on whether a product has novel features regardless of method of origin. In other words, a product is regulated as genetically modified if it carries some trait not previously found in the species whether it was generated using traditional breeding methods (e.g., selective breeding, cell fusion, mutation breeding) or genetic engineering.[13][14][15] Within the scientific community, the term *genetic engineering* is not commonly used; more specific terms such as *transgenic* are preferred.

1.2 Genetically modified organisms

Main article: Genetically modified organism

Plants, animals or micro organisms that have changed through genetic engineering are termed genetically modified organisms or GMOs.[16] Bacteria were the first organisms to be genetically modified. Plasmid DNA containing new genes can be inserted into the bacterial cell and the bacteria will then express those genes. These genes can code for medicines or enzymes that process food and other substrates.[17][18] Plants have been modified for insect protection, herbicide resistance, virus resistance, enhanced nutrition, tolerance to environmental pressures and the production of edible vaccines.[19] Most commercialised GMO's are insect resistant and/or herbicide tolerant crop plants.[20] Genetically modified animals have been used for research, model animals and the production of agricultural or pharmaceutical products.

The genetically modified animals include animals with genes knocked out, increased susceptibility to disease, hormones for extra growth and the ability to express proteins in their milk.[21]

1.3 History

Main article: History of genetic engineering

Humans have altered the genomes of species for thousands of years through selective breeding, or artificial selection[22]:1[23]:1 as contrasted with natural selection, and more recently through mutagenesis. Genetic engineering as the direct manipulation of DNA by humans outside breeding and mutations has only existed since the 1970s. The term "genetic engineering" was first coined by Jack Williamson in his science fiction novel *Dragon's Island*, published in 1951[24] – one year before DNA's role in heredity was confirmed by Alfred Hershey and Martha Chase,[25] and two years before James Watson and Francis Crick showed that the DNA molecule has a double-helix structure – though the general concept of direct genetic manipulation was explored in rudimentary form in Stanley G. Weinbaum's 1936 science fiction story *Proteus Island*.[26][27]

In 1974 Rudolf Jaenisch created the first GM animal.

In 1972, Paul Berg created the first recombinant DNA molecules by combining DNA from the monkey virus SV40 with that of the lambda virus.[28] In 1973 Herbert Boyer and Stanley Cohen created the first transgenic organism by inserting antibiotic resistance genes into the plasmid of an *E. coli* bacterium.[29][30] A year later Rudolf Jaenisch created a transgenic mouse by introducing foreign DNA into its embryo, making it the world's first transgenic animal.[31] These achievements led to concerns in the scientific community about potential risks from genetic engineering, which were first discussed in depth at the Asilomar Conference in 1975. One of the main recommendations from this meeting was that government oversight of recombinant DNA research should be established until the technology was deemed safe.[32][33]

In 1976 Genentech, the first genetic engineering company, was founded by Herbert Boyer and Robert Swanson and a year later the company produced a human protein (somatostatin) in *E.coli*. Genentech announced the production of genetically engineered human insulin in 1978.[34] In 1980, the U.S. Supreme Court in the *Diamond v. Chakrabarty* case ruled that genetically altered life could be patented.[35] The insulin produced by bacteria, branded humulin, was approved for release by the Food and Drug Administration in 1982.[36]

In the 1970s graduate student Steven Lindow of the University of Wisconsin–Madison with D.C. Arny and C. Upper found a bacterium he identified as *P. syringae* that played a role in ice nucleation, and in 1977 he discovered a mutant ice-minus strain. Dr. Lindow (who is now a plant pathologist at the University of California-Berkeley) later successfully created a recombinant ice-minus strain.[37] In 1983, a biotech company, Advanced Genetic Sciences (AGS) applied for U.S. government authorization to perform field tests with the ice-minus strain of *P. syringae* to protect crops from frost, but environmental groups and protestors delayed the field tests for four years with legal challenges.[38] In 1987, the ice-minus strain of *P. syringae* became the first genetically modified organism (GMO) to be released into the environment[39] when a strawberry field and a potato field in California were sprayed with it.[40] Both test fields were attacked by activist groups the night before the tests occurred: "The world's first trial site attracted the world's first field trasher".[39]

The first field trials of genetically engineered plants occurred in France and the USA in 1986, tobacco plants were engineered to be resistant to herbicides.[41] The People's Republic of China was the first country to commercialize transgenic plants, introducing a virus-resistant tobacco in 1992.[42] In 1994 Calgene attained approval to commercially release the Flavr Savr tomato, a tomato engineered to have a longer shelf life.[43] In 1994, the European Union approved tobacco engineered to be resistant to the herbicide bromoxynil, making it the first genetically engineered crop commercialized in Europe.[44] In 1995, Bt Potato was approved safe by the Environmental Protection Agency, after having been approved by the FDA, making it the first pesticide producing crop to be approved in the USA.[45] In 2009 11 transgenic crops were grown commercially in 25 countries, the largest of which by area grown were the USA, Brazil, Argentina, India, Canada, China, Paraguay and South Africa.[46]

In 2010, scientists at the J. Craig Venter Institute created the first synthetic life form by adding a synthetic genome to an empty bacterial cell. The resulting bacterium was named Synthia.[47][48] In 2014, a bacterium was developed that replicated a plasmid containing a unique base pair, creating the first organism engineered to use an expanded genetic alphabet.[49][50]

1.4 Process

Main article: Genetic engineering techniques

The first step is to choose and isolate the gene that will be inserted into the genetically modified organism. As of 2012, most commercialised GM plants have genes transferred into them that provide protection against insects or tolerance to herbicides.[51] The gene can be isolated using restriction enzymes to cut DNA into fragments and gel electrophoresis to separate them out according to length.[52] Polymerase chain reaction (PCR) can also be used to amplify up a gene segment, which can then be isolated through gel electrophoresis.[53] If the chosen gene or the donor organism's genome has been well studied it may be present in a genetic library. If the DNA sequence is known, but no copies of the gene are available, it can be artificially synthesized.[54]

The gene to be inserted into the genetically modified organism must be combined with other genetic elements in order for it to work properly. The gene can also be modified at this stage for better expression or effectiveness. As well as the gene to be inserted most constructs contain a promoter and terminator region as well as a selectable marker gene. The promoter region initiates transcription of the gene and can be used to control the location and level of gene expression, while the terminator region ends transcription. The selectable marker, which in most cases confers antibiotic resistance to the organism it is expressed in, is needed to determine which cells are transformed with the new gene. The constructs are made using recombinant DNA techniques, such as restriction digests, ligations and molecular cloning.[55] The manipulation of the DNA generally occurs within a plasmid.

The most common form of genetic engineering involves inserting new genetic material randomly within the host genome. Other techniques allow new genetic material to be inserted at a specific location in the host genome or generate mutations at desired genomic loci capable of knocking out endogenous genes. The technique of gene targeting uses homologous recombination to target desired changes to a specific endogenous gene. This tends to occur at a relatively low frequency in plants and animals and generally requires the use of selectable markers. The frequency of gene targeting can be greatly enhanced with the use of engineered nucleases such as zinc finger nucleases,[56][57] engineered homing endonucleases,[58][59] or nucleases created from TAL effectors.[60][61]

In addition to enhancing gene targeting, engineered nucleases can also be used to introduce mutations at endogenous genes that generate a gene knockout.[62][63]

1.4.1 Transformation

Main article: Transformation (genetics)
Only about 1% of bacteria are naturally capable of taking

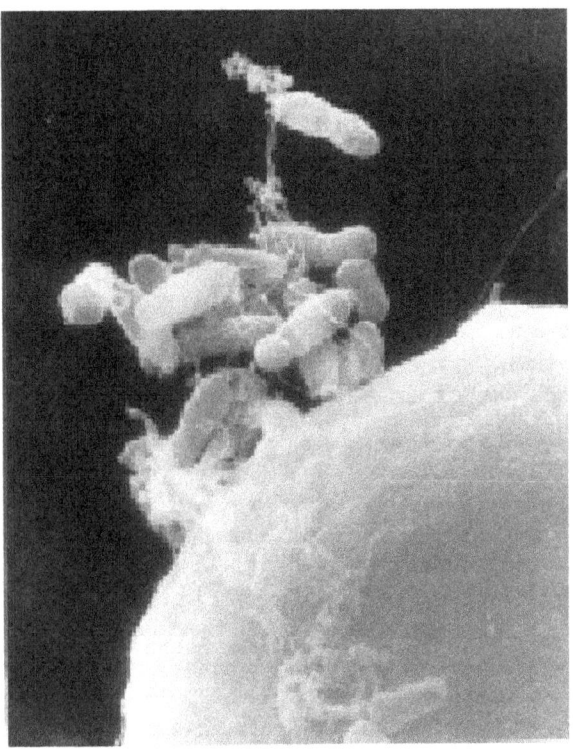

A. tumefaciens *attaching itself to a carrot cell*

up foreign DNA. However, this ability can be induced in other bacteria via stress (e.g. thermal or electric shock), thereby increasing the cell membrane's permeability to DNA; up-taken DNA can either integrate with the genome or exist as extrachromosomal DNA. DNA is generally inserted into animal cells using microinjection, where it can be injected through the cell's nuclear envelope directly into the nucleus or through the use of viral vectors.[64] In plants the DNA is generally inserted using *Agrobacterium*-mediated recombination or biolistics.[65]

In *Agrobacterium*-mediated recombination, the plasmid construct contains T-DNA, DNA which is responsible for insertion of the DNA into the host plants genome. This plasmid is transformed into *Agrobacterium* containing no plasmids prior to infecting the plant cells. The *Agrobacterium* will then naturally insert the genetic material into the plant cells.[66] In biolistics transformation particles of gold or tungsten are coated with DNA and then shot into young plant cells or plant embryos. Some genetic material will enter the cells and transform them. This method can be used on plants that are not susceptible to *Agrobacterium* infection and also allows transformation of plant plastids. Another transformation method for plant and animal cells is electroporation. Electroporation involves subjecting the plant or animal cell to an electric shock, which can make the cell membrane permeable to plasmid DNA. In some cases the electroporated cells will incorporate the DNA into their genome. Due to the damage caused to the cells and DNA the transformation efficiency of biolistics and electroporation is lower than agrobacterial mediated transformation and microinjection.[67]

As often only a single cell is transformed with genetic material the organism must be regenerated from that single cell. As bacteria consist of a single cell and reproduce clonally regeneration is not necessary. In plants this is accomplished through the use of tissue culture. Each plant species has different requirements for successful regeneration through tissue culture. If successful an adult plant is produced that contains the transgene in every cell. In animals it is necessary to ensure that the inserted DNA is present in the embryonic stem cells. Selectable markers are used to easily differentiate transformed from untransformed cells. These markers are usually present in the transgenic organism, although a number of strategies have been developed that can remove the selectable marker from the mature transgenic plant.[68] When the offspring is produced they can be screened for the presence of the gene. All offspring from the first generation will be heterozygous for the inserted gene and must be mated together to produce a homozygous animal.

Further testing uses PCR, Southern hybridization, and DNA sequencing is conducted to confirm that an organism contains the new gene. These tests can also confirm the chromosomal location and copy number of the inserted gene. The presence of the gene does not guarantee it will be expressed at appropriate levels in the target tissue so methods that look for and measure the gene products (RNA and protein) are also used. These include northern hybridization, quantitative RT-PCR, Western blot, immunofluorescence, ELISA and phenotypic analysis. For stable transformation the gene should be passed to the offspring in a Mendelian inheritance pattern, so the organism's offspring are also studied.

1.4.2 Genome editing

Main article: Genome editing

Genome editing is a type of genetic engineering in which DNA is inserted, replaced, or removed from a genome using artificially engineered nucleases, or "molecular scissors." The nucleases create specific double-stranded breaks (DSBs) at desired locations in the genome, and harness the cell's endogenous mechanisms to repair the induced break by natural processes of homologous recombination (HR) and nonhomologous end-joining (NHEJ). There are currently four families of engineered nucleases: meganucleases, zinc finger nucleases (ZFNs), transcription activator-like effector nucleases (TALENs), and the Cas9-guideRNA system (adapted from the CRISPR prokarotic immune system).[69][70] In contrast to artificial genome editing natural genome editing occurs through viral and subviral agents competent in identification of genetic syntax structures for insertion/deletion processes with the result of conserved selection processes.[71]

1.5 Applications

Genetic engineering has applications in medicine, research, industry and agriculture and can be used on a wide range of plants, animals and micro organisms.

1.5.1 Medicine

In medicine, genetic engineering has been used in manufacturing drugs, to create model animals and do laboratory research, and in gene therapy.

Manufacturing

Main article: Industrial fermentation

Genetic engineering is used to mass-produce insulin, human growth hormones, follistim (for treating infertility), human albumin, monoclonal antibodies, antihemophilic factors, vaccines and many other drugs.[72][73] Mouse hybridomas, cells fused together to create monoclonal antibodies, have been humanised through genetic engineering to create human monoclonal antibodies.[74] Genetically engineered viruses are being developed that can still confer immunity, but lack the infectious sequences.[75]

Research

Main article: Genetically modified organism

Genetic engineering is used to create animal models of human diseases. Genetically modified mice are the most common genetically engineered animal model.[76] They have been used to study and model cancer (the oncomouse), obesity, heart disease, diabetes, arthritis, substance abuse, anxiety, aging and Parkinson disease.[77] Potential cures can be tested against these mouse models. Also genetically modified pigs have been bred with the aim of increasing the success of pig to human organ transplantation.[78]

Gene therapy

Main article: Gene therapy

Gene therapy is the genetic engineering of humans, generally by replacing defective genes with effective ones. This can occur in somatic tissue or germline tissue.

Somatic gene therapy has been studied in clinical research in several diseases, including X-linked SCID,[79] chronic lymphocytic leukemia (CLL),[80] and Parkinson's disease.[81] In 2012, Glybera became the first gene therapy treatment to be approved for clinical use in either Europe or the United States after its endorsement by the European Commission.[82][83]

With regard to germline gene therapy, the scientific community has been opposed to attempts to alter genes in humans in inheritable ways using biotechnology since the technology was first introduced,[84] and the caution has continued as the technology has progressed.[85] With the advent of new techniques like CRISPR, in March 2015 scientists urged a worldwide ban on clinical use of gene editing technologies to edit the human genome in a way that can be inherited.[86][87][88][89] In April 2015, Chinese researchers sparked controversy when they reported results of basic research experiments in which they edited the DNA of non-viable human embryos using CRISPR.[90][91] In December 2015, scientists of major world academies called for a moratorium on inheritable human genome edits, including those related to CRISPR-Cas9 technologies.[92]

There are also ethical concerns should the technology be used not just for treatment, but for enhancement, modification or alteration of a human beings' appearance, adaptability, intelligence, character or behavior.[93] The distinction between cure and enhancement can also be difficult to establish.[94] Transhumanists consider the enhancement of humans desirable.

1.5.2   Research

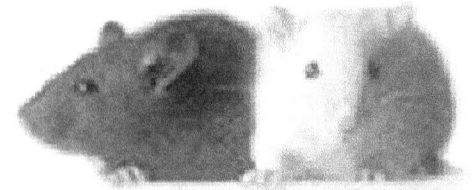

Knockout mice

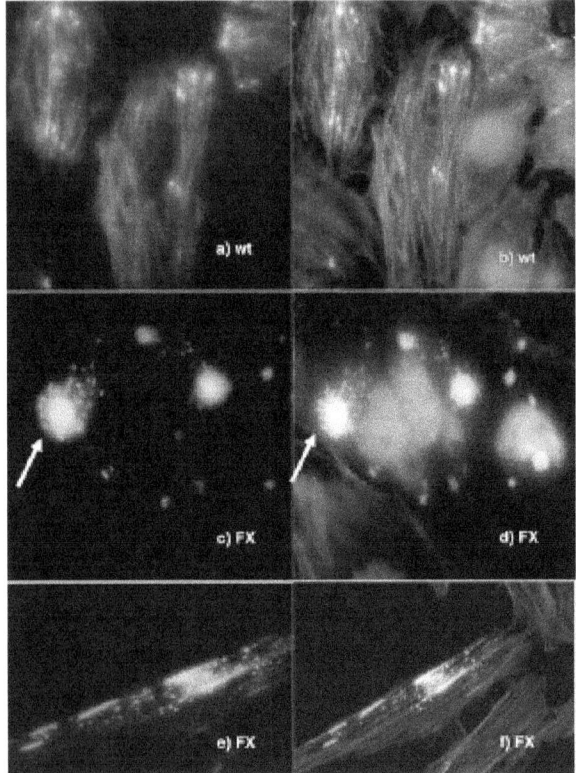

Human cells in which some proteins are fused with green fluorescent protein to allow them to be visualised

Genetic engineering is an important tool for natural scientists. Genes and other genetic information from a wide range of organisms are transformed into bacteria for storage and modification, creating genetically modified bacteria in the process. Bacteria are cheap, easy to grow, clonal, multiply quickly, relatively easy to transform and can be stored at −80 °C almost indefinitely. Once a gene is isolated it can be stored inside the bacteria providing an unlimited supply for research.

Organisms are genetically engineered to discover the functions of certain genes. This could be the effect on the phe-

notype of the organism, where the gene is expressed or what other genes it interacts with. These experiments generally involve loss of function, gain of function, tracking and expression.

- **Loss of function experiments**, such as in a gene knockout experiment, in which an organism is engineered to lack the activity of one or more genes. A knockout experiment involves the creation and manipulation of a DNA construct *in vitro*, which, in a simple knockout, consists of a copy of the desired gene, which has been altered such that it is non-functional. Embryonic stem cells incorporate the altered gene, which replaces the already present functional copy. These stem cells are injected into blastocysts, which are implanted into surrogate mothers. This allows the experimenter to analyze the defects caused by this mutation and thereby determine the role of particular genes. It is used especially frequently in developmental biology. Another method, useful in organisms such as Drosophila (fruit fly), is to induce mutations in a large population and then screen the progeny for the desired mutation. A similar process can be used in both plants and prokaryotes. Loss of function tells whether or not a protein is required for a function, but it does not always mean it's sufficient, especially if a function requires multiple proteins and lose the said function if one protein is missing.

- **Gain of function experiments**, the logical counterpart of knockouts. These are sometimes performed in conjunction with knockout experiments to more finely establish the function of the desired gene. The process is much the same as that in knockout engineering, except that the construct is designed to increase the function of the gene, usually by providing extra copies of the gene or inducing synthesis of the protein more frequently. Gain of function is used to tell whether or not a protein is sufficient for a function, but it does not always mean it's required. Especially when dealing with genetic/functional redundancy.

- **Tracking experiments**, which seek to gain information about the localization and interaction of the desired protein. One way to do this is to replace the wild-type gene with a 'fusion' gene, which is a juxtaposition of the wild-type gene with a reporting element such as green fluorescent protein (GFP) that will allow easy visualization of the products of the genetic modification. While this is a useful technique, the manipulation can destroy the function of the gene, creating secondary effects and possibly calling into question the results of the experiment. More sophisticated techniques are now in development that can track protein products without mitigating their function, such as the addition of small sequences that will serve as binding motifs to monoclonal antibodies.

- **Expression studies** aim to discover where and when specific proteins are produced. In these experiments, the DNA sequence before the DNA that codes for a protein, known as a gene's promoter, is reintroduced into an organism with the protein coding region replaced by a reporter gene such as GFP or an enzyme that catalyzes the production of a dye. Thus the time and place where a particular protein is produced can be observed. Expression studies can be taken a step further by altering the promoter to find which pieces are crucial for the proper expression of the gene and are actually bound by transcription factor proteins; this process is known as promoter bashing.

1.5.3 Industrial

Using genetic engineering techniques one can transform microorganisms such as bacteria or yeast, or transform cells from multicellular organisms such as insects or mammals, with a gene coding for a useful protein, such as an enzyme, so that the transformed organism will overexpress the desired protein. One can manufacture mass quantities of the protein by growing the transformed organism in bioreactor equipment using techniques of industrial fermentation, and then purifying the protein.[95] Some genes do not work well in bacteria, so yeast, insect cells, or mammalians cells, each a eukaryote, can also be used.[96] These techniques are used to produce medicines such as insulin, human growth hormone, and vaccines, supplements such as tryptophan, aid in the production of food (chymosin in cheese making) and fuels.[97] Other applications involving genetically engineered bacteria being investigated involve making the bacteria perform tasks outside their natural cycle, such as making biofuels,[98] cleaning up oil spills, carbon and other toxic waste[99] and detecting arsenic in drinking water.[100] Certain genetically modified microbes can also be used in biomining and bioremediation, due to their ability to extract heavy metals from their environment and incorporate them into compounds that are more easily recoverable.[101]

Experimental, lab scale industrial applications

In materials science, a genetically modified virus has been used in an academic lab as a scaffold for assembling a more environmentally friendly lithium-ion battery.[102][103]

Bacteria have been engineered to function as sensors by expressing a fluorescent protein under certain environmental conditions.[104]

1.5.4 Agriculture

Main article: Genetically modified crops
One of the best-known and controversial applications of

Bt-toxins present in peanut leaves (bottom image) protect it from extensive damage caused by European corn borer larvae (top image).[105]

genetic engineering is the creation and use of genetically modified crops or genetically modified organisms, such as genetically modified fish, which are used to produce genetically modified food and materials with diverse uses. There are four main goals in generating genetically modified crops.[106]

One goal, and the first to be realized commercially, is to provide protection from environmental threats, such as cold (in the case of Ice-minus bacteria), or pathogens, such as insects or viruses, and/or resistance to herbicides. There are also fungal and virus resistant crops developed or in development.[107][108] They have been developed to make the insect and weed management of crops easier and can indirectly increase crop yield.[109]

Another goal in generating GMOs is to modify the quality of produce by, for instance, increasing the nutritional value or providing more industrially useful qualities or quantities.[110] The Amflora potato, for example, produces a more industrially useful blend of starches. Cows have been engineered to produce more protein in their milk to facilitate cheese production.[111] Soybeans and canola have been genetically modified to produce more healthy oils.[112][113]

Another goal consists of driving the GMO to produce materials that it does not normally make. One example is "pharming", which uses crops as bioreactors to produce vaccines, drug intermediates, or drug themselves; the useful product is purified from the harvest and then used in the standard pharmaceutical production process.[114] Cows and goats have been engineered to express drugs and other proteins in their milk, and in 2009 the FDA approved a drug produced in goat milk.[115][116]

Another goal in generating GMOs, is to directly improve yield by accelerating growth, or making the organism more hardy (for plants, by improving salt, cold or drought tolerance).[110] Some agriculturally important animals have been genetically modified with growth hormones to increase their size.[117]

The genetic engineering of agricultural crops can increase the growth rates and resistance to different diseases caused by pathogens and parasites.[118] This is beneficial as it can greatly increase the production of food sources with the usage of fewer resources that would be required to host the world's growing populations. These modified crops would also reduce the usage of chemicals, such as fertilizers and pesticides, and therefore decrease the severity and frequency of the damages produced by these chemical pollution.[118][119]

Ethical and safety concerns have been raised around the use of genetically modified food.[120] A major safety concern relates to the human health implications of eating genetically modified food, in particular whether toxic or allergic reactions could occur.[121] Gene flow into related non-transgenic crops, off target effects on beneficial organisms and the impact on biodiversity are important environmental issues.[122] Ethical concerns involve religious issues, corporate control of the food supply, intellectual property rights and the level of labeling needed on genetically modified products.

1.5.5   BioArt and entertainment

Genetic engineering is also being used to create BioArt.[123] Some bacteria have been genetically engineered to create black and white photographs.[124]

Genetic engineering has also been used to create novelty items such as lavender-colored carnations,[125] blue roses,[126] and glowing fish.[127][128]

1.6   Regulation

Main articles: Regulation of genetic engineering and Regulation of the release of genetically modified organisms

The regulation of genetic engineering concerns the approaches taken by governments to assess and manage the risks associated with the development and release of genetically modified crops. There are differences in the regulation of GM crops between countries, with some of the most marked differences occurring between the USA and Europe. Regulation varies in a given country depending on the intended use of the products of the genetic engineering. For example, a crop not intended for food use is generally not reviewed by authorities responsible for food safety. Starting in the late 1980s, guidance on assessing the safety of genetically engineered plants and food emerged from organizations including the FAO and WHO.[129][130][131][132]

1.7   Controversy

Main articles: Genetically modified food controversies and Human genetic engineering

Critics have objected to use of genetic engineering per se on several grounds, including ethical concerns, ecological concerns, and economic concerns raised by the fact GM techniques and GM organisms are subject to intellectual property law. GMOs also are involved in controversies over GM food with respect to whether food produced from GM crops is safe, whether it should be labeled, and whether GM crops are needed to address the world's food needs. See the genetically modified food controversies article for discussion of issues about GM crops and GM food. These controversies have led to litigation, international trade disputes, and protests, and to restrictive regulation of commercial products in some countries.

1.8 See also

- Biological engineering
- EHA101
- Gene patent
- Gene drive
- Genetic engineering in the United States
- Genetically modified crops
- Genetically modified fish
- Genetically modified food
- Genetically modified food controversies
- Genetically modified livestock
- Genetically modified organisms
- Induced stem cells
- Marker assisted selection
- Paratransgenesis
- Regulation of the release of genetic modified organisms

1.9 References

[1] "First transgenic pet. 'GloFish'. sold to US public". *PHG Foundation*. 9 January 2004.

[2] "Terms and Acronyms". *U.S. Environmental Protection Agency online*. Retrieved 16 July 2015.

[3] Vert, Michel; Doi, Yoshiharu; Hellwich, Karl-Heinz; Hess, Michael; Hodge, Philip; Kubisa, Przemyslaw; Rinaudo, Marguerite; Schué, François (2012). "Terminology for biorelated polymers and applications (IUPAC Recommendations 2012)" (PDF). *Pure and Applied Chemistry* **84** (2): 377–410. doi:10.1351/PAC-REC-10-12-04.

[4] The European Parliament and the council of the European Union (12 March 2001). "Directive on the release of genetically modified organisms (GMOs) Directive 2001/18/EC ANNEX 1 A". Official Journal of the European Communities: 17.

[5] Staff Economic Impacts of Genetically Modified Crops on the Agri-Food Sector; P. 42 Glossary - Term and Definitions The European Commission Directorate-General for Agriculture. "Genetic engineering: The manipulation of an organism's genetic endowment by introducing or eliminating specific genes through modern molecular biology techniques. A broad definition of genetic engineering also includes selective breeding and other means of artificial selection.". Retrieved 5 November 2012

[6] Van Eenennaam. Alison. "Is Livestock Cloning Another Form of Genetic Engineering?" (PDF). agbiotech. Archived from the original (PDF) on 11 May 2011.

[7] Suter, David M.; Dubois-Dauphin, Michel; Krause, Karl-Heinz (2006). "Genetic engineering of embryonic stem cells" (PDF). *Swiss Med Wkly* **136** (27–28): 413–415. PMID 16897894.

[8] Andrianantoandro, Ernesto; Basu, Subhayu; Kariga, David K.; Weiss, Ron (16 May 2006). "Synthetic biology: new engineering rules for an emerging discipline". *Molecular Systems Biology* **2** (2006.0028): 2006.0028. doi:10.1038/msb4100073. PMC 1681505. PMID 16738572.

[9] Jacobsen, E.; Schouten, H. J. (2008). "Cisgenesis, a New Tool for Traditional Plant Breeding. Should be Exempted from the Regulation on Genetically Modified Organisms in a Step by Step Approach". *Potato Research* **51**: 75–88. doi:10.1007/s11540-008-9097-y.

[10] Capecchi, Mario R. (2001). "Generating mice with targeted mutations". *Nature Medicine* **7** (10): 1086–90. doi:10.1038/nm1001-1086. PMID 11590420.

[11] Staff Biotechnology - Glossary of Agricultural Biotechnology Terms United States Department of Agriculture. "Genetic modification: The production of heritable improvements in plants or animals for specific uses, via either genetic engineering or other more traditional methods. Some countries other than the United States use this term to refer specifically to genetic engineering.", Retrieved 5 November 2012

[12] Maryanski, James H. (19 October 1999). "Genetically Engineered Foods". Center for Food Safety and Applied Nutrition at the Food and Drug Administration.

[13] Evans, Brent and Lupescu, Mihai (15 July 2012) Canada - Agricultural Biotechnology Annual – 2012 GAIN (Global Agricultural Information Network) report CA12029, United States Department of Agriculture, Foreign Agricultural Service. Retrieved 5 November 2012

[14] McHugen, Alan (14 September 2000). "Chapter 1: Hors-d'oeuvres and entrees/What is genetic modification? What are GMOs?". *Pandora's Picnic Basket*. Oxford University Press. ISBN 978-0198506744.

[15] Staff (28 November 2005) Health Canada - The Regulation of Genetically Modified Food Glossary definition of Genetically Modified: "An organism, such as a plant, animal or bacterium, is considered genetically modified if its genetic material has been altered through any method, including conventional breeding. A 'GMO' is a genetically modified organism.", Retrieved 5 November 2012

[16] "What is genetic modification (GM)?". CSIRO.

[17] "Genetic Modification of Bacteria". Annenberg Foundation.

[18] Panesar, Pamit et al (2010) "Enzymes in Food Processing: Fundamentals and Potential Applications", Chapter 10, I K International Publishing House, ISBN 978-9380026336

[19] "GM traits list". International Service for the Acquisition of Agri-Biotech Applications.

[20] "ISAAA Brief 43-2011: Executive Summary". International Service for the Acquisition of Agri-Biotech Applications.

[21] Connor, Steve (2 November 2007). "The mouse that shook the world". *The Independent*.

[22] Root, Clive (2007). *Domestication*. Greenwood Publishing Groups.

[23] Zohary, Daniel; Hopf, Maria; Weiss, Ehud (2012). *Domestication of Plants in the Old World: The origin and spread of plants in the old world*. Oxford University Press.

[24] Stableford, Brian M. (2004). *Historical dictionary of science fiction literature*. p. 133. ISBN 9780810849389.

[25] A. Hershey; Chase, M. (1952). "Independent functions of viral protein and nucleic acid in growth of bacteriophage" (PDF). *J Gen Physiol* **36** (1): 39–56. doi:10.1085/jgp.36.1.39. PMC 2147348. PMID 12981234.

[26] "Genetic Engineering". Encyclopedia of Science Fiction. April 2. 2015.

[27] Shiv Kant Prasad. Ajay Dash (2008). *Modern Concepts in Nanotechnology. Volume 5*. Discovery Publishing House. ISBN 9788183562966.

[28] Jackson, DA; Symons, RH; Berg, P (1 October 1972). "Biochemical Method for Inserting New Genetic Information into DNA of Simian Virus 40: Circular SV40 DNA Molecules Containing Lambda Phage Genes and the Galactose Operon of Escherichia coli". *PNAS* **69** (10): 2904–2909. Bibcode:1972PNAS...69.2904J. doi:10.1073/pnas.69.10.2904. PMC 389671. PMID 4342968.

[29] Arnold, Paul (2009). "History of Genetics: Genetic Engineering Timeline".

[30] Cohen, Stanley N.; Chang, Annie C. Y. (1 May 1973). "Recircularization and Autonomous Replication of a Sheared R-Factor DNA Segment in Escherichia coli Transformants — PNAS". Pnas.org. Retrieved 17 July 2010.

[31] Jaenisch, R. and Mintz, B. (1974) Simian virus 40 DNA sequences in DNA of healthy adult mice derived from preimplantation blastocysts injected with viral DNA. Proc. Natl. Acad. 71(4) 1250–1254

[32] Berg P; Baltimore, D; Brenner, S; Roblin, RO; Singer, MF; et al. (1975). "Summary statement of the Asilomar Conference on recombinant DNA molecules" (PDF). *Proc. Natl. Acad. Sci. U.S.A* **72** (6): 1981–4. Bibcode:1975PNAS...72.1981B. doi:10.1073/pnas.72.6.1981. PMC 432675. PMID 806076.

[33] NIH Guidelines for research involving recombinant DNA molecules

[34] Goeddel, David; Kleid, Dennis G.; Bolivar, Francisco; Heyneker, Herbert L.; Yansura, Daniel G.; Crea, Roberto; Hirose, Tadaaki; Kraszewski, Adam; Itakura, Keiichi; Riggs, Arthur D. (January 1979). "Expression in Escherichia coli of chemically synthesized genes for human insulin" (PDF). *PNAS* **76** (1): 106–110. Bibcode:1979PNAS...76..106G. doi:10.1073/pnas.76.1.106. PMC 382885. PMID 85300.

[35] US Supreme Court Cases from Justia & Oyez (16 June 1980). "Diamond V Chakrabarty" **447** (303). Supreme.justia.com. Retrieved 17 July 2010.

[36] "Artificial Genes". *TIME*. 15 November 1982. Retrieved 17 July 2010.

[37] H. Patricia Hynes. (1989) Biotechnology in agriculture: an analysis of selected technologies and policy in the United States. Reproductive and Genetic Engineering (2)1:39–49

[38] Bratspies, Rebecca (2007). "Some Thoughts on the American Approach to Regulating Genetically Modified Organisms" (PDF). *Kansas Journal of Law and Public Policy* **16** (3): 101–131.

[39] BBC News 14 June 2002 GM crops: A bitter harvest?

[40] Thomas H. Maugh II for the Los Angeles Times. 9 June 1987. Altered Bacterium Does Its Job : Frost Failed to Damage Sprayed Test Crop, Company Says

[41] James, Clive (1996). "Global Review of the Field Testing and Commercialization of Transgenic Plants: 1986 to 1995" (PDF). The International Service for the Acquisition of Agri-biotech Applications. Retrieved 17 July 2010.

[42] James, Clive (1997). "Global Status of Transgenic Crops in 1997" (PDF). *ISAAA Briefs No. 5*.: 31.

[43] Bruening, G.; Lyons, J.M. (2000). "The case of the FLAVR SAVR tomato". *California Agriculture* **54** (4): 6–7. doi:10.3733/ca.v054n04p6.

[44] MacKenzie, Debora (18 June 1994). "Transgenic tobacco is European first". New Scientist.

[45] Genetically Altered Potato Ok'd For Crops Lawrence Journal-World - 6 May 1995

[46] Global Status of Commercialized Biotech/GM Crops: 2009 ISAAA Brief 41-2009, 23 February 2010. Retrieved 10 August 2010

[47] Pennisi, Elizabeth (2010-05-21). "Synthetic Genome Brings New Life to Bacterium". *Science* **328** (5981): 958–959. doi:10.1126/science.328.5981.958. ISSN 0036-8075. PMID 20488994.

[48] Gibson, D. G.; Glass, J. I.; Lartigue, C.; Noskov, V. N.; Chuang, R.-Y.; Algire, M. A.; Benders, G. A.; Montague, M. G.; Ma, L.; Moodie, M. M.; Merryman, C.; Vashee, S.; Krishnakumar, R.; Assad-Garcia, N.; Andrews-Pfannkoch, C.; Denisova, E. A.; Young, L.; Qi, Z.-Q.; Segall-Shapiro, T. H.; Calvey, C. H.; Parmar, P. P.; Hutchison Ca, C. A.; Smith, H. O.; Venter, J. C. (2010). "Creation of a Bacterial Cell Controlled by a Chemically Synthesized Genome". *Science* **329** (5987): 52–6. doi:10.1126/science.1190719. PMID 20488990.

[49] Malyshev, Denis A.; Dhami, Kirandeep; Lavergne, Thomas; Chen, Tingjian; Dai, Nan; Foster, Jeremy M.; Corrêa, Ivan R.; Romesberg, Floyd E. (2014-05-15). "A semi-synthetic organism with an expanded genetic alphabet". *Nature* **509** (7500): 385–388. doi:10.1038/nature13314. ISSN 0028-0836. PMC 4058825. PMID 24805238.

[50] Thyer, Ross; Ellefson, Jared (2014-05-15). "Synthetic biology: New letters for life's alphabet". *Nature* **509** (7500): 291–292. doi:10.1038/nature13335. ISSN 0028-0836.

[51] James, Clive (2012). "Global Status of Commercilized Biotech/GM Crops:2012". *ISSA Brief No. 44*.

[52] Alberts B, Johnson A, Lewis J, et al. (2002). "8". *Isolating, Cloning, and Sequencing DNA*. (4th ed.). New York: Garland Science.

[53] Kaufman, R I; Nixon, B T (1996). "Use of PCR to isolate genes encoding sigma54-dependent activators from diverse bacteria". *J Bacteriol* **178** (13): 3967–3970. PMC 232662. PMID 8682806.

[54] Liang, Jing; Luo, Yunzi; Zhao, Huimin (2011). "Synthetic biology: Putting synthesis into biology". *Wiley Interdisciplinary Reviews: Systems Biology and Medicine* **3**: 7–20. doi:10.1002/wsbm.104.

[55] Berg, P.; Mertz, J. E. (2010). "Personal Reflections on the Origins and Emergence of Recombinant DNA Technology". *Genetics* **184** (1): 9–17. doi:10.1534/genetics.109.112144. PMC 2815933. PMID 20061565.

[56] Townsend JA, Wright DA, Winfrey RJ, et al. (May 2009). "High-frequency modification of plant genes using engineered zinc-finger nucleases". *Nature* **459** (7245): 442–5. Bibcode:2009Natur.459..442T. doi:10.1038/nature07845. PMC 2743854. PMID 19404258.

[57] Shukla VK, Doyon Y, Miller JC, et al. (May 2009). "Precise genome modification in the crop species Zea mays using zinc-finger nucleases". *Nature* **459** (7245): 437–41. Bibcode:2009Natur.459..437S. doi:10.1038/nature07992. PMID 19404259.

[58] Grizot S, Smith J, Daboussi F, et al. (September 2009). "Efficient targeting of a SCID gene by an engineered single-chain homing endonuclease". *Nucleic Acids Res.* **37** (16): 5405–19. doi:10.1093/nar/gkp548. PMC 2760784. PMID 19584299.

[59] Gao H, Smith J, Yang M, et al. (January 2010). "Heritable targeted mutagenesis in maize using a designed endonuclease". *Plant J.* **61** (1): 176–87. doi:10.1111/j.1365-313X.2009.04041.x. PMID 19811621.

[60] Christian M, Cermak T, Doyle EL, et al. (July 2010). "TAL Effector Nucleases Create Targeted DNA Double-strand Breaks". *Genetics* **186** (2): 757–61. doi:10.1534/genetics.110.120717. PMC 2942870. PMID 20660643.

[61] Li T, Huang S, Jiang WZ, et al. (August 2010). "TAL nucleases (TALNs): hybrid proteins composed of TAL effectors and FokI DNA-cleavage domain". *Nucleic Acids Res* **39** (1): 359–72. doi:10.1093/nar/gkq704. PMC 3017587. PMID 20699274.

[62] Ekker, S.C. (2008). "Zinc finger-based knockout punches for zebrafish genes". *Zebrafish* **5** (2): 1121–3. doi:10.1089/zeb.2008.9988. PMC 2849655. PMID 18554175.

[63] Geurts AM, Cost GJ, Freyvert Y, et al. (July 2009). "Knockout rats via embryo microinjection of zinc-finger nucleases". *Science* **325** (5939): 433. Bibcode:2009Sci...325..433G. doi:10.1126/science.1172447. PMC 2831805. PMID 19628861.

[64] Chen, I; Dubnau, D (2004). "DNA uptake during bacterial transformation". *Nat. Rev. Microbiol.* **2** (3): 241–9. doi:10.1038/nrmicro844. PMID 15083159.

[65] Head, Graham; Hull, Roger H; Tzotzos, George T. (2009). *Genetically Modified Plants: Assessing Safety and Managing Risk*. London: Academic Pr. p. 244. ISBN 0-12-374106-8.

[66] Gelvin, S. B. (2003). "Agrobacterium-Mediated Plant Transformation: The Biology behind the "Gene-Jockeying" Tool". *Microbiology and Molecular Biology Reviews* **67** (1): 16–37, table of contents. doi:10.1128/MMBR.67.1.16-37.2003. PMC 150518. PMID 12626681.

[67] Darbani, Behrooz; Farajnia, Safar; Toorchi, Mahmoud; Zakerbostanabad, Saeed; Noeparvar, Shahin; Stewart, Jr., C. Neal (2010). "DNA-Delivery Methods to Produce Transgenic Plants". Science Alert.

[68] Hohn, Barbara; Levy, Avraham A; Puchta, Holger (2001). "Elimination of selection markers from transgenic plants". *Current Opinion in Biotechnology* **12** (2): 139–43. doi:10.1016/S0958-1669(00)00188-9. PMID 11287227.

[69] Esvelt, KM.; Wang, HH. (2013). "Genome-scale engineering for systems and synthetic biology". *Mol Syst Biol* **9**: 641. doi:10.1038/msb.2012.66. PMC 3564264. PMID 23340847.

[70] Tan, WS.; Carlson, DF.; Walton, MW.; Fahrenkrug, SC.; Hackett, PB. (2012). "Precision editing of large animal genomes". *Adv Genet.* Advances in Genetics **80**: 37–97. doi:10.1016/B978-0-12-404742-6.00002-8. ISBN 9780124047426. PMC 3683964. PMID 23084873.

[71] "Natural Genetic Engineering and Natural Genome Editing". *Annals of the New York Academy of Sciences* **1178**. 2009.

[72] Avise, John C. (2004). *The hope, hype & reality of genetic engineering: remarkable stories from agriculture, industry, medicine, and the environment*. Oxford University Press US. p. 22. ISBN 978-0-19-516950-8.

[73] "Engineering algae to make complex anti-cancer 'designer' drug". PhysOrg. 10 December 2012. Retrieved 15 April 2013.

[74] =Roque, AC; Lowe, CR; Taipa, MA. (2004). "Antibodies and genetically engineered related molecules: production and purification". *Biotechnol Progress* **20** (3): 639–54. doi:10.1021/bp030070k. PMID 15176864.

[75] Rodriguez, Luis L.; Grubman, Marvin J. (2009). "Foot and mouth disease virus vaccines". *Vaccine* **27**: D90–4. doi:10.1016/j.vaccine.2009.08.039. PMID 19837296.

[76] "Background: Cloned and Genetically Modified Animals". Center for Genetics and Society. 14 April 2005.

[77] "Knockout Mice". Nation Human Genome Research Institute. 2009.

[78] "GM pigs best bet for organ transplant". Medical News Today. 21 September 2003.

[79] Fischer, Alain; Hacein-Bey-Abina, Salima; Cavazzana-Calvo, Marina (2010). "20 years of gene therapy for SCID". *Nature Immunology* **11** (6): 457–60. doi:10.1038/ni0610-457. PMID 20485269.

[80] Ledford, Heidi (2011). "Cell therapy fights leukaemia". *Nature*. doi:10.1038/news.2011.472.

[81] Lewitt, Peter A; Rezai, Ali R; Leehey, Maureen A; Ojemann, Steven G; Flaherty, Alice W; Eskandar, Emad N; Kostyk, Sandra K; Thomas, Karen; Sarkar, Atom; Siddiqui, Mustafa S; Tatter, Stephen B; Schwalb, Jason M; Poston, Kathleen L; Henderson, Jaimie M; Kurlan, Roger M; Richard, Irene H; Van Meter, Lori; Sapan, Christine V; During, Matthew J; Kaplitt, Michael G; Feigin, Andrew (2011). "AAV2-GAD gene therapy for advanced Parkinson's disease: A double-blind, sham-surgery controlled, randomised trial". *The Lancet Neurology* **10** (4): 309–19. doi:10.1016/S1474-4422(11)70039-4. PMID 21419704.

[82] Gallagher, James. (2 November 2012) BBC News – Gene therapy: Glybera approved by European Commission. Bbc.co.uk. Retrieved on 15 December 2012.

[83] Richards, Sabrina. "Gene Therapy Arrives in Europe". The Scientist. Retrieved 16 November 2012.

[84] The Declaration of Inuyama: Human Genome Mapping. Genetic Screening and Gene Therapy

[85] Smith KR, Chan S, Harris J. Human germline genetic modification: scientific and bioethical perspectives. Arch Med Res. 2012 Oct;43(7):491-513. doi: 10.1016/j.arcmed.2012.09.003. PMID 23072719

[86] Wade, Nicholas (19 March 2015). "Scientists Seek Ban on Method of Editing the Human Genome". *New York Times*. Retrieved 20 March 2015.

[87] Pollack, Andrew (3 March 2015). "A Powerful New Way to Edit DNA". *New York Times*. Retrieved 20 March 2015.

[88] Baltimore, David; Berg, Paul; Botchan, Dana; Charo, R. Alta; Church, George; Corn, Jacob E.; Daley, George Q.; Doudna, Jennifer A.; Fenner, Marsha; Greely, Henry T.; Jinek, Martin; Martin, G. Steven; Penhoet, Edward; Puck, Jennifer; Sternberg, Samuel H.; Weissman, Jonathan S.; Yamamoto, Keith R. (19 March 2015). "A prudent path forward for genomic engineering and germline gene modification". *Science* **348**: 36–8. doi:10.1126/science.aab1028. PMID 25791083. Retrieved 20 March 2015.

[89] Lanphier, Edward; Urnov, Fyodor; Haecker, Sarah Ehlen; Werner, Michael; Smolenski, Joanna (26 March 2015). "Don't edit the human germ line". *Nature* **519**: 410–411. doi:10.1038/519410a. PMID 25810189. Retrieved 20 March 2015.

[90] Kolata, Gina (23 April 2015). "Chinese Scientists Edit Genes of Human Embryos, Raising Concerns". *New York Times*. Retrieved 24 April 2015.

[91] Liang, Puping; et al. (18 April 2015). "CRISPR/Cas9-mediated gene editing in human tripronuclear zygotes". *Protein & Cell* **6**: 363–72. doi:10.1007/s13238-015-0153-5. PMC 4417674. PMID 25894090. Retrieved 24 April 2015.

[92] Wade, Nicholas (3 December 2015). "Scientists Place Moratorium on Edits to Human Genome That Could Be Inherited". *New York Times*. Retrieved 3 December 2015.

[93] Bergeson, Emilie R. (1997). "The Ethics of Gene Therapy".

[94] Hanna, Kathi E. "Genetic Enhancement". National Human Genome Research Institute.

[95] "Applications of Genetic Engineering". Microbiologyprocedure. Retrieved 9 July 2010.

[96] "Biotech: What are transgenic organisms?". Easyscience. 2002. Retrieved 9 July 2010.

[97] Savage, Neil (1 August 2007). "Making Gasoline from Bacteria: A biotech startup wants to coax fuels from engineered microbes". Technology Review. Retrieved 16 July 2015.

[98] Summers, Rebecca (24 April 2013) Bacteria churn out first ever petrol-like biofuel New Scientist. Retrieved 27 April 2013

[99] "Applications of Some Genetically Engineered Bacteria". Retrieved 9 July 2010.

[100] Sanderson, Katherine (24 February 2012) New Portable Kit Detects Arsenic In Wells Chemical and Engineering News. Retrieved 23 January 2013

[101] Reece, Jane B.; Urry, Lisa A.; Cain, Michael L.; Wasserman, Steven A.; Minorsky, Peter V.; Jackson, Robert B. (2011). *Campbell Biology Ninth Edition*. San Francisco: Pearson Benjamin Cummings. p. 421. ISBN 0-321-55823-5.

[102] "New virus-built battery could power cars, electronic devices". Web.mit.edu. 2 April 2009. Retrieved 17 July 2010.

[103] "Hidden Ingredient In New, Greener Battery: A Virus". Npr.org. Retrieved 17 July 2010.

[104] "Researchers Synchronize Blinking 'Genetic Clocks' -- Genetically Engineered Bacteria That Keep Track of Time". ScienceDaily. 24 January 2010.

[105] Suszkiw, Jan (November 1999). "Tifton, Georgia: A Peanut Pest Showdown". *Agricultural Research magazine*. Retrieved 23 November 2008.

[106] Magaña-Gómez, JA; de la Barca, A.M. (2009). "Risk assessment of genetically modified crops for nutrition and health". *Nutr. Rev.* **67** (1): 1–16. doi:10.1111/j.1753-4887.2008.00130.x. PMID 19146501.

[107] Islam, Aparna (2008). "Fungus Resistant Transgenic Plants: Strategies, Progress and Lessons Learnt". *Plant Tissue Culture and Biotechnology* **16** (2): 117–38. doi:10.3329/ptcb.v16i2.1113.

[108] "Disease resistant crops". GMO Compass.

[109] Demont, M; Tollens, E (2004). "First impact of biotechnology in the EU: Bt maize adoption in Spain". *Annals of Applied Biology* **145** (2): 197–207. doi:10.1111/j.1744-7348.2004.tb00376.x.

[110] Whitman, Deborah B. (2000). "Genetically Modified Foods: Harmful or Helpful?".

[111] Young, Emma (2003). "GM cows to please cheese-makers". *New Scientist*.

[112] Rapeseed (canola) has been genetically engineered to modify its oil content with a gene encoding a "12:0 thioesterase" (TE) enzyme from the California bay plant (Umbellularia californica) to increase medium length fatty acids, see: Geopie.cornell.edu

[113] Bomgardner Melody M (2012). "Replacing Trans Fat: New crops from Dow Chemical and DuPont target food makers looking for stable, heart-healthy oils". *Chemical and Engineering News* **90** (11): 30–32.

[114] Marvier, Michelle (2008). "Pharmaceutical crops in California, benefits and risks. A review". *Agronomy for Sustainable Development* **28** (1): 1–9. doi:10.1051/agro:2007050.

[115] "FDA Approves First Human Biologic Produced by GE Animals". US Food and Drug Administration.

[116] Rebêlo, Paulo (15 July 2004). "GM cow milk 'could provide treatment for blood disease'". SciDev.

[117] "Giant GM salmon on the way". BBC News. 11 April 2000.

[118] Chivian, Eric; Bernstein, Aaron (2008). *Sustaining Life*. Oxford University Press, Inc. ISBN 978-0-19-517509-7.

[119] Carrington, Damien (13 June 2012) GM crops good for environment, study finds The Guardian. Retrieved 16 June 2012

[120] Pickrell, John (4 September 2006). "Introduction: GM Organisms". New Scientist.

[121] "20 questions on genetically modified foods". World Health Organization. 2010.

[122] "Can GM crops harm the environment?". National Environment Research Council (NERC).

[123] Pasko, Jessica M. (3/4/2007). "Bio-artists bridge gap between arts, sciences: Use of living organisms is attracting attention and controversy". msnbc. Check date values in: |date= (help)

[124] Jackson, Joab (6 December 2005). "Genetically Modified Bacteria Produce Living Photographs". National Geographic News.

[125] Phys.Org website. 4 April 2005 Plant gene replacement results in the world's only blue rose

[126] Katsumoto, Yukihisa; Fukuchi-Mizutani, Masako; Fukui, Yuko; Brugliera, Filippa; Holton, Timothy A.; Karan, Mirko; Nakamura, Noriko; Yonekura-Sakakibara, Keiko; Togami, Junichi; Pigeaire, Alix; Tao, Guo-Qing; Nehra, Narender S.; Lu, Chin-Yi; Dyson, Barry K.; Tsuda, Shinzo; Ashikari, Toshihiko; Kusumi, Takaaki; Mason, John G.; Tanaka, Yoshikazu (2007). "Engineering of the Rose Flavonoid Biosynthetic Pathway Successfully Generated Blue-Hued Flowers Accumulating Delphinidin". *Plant and Cell Physiology* **48** (11): 1589–600. doi:10.1093/pcp/pcm131. PMID 17925311.

[127] Published PCT Application WO2000049150 "Chimeric Gene Constructs for Generation of Fluorescent Transgenic Ornamental Fish." National University of Singapore

[128] Stewart, C. Neal (2006). "Go with the glow: Fluorescent proteins to light transgenic organisms". *Trends in Biotechnology* **24** (4): 155–62. doi:10.1016/j.tibtech.2006.02.002. PMID 16488034.

[129] WHO (1987): Principles for the Safety Assessment of Food Additives and Contaminants in Food. Environmental Health Criteria 70. World Health Organization, Geneva

[130] WHO (1991): Strategies for assessing the safety of foods produced by biotechnology, Report of a Joint FAO/WHO Consultation. World Health Organization, Geneva ISBN 9789241561457. PDF download library.health.go.ug/download/file/fid/790 here]

[131] WHO (1993): Health aspects of marker genes in genetically modified plants. Report of a WHO Workshop. World Health Organization, Geneva

[132] WHO (1995): Application of the principle of substantial equivalence to the safety evaluation of foods or food components from plants derived by modern biotechnology. Report of a WHO Workshop. World Health Organization, Geneva

1.10   Further reading

- British Medical Association (1999). *The Impact of Genetic Modification on Agriculture, Food and Health.* BMJ Books. ISBN 0-7279-1431-6.

- Donnellan, Craig (2004). *Genetic Modification (Issues).* Independence Educational Publishers. ISBN 1-86168-288-3.

- Morgan, Sally (1 January 2009). *Superfoods: Genetic Modification of Foods.* Heinemann Library. ISBN 978-1-4329-2455-3.

- Smiley, Sophie (2005). *Genetic Modification: Study Guide (Exploring the Issues).* Independence Educational Publishers. ISBN 1-86168-307-3.

- James D., Watson (2007). *Recombinant DNA: Genes and Genomes: A Short Course.* San Francisco: W.H. Freeman. ISBN 0-7167-2866-4.

- Weaver, Sean; Michael, Morris (2003). "An Annotated Bibliography of Scientific Publications on the Risks Associated with Genetic Modification". Wellington, N.Z.: Victoria University

- Zaid, A; Hughes, H.G.; Porceddu, E.; Nicholas, F. (2001). *Glossary of Biotechnology for Food and Agriculture - A Revised and Augmented Edition of the Glossary of Biotechnology and Genetic Engineering.* Rome, Italy: FAO. ISBN 92-5-104683-2.

1.11   External links

- GMO Safety - Information about research projects on the biological safety of genetically modified plants.

- GMO-compass, news on GMO en EU

Chapter 2

History of genetic engineering

Herbert Boyer

Stanley Cohen
Herbert Boyer and Stanley Cohen created the first genetically modified organism in 1973

Genetic modification caused by human activity has been occurring since around 12,000 BC, when humans first began to domesticate organisms. Genetic engineering as the direct transfer of DNA from one organism to another was first accomplished by Herbert Boyer and Stanley Cohen in 1973. The first genetically modified animal was a mouse created in 1973 by Rudolf Jaenisch. In 1983 an antibiotic resistant gene was inserted into tobacco, leading to the first

genetically engineered plant. Advances followed that allowed scientists to manipulate and add genes to a variety of different organism and induce a range of different effects.

In 1976 the technology was commercialised, with the advent of genetically modified bacteria that produced somatostatin, followed by insulin in 1978. Plants were first commercialised with virus resistant tobacco released in China in 1992. The first genetically modified food was the Flavr Savr tomato marketed in 1994. By 2010, 29 countries had planted commercialized biotech crops. In 2000 a paper published in *Science* introduced golden rice, the first food developed with increased nutrient value.

2.1 Agriculture

Main article: History of agriculture
Genetic engineering is the direct manipulation of an or-

DNA studies suggested that the dog most likely arose from a common ancestor with the grey wolf.[1]

ganism's genome using certain biotechnology techniques that have only existed since the 1970s.[2] Human directed genetic manipulation was occurring much earlier, beginning with the domestication of plants and animals through artificial selection. The dog is believed to be the first animal domesticated, possibly arising from a common ancestor of the grey wolf,[1] with archeologically evidence dating to about 12,000 BC.[3] Other carnivores domesticated in prehistoric times include the cat, which cohabited with human 9 500 years ago.[4] Archeologically evidence suggests sheep, cattle, pigs and goats were domesticated between 9 000 BC and 8 000 BC in the Fertile Crescent.[5]

The first evidence of plant domestication comes from emmer and einkorn wheat found in pre-Pottery Neolithic A villages in Southwest Asia dated about 10.500 to 10.100 BC.[6] The Fertile Crescent of Western Asia, Egypt, and India were sites of the earliest planned sowing and harvesting of plants that had previously been gathered in the wild. Independent development of agriculture occurred in northern and southern China, Africa's Sahel, New Guinea and several regions of the Americas.[7] The eight Neolithic founder crops (emmer wheat, einkorn wheat, barley, peas, lentils, bitter vetch, chick peas and flax) had all appeared by about 7000 BC.[8] Horticulture first appears in the Levant during the Chalcolithic period about 6 800 to 6,300 BC.[9] Due to the soft tissues, archeological evidence for early vegetables is scarce. The earliest vegetable remains have been found in Egyptian caves that date back to the 2nd millennium BC.[10]

Selective breeding of domesticated plants was once the main way early farmers shaped organisms to suit their needs. Charles Darwin described three types of selection: methodical selection, wherein humans deliberately select for particular characteristics; unconscious selection, wherein a characteristic is selected simply because it is desirable; and natural selection, wherein a trait that helps an organism survive better is passed on.[11]:25 Early breeding relied on unconscious and natural selection. The introduction of methodical selection is unknown.[11]:25 Common characteristics that were bred into domesticated plants include grains that did not shatter to allow easier harvesting, uniform ripening, shorter lifespans that translate to faster growing, loss of toxic compounds, and productivity.[11]:27-30 Some plants, like the Banana, were able to be propagated by vegetative cloning. Offspring often did not contain seeds, and therefore sterile. However, these offspring were usually juicier and larger. Propagation through cloning allows these mutant varieties to be cultivated despite their lack of seeds.[11]:31

Hybridization was another way that rapid changes in plant's makeup were introduced. It often increased vigor in plants, and combined desirable traits together. Hybridization most likely first occurred when humans first grew similar, yet slightly different plants in close proximity.[11]:32 Triticum aestivum, wheat used in baking bread, is an allopolyploid. Its creation is the result of two separate hybridization events.[12]

X-rays were first used to deliberately mutate plants in 1927. Between 1927 and 2007, more than 2.540 genetically mutated plant varieties had been produced using x-rays.[13]

2.2 Genetics

Main article: History of genetics
Various genetic discoveries have been essential in the de-

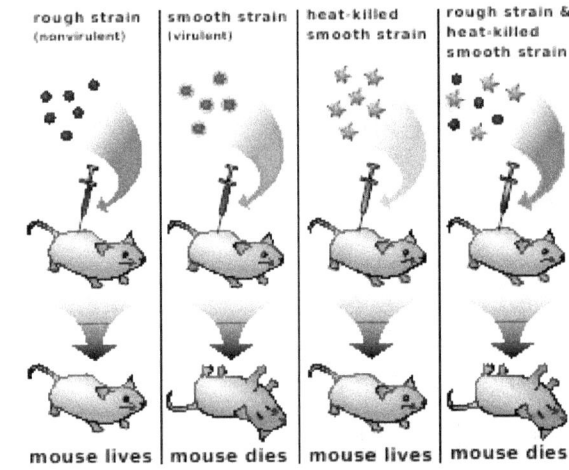

Griffith proved the existence of a "transforming principle", which Avery, MacLeod and McCarty later showed to be DNA

velopment of genetic engineering. Genetic inheritance was first discovered by Gregor Mendel in 1865 following experiments crossing peas. Although largely ignored for 34 years he provided the first evidence of hereditary segregation and independent assortment.[14] In 1889 Hugo de Vries came up with the name "(pan)gene" after postulating that particles are responsible for inheritance of characteristics[15] and the term "genetics" was coined by William Bateson in 1905.[16] In 1928 Frederick Griffith proved the existence of a "transforming principle" involved in inheritance, which Avery, MacLeod and McCarty later (1944) identified as DNA. Edward Lawrie Tatum and George Wells Beadle developed the central dogma that genes code for proteins in 1941. The double helix structure of DNA was identified by James Watson and Francis Crick in 1953.

As well as discovering how DNA works, tools had to be developed that allowed it to be manipulated. In 1970 Hamilton Smiths lab discovered restriction enzymes that allowed DNA to be cut at specific places and separated out on an electrophoresis gel. This enabled scientists to isolate genes from an organism's genome.[17] DNA ligases, that join broken DNA together, had been discovered earlier in 1967[18] and by combining the two enzymes it was possible to "cut and paste" DNA sequences to create recombinant

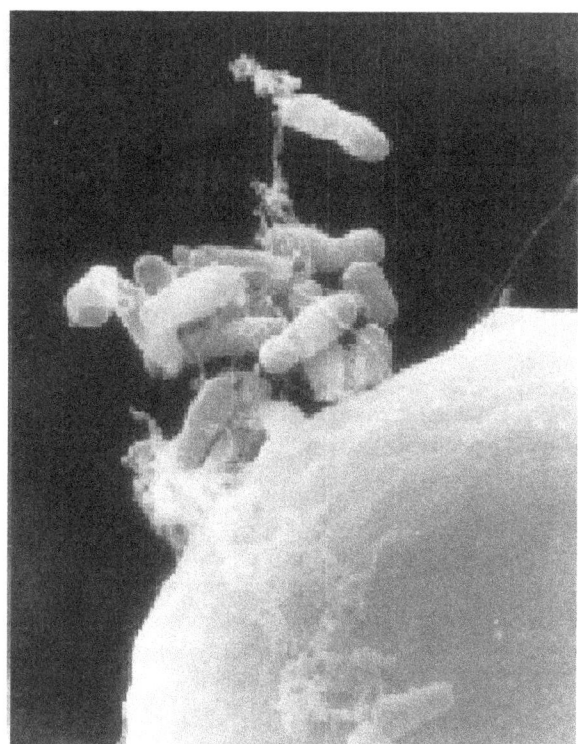

The bacterium Agrobacterium tumefaciens *inserts T-DNA into infected plant cells, which is then incorporated into the plants genome.*

DNA. Plasmids, discovered in 1952,[19] became important tools for transferring information between cells and replicating DNA sequences. Frederick Sanger developed a method for sequencing DNA in 1977, greatly increasing the genetic information available to researchers. Polymerase chain reaction (PCR), developed by Kary Mullis in 1983, allowed small sections of DNA to be amplified and aided identification and isolation of genetic material.

As well as manipulating the DNA, techniques had to be developed for its insertion (known as transformation) into an organism's genome. Griffiths experiment had already shown that some bacteria had the ability to naturally uptake and express foreign DNA. Artificial competence was induced in *Escherichia coli* in 1970 when Morton Mandel and Akiko Higa showed that it could take up bacteriophage λ after treatment with calcium chloride solution ($CaCl_2$).[20] Two years later, Stanley Cohen showed that $CaCl_2$ treatment was also effective for uptake of plasmid DNA.[21] Transformation using electroporation was developed in the late 1980s, increasing the efficiency and bacterial range.[22] In 1907 a bacterium that caused plant tumors, *Agrobacterium tumefaciens*, was discovered and in the early 1970s the tumor inducing agent was found to be a DNA plasmid called the Ti plasmid.[23] By removing the genes in the plasmid that caused the tumor and adding in novel genes researchers were able to infect plants with *A.*

tumefaciens and let the bacteria insert their chosen DNA into the genomes of the plants.[24]

2.3 Early genetically modified organisms

Paul Berg created the first recombinant DNA molecules in 1972.

In 1972 Paul Berg utilised restriction enzymes and DNA ligases to create the first recombinant DNA molecules. He combined DNA from the monkey virus SV40 with that of the lambda virus.[25] Herbert Boyer and Stanley N. Cohen took Berg's work a step further and introduced recombinant DNA into a bacterial cell. Cohen was researching plasmids, while Boyers work involved restriction enzymes. They recognised the complementary nature of their work and teamed up in 1972. Together they found a restriction enzyme that cut the pSC101 plasmid at a single point and were able to insert and ligate a gene that conferred resistance to the kanamycin antibiotic into the gap. Cohen had previously devised a method where bacteria could be induced to take up a plasmid and using this they were able to create a bacteria that survived in the presence of the kanamycin. This represented the first genetically modified organism. They repeated experiments showing that other genes could be expressed in bacteria, including one from the toad Xenopus laevis, the first cross kingdom transformation.[26][27][28]

In 1973 Rudolf Jaenisch created the first GM animal.

In 1973 Rudolf Jaenisch created a transgenic mouse by introducing foreign DNA into its embryo, making it the world's first transgenic animal.[29] Jaenisch was studying mammalian cells infected with simian virus 40 (SV40) when he happened to read a paper from Beatrice Mintz describing the generation of chimera mice. He took his SV40 samples to Mintz's lab and injected them into early mouse embryos expecting tumours to develop. The mice appeared normal, but after using radioactive probes he discovered that the virus had integrated itself into the mice genome.[30] However the mice did not pass the transgene to their offspring. In 1981 the laboratories of Frank Ruddle, Frank Constantini and Elizabeth Lacy injected purified DNA into a single-cell mouse embryo and showed transmission of the genetic material to subsequent generations.[31][32]

The first genetically engineered plant was tobacco, reported in 1983.[33] It was developed by Michael W. Bevan, Richard B. Flavell and Mary-Dell Chilton by creating a chimeric gene that joined an antibiotic resistant gene to the T1 plasmid from *Agrobacterium*. The tobacco was infected with *Agrobacterium* transformed with this plasmid resulting in the chimeric gene being inserted into the plant. Through tissue culture techniques a single tobacco cell was selected that contained the gene and a new plant grown from it.[34]

2.4   Regulation

Main article: Regulation of genetic engineering

The development of genetic engineering technology led to

concerns in the scientific community about potential risks. The development of a regulatory framework concerning genetic engineering began in 1975, at Asilomar, California. The Asilomar meeting recommended a set of guidelines regarding the cautious use of recombinant technology and any products resulting from that technology.[35] The Asilomar recommendations were voluntary, but in 1976 the US National Institute of Health (NIH) formed a recombinant DNA advisory committee.[36] This was followed by other regulatory offices (the United States Department of Agriculture (USDA), Environmental Protection Agency (EPA) and Food and Drug Administration (FDA)), effectively making all recombinant DNA research tightly regulated in the USA.[37]

In 1982 the Organization for Economic Co-operation and Development (OECD) released a report into the potential hazards of releasing genetically modified organisms into the environment as the first transgenic plants were being developed.[38] As the technology improved and genetically organisms moved from model organisms to potential commercial products the USA established a committee at the Office of Science and Technology (OSTP) to develop mechanisms to regulate the developing technology.[37] In 1986 the OSTP assigned regulatory approval of genetically modified plants in the US to the USDA, FDA and EPA.[39] In the late 1980s and early 1990s, guidance on assessing the safety of genetically engineered plants and food emerged from organizations including the FAO and WHO.[40][41][42][43]

The European Union first introduced laws requiring GMO's to be labelled in 1997.[44] In 2013 Connecticut became the first state to enacted a labeling law in the USA, although it would not take effect until other states followed suit.[45]

2.5   Research and Medicine

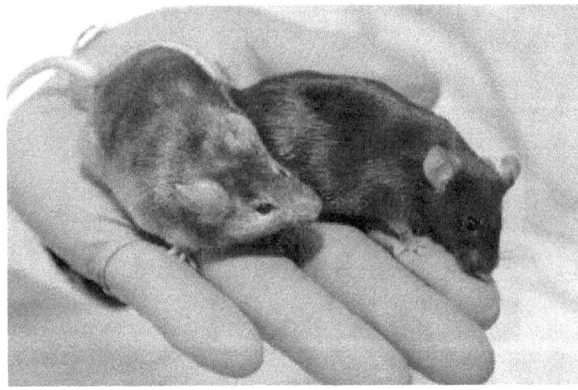

A laboratory mouse in which a gene affecting hair growth has been knocked out (left), is shown next to a normal lab mouse.

The ability to insert, alter or remove genes in model organisms allowed scientists to study the genetic elements of human diseases.[46] Genetically modified mice were created in 1984 that carried cloned oncogenes that predisposed them to developing cancer.[47] The technology has also been used to generate mice with genes knocked out. The first recorded knockout mouse was created by Mario R. Capecchi, Martin Evans and Oliver Smithies in 1989. In 1992 oncomice with tumor suppressor genes knocked out were generated.[47] Creating Knockout rats is much harder and only became possible in 2003.[48][49]

After the discovery of microRNA in 1993,[50] RNA interference (RNAi) has been used to silence an organism's genes.[51] By modifying an organism to express mircoRNA targeted to its endogenous genes, researchers have been able to knockout or partially reduce gene function in a range of species. The ability to partially reduce gene function has allowed the study of genes that are lethal when completely knocked out. Other advantages of using RNAi include the availability of inducible and tissue specific knockout.[52] In 2007 microRNA targeted to insect and nematode genes was expressed in plants, leading to suppression when they fed on the transgenic plant, potentially creating a new way to control pests.[53] Targeting endogenous microRNA expression has allowed further fine tuning of gene expression, supplementing the more traditional gene knock out approach.[54]

Genetic engineering has been used to produce proteins derived from humans and other sources in organisms that normally cannot synthesize these proteins. Human insulin-synthesising bacteria were developed in 1979 and were first used as a treatment in 1982.[55] In 1988 the first human antibodies were produced in plants.[56] In 2000 Vitamin A-enriched golden rice, was the first food with increased nutrient value.[57]

2.6 Further advances

As not all plant cells were susceptible to infection by *A. tumefaciens* other methods were developed, including electroporation, micro-injection[58] and particle bombardment with a gene gun (invented in 1987).[59][60] In the 1980s techniques were developed to introduce isolated chloroplasts back into a plant cell that had its cell wall removed. With the introduction of the gene gun in 1987 it became possible to integrate foreign genes into a chloroplast.[61]

Genetic transformation has become very efficient in some model organism. In 2008 genetically modified seeds were produced in *Arabidopsis thaliana* by simply dipping the flowers in an *Agrobacterium* solution.[62] The range of plants that can be transformed has increased as tissue culture techniques have been developed for different species.

The first transgenic livestock were produced in 1985,[63] by micro-injecting foreign DNA into rabbit, sheep and pig eggs.[64] The first animal to synthesise transgenic proteins in their milk were mice,[65] engineered to produce human tissue plasminogen activator.[66] This technology was applied to sheep, pigs, cows and other livestock.[65]

In 2010 scientists at the J. Craig Venter Institute announced that they had created the first synthetic bacterial genome. The researchers added the new genome to bacterial cells and selected for cells that contained the new genome. To do this the cells undergoes a process called resolution, where during bacterial cell division one new cell receives the original DNA genome of the bacteria, whilst the other receives the new synthetic genome. When this cell replicates it uses the synthetic genome as its template. The resulting bacterium the researchers developed, named Synthia, was the world's first synthetic life form.[67][68]

In 2014 a bacteria was developed that replicated a plasmid containing an unnatural base pair. This required altering the bacterium so it could import the unnatural nucleotides and then efficiently replicate them. The plasmid retained the unnatural base pairs when it doubled an estimated 99.4% of the time.[69] This is the first organism engineered to use an expanded genetic alphabet.[70]

In 2015 CRISPR and TALENs was used to modify plant genomes. Chinese labs used it to create a fungus-resistant wheat and boost rice yields, while a U.K. group used it to tweak a barley gene that could help produce drought-resistant varieties. When used to precisely remove material from DNA without adding genes from other species, the result is not subject the lengthy and expensive regulatory process associated with GMOs. While CRISPR may use foreign DNA to aid the editing process, the second generation of edited plants contain none of that DNA. Researchers celebrated the acceleration because it may allow them to "keep up" with rapidly evolving pathogens. The U.S. Department of Agriculture stated that some examples of gene-edited corn, potatoes and soybeans are not subject to existing regulations. As of 2016 other review bodies had yet to make statements.[71]

2.7 Commercialisation

In 1976 Genentech, the first genetic engineering company was founded by Herbert Boyer and Robert Swanson and a year later and the company produced a human protein (somatostatin) in *E.coli*. Genentech announced the production of genetically engineered human insulin in 1978.[72] In 1980 the U.S. Supreme Court in the Diamond v. Chakrabarty case ruled that genetically altered life could be patented.[73] The insulin produced by bacteria, branded humulin, was approved for release by the Food and Drug Administration in 1982.[74]

In 1983 a biotech company, Advanced Genetic Sciences (AGS) applied for U.S. government authorization to perform field tests with the ice-minus strain of *P. syringae* to protect crops from frost, but environmental groups and protestors delayed the field tests for four years with legal challenges.[75] In 1987 the ice-minus strain of *P. syringae* became the first genetically modified organism (GMO) to be released into the environment[76] when a strawberry field and a potato field in California were sprayed with it.[77] Both test fields were attacked by activist groups the night before the tests occurred: "The world's first trial site attracted the world's first field trasher".[76]

The first genetically modified crop plant was produced in 1982, an antibiotic-resistant tobacco plant.[78] The first field trials of genetically engineered plants occurred in France and the USA in 1986, tobacco plants were engineered to be resistant to herbicides.[79] In 1987 Plant Genetic Systems, founded by Marc Van Montagu and Jeff Schell, was the first company to genetically engineer insect-resistant plants by incorporating genes that produced insecticidal proteins from Bacillus thuringiensis (Bt) into tobacco.[80]

Genetically modified microbial enzymes were the first application of genetically modified organisms in food production and were approved in 1988 by the US Food and Drug Administration.[81] In the early 1990s, recombinant chymosin was approved for use in several countries.[81][82] Cheese had typically been made using the enzyme complex rennet that had been extracted from cows' stomach lining. Scientists modified bacteria to produce chymosin, which was also able to clot milk, resulting in cheese curds.[83] The People's Republic of China was the first country to commercialize transgenic plants, introducing a virus-resistant tobacco in 1992.[84] In 1994 Calgene attained approval to commercially release the Flavr Savr tomato, a tomato engineered to have a longer shelf life.[85] Also in 1994, the European Union approved tobacco engineered to be resistant to the herbicide bromoxynil, making it the first genetically engineered crop commercialized in Europe.[86] In 1995 Bt Potato was approved safe by the Environmental Protection Agency, after having been approved by the FDA, making it the first pesticide producing crop to be approved in the USA.[87] In 1996 a total of 35 approvals had been granted to commercially grow 8 transgenic crops and one flower crop (carnation), with 8 different traits in 6 countries plus the EU.[79]

By 2010, 29 countries had planted commercialized biotech crops and a further 31 countries had granted regulatory approval for transgenic crops to be imported.[88] In 2013 Robert Fraley (Monsanto's executive vice president and chief technology officer), Marc Van Montagu and Mary-Dell Chilton were awarded the World Food Prize for improving the "quality, quantity or availability" of food in the world.[89]

The first genetically modified animal to be commercialised was the GloFish, a Zebra fish with a fluorescent gene added that allows it to glow in the dark under ultraviolet light.[90] The first genetically modified animal to be approved for food use was AquAdvantage salmon in 2015.[91] The salmon were transformed with a growth hormone-regulating gene from a Pacific Chinook salmon and a promoter from an ocean pout enabling it to grow year-round instead of only during spring and summer.[92]

2.8   Opposition

Opposition and support for the use of genetic engineering has existed since the technology was developed.[76] After Arpad Pusztai went public with research he was conducting in 1998 the public opposition to genetically modified food increased.[93] Opposition continued following controversial and publicly debated papers published in 1999 and 2013 that claimed negative environmental and health impacts from genetically modified crops.[94][95]

2.9   References

[1] Skoglund, Pontus; Ersmark, Erik; Palkopoulou, Eleftheria; Dalén, Love (2015-06-01). "Ancient Wolf Genome Reveals an Early Divergence of Domestic Dog Ancestors and Admixture into High-Latitude Breeds". *Current Biology* **25** (11): 1515–1519. doi:10.1016/j.cub.2015.04.019. ISSN 0960-9822. PMID 26004765.

[2] Jackson, DA; Symons, RH; Berg, P (1 October 1972). "Biochemical Method for Inserting New Genetic Information into DNA of Simian Virus 40: Circular SV40 DNA Molecules Containing Lambda Phage Genes and the Galactose Operon of Escherichia coli". *PNAS* **69** (10): 2904–2909. Bibcode:1972PNAS...69.2904J. doi:10.1073/pnas.69.10.2904. PMC 389671. PMID 4342968.

[3] Larson, Greger; Karlsson, Elinor K.; Perri, Angela; Webster, Matthew T.; Ho, Simon Y. W.; Peters, Joris; Stahl, Peter W.; Piper, Philip J.; Lingaas, Frode (2012-06-05). "Rethinking dog domestication by integrating genetics, archeology, and biogeography". *Proceedings of the National Academy of Sciences* **109** (23): 8878–8883. doi:10.1073/pnas.1203005109. ISSN 0027-8424. PMC 3384140. PMID 22615366.

[4] Montague, Michael J.; Li, Gang; Gandolfi, Barbara; Khan, Razib; Aken, Bronwen L.; Searle, Steven M. J.; Minx, Patrick; Hillier, LaDeana W.; Koboldt, Daniel

C. (2014-12-02). "Comparative analysis of the domestic cat genome reveals genetic signatures underlying feline biology and domestication". *Proceedings of the National Academy of Sciences* **111** (48): 17230–17235. doi:10.1073/pnas.1410083111. ISSN 0027-8424. PMC 4260561. PMID 25385592.

[5] Zeder, Melinda A. (2008-08-19). "Domestication and early agriculture in the Mediterranean Basin: Origins, diffusion, and impact". *Proceedings of the National Academy of Sciences* **105** (33): 11597–11604. doi:10.1073/pnas.0801317105. ISSN 0027-8424. PMC 2575338. PMID 18697943.

[6] Zohary, & Hopf Weiss, p. 1.

[7] the history of maize cultivation in southern Mexico dates back 9000 years. *New York Times*, accessdate=2010-5-4

[8] Sue Colledge and James Conolly (2007). *The Origins and Spread of Domestic Plants in Southwest Asia and Europe*, p. 40.

[9] Zohary, & Hopf Weiss, p. 5.

[10] Zohary, & Hopf Weiss, p. 6.

[11] Noel Kingsbury. Hybrid: The History and Science of Plant Breeding University of Chicago Press, Oct 15, 2009

[12] "Evolution of Wheatpublisher=Wheat, the big picture".

[13] Schouten, H. J.; Jacobsen, E. (2007). "Are Mutations in Genetically Modified Plants Dangerous?". *Journal of Biomedicine and Biotechnology* **2007**: 1–2. doi:10.1155/2007/82612.

[14] D. L. Hartl and V. Orel (1992). "What Did Gregor Mendel Think He Discovered?". *Genetics* **131** (2): 245–25.

[15] Vries, H. de (1889) *Intracellular Pangenesis* ("pan-gene" definition on page 7 and 40 of this 1910 translation in English)

[16] Creative Sponge. "The Bateson Lecture".

[17] Roberts, R. J. (2005). "Classic Perspective: How restriction enzymes became the workhorses of molecular biology". *Proceedings of the National Academy of Sciences* **102** (17): 5905–5908. doi:10.1073/pnas.0500923102. PMC 1087929. PMID 15840723.

[18] Weiss, B.; Richardson, C. C. (1967). "Enzymatic breakage and joining of deoxyribonucleic acid, I. Repair of single-strand breaks in DNA by an enzyme system from Escherichia coli infected with T4 bacteriophage". *Proceedings of the National Academy of Sciences* **57** (4): 1021–8. doi:10.1073/pnas.57.4.1021. PMC 224649. PMID 5340583.

[19] Lederberg, J (1952). "Cell genetics and hereditary symbiosis". *Physiological reviews* **32** (4): 403–30. PMID 13003535.

[20] Mandel, Morton; Higa, Akiko (1970). "Calcium-dependent bacteriophage DNA infection". *Journal of Molecular Biology* **53** (1): 159–162. doi:10.1016/0022-2836(70)90051-3. PMID 4922220.

[21] Cohen, S. N.; Chang, A. C. Y.; Hsu, L. (1972). "Nonchromosomal Antibiotic Resistance in Bacteria: Genetic Transformation of Escherichia coli by R-Factor DNA". *Proceedings of the National Academy of Sciences* **69** (8): 2110–4. doi:10.1073/pnas.69.8.2110. PMC 426879. PMID 4559594.

[22] Wirth, Reinhard; Friesenegger, Anita; Fiedlerand, Stefan (1989). "Transformation of various species of gram-negative bacteria belonging to 11 different genera by electroporation". *Molecular and General Genetics MGG* **216**: 175–177. doi:10.1007/BF00332248.

[23] Nester, Eugene. "Agrobacterium: The Natural Genetic Engineer (100 Years Later)". Retrieved 14 January 2011.

[24] Zambryski, P.; Joos, H.; Genetello, C.; Leemans, J.; Montagu, M. V.; Schell, J. (1983). "Ti plasmid vector for the introduction of DNA into plant cells without alteration of their normal regeneration capacity". *The EMBO Journal* **2** (12): 2143–2150. PMC 555426. PMID 16453482.

[25] Jackson, D. A.; Symons, R. H.; Berg, P. (1972). "Biochemical Method for Inserting New Genetic Information into DNA of Simian Virus 40: Circular SV40 DNA Molecules Containing Lambda Phage Genes and the Galactose Operon of Escherichia coli". *Proceedings of the National Academy of Sciences* **69** (10): 2904–9. doi:10.1073/pnas.69.10.2904. PMC 389671. PMID 4342968.

[26] "Genome and genetics timeline - 1973". Genome news network.

[27] Arnold, Paul (2009). "History of Genetics: Genetic Engineering Timeline".

[28] Stanley N. Cohen and Annie C. Y. Chang (1 May 1973). "Recircularization and Autonomous Replication of a Sheared R-Factor DNA Segment in Escherichia coli Transformants — PNAS". Pnas.org. Retrieved 17 July 2010.

[29] Jaenisch, R. and Mintz, B. (1974) Simian virus 40 DNA sequences in DNA of healthy adult mice derived from preimplantation blastocysts injected with viral DNA. Proc. Natl. Acad. 71(4):1250–1254

[30] Brownlee, C. (2004). "Inaugural Article: Biography of Rudolf Jaenisch". *Proceedings of the National Academy of Sciences* **101** (39): 13982–13984. doi:10.1073/pnas.0406416101. PMC 521108. PMID 15383657.

[31] Gordon, J.; Ruddle, F. (1981). "Integration and stable germ line transmission of genes injected into mouse pronuclei". *Science* **214** (4526): 1244–6. Bibcode:1981Sci...214.1244G. doi:10.1126/science.6272397. PMID 6272397.

[32] Costantini, F.; Lacy, E. (1981). "Introduction of a rabbit β-globin gene into the mouse germ line". *Nature* **294** (5836): 92–4. Bibcode:1981Natur.294...92C. doi:10.1038/294092a0. PMID 6945481.

[33] Lemaux, P. (2008). "Genetically Engineered Plants and Foods: A Scientist's Analysis of the Issues (Part I)". *Annual Review of Plant Biology* **59**: 771–812. doi:10.1146/annurev.arplant.58.032806.103840. PMID 18284373.

[34] Bevan, M. W.; Flavell, R. B.; Chilton, M. D. (1983). "A chimaeric antibiotic resistance gene as a selectable marker for plant cell transformation". *Nature* **304** (5922): 184–187. doi:10.1038/304184a0.

[35] Berg, P.; Baltimore, D.; Brenner, S.; Roblin, R. O.; Singer, M. F. (1975). "Summary statement of the Asilomar conference on recombinant DNA molecules". *Proceedings of the National Academy of Sciences* **72** (6): 1981–4. doi:10.1073/pnas.72.6.1981. PMC 432675. PMID 806076.

[36] Hutt, P. B. (1978). "Research on recombinant DNA molecules: The regulatory issues". *Southern California law review* **51** (6): 1435–50. PMID 11661661.

[37] McHughen A. Smyth S (2008). "US regulatory system for genetically modified [genetically modified organism (GMO), rDNA or transgenic] crop cultivars". *Plant biotechnology journal* **6** (1): 2–12. doi:10.1111/j.1467-7652.2007.00300.x. PMID 17956539.

[38] Bull, A.T., Holt, G. and Lilly, M.D. (1982). *Biotechnology : international trends and perspectives* (PDF). Paris: Organisation for Economic Co-operation and Development.

[39] U.S. Office of Science and Technology Policy (1986). "Coordinated framework for regulation of biotechnology; announcement of policy; notice for public comment". *Federal register* **51** (123): 23302–50. PMID 11655807.

[40] WHO (1987): Principles for the Safety Assessment of Food Additives and Contaminants in Food, Environmental Health Criteria 70. World Health Organization, Geneva

[41] WHO (1991): Strategies for assessing the safety of foods produced by biotechnology, Report of a Joint FAO/WHO Consultation. World Health Organization, Geneva

[42] WHO (1993): Health aspects of marker genes in genetically modified plants, Report of a WHO Workshop. World Health Organization, Geneva

[43] WHO (1995): Application of the principle of substantial equivalence to the safety evaluation of foods or food components from plants derived by modern biotechnology, Report of a WHO Workshop. World Health Organization, Geneva

[44] Gruère, Colin A. Carter and Guillaume P. (2003-12-15). "Mandatory Labeling of Genetically Modified Foods: Does it Really Provide Consumer Choice?". *www.agbioforum.org*. Retrieved 2016-01-21.

[45] Strom, Stephanie (2013-06-03). "Connecticut Approves Qualified Genetic Labeling". *The New York Times*. ISSN 0362-4331. Retrieved 2016-01-21.

[46] "Knockout Mice". National Human Genome Research Institute.

[47] Hanahan, D.; Wagner, E. F.; Palmiter, R. D. (2007). "The origins of oncomice: A history of the first transgenic mice genetically engineered to develop cancer". *Genes & Development* **21** (18): 2258–2270. doi:10.1101/gad.1583307. PMID 17875663.

[48] Helen R. Pilcher (2003). "It's a knockout: First rat to have key genes altered". *Nature*. doi:10.1038/news030512-17 (inactive 2015-01-09).

[49] Zan, Y; Haag, J. D.; Chen, K. S.; Shepel, L. A.; Wigington, D; Wang, Y. R.; Hu, R; Lopez-Guajardo, C. C.; Brose, H. L.; Porter, K. I.; Leonard, R. A.; Hitt, A. A.; Schommer, S. L.; Elegbede, A. F.; Gould, M. N. (2003). "Production of knockout rats using ENU mutagenesis and a yeast-based screening assay". *Nature Biotechnology* **21** (6): 645–51. doi:10.1038/nbt830. PMID 12754522.

[50] Lee, R.C.; Ambros, V. (1993). "The C. elegans heterochronic gene lin-4 encodes small RNAs with antisense complementarity to lin-14.". *Cell* **75**: 843–854. doi:10.1016/0092-8674(93)90529-y.

[51] Fire, A.; Xu, S.; Montgomery, M. K.; Kostas, S. A.; Driver, S. E.; Mello, C. C. (1998). "Potent and specific genetic interference by double-stranded RNA in Caenorhabditis elegans". *Nature* **391** (6669): 806–811. doi:10.1038/35888. PMID 9486653.

[52] Schwab, Rebecca; Ossowski, Stephan; Warthmann, Norman; Weigel, Detlef (2010-01-01). Meyers, Blake C.; Green, Pamela J., eds. *Directed Gene Silencing with Artificial MicroRNAs*. Methods in Molecular Biology. Humana Press. pp. 71–88. ISBN 9781603270045.

[53] Vaucheret, H.; Chupeau, Y. (2011). "Ingested plant miRNAs regulate gene expression in animals". *Cell Research* **22** (1): 3–5. doi:10.1038/cr.2011.164. PMC 3351922. PMID 22025251.

[54] Gentner, B.; Naldini, L. (2012-11-01). "Exploiting microRNA regulation for genetic engineering". *Tissue Antigens* **80** (5): 393–403. doi:10.1111/tan.12002. ISSN 1399-0039.

[55] Ladisch, M. R.; Kohlmann, K. L. (1992). "Recombinant human insulin". *Biotechnology Progress* **8** (6): 469–478. doi:10.1021/bp00018a001. PMID 1369033.

[56] Woodard, S. L.; Woodard, J. A.; Howard, M. E. (2004). "Plant molecular farming: Systems and products". *Plant Cell Reports* **22** (10): 711–720. doi:10.1007/s00299-004-0767-1. PMID 14997337.

[57] Ye, Xudong; Al-Babili, Salim; Klöti, Andreas; Zhang, Jing; Lucca, Paola; Beyer, Peter; Potrykus, Ingo (2000-01-14). "Engineering the Provitamin A (β-Carotene) Biosynthetic Pathway into (Carotenoid-Free) Rice Endosperm". *Science* **287** (5451): 303–305. doi:10.1126/science.287.5451.303. ISSN 0036-8075. PMID 10634784.

[58] Peters, Pamela. "Transforming Plants - Basic Genetic Engineering Techniques". Retrieved 28 January 2010.

[59] Voiland, Michael; McCandless, Linda. "Development Of The "Gene Gun" At Cornell". Archived from the original on May 1, 2008. Retrieved January 19, 2013.

[60] Roger Segelken for the Cornell Chronicle. Mary 14, 1987. Biologists Invent Gun for Shooting Cells with DNA Issue available as pdf download here, page 3

[61] http://web.archive.org/web/20130330150255/http://www.lifesciencesfoundation.org/events-item-799.html. Archived from the original on March 30, 2013. Retrieved January 18, 2013. Missing or empty |title= (help)

[62] Clough, S. J.; Bent, A. F. (1998). "Floral dip: A simplified method for Agrobacterium-mediated transformation of Arabidopsis thaliana". *The Plant Journal* **16** (6): 735–743. doi:10.1046/j.1365-313x.1998.00343.x. PMID 10069079.

[63] Brophy, B.; Smolenski, G.; Wheeler, T.; Wells, D.; l'Huillier, P.; Laible, G. T. (2003). "Cloned transgenic cattle produce milk with higher levels of β-casein and κ-casein". *Nature Biotechnology* **21** (2): 157–162. doi:10.1038/nbt783. PMID 12548290.

[64] Hammer, R. E.; Pursel, V. G.; Rexroad, C. E.; Wall, R. J.; Bolt, D. J.; Ebert, K. M.; Palmiter, R. D.; Brinster, R. L. (1985). "Production of transgenic rabbits, sheep and pigs by microinjection". *Nature* **315** (6021): 680–683. doi:10.1038/315680a0. PMID 3892305.

[65] A. John Clark. "The Mammary Gland as a Bioreactor: Expression, Processing, and Production of Recombinant Proteins". *Journal of Mammary Gland Biology and Neoplasia* **3** (3): 337–350. doi:10.1023/a:1018723712996.

[66] K. Gordon, E. Lee, J. Vitale, A. Smith, H. Westphal, and L. Hennighausen (1987). "Production of human tissue plasminogen activator in transgenic mouse milk". *Biotechnology* **5**: 1183±1187. doi:10.1038/nbt1187-1183.

[67] Gibson, D. G.; Glass, J. I.; Lartigue, C.; Noskov, V. N.; Chuang, R.-Y.; Algire, M. A.; Benders, G. A.; Montague, M. G.; Ma, L.; Moodie, M. M.; Merryman, C.; Vashee, S.; Krishnakumar, R.; Assad-Garcia, N.; Andrews-Pfannkoch, C.; Denisova, E. A.; Young, L.; Qi, Z.-Q.; Segall-Shapiro, T. H.; Calvey, C. H.; Parmar, P. P.; Hutchison Ca, C. A.; Smith, H. O.; Venter, J. C. (2010). "Creation of a Bacterial Cell Controlled by a Chemically Synthesized Genome". *Science* **329** (5987): 52–6. doi:10.1126/science.1190719. PMID 20488990.

[68] Sample, Ian (20 May 2010). "Craig Venter creates synthetic life form". London: guardian.co.uk.

[69] Malyshev, Denis A.; Dhami, Kirandeep; Lavergne, Thomas; Chen, Tingjian; Dai, Nan; Foster, Jeremy M.; Corrêa, Ivan R.; Romesberg, Floyd E. (2014-05-15). "A semi-synthetic organism with an expanded genetic alphabet". *Nature* **509** (7500): 385–388. doi:10.1038/nature13314. ISSN 0028-0836. PMC 4058825. PMID 24805238.

[70] Thyer, Ross; Ellefson, Jared (2014-05-15). "Synthetic biology: New letters for life's alphabet". *Nature* **509** (7500): 291–292. doi:10.1038/nature13335. ISSN 0028-0836.

[71] Talbot, David (2016-03). "10 Breakthrough Technologies 2016: Precise Gene Editing in Plants". *MIT Technology Review*. Retrieved 2016-03-08. Check date values in: |date= (help)

[72] Goeddel, D. V.; Kleid, D. G.; Bolivar, F.; Heyneker, H. L.; Yansura, D. G.; Crea, R.; Hirose, T.; Kraszewski, A.; Itakura, K.; Riggs, A. D. (1979). "Expression in Escherichia coli of chemically synthesized genes for human insulin". *Proceedings of the National Academy of Sciences* **76** (1): 106–10. doi:10.1073/pnas.76.1.106. PMC 382885. PMID 85300.

[73] US Supreme Court Cases from Justia & Oyez (16 June 1980). "Diamond V Chakrabarty" **447** (303). Supreme.justia.com. Retrieved 17 July 2010.

[74] "Artificial Genes". TIME. 15 November 1982. Retrieved 17 July 2010.

[75] Rebecca Bratspies (2007) Some Thoughts on the American Approach to Regulating Genetically Modified Organisms. Kansas Journal of Law and Public Policy 16:393

[76] BBC News 14 June 2002 GM crops: A bitter harvest?

[77] Thomas H. Maugh II for the Los Angeles Times. June 9, 1987. Altered Bacterium Does Its Job : Frost Failed to Damage Sprayed Test Crop, Company Says

[78] Fraley, RT et al. (1983) Expression of bacterial genes in plant cells. Proc. Natl. Acad. Sci. USA 80: 4803–4807

[79] James, Clive (1996). "Global Review of the Field Testing and Commercialization of Transgenic Plants: 1986 to 1995" (PDF). The International Service for the Acquisition of Agri-biotech Applications. Retrieved 17 July 2010.

[80] Vaeck, M et al. (1987) Transgenic plants protected from insect attack. *Nature* 328, 33–37 Transgenic plants protected from insect attack

[81] "FDA Approves 1st Genetically Engineered Product for Food". *Los Angeles Times*. 24 March 1990. Retrieved 1 May 2014.

[82] Staff, National Centre for Biotechnology Education, 2006. Case Study: Chymosin

[83] Campbell-Platt, Geoffrey (26 August 2011). *Food Science and Technology*. John Wiley & Sons. ISBN 978-1-4443-5782-0.

[84] James, Clive (1997). "Global Status of Transgenic Crops in 1997" (PDF). *ISAAA Briefs No. 5.*: 31.

[85] Bruening, G.; Lyons, J. M. (2000). "The case of the FLAVR SAVR tomato". *California Agriculture* **54** (4): 6–7. doi:10.3733/ca.v054n04p6.

[86] Debora MacKenzie (18 June 1994). "Transgenic tobacco is European first". New Scientist.

[87] Genetically Altered Potato Ok'd For Crops Lawrence Journal-World - 6 May 1995

[88] Global Status of Commercialized Biotech/GM Crops: 2011 ISAAA Brief ISAAA Brief 43-2011. Retrieved 14 October 2012

[89] Andrew Pollack (19 June 2013). "Executive at Monsanto wins global food honor". *The New York Times*. Retrieved 20 June 2013.

[90] Vàzquez-Salat, Núria; Salter, Brian; Smets, Greet; Houdebine, Louis-Marie (2012-11-01). "The current state of GMO governance: Are we ready for GM animals?". *Biotechnology Advances*. Special issue on ACB 2011 **30** (6): 1336–1343. doi:10.1016/j.biotechadv.2012.02.006.

[91] "AQUABOUNTY CLEARED TO SELL SALMON IN USA FOR COMMERCIAL PURPOSES".

[92] Bodnar, Anastasia (October 2010). "Risk Assessment and Mitigation of AquAdvantage Salmon" (PDF). ISB News Report.

[93] Arpad Pusztai: Biological divide James Randerson The Guardian January 15, 2008

[94] Waltz, Emily (2009-09-02). "GM crops: Battlefield". *Nature News* **461** (7260): 27–32. doi:10.1038/461027a.

[95] "Rat study sparks GM furore". *Nature News & Comment*. Retrieved 2016-01-21.

2.10   Sources

- Zohary, Daniel; Hopf, Maria; Weiss, Ehud (1 March 2012). *Domestication of Plants in the Old World: The Origin and Spread of Domesticated Plants in Southwest Asia, Europe, and the Mediterranean Basin*. OUP Oxford. ISBN 978-0-19-954906-1.

Chapter 3

Introduction to genetics

This article is a non-technical introduction to the subject. For the main encyclopedia article, see Genetics.
Genetics glossary
DNA

A long molecule that looks like a twisted ladder. It is made of four types of simple units and the sequence of these units carries information, just as the sequence of letters carries information on a page.

Nucleotides

They form the rungs of the DNA ladder and are the repeating units in DNA. There are four types of nucleotides (A, T, G and C) and it is the sequence of these nucleotides that carries information.

Chromosome

A package for carrying DNA in the cells. They contain a single long piece of DNA that is wound up and bunched together into a compact structure. Different species of plants and animals have different numbers and sizes of chromosomes.

Gene

A segment of DNA. Genes are like sentences made of the "letters" of the nucleotide alphabet, between them genes direct the physical development and behavior of an organism. Genes are like a recipe or instruction book, providing information that an organism needs so it can build or do something - like making an eye or a leg, or repairing a wound.

Allele

The different forms of a given gene that an organism may possess. For example, in humans, one allele of the eye-color gene produces green eyes and another allele of the eye-color gene produces brown eyes.

Genome

The complete set of genes in a particular organism.

Genetic engineering

When people change an organism by adding new genes, or deleting genes from its genome.

Mutation

An event that changes the sequence of the DNA in a gene.

Genetics is the study of genes — what they are, what they do, and how they work. Genes are made up of molecules inside the nucleus of a cell that are strung together in such a way that the sequence carries information: that information determines how living organisms inherit phenotypic traits, (features) determined by the genes they received from their parents and thereby going back through the generations. For example, offspring produced by sexual reproduction usually look similar to each of their parents because they have inherited some of each of their parents' genes. Genetics identifies which features are inherited, and explains how these features pass from generation to generation. In addition to inheritance, genetics studies how genes are turned on and off to control what substances are made in a cell - gene expression; and how a cell divides - mitosis or meiosis.

Some phenotypic traits can be seen, such as eye color while others can only be detected, such as blood type or intelligence. Traits determined by genes can be modified by the animal's surroundings (environment): for example, the general design of a tiger's stripes is inherited, but the specific stripe pattern is determined by the tiger's surroundings. Another example is a person's height: it is determined by both genetics and nutrition.

Genes are made of DNA, which is divided into separate pieces called chromosomes. Humans have 46: 23 pairs, though this number varies between species, for example many primates have 24 pairs. Meiosis creates special cells, sperm in males and eggs in females, which only have 23 chromosomes. These two cells merge into one during the fertilization stage of sexual reproduction, creating a zygote in which a nucleic acid double helix divides, with each single helix occupying one of the daughter cells, resulting in half the normal number of genes. The zygote then divides into four daughter cells by which time genetic recombination has created a new embryo with 23 pairs of chromosomes, half from each parent. Mating and resultant mate choice result in sexual selection. In normal cell division (mitosis) is possible when the double helix separates, and a complement of each separated half is made, resulting in two identical double helices in one cell, with each occupying one of the two new daughter cells created when the cell divides.

Chromosomes all contain four nucleotides, abbreviated C (cytosine), G (guanine), A (adenine), or T (thymine), which line up in a particular sequence and make a long string. There are two strings of nucleotides coiled around one another in each chromosome: a double helix. C on one string is always opposite from G on the other string; A is always opposite T. There are about 3.2 billion nucleotide pairs on all the human chromosomes: this is the human genome. The order of the nucleotides carries genetic information, whose rules are defined by the genetic code, similar to how the order of letters on a page of text carries information. Three nucleotides in a row - a triplet - carry one unit of information: a codon.

The genetic code not only controls inheritance: it also controls gene expression, which occurs when a portion of the double helix is uncoiled, exposing a series of the nucleotides, which are within the interior of the DNA. This series of exposed triplets (codons) carries the information to allow machinery in the cell to "read" the codons on the exposed DNA, which results in the making of RNA molecules. RNA in turn makes either amino acids or microRNA, which are responsible for all of the structure and function of a living organism; i.e. they determine all the features of the cell and thus the entire individual. Closing the uncoiled segment turns off the gene.

Heritability means the information in a given gene is not always exactly the same in every individual in that species, so the same gene in different individuals does not give exactly the same instructions. Each unique form of a single gene is called an allele; different forms are collectively called polymorphisms. As an example, one allele for the gene for hair color and skin cell pigmentation could instruct the body to produce black pigment, producing black hair and pigmented skin; while a different allele of the same gene in a different individual could give garbled instructions that would result in a failure to produce any pigment, giving white hair and no pigmented skin: albinism. Mutations are random changes in genes creating new alleles, which in

turn produce new traits, which could help, harm, or have no new effect on the individual's likelihood of survival; thus, mutations are the basis for evolution.

3.1 Inheritance in biology

3.1.1 Genes and inheritance

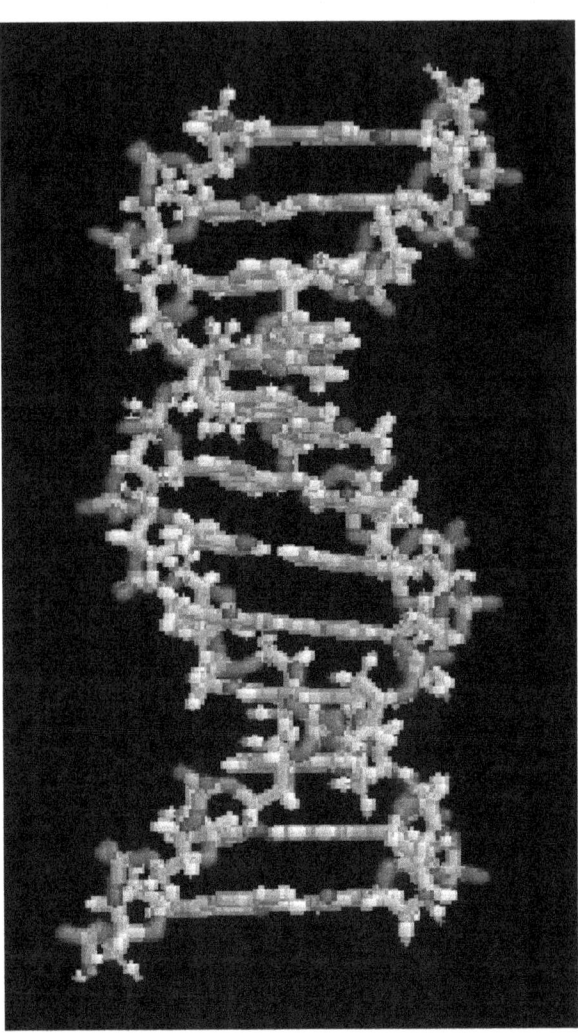

A section of DNA: the sequence of the plate-like units (nucleotides) in the center carries information.

Genes are pieces of DNA that contain information for synthesis of ribonucleic acids (RNAs) or polypeptides. Genes are inherited as units, with two parents dividing out copies of their genes to their offspring. This process can be compared with mixing two hands of cards, shuffling them, and then dealing them out again. Humans have two copies of each of their genes, and make copies that are found in eggs or sperm—but they only include *one* copy of each type of gene. An egg and sperm join to form a complete set

Red hair is a recessive trait.

of genes. The eventually resulting offspring has the same number of genes as their parents, but for any gene one of their two copies comes from their father, and one from their mother.[1]

The effects of this mixing depend on the types (the alleles) of the gene. If the father has two copies of an allele for red hair, and the mother has two copies for brown hair, all their children get the two alleles that give different instructions, one for red hair and one for brown. The hair color of these children depends on how these alleles work together. If one allele dominates the instructions from another, it is called the *dominant* allele, and the allele that is overridden is called the *recessive* allele. In the case of a daughter with alleles for both red and brown hair, brown is dominant and she ends up with brown hair.[2]

Although the red color allele is still there in this brown-haired girl, it doesn't show. This is a difference between what you see on the surface (the traits of an organism, called its phenotype) and the genes within the organism (its genotype). In this example you can call the allele for brown "B" and the allele for red "b". (It is normal to write dominant alleles with capital letters and recessive ones with lower-case letters.) The brown hair daughter has the "brown hair phenotype" but her genotype is Bb, with one copy of the B allele, and one of the b allele.

Now imagine that this woman grows up and has children with a brown-haired man who also has a Bb genotype. Her eggs will be a mixture of two types, one sort containing the B allele, and one sort the b allele. Similarly, her partner will produce a mix of two types of sperm containing one or the other of these two alleles. When the transmitted genes are joined up in their offspring, these children have a chance of getting either brown or red hair, since they could get a genotype of BB = brown hair, Bb = brown hair or bb = red hair. In this generation, there is therefore a chance of the recessive allele showing itself in the phenotype of the children - some of them may have red hair like their grandfather.[2]

Many traits are inherited in a more complicated way than the example above. This can happen when there are several genes involved, each contributing a small part to the end result. Tall people tend to have tall children because their children get a package of many alleles that each contribute a bit to how much they grow. However, there are not clear groups of "short people" and "tall people", like there are groups of people with brown or red hair. This is because of the large number of genes involved; this makes the trait very variable and people are of many different heights.[3] Despite a common misconception, the green/blue eye traits are also inherited in this complex inheritance model.[4] Inheritance can also be complicated when the trait depends on interaction between genetics and environment. For example, malnutrition does not change traits like eye color, but can stunt growth.[5]

3.1.2 Inherited diseases

Some diseases are hereditary and run in families; others, such as infectious diseases, are caused by the environment. Other diseases come from a combination of genes and the environment.[6] Genetic disorders are diseases that are caused by a single allele of a gene and are inherited in families. These include Huntington's disease, Cystic fibrosis or Duchenne muscular dystrophy. Cystic fibrosis, for example, is caused by mutations in a single gene called *CFTR* and is inherited as a recessive trait.[7]

Other diseases are influenced by genetics, but the genes a person gets from their parents only change their risk of getting a disease. Most of these diseases are inherited in a complex way, with either multiple genes involved, or coming from both genes and the environment. As an example, the risk of breast cancer is 50 times higher in the families most at risk, compared to the families least at risk. This variation is probably due to a large number of alleles, each changing the risk a little bit.[8] Several of the genes have been identified, such as *BRCA1* and *BRCA2*, but not all of them. However, although some of the risk is genetic, the risk of this cancer is also increased by being overweight, drinking a lot of alcohol and not exercising.[9] A woman's risk of breast cancer therefore comes from a large number of alleles interacting with her environment, so it is very hard to predict.

3.2 How genes work

3.2.1 Genes make proteins

Main article: Genetic code

The function of genes is to provide the information needed to make molecules called proteins in cells.[1] Cells are the smallest independent parts of organisms: the human body contains about 100 trillion cells, while very small organisms like bacteria are just one single cell. A cell is like a miniature and very complex factory that can make all the parts needed to produce a copy of itself, which happens when cells divide. There is a simple division of labor in cells - genes give instructions and proteins carry out these instructions, tasks like building a new copy of a cell, or repairing damage.[10] Each type of protein is a specialist that only does one job, so if a cell needs to do something new, it must make a new protein to do this job. Similarly, if a cell needs to do something faster or slower than before, it makes more or less of the protein responsible. Genes tell cells what to do by telling them which proteins to make and in what amounts.

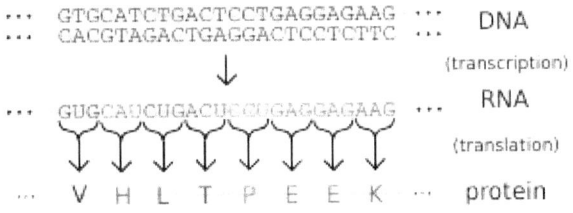

Genes are expressed by being transcribed into RNA, and this RNA then translated into protein.

Proteins are made of a chain of 20 different types of amino acid molecules. This chain folds up into a compact shape, rather like an untidy ball of string. The shape of the protein is determined by the sequence of amino acids along its chain and it is this shape that, in turn, determines what the protein does.[10] For example, some proteins have parts of their surface that perfectly match the shape of another molecule, allowing the protein to bind to this molecule very tightly. Other proteins are enzymes, which are like tiny machines that alter other molecules.[11]

The information in DNA is held in the sequence of the repeating units along the DNA chain.[12] These units are

four types of nucleotides (A,T,G and C) and the sequence of nucleotides stores information in an alphabet called the genetic code. When a gene is read by a cell the DNA sequence is copied into a very similar molecule called RNA (this process is called transcription). Transcription is controlled by other DNA sequences (such as promoters), which show a cell where genes are, and control how often they are copied. The RNA copy made from a gene is then fed through a structure called a ribosome, which translates the sequence of nucleotides in the RNA into the correct sequence of amino acids and joins these amino acids together to make a complete protein chain. The new protein then folds up into its active form. The process of moving information from the language of RNA into the language of amino acids is called translation.[13]

If the sequence of the nucleotides in a gene changes, the sequence of the amino acids in the protein it produces may also change - if part of a gene is deleted, the protein produced is shorter and may not work any more.[10] This is the reason why different alleles of a gene can have different effects in an organism. As an example, hair color depends on how much of a dark substance called melanin is put into the hair as it grows. If a person has a normal set of the genes involved in making melanin, they make all the proteins needed and they grow dark hair. However, if the alleles for a particular protein have different sequences and produce proteins that can't do their jobs, no melanin is produced and the person has white skin and hair (albinism).[14]

3.2.2 Genes are copied

Main article: DNA replication

Genes are copied each time a cell divides into two new cells. The process that copies DNA is called DNA replication.[12] It is through a similar process that a child inherits genes from its parents, when a copy from the mother is mixed with a copy from the father.

DNA can be copied very easily and accurately because each piece of DNA can direct the creation of a new copy of its information. This is because DNA is made of two strands that pair together like the two sides of a zipper. The nucleotides are in the center, like the teeth in the zipper, and pair up to hold the two strands together. Importantly, the four different sorts of nucleotides are different shapes, so for the strands to close up properly, an **A** nucleotide must go opposite a **T** nucleotide, and a **G** opposite a **C**. This exact pairing is called base pairing.[12]

When DNA is copied, the two strands of the old DNA are pulled apart by enzymes; then they pair up with new nucleotides and then close. This produces two new pieces of

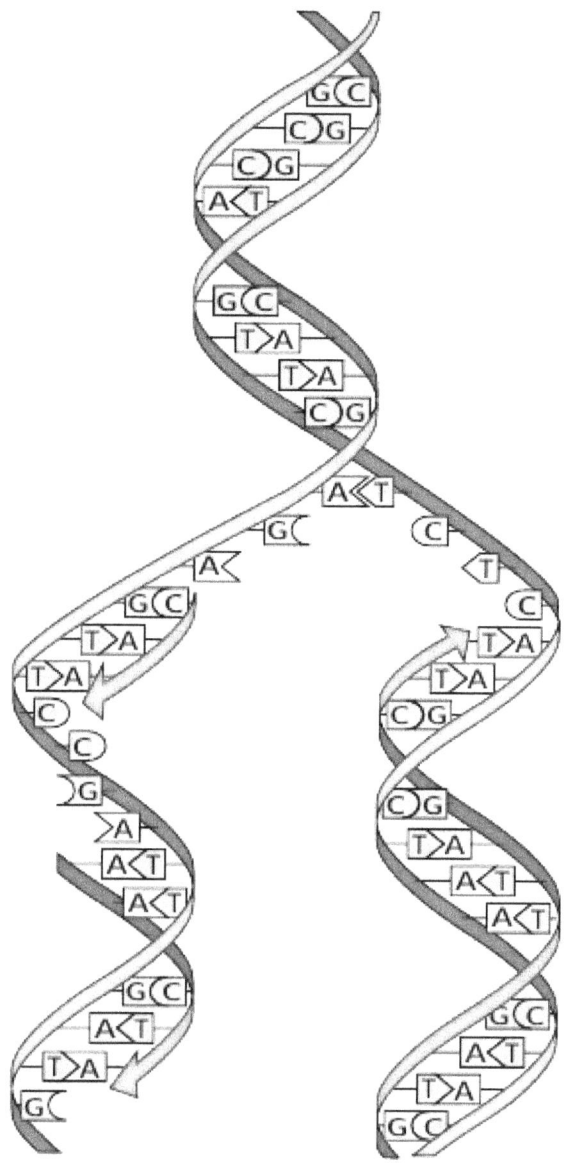

DNA replication. DNA is unwound and nucleotides are matched to make two new strands.

3.3 Genes and evolution

Further information: Evolution, Introduction to evolution, and History of evolutionary thought
A population of organisms evolves when an inherited trait

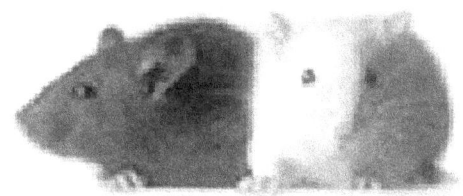

Mice with different coat colors.

becomes more common or less common over time.[16] For instance, all the mice living on an island would be a single population of mice: some with white fur, some gray. If over generations, white mice became more frequent and gray mice less frequent, then the color of the fur in this population of mice would be evolving. In terms of genetics, this is called an increase in allele frequency.

Alleles become more or less common either by chance in a process called genetic drift, or by natural selection.[17] In natural selection, if an allele makes it more likely for an organism to survive and reproduce, then over time this allele becomes more common. But if an allele is harmful, natural selection makes it less common. In the above example, if the island were getting colder each year and snow became present for much of the time, then the allele for white fur would favor survival, since predators would be less likely to see them against the snow, and more likely to see the gray mice. Over time white mice would become more and more frequent, while gray mice less and less.

Mutations create new alleles. These alleles have new DNA sequences and can produce proteins with new properties.[18] So if an island was populated entirely by black mice, mutations could happen creating alleles for white fur. The combination of mutations creating new alleles at random, and natural selection picking out those that are useful, causes adaptation. This is when organisms change in ways that help them to survive and reproduce.

3.4 Genetic engineering

Main article: Genetic engineering

Since traits come from the genes in a cell, putting a new piece of DNA into a cell can produce a new trait. This is

DNA, each containing one strand from the old DNA and one newly made strand. This process is not predictably perfect as proteins attach to a nucleotide while they are building and cause a change in the sequence of that gene. These changes in DNA sequence are called mutations.[15] Mutations produce new alleles of genes. Sometimes these changes stop the functioning of that gene or make it serve another advantageous function, such as the melanin genes discussed above. These mutations and their effects on the traits of organisms are one of the causes of evolution.[16]

how genetic engineering works. For example, rice can be given genes from a maize and a soil bacteria so the rice produces beta-carotene, which the body converts to Vitamin A.[19] This can help children suffering from Vitamin A deficiency. Another gene being put into some crops comes from the bacterium *Bacillus thuringiensis*; the gene makes a protein that is an insecticide. The insecticide kills insects that eat the plants, but is harmless to people.[20] In these plants, the new genes are put into the plant before it is grown, so the genes are in every part of the plant, including its seeds.[21] The plant's offspring inherit the new genes, which has led to concern about the spread of new traits into wild plants.[22]

The kind of technology used in genetic engineering is also being developed to treat people with genetic disorders in an experimental medical technique called gene therapy.[23] However, here the new gene is put in after the person has grown up and become ill, so any new gene is not inherited by their children. Gene therapy works by trying to replace the allele that causes the disease with an allele that works properly.

3.5   See also

- Common misunderstandings of genetics
- Epigenetics
- Full genome sequencing
- History of genetics
- Genetics in simple English
- List of basic genetics topics
- Molecular genetics
- Predictive medicine
- Timeline of the history of genetics

3.6   References

[1] *University of Utah Genetics Learning Center animated tour of the basics of genetics.* Howstuffworks.com. Retrieved 2008-01-24.

[2] MELANOCORTIN 1 RECEPTOR, Accessed 27 November 2010

[3] Multifactorial Inheritance Health Library, Morgan Stanley Children's Hospital. Accessed 20 May 2008

[4] Eye color is more complex than two genes, Athro Limited. Accessed 27 November 2010

[5] "Low income kids' height doesn't measure up by age 1". University of Michigan Health System. Retrieved May 20, 2008.

[6] Frequently Asked Questions About Genetic Disorders NIH. Accessed 20 May 2008

[7] Cystic fibrosis Genetics Home Reference, NIH, Accessed 16 May 2008

[8] Peto J (June 2002). "Breast cancer susceptibility-A new look at an old model". *Cancer Cell* 1 (5): 411–2. doi:10.1016/S1535-6108(02)00079-X. ISSN 1535-6108. PMID 12124169.

[9] What Are the Risk Factors for Breast Cancer? American Cancer Society. Accessed 16 May 2008

[10] The Structures of Life National Institute of General Medical Sciences, Accessed 20 May 2008

[11] Enzymes HowStuffWorks, Accessed 20 May 2008

[12] What is DNA? Genetics Home Reference, Accessed 16 May 2008

[13] DNA-RNA-Protein Nobelprize.org, Accessed 20 May 2008

[14] What is Albinism? The National Organization for Albinism and Hypopigmentation, Accessed 20 May 2008

[15] Mutations The University of Utah, Genetic Science Learning Center, Accessed 20 May 2008

[16] Brain, Marshall. "How Evolution Works", *How Stuff Works: Evolution Library.* Howstuffworks.com. Retrieved 2008-01-24.

[17] Mechanisms: The Processes of Evolution Understanding Evolution, Accessed 20 May 2008

[18] Genetic Variation Understanding Evolution, Accessed 20 May 2008

[19] Staff Golden Rice Project Retrieved 5 November 2012

[20] Tifton, Georgia: A Peanut Pest Showdown USDA, accessed 16 May 2008

[21] Genetic engineering: Bacterial arsenal to combat chewing insects GMO Safety, Jul 2010

[22] Genetically engineered organisms public issues education Cornell University, Accessed 16 May 2008

[23] Staff (November 18, 2005). "Gene Therapy" (FAQ). *Human Genome Project Information.* Oak Ridge National Laboratory. Retrieved 2006-05-28.

3.7   External links

- Introduction to Genetics, University of Utah

- Introduction to Genes and Disease, NCBI open book

- Genetics glossary, A talking glossary of genetic terms.

- Animated guide to cloning

- Khan Academy on YouTube

- What Color Eyes Would Your Children Have? Genetics of human eye color: An interactive introduction

- Double Helix Game from the Nobel Prize website. Match CATG bases with each other, and other games

- Transcribe and translate a gene, University of Utah

- StarGenetics software simulates mating experiments between organisms that are genetically different across a range of traits

Chapter 4

Genetics

This article is about the general scientific term. For the scientific journal, see Genetics (journal).
For a more accessible and less technical introduction to this topic, see Introduction to genetics.

Genetics is the study of genes, genetic variation, and heredity in living organisms.[1][2] It is generally considered a field of biology, but it intersects frequently with many of the life sciences and is strongly linked with the study of information systems.

The modern science of genetics, seeking to understand this process, began with the work of Imre Festetics, a Hungarian noble, who lived in Brno (Brünn) before Gregor Mendel. Imre Festetics was the first who used the word "genetics", more than 80 years earlier than William Bateson. He described several rules of genetic inheritance in his work *The genetic law of the Nature* (Die genetische Gesätze der Natur, 1819). His second law is the same as what Mendel published. In his third law, he developed the basic principles of mutation (he can be considered a forerunner of Hugo de Vries.[3]

The father of genetics is Gregor Mendel, a late 19th-century scientist and Augustinian friar. Mendel studied 'trait inheritance', patterns in the way traits were handed down from parents to offspring. He observed that organisms (pea plants) inherit traits by way of discrete "units of inheritance". This term, still used today, is a somewhat ambiguous definition of what is referred to as a gene.

Trait inheritance and molecular inheritance mechanisms of genes are still primary principles of genetics in the 21st century, but modern genetics has expanded beyond inheritance to studying the function and behavior of genes. Gene structure and function, variation, and distribution are studied within the context of the cell, the organism (e.g. dominance) and within the context of a population. Genetics has given rise to a number of sub-fields including epigenetics and population genetics. Organisms studied within the broad field span the domain of life, including bacteria, plants, animals, and humans.

Genetic processes work in combination with an organism's environment and experiences to influence development and behavior, often referred to as nature versus nurture. The intra- or extra-cellular environment of a cell or organism may switch gene transcription on or off. A classic example is two seeds of genetically identical corn, one placed in a temperate climate and one in an arid climate. While the average height of the two corn stalks may be genetically determined to be equal, the one in the arid climate only grows to half the height of the one in the temperate climate due to lack of water and nutrients in its environment.

4.1 Etymology

The word genetics stems from the Ancient Greek γενετικός *genetikos* meaning "genitive"/"generative", which in turn derives from γένεσις *genesis* meaning "origin".[4][5][6]

4.2 The gene

The modern working definition of a gene is a portion (or sequence) of DNA that codes for a known cellular function or process (e.g. the function "make melanin molecules"). A single 'gene' is most similar to a single 'word' in the English language. The nucleotides (molecules) that make up genes can be seen as 'letters' in the English language. Nucleotides are named according to which of the four nitrogenous bases they contain. The four bases are cytosine, guanine, adenine, and thymine. A single gene may have a small number of nucleotides or a large number of nucleotides, in the same way that a word may be small or large (e.g. 'cell' vs. 'electrophysiology'). A single gene often interacts with neighboring genes to produce a cellular function and can even be ineffectual without those neighboring genes. This can be seen in the same way that a 'word' may have meaning only in the context of a 'sentence.' A series of nucleotides can be put together without forming a gene (non coding regions of DNA), like a string of letters can be put together without forming a word (e.g. udkslk). Nonetheless, all words have letters, like all genes must have nucleotides.

A quick heuristic that is often used (but not always true) is "one gene, one protein" meaning a singular gene codes for a singular protein type in a cell (enzyme, transcription factor, etc.).

The sequence of nucleotides in a gene is read and translated by a cell to produce a chain of amino acids which in turn folds into a protein. The order of amino acids in a protein corresponds to the order of nucleotides in the gene. This relationship between nucleotide sequence and amino acid sequence is known as the genetic code. The amino acids in a protein determine how it folds into its unique three-dimensional shape, a structure that is ultimately responsible for the protein's function. Proteins carry out many of the functions needed for cells to live. A change to the DNA in a gene can alter a protein's amino acid sequence, thereby changing its shape and function and rendering the protein ineffective or even malignant (e.g. sickle cell anemia). Changes to genes are called mutations.

4.3 History

Main article: History of genetics

The observation that living things inherit traits from their parents has been used since prehistoric times to improve crop plants and animals through selective breeding.[7] The modern science of genetics, seeking to understand this process, began with the work of Gregor Mendel in the mid-19th century.[8]

Although the science of genetics began with the applied and theoretical work of Mendel, other theories of inheritance preceded his work. A popular theory during Mendel's time was the concept of blending inheritance: the idea that individuals inherit a smooth blend of traits from their parents.[9] Mendel's work provided examples where traits were definitely not blended after hybridization, showing that traits are produced by combinations of distinct genes rather than a continuous blend. Blending of traits in the progeny is now explained by the action of multiple genes with quantitative effects. Another theory that had some support at that time was the inheritance of acquired characteristics: the belief that individuals inherit traits strengthened by their parents. This theory (commonly associated with Jean-Baptiste Lamarck) is now known to be wrong—the experiences of individuals do not affect the genes they pass to their children,[10] although evidence in the field of epigenetics has revived some aspects of Lamarck's theory.[11] Other theories included the pangenesis of Charles Darwin (which had both acquired and inherited aspects) and Francis Galton's reformulation of pangenesis as both particulate and

inherited.[12]

4.3.1 Mendelian and classical genetics

Modern genetics started with Gregor Johann Mendel, a scientist and Augustinian friar who studied the nature of inheritance in plants. In his paper "Versuche über Pflanzenhybriden" ("Experiments on Plant Hybridization"), presented in 1865 to the Naturforschender Verein (Society for Research in Nature) in Brünn, Mendel traced the inheritance patterns of certain traits in pea plants and described them mathematically.[13] Although this pattern of inheritance could only be observed for a few traits, Mendel's work suggested that heredity was particulate, not acquired, and that the inheritance patterns of many traits could be explained through simple rules and ratios.

The importance of Mendel's work did not gain wide understanding until the 1890s, after his death, when other scientists working on similar problems re-discovered his research. William Bateson, a proponent of Mendel's work, coined the word genetics in 1905.[14][15] (The adjective genetic, derived from the Greek word genesis—γένεσις, "origin", predates the noun and was first used in a biological sense in 1860.)[16] Bateson both acted as a mentor and was aided significantly by the work of women scientists from Newnham College at Cambridge, specifically the work of Becky Saunders, Nora Darwin Barlow, and Muriel Wheldale Onslow.[17] Bateson popularized the usage of the word genetics to describe the study of inheritance in his inaugural address to the Third International Conference on Plant Hybridization in London, England, in 1906.[18]

After the rediscovery of Mendel's work, scientists tried to determine which molecules in the cell were responsible for inheritance. In 1911, Thomas Hunt Morgan argued that genes are on chromosomes, based on observations of a sex-linked white eye mutation in fruit flies.[19] In 1913, his student Alfred Sturtevant used the phenomenon of genetic linkage to show that genes are arranged linearly on the chromosome.[20]

4.3.2 Molecular genetics

Although genes were known to exist on chromosomes, chromosomes are composed of both protein and DNA, and scientists did not know which of the two is responsible for inheritance. In 1928, Frederick Griffith discovered the phenomenon of transformation (see Griffith's experiment): dead bacteria could transfer genetic material to "transform" other still-living bacteria. Sixteen years later, in 1944, the Avery–MacLeod–McCarty experiment identified DNA as the molecule responsible for transformation.[21] The role of the nucleus as the repository of genetic information in eukaryotes had been established by Hämmerling in 1943 in his work on the single celled alga *Acetabularia*.[22] The Hershey–Chase experiment in 1952 confirmed that DNA (rather than protein) is the genetic material of the viruses that infect bacteria, providing further evidence that DNA is the molecule responsible for inheritance.[23]

James Watson and Francis Crick determined the structure of DNA in 1953, using the X-ray crystallography work of Rosalind Franklin and Maurice Wilkins that indicated DNA had a helical structure (i.e., shaped like a corkscrew).[24][25] Their double-helix model had two strands of DNA with the nucleotides pointing inward, each matching a complementary nucleotide on the other strand to form what looks like rungs on a twisted ladder.[26] This structure showed that genetic information exists in the sequence of nucleotides on each strand of DNA. The structure also suggested a simple method for replication: if the strands are separated, new partner strands can be reconstructed for each based on the sequence of the old strand. This property is what gives DNA its semi-conservative nature where one strand of new DNA is from an original parent strand.[27]

Although the structure of DNA showed how inheritance works, it was still not known how DNA influences the behavior of cells. In the following years, scientists tried to understand how DNA controls the process of protein production.[28] It was discovered that the cell uses DNA as a template to create matching messenger RNA, molecules with nucleotides very similar to DNA. The nucleotide sequence of a messenger RNA is used to create an amino acid sequence in protein; this translation between nucleotide sequences and amino acid sequences is known as the genetic code.[29]

With the newfound molecular understanding of inheritance came an explosion of research.[30] A notable theory arose from Tomoko Ohta in 1973 with her amendment to the neutral theory of molecular evolution through publishing the nearly neutral theory of molecular evolution. In this theory, Ohta stressed the importance of natural selection and the environment to the rate at which genetic evolution occurs.[31] One important development was chain-termination DNA sequencing in 1977 by Frederick Sanger. This technology allows scientists to read the nucleotide sequence of a DNA molecule.[32] In 1983, Kary Banks Mullis developed the polymerase chain reaction, providing a quick way to isolate and amplify a specific section of DNA from a mixture.[33] The efforts of the Human Genome Project, Department of Energy, NIH, and parallel private efforts by Celera Genomics led to the sequencing of the human genome in 2003.[34]

4.4 Features of inheritance

4.4.1 Discrete inheritance and Mendel's laws

Main article: Mendelian inheritance

At its most fundamental level, inheritance in organisms occurs by passing discrete heritable units, called genes, from parents to progeny.[35] This property was first observed by Gregor Mendel, who studied the segregation of heritable traits in pea plants.[13][36] In his experiments studying the trait for flower color, Mendel observed that the flowers of each pea plant were either purple or white—but never an intermediate between the two colors. These different, discrete versions of the same gene are called alleles.

In the case of the pea, which is a diploid species, each individual plant has two copies of each gene, one copy inherited from each parent.[37] Many species, including humans, have this pattern of inheritance. Diploid organisms with two copies of the same allele of a given gene are called homozygous at that gene locus, while organisms with two different alleles of a given gene are called heterozygous.

The set of alleles for a given organism is called its genotype, while the observable traits of the organism are called its phenotype. When organisms are heterozygous at a gene, often one allele is called dominant as its qualities dominate the phenotype of the organism, while the other allele is called recessive as its qualities recede and are not observed. Some alleles do not have complete dominance and instead have incomplete dominance by expressing an intermediate phenotype, or codominance by expressing both alleles at once.[38]

When a pair of organisms reproduce sexually, their offspring randomly inherit one of the two alleles from each parent. These observations of discrete inheritance and the segregation of alleles are collectively known as Mendel's first law or the Law of Segregation.

4.4.2 Notation and diagrams

Geneticists use diagrams and symbols to describe inheritance. A gene is represented by one or a few letters. Often a "+" symbol is used to mark the usual, non-mutant allele for a gene.[39]

In fertilization and breeding experiments (and especially when discussing Mendel's laws) the parents are referred to as the "P" generation and the offspring as the "F1" (first filial) generation. When the F1 offspring mate with each other, the offspring are called the "F2" (second filial) generation. One of the common diagrams used to predict the result of cross-breeding is the Punnett square.

When studying human genetic diseases, geneticists often use pedigree charts to represent the inheritance of traits.[40] These charts map the inheritance of a trait in a family tree.

4.4.3   Multiple gene interactions

Organisms have thousands of genes, and in sexually reproducing organisms these genes generally assort independently of each other. This means that the inheritance of an allele for yellow or green pea color is unrelated to the inheritance of alleles for white or purple flowers. This phenomenon, known as "Mendel's second law" or the "Law of independent assortment", means that the alleles of different genes get shuffled between parents to form offspring with many different combinations. (Some genes do not assort independently, demonstrating genetic linkage, a topic discussed later in this article.)

Often different genes can interact in a way that influences the same trait. In the Blue-eyed Mary (*Omphalodes verna*), for example, there exists a gene with alleles that determine the color of flowers: blue or magenta. Another gene, however, controls whether the flowers have color at all or are white. When a plant has two copies of this white allele, its flowers are white—regardless of whether the first gene has blue or magenta alleles. This interaction between genes is called epistasis, with the second gene epistatic to the first.[41]

Many traits are not discrete features (e.g. purple or white flowers) but are instead continuous features (e.g. human height and skin color). These complex traits are products of many genes.[42] The influence of these genes is mediated, to varying degrees, by the environment an organism has experienced. The degree to which an organism's genes contribute to a complex trait is called heritability.[43] Measurement of the heritability of a trait is relative—in a more variable environment, the environment has a bigger influence on the total variation of the trait. For example, human height is a trait with complex causes. It has a heritability of 89% in the United States. In Nigeria, however, where people experience a more variable access to good nutrition and health care, height has a heritability of only 62%.[44]

4.5   Molecular basis for inheritance

4.5.1   DNA and chromosomes

Main articles: DNA and Chromosome
 The molecular basis for genes is deoxyribonucleic acid (DNA). DNA is composed of a chain of nucleotides, of which there are four types: adenine (A), cytosine (C), guanine (G), and thymine (T). Genetic information exists in the sequence of these nucleotides, and genes exist as stretches of sequence along the DNA chain.[45] Viruses are the only exception to this rule—sometimes viruses use the very similar molecule, RNA, instead of DNA as their genetic material.[46] Viruses cannot reproduce without a host and are unaffected by many genetic processes, so tend not to be considered living organisms.

DNA normally exists as a double-stranded molecule, coiled into the shape of a double helix. Each nucleotide in DNA preferentially pairs with its partner nucleotide on the opposite strand: A pairs with T, and C pairs with G. Thus, in its two-stranded form, each strand effectively contains all necessary information, redundant with its partner strand. This structure of DNA is the physical basis for inheritance: DNA replication duplicates the genetic information by splitting the strands and using each strand as a template for synthesis of a new partner strand.[47]

Genes are arranged linearly along long chains of DNA base-pair sequences. In bacteria, each cell usually contains a single circular genophore, while eukaryotic organisms (such as plants and animals) have their DNA arranged in multiple linear chromosomes. These DNA strands are often extremely long; the largest human chromosome, for example, is about 247 million base pairs in length.[48] The DNA of a chromosome is associated with structural proteins that organize, compact and control access to the DNA, forming a material called chromatin; in eukaryotes, chromatin is usually composed of nucleosomes, segments of DNA wound around cores of histone proteins.[49] The full set of hereditary material in an organism (usually the combined DNA sequences of all chromosomes) is called the genome.

While haploid organisms have only one copy of each chromosome, most animals and many plants are diploid, containing two of each chromosome and thus two copies of every gene.[37] The two alleles for a gene are located on identical loci of the two homologous chromosomes, each allele inherited from a different parent.

Many species have so-called sex chromosomes that determine the gender of each organism.[50] In humans and many other animals, the Y chromosome contains the gene that triggers the development of the specifically male characteristics. In evolution, this chromosome has lost most of its content and also most of its genes, while the X chromosome is similar to the other chromosomes and contains many genes. The X and Y chromosomes form a strongly heterogeneous pair.

4.5.2   Reproduction

Main articles: Asexual reproduction and Sexual reproduction

When cells divide, their full genome is copied and each daughter cell inherits one copy. This process, called mitosis, is the simplest form of reproduction and is the basis for asexual reproduction. Asexual reproduction can also occur in multicellular organisms, producing offspring that inherit their genome from a single parent. Offspring that are genetically identical to their parents are called clones.

Eukaryotic organisms often use sexual reproduction to generate offspring that contain a mixture of genetic material inherited from two different parents. The process of sexual reproduction alternates between forms that contain single copies of the genome (haploid) and double copies (diploid).[37] Haploid cells fuse and combine genetic material to create a diploid cell with paired chromosomes. Diploid organisms form haploids by dividing, without replicating their DNA, to create daughter cells that randomly inherit one of each pair of chromosomes. Most animals and many plants are diploid for most of their lifespan, with the haploid form reduced to single cell gametes such as sperm or eggs.

Although they do not use the haploid/diploid method of sexual reproduction, bacteria have many methods of acquiring new genetic information. Some bacteria can undergo conjugation, transferring a small circular piece of DNA to another bacterium.[51] Bacteria can also take up raw DNA fragments found in the environment and integrate them into their genomes, a phenomenon known as transformation.[52] These processes result in horizontal gene transfer, transmitting fragments of genetic information between organisms that would be otherwise unrelated.

4.5.3   Recombination and genetic linkage

Main articles: Chromosomal crossover and Genetic linkage

The diploid nature of chromosomes allows for genes on different chromosomes to assort independently or be separated from their homologous pair during sexual reproduction wherein haploid gametes are formed. In this way new combinations of genes can occur in the offspring of a mating pair. Genes on the same chromosome would theoretically never recombine. However, they do via the cellular process of chromosomal crossover. During crossover, chromosomes exchange stretches of DNA, effectively shuffling the gene alleles between the chromosomes.[53] This process of chromosomal crossover generally occurs during meiosis, a series of cell divisions that creates haploid cells.

The first cytological demonstration of crossing over was performed by Harriet Creighton and Barbara McClintock in 1931. Their research and experiments on corn provided cytological evidence for the genetic theory that linked genes on paired chromosomes do in fact exchange places from one homolog to the other.

The probability of chromosomal crossover occurring between two given points on the chromosome is related to the distance between the points. For an arbitrarily long distance, the probability of crossover is high enough that the inheritance of the genes is effectively uncorrelated.[54] For genes that are closer together, however, the lower probability of crossover means that the genes demonstrate genetic linkage; alleles for the two genes tend to be inherited together. The amounts of linkage between a series of genes can be combined to form a linear linkage map that roughly describes the arrangement of the genes along the chromosome.[55]

4.6   Gene expression

4.6.1   Genetic code

Main article: Genetic code

Genes generally express their functional effect through the production of proteins, which are complex molecules responsible for most functions in the cell. Proteins are made up of one or more polypeptide chains, each of which is composed of a sequence of amino acids, and the DNA sequence of a gene (through an RNA intermediate) is used to produce a specific amino acid sequence. This process begins with the production of an RNA molecule with a sequence matching the gene's DNA sequence, a process called transcription.

This messenger RNA molecule is then used to produce a corresponding amino acid sequence through a process called translation. Each group of three nucleotides in the sequence, called a codon, corresponds either to one of the twenty possible amino acids in a protein or an instruction to end the amino acid sequence; this correspondence is called the genetic code.[56] The flow of information is unidirectional: information is transferred from nucleotide sequences into the amino acid sequence of proteins, but it never transfers from protein back into the sequence of DNA—a phenomenon Francis Crick called the central dogma of molecular biology.[57]

The specific sequence of amino acids results in a unique three-dimensional structure for that protein, and the three-dimensional structures of proteins are related to their functions.[58][59] Some are simple structural molecules, like the fibers formed by the protein collagen. Proteins can bind to other proteins and simple molecules, sometimes acting as enzymes by facilitating chemical reactions within the bound molecules (without changing the structure of the protein itself). Protein structure is dynamic: the protein hemoglobin bends into slightly different forms as it facilitates the capture, transport, and release of oxygen molecules within mammalian blood.

A single nucleotide difference within DNA can cause a change in the amino acid sequence of a protein. Because protein structures are the result of their amino acid sequences, some changes can dramatically change the properties of a protein by destabilizing the structure or changing the surface of the protein in a way that changes its interaction with other proteins and molecules. For example, sickle-cell anemia is a human genetic disease that results from a single base difference within the coding region for the β-globin section of hemoglobin, causing a single amino acid change that changes hemoglobin's physical properties.[60] Sickle-cell versions of hemoglobin stick to themselves, stacking to form fibers that distort the shape of red blood cells carrying the protein. These sickle-shaped cells no longer flow smoothly through blood vessels, having a tendency to clog or degrade, causing the medical problems associated with this disease.

Some DNA sequences are transcribed into RNA but are not translated into protein products—such RNA molecules are called non-coding RNA. In some cases, these products fold into structures which are involved in critical cell functions (e.g. ribosomal RNA and transfer RNA). RNA can also have regulatory effects through hybridization interactions with other RNA molecules (e.g. microRNA).

4.6.2 Nature and nurture

Main article: Nature and nurture

Although genes contain all the information an organism uses to function, the environment plays an important role in determining the ultimate phenotypes an organism displays. This is the complementary relationship often referred to as "nature and nurture". The phenotype of an organism depends on the interaction of genes and the environment. An interesting example is the coat coloration of the Siamese cat. In this case, the body temperature of the cat plays the role of the environment. The cat's genes code for dark hair, thus the hair-producing cells in the cat make cellular proteins resulting in dark hair. But these dark hair-producing proteins are sensitive to temperature (i.e. have a mutation causing temperature-sensitivity) and denature in higher-temperature environments, failing to produce dark-hair pigment in areas where the cat has a higher body temperature. In a low-temperature environment, however, the protein's structure is stable and produces dark-hair pigment normally. The protein remains functional in areas of skin that are colder – such as its legs, ears, tail and face – so the cat has dark-hair at its extremities.[61]

Environment plays a major role in effects of the human genetic disease phenylketonuria.[62] The mutation that causes phenylketonuria disrupts the ability of the body to break down the amino acid phenylalanine, causing a toxic build-up of an intermediate molecule that, in turn, causes severe symptoms of progressive mental retardation and seizures. However, if someone with the phenylketonuria mutation follows a strict diet that avoids this amino acid, they remain normal and healthy.

A popular method in determining how genes and environment ("nature and nurture") contribute to a phenotype is by studying identical and fraternal twins or siblings of multiple births.[63] Because identical siblings come from the same zygote, they are genetically the same. Fraternal siblings are as genetically different from one another as normal siblings. By analyzing statistics on how often a twin of a set has a certain disorder compared to other sets of twins, scientists can determine whether that disorder is caused by genetic or environmental factors (i.e. whether it has 'nature' or 'nurture' causes). One famous example is the multiple birth study of the Genain quadruplets, who were identical quadruplets all diagnosed with schizophrenia.[64]

4.6.3 Gene regulation

Main article: Regulation of gene expression

The genome of a given organism contains thousands of genes, but not all these genes need to be active at any given moment. A gene is expressed when it is being transcribed into mRNA and there exist many cellular methods of controlling the expression of genes such that proteins are produced only when needed by the cell. Transcription factors are regulatory proteins that bind to DNA, either promoting or inhibiting the transcription of a gene.[65] Within the genome of *Escherichia coli* bacteria, for example, there exists a series of genes necessary for the synthesis of the amino acid tryptophan. However, when tryptophan is already available to the cell, these genes for tryptophan synthesis are no longer needed. The presence of tryptophan directly affects the activity of the genes—tryptophan molecules bind to the tryptophan repressor (a transcription factor), changing the repressor's structure such that the repressor binds to the genes. The tryptophan repressor blocks the transcription and expression of the genes, thereby creating negative feedback regulation of the tryptophan synthesis process.[66]

Differences in gene expression are especially clear within multicellular organisms, where cells all contain the same genome but have very different structures and behaviors due to the expression of different sets of genes. All the cells in a multicellular organism derive from a single cell, differentiating into variant cell types in response to external and intercellular signals and gradually establishing different patterns of gene expression to create different behaviors. As no single gene is responsible for the development of structures within multicellular organisms, these patterns arise from the complex interactions between many cells.

Within eukaryotes, there exist structural features of chromatin that influence the transcription of genes, often in the form of modifications to DNA and chromatin that are stably inherited by daughter cells.[67] These features are called "epigenetic" because they exist "on top" of the DNA sequence and retain inheritance from one cell generation to the next. Because of epigenetic features, different cell types grown within the same medium can retain very different properties. Although epigenetic features are generally dynamic over the course of development, some, like the phenomenon of paramutation, have multigenerational inheritance and exist as rare exceptions to the general rule of DNA as the basis for inheritance.[68]

4.7 Genetic change

4.7.1 Mutations

Main article: Mutation
During the process of DNA replication, errors occasionally occur in the polymerization of the second strand. These errors, called mutations, can have an impact on the phenotype of an organism, especially if they occur within the protein coding sequence of a gene. Error rates are usually very low—1 error in every 10–100 million bases—due to the "proofreading" ability of DNA polymerases.[69][70] Processes that increase the rate of changes in DNA are called mutagenic: mutagenic chemicals promote errors in DNA replication, often by interfering with the structure of base-pairing, while UV radiation induces mutations by causing damage to the DNA structure.[71] Chemical damage to DNA occurs naturally as well and cells use DNA repair mechanisms to repair mismatches and breaks. The repair does not, however, always restore the original sequence.

In organisms that use chromosomal crossover to exchange DNA and recombine genes, errors in alignment during meiosis can also cause mutations.[72] Errors in crossover are especially likely when similar sequences cause partner chromosomes to adopt a mistaken alignment; this makes some regions in genomes more prone to mutating in this way. These errors create large structural changes in DNA sequence - duplications, inversions, deletions of entire regions - or the accidental exchange of whole parts of sequences between different chromosomes (chromosomal translocation).

4.7.2 Natural selection and evolution

Main article: Evolution
Further information: Natural selection

Mutations alter an organism's genotype and occasionally this causes different phenotypes to appear. Most mutations have little effect on an organism's phenotype, health, or reproductive fitness.[73] Mutations that do have an effect are usually deleterious, but occasionally some can be beneficial.[74] Studies in the fly *Drosophila melanogaster* suggest that if a mutation changes a protein produced by a gene, about 70 percent of these mutations will be harmful with the remainder being either neutral or weakly beneficial.[75]

Population genetics studies the distribution of genetic differences within populations and how these distributions change over time.[76] Changes in the frequency of an allele in a population are mainly influenced by natural selection, where a given allele provides a selective or reproductive advantage to the organism,[77] as well as other factors such as mutation, genetic drift, genetic draft,[78] artificial selection and migration.[79]

Over many generations, the genomes of organisms can change significantly, resulting in evolution. In the process called adaptation, selection for beneficial mutations can cause a species to evolve into forms better able to survive in their environment.[80] New species are formed through the process of speciation, often caused by geographical separations that prevent populations from exchanging genes with each other.[81] The application of genetic principles to the study of population biology and evolution is known as the "modern synthesis".

By comparing the homology between different species' genomes, it is possible to calculate the evolutionary distance between them and when they may have diverged. Genetic comparisons are generally considered a more accurate method of characterizing the relatedness between species than the comparison of phenotypic characteristics. The evolutionary distances between species can be used to form evolutionary trees; these trees represent the common descent and divergence of species over time, although they do not show the transfer of genetic material between unrelated species (known as horizontal gene transfer and most common in bacteria).[82]

4.7.3 Model organisms

Although geneticists originally studied inheritance in a wide range of organisms, researchers began to specialize in studying the genetics of a particular subset of organisms. The fact that significant research already existed for a given organism would encourage new researchers to choose it for further study, and so eventually a few model organisms became the basis for most genetics research.[83] Common research topics in model organism genetics include the study of gene regulation and the involvement of genes in development and cancer.

Organisms were chosen, in part, for convenience—short generation times and easy genetic manipulation made some organisms popular genetics research tools. Widely used model organisms include the gut bacterium *Escherichia coli*, the plant *Arabidopsis thaliana*, baker's yeast (*Saccharomyces cerevisiae*), the nematode *Caenorhabditis elegans*, the common fruit fly (*Drosophila melanogaster*), and the common house mouse (*Mus musculus*).

4.7.4 Medicine

Medical genetics seeks to understand how genetic variation relates to human health and disease.[84] When searching for an unknown gene that may be involved in a disease, researchers commonly use genetic linkage and genetic pedigree charts to find the location on the genome associated with the disease. At the population level, researchers take advantage of Mendelian randomization to look for locations in the genome that are associated with diseases, a method especially useful for multigenic traits not clearly defined by a single gene.[85] Once a candidate gene is found, further research is often done on the corresponding gene – the orthologous gene – in model organisms. In addition to studying genetic diseases, the increased availability of genotyping methods has led to the field of pharmacogenetics: the study of how genotype can affect drug responses.[86]

Individuals differ in their inherited tendency to develop cancer,[87] and cancer is a genetic disease.[88] The process of cancer development in the body is a combination of events. Mutations occasionally occur within cells in the body as they divide. Although these mutations will not be inherited by any offspring, they can affect the behavior of cells, sometimes causing them to grow and divide more frequently. There are biological mechanisms that attempt to stop this process; signals are given to inappropriately dividing cells that should trigger cell death, but sometimes additional mutations occur that cause cells to ignore these messages. An internal process of natural selection occurs within the body and eventually mutations accumulate within cells to promote their own growth, creating a cancerous tumor that grows and invades various tissues of the body.

Normally, a cell divides only in response to signals called growth factors and stops growing once in contact with surrounding cells and in response to growth-inhibitory signals. It usually then divides a limited number of times and dies,

staying within the epithelium where it is unable to migrate to other organs. To become a cancer cell, a cell has to accumulate mutations in a number of genes (3–7) that allow it to bypass this regulation: it no longer needs growth factors to divide, it continues growing when making contact to neighbor cells, and ignores inhibitory signals, it will keep growing indefinitely and is immortal, it will escape from the epithelium and ultimately may be able to escape from the primary tumor, cross the endothelium of a blood vessel, be transported by the bloodstream and will colonize a new organ, forming deadly metastasis. Although there are some genetic predispositions in a small fraction of cancers, the major fraction is due to a set of new genetic mutations that originally appear and accumulate in one or a small number of cells that will divide to form the tumor and are not transmitted to the progeny (somatic mutations). The most frequent mutations are a loss of function of p53 protein, a tumor suppressor, or in the p53 pathway, and gain of function mutations in the ras proteins, or in other oncogenes.

4.7.5 Research methods

DNA can be manipulated in the laboratory. Restriction enzymes are commonly used enzymes that cut DNA at specific sequences, producing predictable fragments of DNA.[89] DNA fragments can be visualized through use of gel electrophoresis, which separates fragments according to their length.

The use of ligation enzymes allows DNA fragments to be connected. By binding ("ligating") fragments of DNA together from different sources, researchers can create recombinant DNA, the DNA often associated with genetically modified organisms. Recombinant DNA is commonly used in the context of plasmids: short circular DNA molecules with a few genes on them. In the process known as molecular cloning, researchers can amplify the DNA fragments by inserting plasmids into bacteria and then culturing them on plates of agar (to isolate clones of bacteria cells). ("Cloning" can also refer to the various means of creating cloned ("clonal") organisms.)

DNA can also be amplified using a procedure called the polymerase chain reaction (PCR).[90] By using specific short sequences of DNA, PCR can isolate and exponentially amplify a targeted region of DNA. Because it can amplify from extremely small amounts of DNA, PCR is also often used to detect the presence of specific DNA sequences.

4.7.6 DNA sequencing and genomics

DNA sequencing, one of the most fundamental technologies developed to study genetics, allows researchers to determine the sequence of nucleotides in DNA fragments.

The technique of chain-termination sequencing, developed in 1977 by a team led by Frederick Sanger, is still routinely used to sequence DNA fragments.[91] Using this technology, researchers have been able to study the molecular sequences associated with many human diseases.

As sequencing has become less expensive, researchers have sequenced the genomes of many organisms, using a process called genome assembly, which utilizes computational tools to stitch together sequences from many different fragments.[92] These technologies were used to sequence the human genome in the Human Genome Project completed in 2003.[34] New high-throughput sequencing technologies are dramatically lowering the cost of DNA sequencing, with many researchers hoping to bring the cost of resequencing a human genome down to a thousand dollars.[93]

Next generation sequencing (or high-throughput sequencing) came about due to the ever-increasing demand for low-cost sequencing. These sequencing technologies allow the production of potentially millions of sequences concurrently.[94][95] The large amount of sequence data available has created the field of genomics, research that uses computational tools to search for and analyze patterns in the full genomes of organisms. Genomics can also be considered a subfield of bioinformatics, which uses computational approaches to analyze large sets of biological data. A common problem to these fields of research is how to manage and share data that deals with human subject and personally identifiable information. See also genomics data sharing.

4.8   Society and culture

On 19 March 2015, a leading group of biologists urged a worldwide ban on clinical use of methods, particularly the use of CRISPR and zinc finger, to edit the human genome in a way that can be inherited.[96][97][98][99] In April 2015, Chinese researchers reported results of basic research to edit the DNA of non-viable human embryos using CRISPR.[100][101]

4.9   See also

- Bacterial genome size
- Eugenics
- Embryology
- Evolution
- Genetic disorder

- Genetic engineering
- Genetic enhancement
- Index of genetics articles
- Medical genetics
- Molecular tools for gene study
- Mutation
- Outline of genetics
- Timeline of the history of genetics

4.10   References

[1] Griffiths, Anthony J. F.; Miller, Jeffrey H.; Suzuki, David T.; Lewontin, Richard C.; Gelbart, eds. (2000). "Genetics and the Organism: Introduction". *An Introduction to Genetic Analysis* (7th ed.). New York: W. H. Freeman. ISBN 0-7167-3520-2.

[2] Hartl D, Jones E (2005)

[3] name=Poczai et all. 2014 Poczai P, Bell N, Hyvönen J (2014) Imre Festetics and the Sheep Breeders' Society of Moravia: Mendel's Forgotten "Research Network". PLoS Biol 12(1): e1001772. doi:10.1371/journal.pbio.1001772

[4] "Genetikos (γενετ-ικός)". *Henry George Liddell, Robert Scott, A Greek-English Lexicon*. Perseus Digital Library, Tufts University. Retrieved 20 February 2012.

[5] "Genesis (γένεσις)". *Henry George Liddell, Robert Scott, A Greek-English Lexicon*. Perseus Digital Library, Tufts University. Retrieved 20 February 2012.

[6] "Genetic". Online Etymology Dictionary. Retrieved 20 February 2012.

[7] DK Publishing (2009). *Science: The Definitive Visual Guide*. Penguin. p. 362. ISBN 978-0-7566-6490-9.

[8] Weiling, F (1991). "Historical study: Johann Gregor Mendel 1822–1884.". *American Journal of Medical Genetics* **40** (1): 1–25; discussion 26. doi:10.1002/ajmg.1320400103. PMID 1887835.

[9] Matthew Hamilton (2011). *Population Genetics*. Georgetown University. p. 26. ISBN 978-1-4443-6245-9.

[10] Lamarck, J-B (2008). In Encyclopædia Britannica. Retrieved from Encyclopædia Britannica Online on 16 March 2008.

[11] Singer, Emily (4 February 2009). "A Comeback for Lamarckian Evolution?". *Technology Review*. Retrieved 14 March 2013.

[12] Peter J. Bowler, *The Mendelian Revolution: The Emergency of Hereditarian Concepts in Modern Science and Society* (Baltimore: Johns Hopkins University Press, 1989): chapters 2 & 3.

[13] Blumberg, Roger B. "Mendel's Paper in English".

[14] genetics, *n.*, Oxford English Dictionary, 3rd ed.

[15] Bateson W. "Letter from William Bateson to Alan Sedgwick in 1905". The John Innes Centre. Retrieved 15 March 2008. Note that the letter was to an Adam Sedgwick, a zoologist and "Reader in Animal Morphology" at Trinity College, Cambridge

[16] genetic, *adj.*, Oxford English Dictionary, 3rd ed.

[17] Richmond, Marsha L. (November 2007). "Opportunities for women in early genetics". *Nature Review Genetics* **8**: 897–902. doi:10.1038/nrg2200. Retrieved April 23, 2015.

[18] Bateson, W (1907). "The Progress of Genetic Research". In Wilks, W. *Report of the Third 1906 International Conference on Genetics: Hybridization (the cross-breeding of genera or species), the cross-breeding of varieties, and general plant breeding*. London: Royal Horticultural Society.

> Initially titled the "International Conference on Hybridisation and Plant Breeding", the title was changed as a result of Bateson's speech. See: Cock AG, Forsdyke DR (2008). *Treasure your exceptions: the science and life of William Bateson*. Springer. p. 248. ISBN 978-0-387-75687-5.

[19] Moore, John A. (1983). "Thomas Hunt Morgan—The Geneticist". *Integrative and Comparative Biology* **23** (4): 855–865. doi:10.1093/icb/23.4.855.

[20] Sturtevant AH (1913). "The linear arrangement of six sex-linked factors in Drosophila, as shown by their mode of association" (PDF). *Journal of Experimental Biology* **14**: 43–59. doi:10.1002/jez.1400140104.

[21] Avery, OT; MacLeod, CM; McCarty, M (1944). "Studies on the Chemical Nature of the Substance Inducing Transformation of Pneumococcal Types: Induction of Transformation by a Desoxyribonucleic Acid Fraction Isolated from Pneumococcus Type III". *The Journal of Experimental Medicine* **79** (2): 137–58. doi:10.1084/jem.79.2.137. PMC 2135445. PMID 19871359. Reprint: Avery, OT; MacLeod, CM; McCarty, M (1979). "Studies on the chemical nature of the substance inducing transformation of pneumococcal types. Inductions of transformation by a desoxyribonucleic acid fraction isolated from pneumococcus type III". *The Journal of Experimental Medicine* **149** (2): 297–326. doi:10.1084/jem.149.2.297. PMC 2184805. PMID 33226.

[22] Cell and Molecular Biology", Pragya Khanna. I. K. International Pvt Ltd, 2008. p. 221. ISBN 81-89866-59-1, ISBN 978-81-89866-59-4

[23] Hershey, AD; Chase, M (1952). "Independent functions of viral protein and nucleic acid in growth of bacteriophage". *The Journal of General Physiology* **36** (1): 39–56. doi:10.1085/jgp.36.1.39. PMC 2147348. PMID 12981234.

[24] Judson, Horace (1979). *The Eighth Day of Creation: Makers of the Revolution in Biology*. Cold Spring Harbor Laboratory Press. pp. 51–169. ISBN 0-87969-477-7.

[25] Watson, J. D.; Crick, FH (1953). "Molecular Structure of Nucleic Acids: A Structure for Deoxyribose Nucleic Acid" (PDF). *Nature* **171** (4356): 737–8. Bibcode:1953Natur.171..737W. doi:10.1038/171737a0. PMID 13054692.

[26] Watson, J. D.; Crick, FH (1953). "Genetical Implications of the Structure of Deoxyribonucleic Acid" (PDF). *Nature* **171** (4361): 964–7. Bibcode:1953Natur.171..964W. doi:10.1038/171964b0. PMID 13063483.

[27] Stratmann, S. A. (1 Nov 2013). "DNA replication at the single molecule level". *Chemical Society Reviews* **43** (4): 1201–20. doi:10.1039/c3cs60391a. PMID 24395040.

[28] Frederick Betz (2010). *Managing Science: Methodology and Organization of Research*. Springer. p. 76. ISBN 978-1-4419-7488-4.

[29] Stanley A. Rice (2009). *Encyclopedia of Evolution*. Infobase Publishing. p. 134. ISBN 978-1-4381-1005-9.

[30] Sahotra Sarkar (1998). *Genetics and Reductionism*. Cambridge University Press. p. 140. ISBN 978-0-521-63713-8.

[31] Ohta, Tomoko (1973). "Slightly Deleterious Mutant Substitutions in Evolution". *Nature* **246** (5428): 96–98. Bibcode:1973Natur.246...96O. doi:10.1038/246096a0. PMID 4585855.

[32] Sanger, F; Nicklen, S; Coulson, AR (1977). "DNA sequencing with chain-terminating inhibitors". *Proceedings of the National Academy of Sciences of the United States of America* **74** (12): 5463–7. Bibcode:1977PNAS...74.5463S. doi:10.1073/pnas.74.12.5463. PMC 431765. PMID 271968.

[33] Saiki, RK; Scharf, S; Faloona, F; Mullis, KB; Horn, GT; Erlich, HA; Arnheim, N (1985). "Enzymatic amplification of beta-globin genomic sequences and restriction site analysis for diagnosis of sickle cell anemia". *Science* **230** (4732): 1350–4. Bibcode:1985Sci...230.1350S. doi:10.1126/science.2999980. PMID 2999980.

[34] "Human Genome Project Information". Human Genome Project. Retrieved 15 March 2008.

[35] Griffiths, Anthony J. F.; Miller, Jeffrey H.; Suzuki, David T.; Lewontin, Richard C.; Gelbart, eds. (2000). "Patterns of Inheritance: Introduction". *An Introduction to Genetic Analysis* (7th ed.). New York: W. H. Freeman. ISBN 0-7167-3520-2.

[36] Griffiths, Anthony J. F.; Miller, Jeffrey H.; Suzuki, David T.; Lewontin, Richard C.; Gelbart, eds. (2000). "Mendel's experiments". *An Introduction to Genetic Analysis* (7th ed.). New York: W. H. Freeman. ISBN 0-7167-3520-2.

[37] Griffiths, Anthony J. F.; Miller, Jeffrey H.; Suzuki, David T.; Lewontin, Richard C.; Gelbart, eds. (2000). "Mendelian genetics in eukaryotic life cycles". *An Introduction to Genetic Analysis* (7th ed.). New York: W. H. Freeman. ISBN 0-7167-3520-2.

[38] Griffiths, Anthony J. F.; Miller, Jeffrey H.; Suzuki, David T.; Lewontin, Richard C.; Gelbart, eds. (2000). "Interactions between the alleles of one gene". *An Introduction to Genetic Analysis* (7th ed.). New York: W. H. Freeman. ISBN 0-7167-3520-2.

[39] Cheney, Richard W. "Genetic Notation". Archived from the original on 3 January 2008. Retrieved 18 March 2008.

[40] Griffiths, Anthony J. F.; Miller, Jeffrey H.; Suzuki, David T.; Lewontin, Richard C.; Gelbart, eds. (2000). "Human Genetics". *An Introduction to Genetic Analysis* (7th ed.). New York: W. H. Freeman. ISBN 0-7167-3520-2.

[41] Griffiths, Anthony J. F.; Miller, Jeffrey H.; Suzuki, David T.; Lewontin, Richard C.; Gelbart, eds. (2000). "Gene interaction and modified dihybrid ratios". *An Introduction to Genetic Analysis* (7th ed.). New York: W. H. Freeman. ISBN 0-7167-3520-2.

[42] Mayeux, R (2005). "Mapping the new frontier: complex genetic disorders". *The Journal of Clinical Investigation* **115** (6): 1404–7. doi:10.1172/JCI25421. PMC 1137013. PMID 15931374.

[43] Griffiths, Anthony J. F.; Miller, Jeffrey H.; Suzuki, David T.; Lewontin, Richard C.; Gelbart, eds. (2000). "Quantifying heritability". *An Introduction to Genetic Analysis* (7th ed.). New York: W. H. Freeman. ISBN 0-7167-3520-2.

[44] Luke, A; Guo, X; Adeyemo, AA; Wilks, R; Forrester, T; Lowe Jr, W; Comuzzie, AG; Martin, LJ; Zhu, X; Rotimi, CN; Cooper, RS (2001). "Heritability of obesity-related traits among Nigerians, Jamaicans and US black people". *International journal of obesity and related metabolic disorders* **25** (7): 1034–41. doi:10.1038/sj.ijo.0801650. PMID 11443503.

[45] Pearson, H (2006). "Genetics: what is a gene?". *Nature* **441** (7092): 398–401. Bibcode:2006Natur.441..398P. doi:10.1038/441398a. PMID 16724031.

[46] Prescott, L (1993). *Microbiology*. Wm. C. Brown Publishers. ISBN 0-697-01372-3.

[47] Griffiths, Anthony J. F.; Miller, Jeffrey H.; Suzuki, David T.; Lewontin, Richard C.; Gelbart, eds. (2000). "Mechanism of DNA Replication". *An Introduction to Genetic Analysis* (7th ed.). New York: W. H. Freeman. ISBN 0-7167-3520-2.

[48] Gregory, SG; Barlow, KF; McLay, KE; Kaul, R; Swarbreck, D; Dunham, A; Scott, CE; Howe, KL; et al. (2006). "The DNA sequence and biological annotation of human chromosome 1". *Nature* **441** (7091): 315–21. Bibcode:2006Natur.441..315G. doi:10.1038/nature04727. PMID 16710414.

[49] Alberts et al. (2002), II.4. DNA and chromosomes: Chromosomal DNA and Its Packaging in the Chromatin Fiber

[50] Griffiths, Anthony J. F.; Miller, Jeffrey H.; Suzuki, David T.; Lewontin, Richard C.; Gelbart, eds. (2000). "Sex chromosomes and sex-linked inheritance". *An Introduction to Genetic Analysis* (7th ed.). New York: W. H. Freeman. ISBN 0-7167-3520-2.

[51] Griffiths, Anthony J. F.; Miller, Jeffrey H.; Suzuki, David T.; Lewontin, Richard C.; Gelbart, eds. (2000). "Bacterial conjugation". *An Introduction to Genetic Analysis* (7th ed.). New York: W. H. Freeman. ISBN 0-7167-3520-2.

[52] Griffiths, Anthony J. F.; Miller, Jeffrey H.; Suzuki, David T.; Lewontin, Richard C.; Gelbart, eds. (2000). "Bacterial transformation". *An Introduction to Genetic Analysis* (7th ed.). New York: W. H. Freeman. ISBN 0-7167-3520-2.

[53] Griffiths, Anthony J. F.; Miller, Jeffrey H.; Suzuki, David T.; Lewontin, Richard C.; Gelbart, eds. (2000). "Nature of crossing-over". *An Introduction to Genetic Analysis* (7th ed.). New York: W. H. Freeman. ISBN 0-7167-3520-2.

[54] Jack E. Staub (1994). *Crossover: Concepts and Applications in Genetics, Evolution, and Breeding*. University of Wisconsin Press. p. 55. ISBN 978-0-299-13564-5.

[55] Griffiths, Anthony J. F.; Miller, Jeffrey H.; Suzuki, David T.; Lewontin, Richard C.; Gelbart, eds. (2000). "Linkage maps". *An Introduction to Genetic Analysis* (7th ed.). New York: W. H. Freeman. ISBN 0-7167-3520-2.

[56] Berg JM, Tymoczko JL, Stryer L, Clarke ND (2002). "I. 5. DNA, RNA, and the Flow of Genetic Information: Amino Acids Are Encoded by Groups of Three Bases Starting from a Fixed Point". *Biochemistry* (5th ed.). New York: W. H. Freeman and Company.

[57] Crick, F (1970). "Central dogma of molecular biology" (PDF). *Nature* **227** (5258): 561–3. Bibcode:1970Natur.227..561C. doi:10.1038/227561a0. PMID 4913914.

[58] Alberts et al. (2002), I.3. Proteins: The Shape and Structure of Proteins

[59] Alberts et al. (2002), I.3. Proteins: Protein Function

[60] "How Does Sickle Cell Cause Disease?". Brigham and Women's Hospital: Information Center for Sickle Cell and Thalassemic Disorders. 11 April 2002. Retrieved 23 July 2007.

[61] Imes, DL; Geary, LA; Grahn, RA; Lyons, LA (2006). "Albinism in the domestic cat (*Felis catus*) is associated with a tyrosinase (TYR) mutation". *Animal genetics* **37** (2): 175–8. doi:10.1111/j.1365-2052.2005.01409.x. PMC 1464423. PMID 16573534.

[62] "MedlinePlus: Phenylketonuria". NIH: National Library of Medicine. Retrieved 15 March 2008.

[63] For example, Ridley M (2003). *Nature via nurture: genes, experience and what makes us human.* Fourth Estate. p. 73. ISBN 978-1-84115-745-0.

[64] Rosenthal, David (1964). *The Genain quadruplets: a case study and theoretical analysis of heredity and environment in schizophrenia.* New York: Basic Books. doi:10.1002/bs.3830090407.

[65] Brivanlou, AH; Darnell Jr, JE (2002). "Signal transduction and the control of gene expression". *Science* **295** (5556): 813–8. Bibcode:2002Sci...295..813B. doi:10.1126/science.1066355. PMID 11823631.

[66] Alberts et al. (2002). II.3. Control of Gene Expression – The Tryptophan Repressor Is a Simple Switch That Turns Genes On and Off in Bacteria

[67] Jaenisch, R; Bird, A (2003). "Epigenetic regulation of gene expression: how the genome integrates intrinsic and environmental signals". *Nature Genetics.* 33 Suppl (3s): 245–54. doi:10.1038/ng1089. PMID 12610534.

[68] Chandler, VL (2007). "Paramutation: from maize to mice". *Cell* **128** (4): 641–5. doi:10.1016/j.cell.2007.02.007. PMID 17320501.

[69] Griffiths, Anthony J. F.; Miller, Jeffrey H.; Suzuki, David T.; Lewontin, Richard C.; Gelbart, eds. (2000). "Spontaneous mutations". *An Introduction to Genetic Analysis* (7th ed.). New York: W. H. Freeman. ISBN 0-7167-3520-2.

[70] Freisinger, E; Grollman, AP; Miller, H; Kisker, C (2004). "Lesion (in)tolerance reveals insights into DNA replication fidelity". *The EMBO Journal* **23** (7): 1494–505. doi:10.1038/sj.emboj.7600158. PMC 391067. PMID 15057282.

[71] Griffiths, Anthony J. F.; Miller, Jeffrey H.; Suzuki, David T.; Lewontin, Richard C.; Gelbart, eds. (2000). "Induced mutations". *An Introduction to Genetic Analysis* (7th ed.). New York: W. H. Freeman. ISBN 0-7167-3520-2.

[72] Griffiths, Anthony J. F.; Miller, Jeffrey H.; Suzuki, David T.; Lewontin, Richard C.; Gelbart, eds. (2000). "Chromosome Mutation I: Changes in Chromosome Structure: Introduction". *An Introduction to Genetic Analysis* (7th ed.). New York: W. H. Freeman. ISBN 0-7167-3520-2.

[73] Moselio Schaechter (2009). *Encyclopedia of Microbiology.* Academic Press. p. 551. ISBN 978-0-12-373944-5.

[74] Mike Calver; Alan Lymbery; Jennifer McComb; Mike Bamford (2009). *Environmental Biology.* Cambridge University Press. p. 118. ISBN 978-0-521-67982-4.

[75] Sawyer, SA; Parsch, J; Zhang, Z; Hartl, DL (2007). "Prevalence of positive selection among nearly neutral amino acid replacements in Drosophila". *Proceedings of the National Academy of Sciences of the United States of America* **104** (16): 6504–10. Bibcode:2007PNAS..104.6504S. doi:10.1073/pnas.0701572104. PMC 1871816. PMID 17409186.

[76] Griffiths, Anthony J. F.; Miller, Jeffrey H.; Suzuki, David T.; Lewontin, Richard C.; Gelbart, eds. (2000). "Variation and its modulation". *An Introduction to Genetic Analysis* (7th ed.). New York: W. H. Freeman. ISBN 0-7167-3520-2.

[77] Griffiths, Anthony J. F.; Miller, Jeffrey H.; Suzuki, David T.; Lewontin, Richard C.; Gelbart, eds. (2000). "Selection". *An Introduction to Genetic Analysis* (7th ed.). New York: W. H. Freeman. ISBN 0-7167-3520-2.

[78] Gillespie, John H. (2001). "Is the population size of a species relevant to its evolution?". *Evolution* **55** (11): 2161–2169. doi:10.1111/j.0014-3820.2001.tb00732.x. PMID 11794777.

[79] Griffiths, Anthony J. F.; Miller, Jeffrey H.; Suzuki, David T.; Lewontin, Richard C.; Gelbart, eds. (2000). "Random events". *An Introduction to Genetic Analysis* (7th ed.). New York: W. H. Freeman. ISBN 0-7167-3520-2.

[80] Darwin, Charles (1859). *On the Origin of Species* (1st ed.). London: John Murray. p. 1. ISBN 0-8014-1319-2. Earlier related ideas were acknowledged in Darwin, Charles (1861). *On the Origin of Species* (3rd ed.). London: John Murray. xiii. ISBN 0-8014-1319-2.

[81] Gavrilets, S (2003). "Perspective: models of speciation: what have we learned in 40 years?". *Evolution; international journal of organic evolution* **57** (10): 2197–215. doi:10.1554/02-727. PMID 14628909.

[82] Wolf, YI; Rogozin, IB; Grishin, NV; Koonin, EV (2002). "Genome trees and the tree of life". *Trends in Genetics* **18** (9): 472–9. doi:10.1016/S0168-9525(02)02744-0. PMID 12175808.

[83] "The Use of Model Organisms in Instruction". University of Wisconsin: Wisconsin Outreach Research Modules. Retrieved 15 March 2008.

[84] "NCBI: Genes and Disease". NIH: National Center for Biotechnology Information. Retrieved 15 March 2008.

[85] Davey Smith, G; Ebrahim, S (2003). "'Mendelian randomization': can genetic epidemiology contribute to understanding environmental determinants of disease?". *International Journal of Epidemiology* **32** (1): 1–22. doi:10.1093/ije/dyg070. PMID 12689998.

[86] "Pharmacogenetics Fact Sheet". NIH: National Institute of General Medical Sciences. Retrieved 15 March 2008.

[87] Frank, SA (2004). "Genetic predisposition to cancer – insights from population genetics". *Nature reviews. Genetics* **5** (10): 764–72. doi:10.1038/nrg1450. PMID 15510167.

[88] Strachan T, Read AP (1999). *Human Molecular Genetics 2* (second ed.). John Wiley & Sons Inc. Chapter 18: Cancer Genetics

[89] Lodish et al. (2000), Chapter 7: 7.1. DNA Cloning with Plasmid Vectors

[90] Lodish et al. (2000), Chapter 7: 7.7. Polymerase Chain Reaction: An Alternative to Cloning

[91] Brown TA (2002). "Section 2, Chapter 6: 6.1. The Methodology for DNA Sequencing". *Genomes 2* (2nd ed.). Oxford: Bios. ISBN 1-85996-228-9.

[92] Brown (2002), Section 2, Chapter 6: 6.2. Assembly of a Contiguous DNA Sequence

[93] Service, RF (2006). "Gene sequencing. The race for the $1000 genome". *Science* **311** (5767): 1544–6. doi:10.1126/science.311.5767.1544. PMID 16543431.

[94] Hall, Nell (May 2007). "Advanced sequencing technologies and their wider impact in microbiology". *J. Exp. Biol.* **209** (Pt 9): 1518–1525. doi:10.1242/jeb.001370. PMID 17449817.

[95] Church, George M. (January 2006). "Genomes for all". *Sci. Am.* **294** (1): 46–54. doi:10.1038/scientificamerican0106-46. PMID 16468433.(subscription required)

[96] Wade, Nicholas (19 March 2015). "Scientists Seek Ban on Method of Editing the Human Genome". *New York Times*. Retrieved 20 March 2015.

[97] Pollack, Andrew (3 March 2015). "A Powerful New Way to Edit DNA". *New York Times*. Retrieved 20 March 2015.

[98] Baltimore, David; Berg, Paul; Botchan, Dana; Charo, R. Alta; Church, George; Corn, Jacob E.; Daley, George Q.; Doudna, Jennifer A.; Fenner, Marsha; Greely, Henry T.; Jinek, Martin; Martin, G. Steven; Penhoet, Edward; Puck, Jennifer; Sternberg, Samuel H.; Weissman, Jonathan S.; Yamamoto, Keith R. (19 March 2015). "A prudent path forward for genomic engineering and germline gene modification". *Science* **348**: 36–8. Bibcode:2015Sci...348...36B. doi:10.1126/science.aab1028. PMID 25791083. Retrieved 20 March 2015.

[99] Lanphier, Edward; Urnov, Fyodor; Haecker, Sarah Ehlen; Werner, Michael; Smolenski, Joanna (26 March 2015). "Don't edit the human germ line". *Nature* **519**: 410–411. Bibcode:2015Natur.519..410L. doi:10.1038/519410a. PMID 25810189. Retrieved 20 March 2015.

[100] Kolata, Gina (23 April 2015). "Chinese Scientists Edit Genes of Human Embryos, Raising Concerns". *New York Times*. Retrieved 24 April 2015.

[101] Liang, Puping; et al. (18 April 2015). "CRISPR/Cas9-mediated gene editing in human tripronuclear zygotes". *Protein & Cell* **6**: 363–72. doi:10.1007/s13238-015-0153-5. PMC 4417674. PMID 25894090. Retrieved 24 April 2015.

4.11   Further reading

See also: Bibliography of biology § Genetics

- Bruce Alberts; Dennis Bray; Karen Hopkin; Alexander Johnson; Julian Lewis; Martin Raff; Keith Roberts; Peter Walter (2013). *Essential Cell Biology, 4th Edition*. Garland Science. ISBN 978-1-317-80627-1.

- Griffiths, Anthony J. F.; Miller, Jeffrey H.; Suzuki, David T.; Lewontin, Richard C.; Gelbart. eds. (2000). *An Introduction to Genetic Analysis* (7th ed.). New York: W. H. Freeman. ISBN 0-7167-3520-2.

- Hartl D, Jones E (2005). *Genetics: Analysis of Genes and Genomes* (6th ed.). Jones & Bartlett. ISBN 0-7637-1511-5.

- King, Robert C; Mulligan, Pamela K; Stansfield, William D (2013). *A Dictionary of Genetics* (8th ed.). New York: Oxford University Press. ISBN 0-1997-6644-4.

- Lodish H, Berk A, Zipursky LS, Matsudaira P, Baltimore D, and Darnell J (2000). *Molecular Cell Biology* (4th ed.). New York: Scientific American Books. ISBN 0-7167-3136-3.

4.12   External links

-
- Genetics on *In Our Time* at the BBC. (listen now)
- Genetics at DMOZ

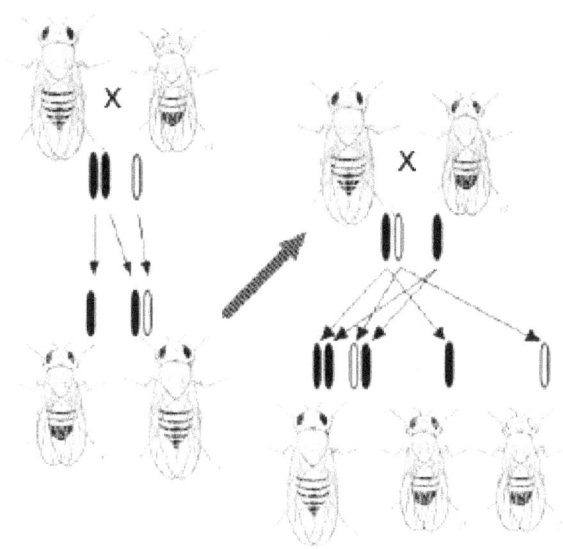

Morgan's observation of sex-linked inheritance of a mutation caus-ing white eyes in Drosophila *led him to the hypothesis that genes are located upon chromosomes.*

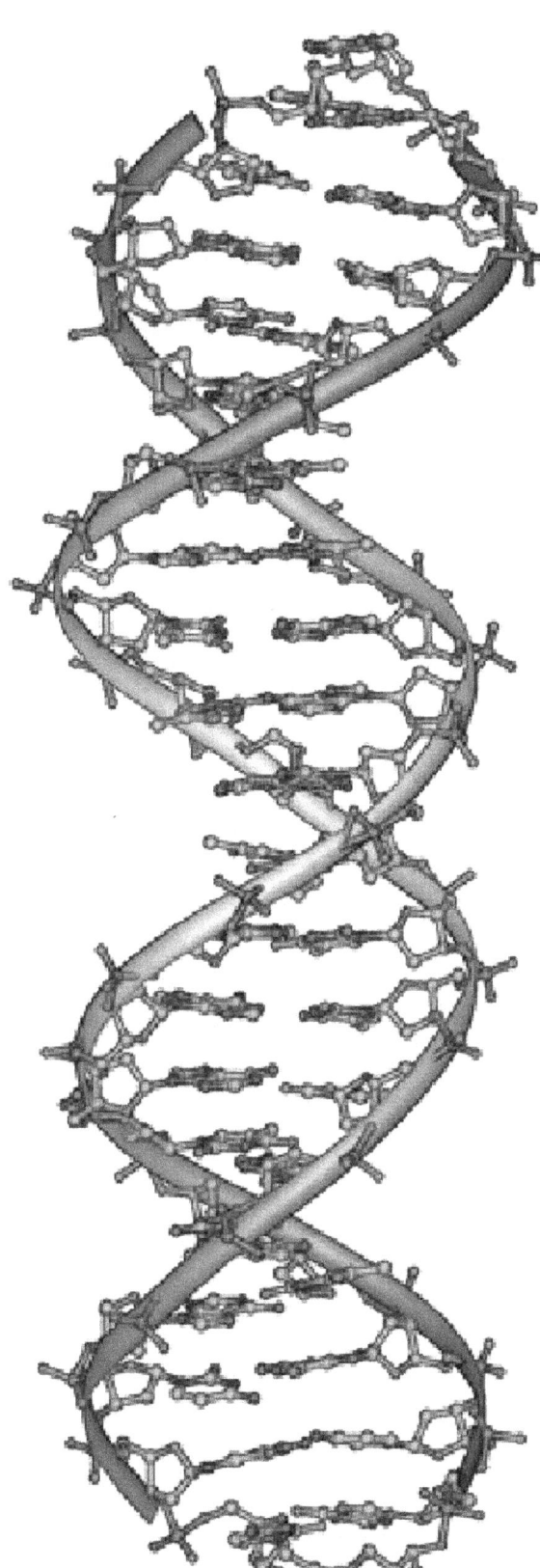

DNA, the molecular basis for biological inheritance. Each strand of DNA is a chain of nucleotides, matching each other in the center to form what look like rungs on a twisted ladder.

		pollen ♂	
		B	**b**
pistil ♀	**B**	**BB**	**Bb**
	b	**Bb**	**bb**

A Punnett square depicting a cross between two pea plants heterozygous for purple (B) and white (b) blossoms.

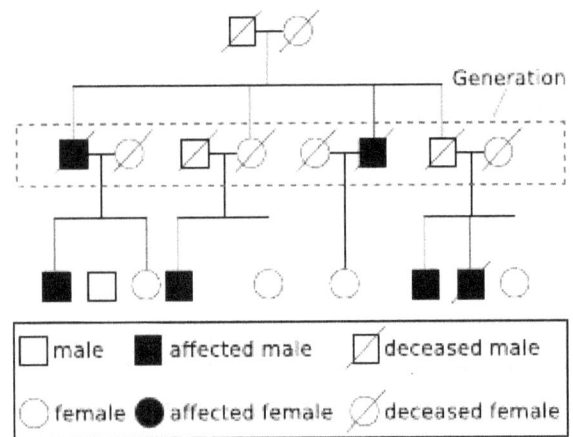

Genetic pedigree charts help track the inheritance patterns of traits.

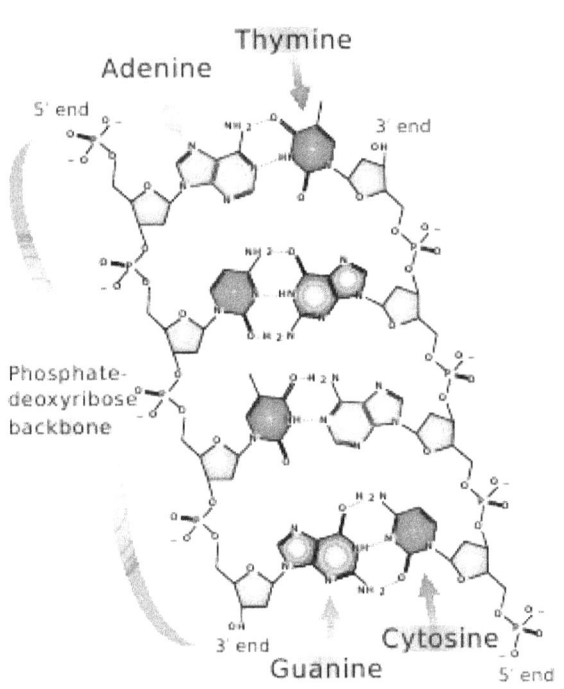

The molecular structure of DNA. Bases pair through the arrangement of hydrogen bonding between the strands.

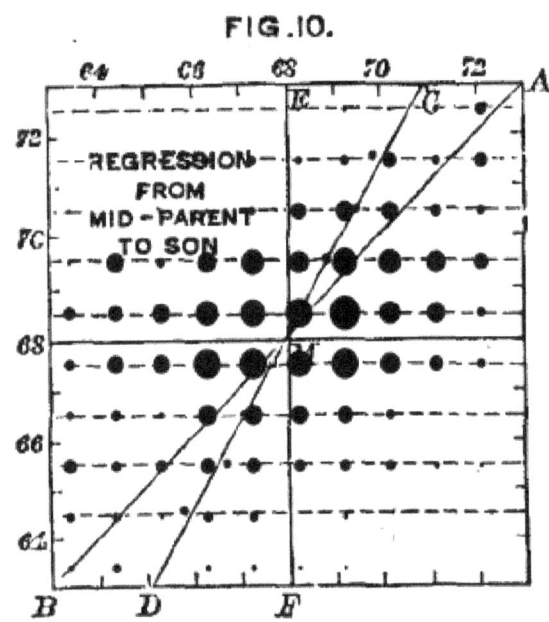

Human height is a trait with complex genetic causes. Francis Galton's data from 1889 shows the relationship between offspring height as a function of mean parent height. While correlated, remaining variation in offspring heights indicates environment is also an important factor in this trait.

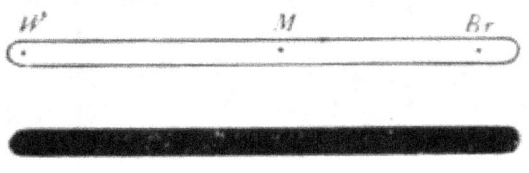

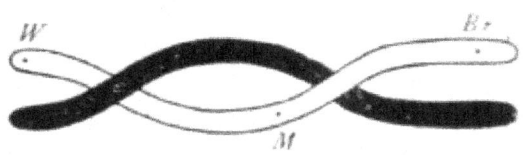

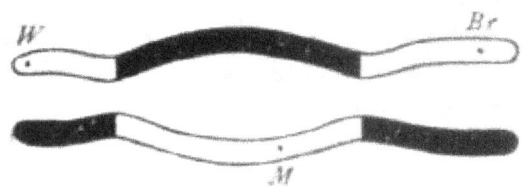

Thomas Hunt Morgan's 1916 illustration of a double crossover between chromosomes.

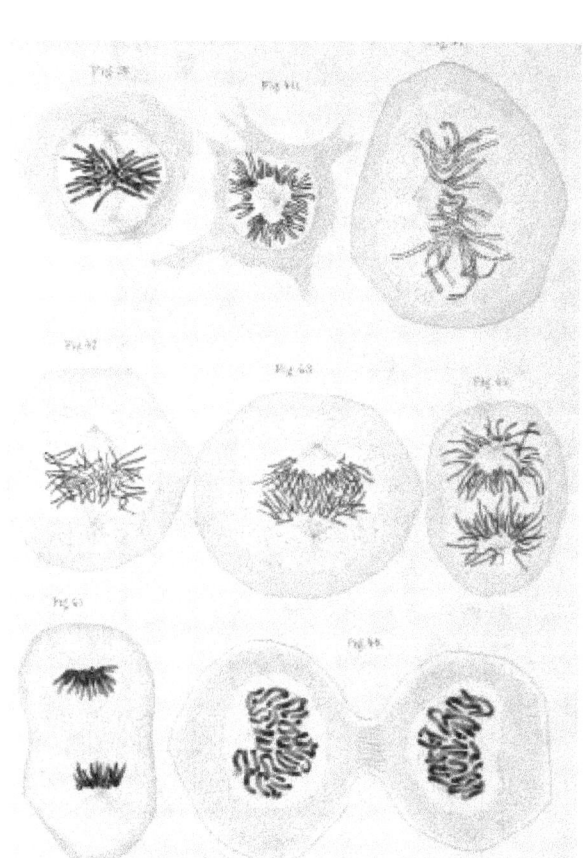

Walther Flemming's 1882 diagram of eukaryotic cell division. Chromosomes are copied, condensed, and organized. Then, as the cell divides, chromosome copies separate into the daughter cells.

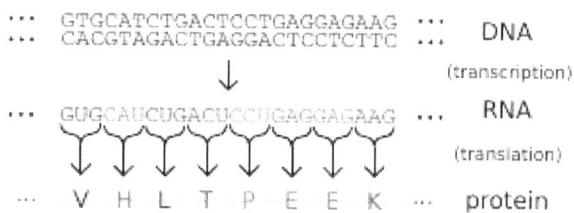

The genetic code: Using a triplet code, DNA, through a messenger RNA intermediary, specifies a protein.

Siamese cats have a temperature-sensitive pigment-production mutation.

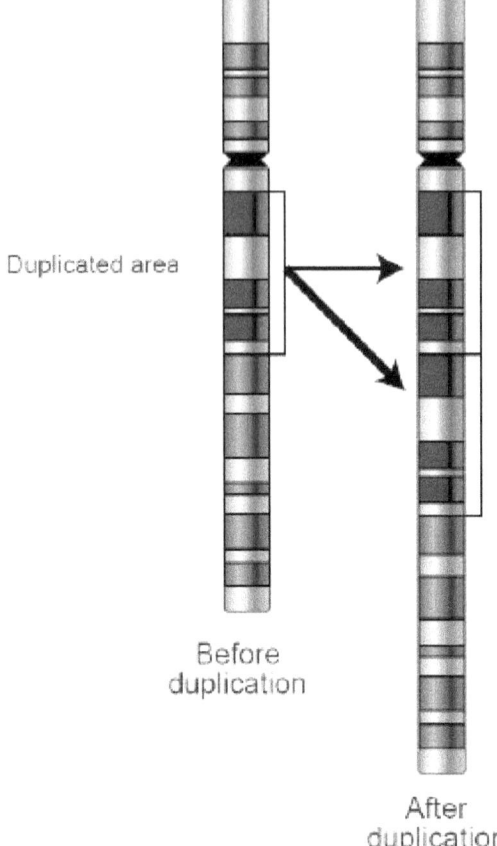

Gene duplication allows diversification by providing redundancy: one gene can mutate and lose its original function without harming the organism.

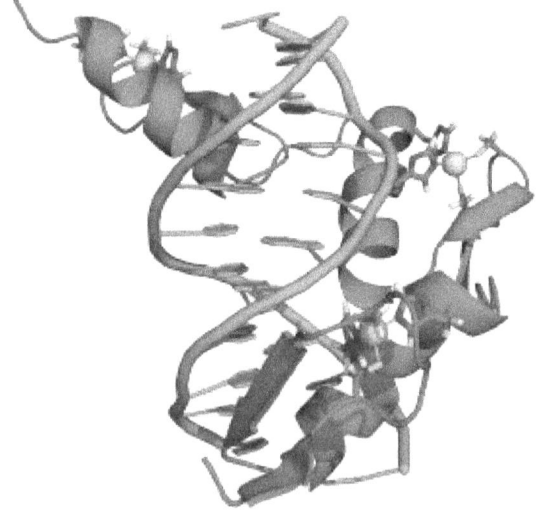

Transcription factors bind to DNA, influencing the transcription of associated genes.

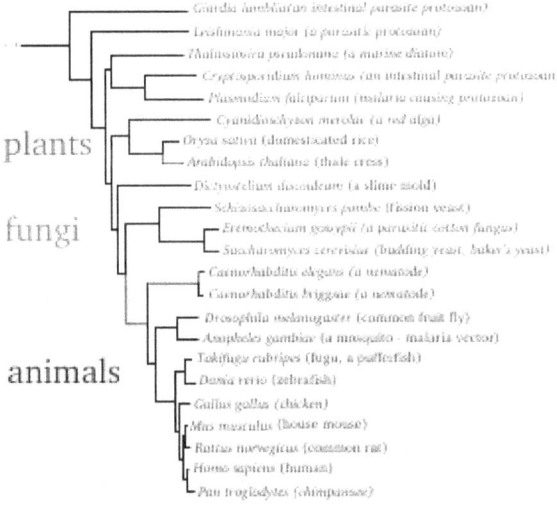

An evolutionary tree of eukaryotic organisms, constructed by the comparison of several orthologous gene sequences.

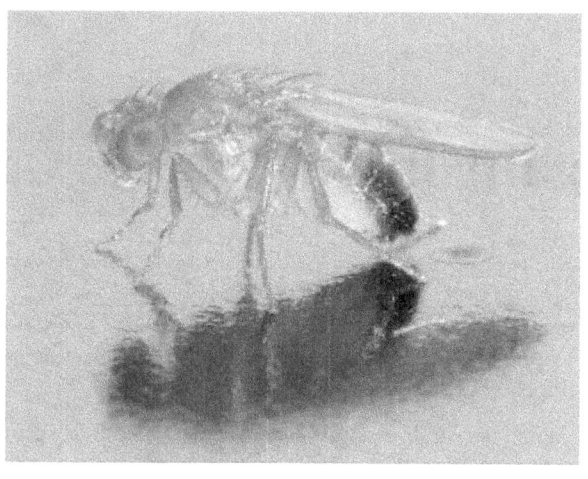

The common fruit fly (Drosophila melanogaster) is a popular model organism in genetics research.

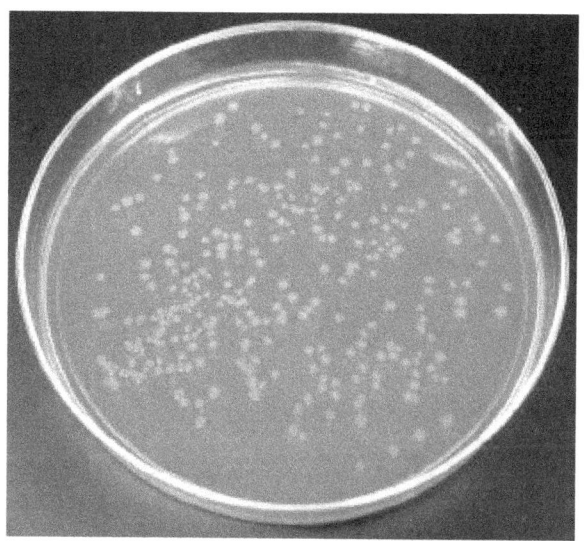

Colonies of E. coli produced by cellular cloning. A similar methodology is often used in molecular cloning.

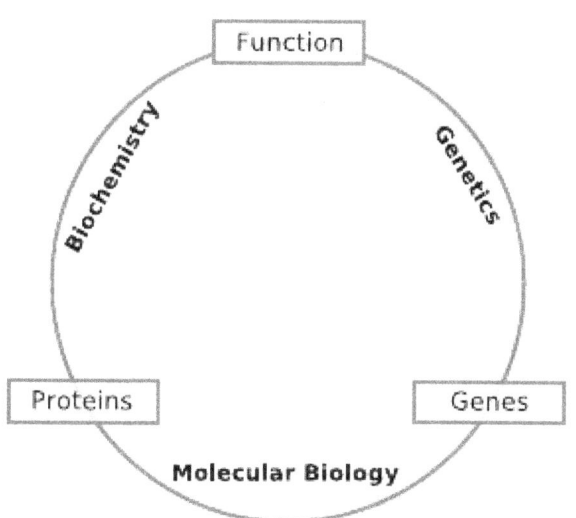

Schematic relationship between biochemistry, genetics and molecular biology.

Chapter 5

Genetically modified organism

"GMO" redirects here. For other uses, see GMO (disambiguation).
For related content, see genetically modified food, genetically modified crops, and genetic engineering.

GloFish, the first genetically modified animal to be sold as a pet

A **genetically modified organism** (**GMO**) is any organism whose genetic material has been altered using genetic engineering techniques (i.e. genetically *engineered* organism). GMOs are the source of medicines and genetically modified foods and are also widely used in scientific research and to produce other goods. The term GMO is very close to the technical legal term, 'living modified organism', defined in the Cartagena Protocol on Biosafety, which regulates international trade in living GMOs (specifically, "any living organism that possesses a novel combination of genetic material obtained through the use of modern biotechnology").

A more specifically defined type of GMO is a "Transgenic Organism". This is an organism whose genetic makeup has been altered by the addition of genetic material from another, unrelated organism. This should not be confused with the more general way in which "GMO" is used to classify genetically altered organisms, as typically GMOs are organisms whose genetic makeup has been altered without the addition of genetic material from an unrelated organism.

The first genetically modified mouse was in 1973 and the first plant was produced in 1983.[1]

5.1 Production

Further information: Genetic engineering, Modifications (genetics), Horizontal gene transfer, Molecular cloning, Recombinant DNA and Transformation (genetics)

Genetic modification involves the mutation, insertion, or deletion of genes. Inserted genes usually come from a different species in a form of horizontal gene-transfer. In nature this can occur when exogenous DNA penetrates the cell membrane for any reason. This can be accomplished artificially by:

- attaching the genes to a virus.

- physically inserting the extra DNA into the nucleus of the intended host with a very small syringe.

- using electroporation (that is, introducing DNA from one organism into the cell of another by use of an electric pulse).

- firing small particles from a gene gun.[2][3][4]

Other methods exploit natural forms of gene transfer, such as the ability of *Agrobacterium* to transfer genetic material to plants,[5] or the ability of lentiviruses to transfer genes to animal cells.[6]

5.2 History

Main article: History of genetic engineering

Herbert Boyer

Stanley Cohen

Herbert Boyer and Stanley Cohen created the first genetically modified organism in 1973

Humans have domesticated plants and animals since around 12,000 BCE, using selective breeding or artificial selection (as contrasted with natural selection).[7]:25 The process of selective breeding, in which organisms with desired traits (and thus with the desired genes) are used to breed the next generation and organisms lacking the trait are not bred, is a precursor to the modern concept of genetic modification.[8]:1[9]:1 Various advancements in genetics allowed humans to directly alter the DNA and therefore genes of organisms. In 1972 Paul Berg created the first recombinant DNA molecule when he combined DNA from a monkey virus with that of the lambda virus.[10][11]

Herbert Boyer and Stanley Cohen made the first genetically modified organism (GMO) in 1973. They took a gene from a bacterium that provided resistance to the antibiotic kanamycin, inserted it into a plasmid and then induced another bacteria to uptake the plasmid. The bacteria was then able to survive in the presence of kanamycin.[12] Boyer and Cohen expressed other genes in bacteria. This included genes from the toad Xenopus laevis in 1974, creating the first GMO expressing a gene from an organism from different kingdom.[13]

In 1973 Rudolf Jaenisch created the first GM animal.

In 1973 Rudolf Jaenisch created a transgenic mouse by introducing foreign DNA into its embryo, making it the world's first transgenic animal.[14] However it took another eight years before transgenic mice were developed that passed the transgene to their offspring.[15][16] Genetically modified mice were created in 1984 that carried cloned oncogenes, predisposed them to developing cancer.[17] Mice with genes knocked out (knockout mouse) were created in 1989. The first transgenic livestock were produced in 1985[18] and the first animal to synthesise transgenic proteins in their milk were mice.[19] engineered to produce human tissue plasminogen activator in 1987.[20]

In 1983 the first genetically engineered plant was developed by Michael W. Bevan, Richard B. Flavell and Mary-Dell Chilton. They infected tobacco with *Agrobacterium* transformed with an antibiotic resistance gene and through tissue culture techniques were able to grow a new plant containing the resistance gene.[21] The gene gun was invented in 1987, allowing transformation of plants not susceptible to *Agrobacterium* infection.[22] In 2000, Vitamin A-enriched golden rice, was the first plant developed with increased nutrient value.[23]

In 1976 Genentech, the first genetic engineering company was founded by Herbert Boyer and Robert Swanson and a year later and the company produced a human protein (somatostatin) in *E.coli*. Genentech announced the production of genetically engineered human insulin in 1978.[24] The insulin produced by bacteria, branded humulin, was approved for release by the Food and Drug Administration in 1982.[25] In 1988 the first human antibodies were produced in plants.[26] In 1987, the ice-minus strain of *P. syringae* became the first genetically modified organism to be released into the environment[27] when a strawberry field and a potato field in California were sprayed with it.[28]

The first genetically modified crop, an antibiotic-resistant tobacco plant, was produced in 1982.[29] China was the first country to commercialize transgenic plants, introducing a virus-resistant tobacco in 1992.[30] In 1994 Calgene attained approval to commercially release the Flavr Savr tomato, the first genetically modified food.[31] Also in 1994, the European Union approved tobacco engineered to be resistant to the herbicide bromoxynil, making it the first genetically engineered crop commercialized in Europe.[32] An insect resistant Potato was approved for release in the USA in 1995,[33] and by 1996 approval had been granted to commercially grow 8 transgenic crops and one flower crop (carnation) in 6 countries plus the EU.[34]

In 2010, scientists at the J. Craig Venter Institute, announced that they had created the first synthetic bacterial genome. They named it Synthia and it was the world's first synthetic life form.[35][36]

The first genetically modified animal to be commercialised was the GloFish, a Zebra fish with a fluorescent gene added that allows it to glow in the dark under ultraviolet light.[37] The first genetically modified animal to be approved for food use was AquAdvantage salmon in 2015.[38] The salmon were transformed with a growth hormone-regulating gene from a Pacific Chinook salmon and a promoter from an ocean pout enabling it to grow year-round instead of only during spring and summer.[39]

5.3 Uses

GMOs are used in biological and medical research, production of pharmaceutical drugs,[40] experimental medicine (e.g. gene therapy), and agriculture (e.g. golden rice, resistance to herbicides). The term "genetically modified organism" does not always imply, but can include, targeted insertions of genes from one species into another. For example, a gene from a jellyfish, encoding a fluorescent protein called GFP, or green fluorescent protein, can be physically linked and thus co-expressed with mammalian genes to identify the location of the protein encoded by the GFP-tagged gene in the mammalian cell. Such methods are useful tools for biologists in many areas of research, including those who study the mechanisms of human and other diseases or fundamental biological processes in eukaryotic or prokaryotic

cells.

5.3.1 Microbes

Bacteria were the first organisms to be modified in the laboratory, due to the relative ease of modifying their genetics.[41]

They continue to be important model organisms for experiments in genetic engineering. In the field of synthetic biology, they have been used to test various synthetic approaches, from synthesizing genomes to creating novel nucleotides.[42][43][44]

These organisms are now used for several purposes, and are particularly important in producing large amounts of pure human proteins for use in medicine.[45]

Genetically modified bacteria are used to produce the protein insulin to treat diabetes.[46] Similar bacteria have been used to produce biofuels,[47] clotting factors to treat haemophilia,[48] and human growth hormone to treat various forms of dwarfism.[49][50]

In addition, various genetically engineered microorganisms are routinely used as sources of enzymes for the manufacture of a variety of processed foods. These include alpha-amylase from bacteria, which converts starch to simple sugars, chymosin from bacteria or fungi, which clots milk protein for cheese making, and pectinesterase from fungi, which improves fruit juice clarity.[51]

5.3.2 Plants

Transgenic plants

Kenyans examining insect-resistant transgenic Bt corn

Transgenic plants have been engineered for scientific research, to create new colours in plants, and to create different crops.

In research, plants are engineered to help discover the functions of certain genes. One way to do this is to knock out the gene of interest and see what phenotype develops. Another strategy is to attach the gene to a strong promoter and see what happens when it is over expressed. A common technique used to find out where the gene is expressed is to attach it to GUS or a similar reporter gene that allows visualisation of the location.[52]

Suntory "blue" rose

After thirteen years of collaborative research, an Australian company – Florigene, and a Japanese company – Suntory, created a blue rose (actually lavender or mauve) in 2004.[53] The genetic engineering involved three alterations – adding two genes, and interfering with another. One of the added genes was for the blue plant pigment delphinidin cloned from the pansy.[54] The researchers then used RNA interference (RNAi) technology to depress all color production by endogenous genes by blocking a crucial protein in color production, called dihydroflavonol 4-reductase) (DFR), and adding a variant of that protein that would not be blocked by the RNAi but that would allow the delphinidin to work.[54] The roses are sold in Japan, the United States, and Canada.[55][56] Florigene has also created and sells lavender-colored carnations that are genetically engineered in a similar way.[54]

Simple plants and plant cells have been genetically engineered for production of biopharmaceuticals in bioreactors as opposed to cultivating plants in open fields. Work has been done with duckweed *Lemna minor*,[57] the algae *Chlamydomonas reinhardtii*[58] and the moss *Physcomitrella patens*.[59][60] An Israeli company, Protalix, has developed a method to produce therapeutics in cultured transgenic carrot and tobacco cells.[61] Protalix and its partner, Pfizer, received FDA approval to market its drug Elelyso, a treatment for Gaucher's disease, in 2012.[62]

Genetically modified crops   Main article: Genetically modified crops

Genetically modified crops (GM crops, or biotech crops) are plants used in agriculture, the DNA of which has been modified using genetic engineering techniques. In most cases the aim is to introduce a new trait to the plant which does not occur naturally in the species. Examples in food crops include resistance to certain pests, diseases, or environmental conditions, reduction of spoilage, or resistance to chemical treatments (e.g. resistance to a herbicide), or improving the nutrient profile of the crop. Examples in non-food crops include production of pharmaceutical agents, biofuels, and other industrially useful goods, as well as for bioremediation.[63]

Farmers have widely adopted GM technology. Between 1996 and 2013, the total surface area of land cultivated with GM crops increased by a factor of 100, from 17,000 square kilometers (4,200,000 acres) to 1,750,000 km^2 (432 million acres).[63] 10% of the world's croplands were planted with GM crops in 2010.[64] In the US, by 2014, 94% of the planted area of soybeans, 96% of cotton and 93% of corn were genetically modified varieties.[65] In recent years GM crops expanded rapidly in developing countries. In 2013 approximately 18 million farmers grew 54% of worldwide GM crops in developing countries.[63]

For discussions of issues about GM crops and GM food, see the Controversies section below and the article on genetically modified food controversies.

Cisgenic plants

Cisgenesis, sometimes also called intragenesis, is a product designation for a category of genetically engineered plants. A variety of classification schemes have been proposed[66] that order genetically modified organisms based on the nature of introduced genotypical changes rather than the process of genetic engineering.

While some genetically modified plants are developed by the introduction of a gene originating from distant, sexually incompatible species into the host genome, cisgenic plants contain genes that have been isolated either directly from the host species or from sexually compatible species. The new genes are introduced using recombinant DNA methods and gene transfer. Some scientists hope that the approval process of cisgenic plants might be simpler than that of proper transgenics,[67] but it remains to be seen.[68]

5.3.3   Mammals

Genetically modified mammals are an important category

Some chimeras, like the blotched mouse shown, are created through genetic modification techniques like gene targeting.

Research use

Dolly was a female domestic sheep and the first animal to be cloned from an adult somatic cell

of genetically modified organisms.[69] Ralph L. Brinster and Richard Palmiter developed the techniques responsible for transgenic mice, rats, rabbits, sheep, and pigs in the early 1980s, and established many of the first transgenic models of human disease, including the first carcinoma caused by a transgene. The process of genetically engineering animals is a slow, tedious, and expensive process. However, new technologies are making genetic modifications easier and more precise.[70]

The first transgenic (genetically modified) animal was produced by injecting DNA into mouse embryos then implanting the embryos in female mice.[71]

Genetically modified animals currently being developed can be placed into six different broad classes based on the intended purpose of the genetic modification:

1. to research human diseases (for example, to develop animal models for these diseases);

2. to produce industrial or consumer products (fibres for multiple uses);

3. to produce products intended for human therapeutic use (pharmaceutical products or tissue for implantation);

4. to enrich or enhance the animals' interactions with humans (hypo-allergenic pets);

5. to enhance production or food quality traits (faster growing fish, pigs that digest food more efficiently);

6. to improve animal health (disease resistance)[72]

Transgenic animals are used as experimental models to perform phenotypic and for testing in biomedical research.[73]

Genetically modified (genetically engineered) animals are becoming more vital to the discovery and development of cures and treatments for many serious diseases. By altering the DNA or transferring DNA to an animal, we can develop certain proteins that may be used in medical treatment. Stable expressions of human proteins have been developed in many animals, including sheep, pigs, and rats. Human-alpha-1-antitrypsin,[74] which has been tested in sheep and is used in treating humans with this deficiency and transgenic pigs with human-histo-compatibility have been studied in the hopes that the organs will be suitable for transplant with less chances of rejection.

Scientists have genetically engineered several organisms, including some mammals, to include green fluorescent protein (GFP), first observed in the jellyfish, *Aequorea victoria* in 1962, for medical research purposes (Chalfie, Shimoura, and Tsien were awarded the Nobel prize in Chemistry in 2008 for the discovery and development of GFP [75]). For example, fluorescent pigs have been bred to study human organ transplants (xenotransplantation), regenerating ocular photoreceptor cells, and other topics.[76] In 2011 a Japanese-American team created green-fluorescent cats to find therapies for HIV/AIDS and other diseases[77] as feline immunodeficiency virus (FIV) is related to HIV.[78]

In 2009, scientists in Japan announced that they had successfully transferred a gene into a primate species (marmosets) and produced a stable line of breeding transgenic primates for the first time.[79][80] Their first research target for these marmosets was Parkinson's disease, but they were also considering amyotrophic lateral sclerosis and

Huntington's disease.[81]

Producing human therapeutics

Herman the Bull, Naturalis, for the production of lactoferrin enhanced milk.

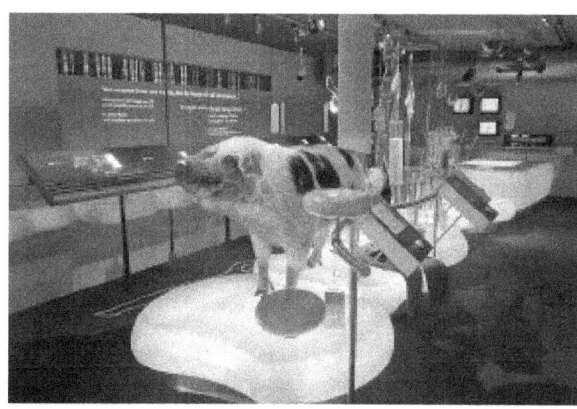

Transgenic pig for cheese production.

Within the field known as pharming, intensive research has been conducted to develop transgenic animals that produce biotherapeutics.[82] On 6 February 2009, the U.S. Food and Drug Administration approved the first human biological drug produced from such an animal, a goat. The drug, ATryn, is an anticoagulant which reduces the probability of blood clots during surgery or childbirth. It is extracted from the goat's milk.[83]

Production or food quality traits

In 2006, a pig was engineered to produce omega-3 fatty acids through the expression of a roundworm gene.[84]

Enviropig was a genetically enhanced line of Yorkshire pigs in Canada created with the capability of digesting plant phosphorus more efficiently than conventional Yorkshire pigs. The project ended in 2012.[85][86] These pigs produced the enzyme phytase, which breaks down the indigestible phosphorus, in their saliva. The enzyme was introduced into the pig chromosome by pronuclear microinjection. With this enzyme, the animal is able to digest cereal grain phosphorus.[85][87] The use of these pigs would reduce the potential of water pollution since they excrete from 30 to 70.7% less phosphorus in manure depending upon the age and diet.[85][87] The lower concentrations of phosphorus in surface runoff reduces algal growth, because phosphorus is the limiting nutrient for algae.[85] Because algae consume large amounts of oxygen, it can result in dead zones for fish.

In 2011, Chinese scientists generated dairy cows genetically engineered with genes from human beings to produce milk that would be the same as human breast milk.[88] This could potentially benefit mothers who cannot produce breast milk but want their children to have breast milk rather than formula. Aside from milk production, the researchers claim these transgenic cows to be identical to regular cows.[89] Two months later scientists from Argentina presented Rosita, a transgenic cow incorporating two human genes, to produce milk with similar properties as human breast milk.[90] In 2012, researchers from New Zealand also developed a genetically engineered cow that produced allergy-free milk.[91]

Goats have been genetically engineered to produce milk with strong spiderweb-like silk proteins in their milk.[92]

Human gene therapy

Gene therapy,[93] uses genetically modified viruses to deliver genes that can cure disease in humans. Although gene therapy is still relatively new, it has had some successes. It has been used to treat genetic disorders such as severe combined immunodeficiency,[94] and Leber's congenital amaurosis.[95] Treatments are also being developed for a range of other currently incurable diseases, such as cystic fibrosis,[96] sickle cell anemia,[97] Parkinson's disease,[98][99] cancer,[100][101][102] diabetes,[103] heart disease[104] and muscular dystrophy.[105]

5.3.4 Fish

Genetically modified fish are used for scientific research and as pets, and are being considered for use as food and as aquatic pollution sensors.

GM fish are widely used in basic research in genetics and development. Two species of fish, zebrafish and medaka, are most commonly modified because they have optically clear chorions (membranes in the egg), rapidly develop, and the 1-cell embryo is easy to see and microinject with transgenic DNA.[106]

The GloFish is a patented[107] brand of genetically modified (GM) fluorescent zebrafish with bright red, green, and orange fluorescent color. Although not originally developed for the ornamental fish trade, it became the first genetically modified animal to become publicly available as a pet when it was introduced for sale in 2003.[108] They were quickly banned for sale in California.[109]

GM fish have been developed with promoters driving an over-production of "all fish" growth hormone for use in the aquaculture industry to increase the speed of development and potentially reduce fishing pressure on wild stocks. This has resulted in dramatic growth enhancement in several species, including salmon,[110] trout[111] and tilapia.[112] AquaBounty Technologies, a biotechnology company working on bringing a GM salmon to market, claims that their GM AquAdvantage salmon can mature in half the time as wild salmon.[113] AquaBounty applied for regulatory approval to market their GM salmon in the US, and was approved in November 2015.[114] On 25 November 2013 Canada approved commercial scale production and export of GM Salmon eggs but they are not approved for human consumption in Canada.[115]

Several academic groups have been developing GM zebrafish to detect aquatic pollution. The lab that originated the GloFish discussed above originally developed them to change color in the presence of pollutants, to be used as environmental sensors.[116][117] A lab at University of Cincinnati has been developing GM zebrafish for the same purpose,[118][119] as has a lab at Tulane University.[120]

Recent research on pain in fish has resulted in concerns being raised that genetic-modifications induced for scientific research may have detrimental effects on the welfare of fish.[121]

5.3.5 Frogs

Genetically modified frogs are used for scientific research and being considered are widely used in basic research including genetics and early development. Two species of frog, *Xenopus laevis* and *Xenopus tropicalis*, are most commonly used.

GM frogs are also being used as pollution sensors, especially for endocrine disrupting chemicals.[122]

5.3.6 Invertebrates

See also: Genetically modified insect

Fruit flies

In biological research, transgenic fruit flies (*Drosophila melanogaster*) are model organisms used to study the effects of genetic changes on development.[123] Fruit flies are often preferred over other animals due to their short life cycle, low maintenance requirements, and relatively simple genome compared to many vertebrates.

Mosquitoes

In 2010, scientists created "malaria-resistant mosquitoes" in the laboratory.[124][125][126] The World Health Organization estimated that malaria killed almost one million people in 2008.[127] Genetically modified male mosquitoes containing a lethal gene have been developed to combat the spread of dengue fever[128] and the Zika virus.[129] *Aedes aegypti* mosquitoes, the single most important carrier of dengue fever and the Zika virus, were reduced by 80% in a 2010 trial of these GM mosquitoes in the Cayman Islands[130][131] and by 90% in a 2015 trial in Bahia, Brazil.[129] Between 50 and 100 million people are affected by dengue fever every year and 40,000 people die from it.[132] The Zika virus causes babies to be born with shrunken heads and spread rapidly in the Americas in 2015 and 2016.[129]

Bollworms

A strain of *Pectinophora gossypiella* (Pink bollworm) has been genetically engineered to express a red fluorescent protein. This allows researchers to monitor bollworms that have been sterilized by radiation and released to reduce bollworm infestation. The strain has been field tested for over three years and has been approved for release.[132][133][134]

Cnidarians

Cnidarians such as *Hydra* and the sea anemone *Nematostella vectensis* have become attractive model organisms to study the evolution of immunity and certain developmental processes. An important technical breakthrough was the development of procedures for generation of stably transgenic hydras and sea anemones by embryo microinjection.[135]

5.4 Regulation

Main articles: Regulation of genetic engineering and Regulation of the release of genetically modified organisms

The regulation of genetic engineering concerns the approaches taken by governments to assess and manage the risks associated with the use of genetic engineering technology and the development and release of genetically modified organisms (GMO), including genetically modified crops and genetically modified fish. There are differences in the regulation of GMOs between countries, with some of the most marked differences occurring between the USA and Europe.[136] Regulation varies in a given country depending on the intended use of the products of the genetic engineering. For example, a crop not intended for food use is generally not reviewed by authorities responsible for food safety.[137] The European Union differentiates between approval for cultivation within the EU and approval for import and processing. While only a few GMOs have been approved for cultivation in the EU a number of GMOs have been approved for import and processing.[138] The cultivation of GMOs has triggered a debate about coexistence of GM and nonGM crops. Depending on the coexistence regulations, incentives for cultivation of GM crops differ.[139]

5.5 Controversy

See also: Genetically modified food controversies

There is controversy over GMOs, especially with regard to their use in producing food. The dispute involves buyers, biotechnology companies, governmental regulators, nongovernmental organizations, and scientists. The key areas of controversy related to GMO food are whether GM food should be labeled, the role of government regulators, the effect of GM crops on health and the environment, the effect on pesticide resistance, the impact of GM crops for farmers, and the role of GM crops in feeding the world population. In 2014, sales of products which had been labeled as non-GMO grew 30 percent to $1.1 billion.[140]

There is a general scientific agreement that food from genetically modified crops is not inherently riskier to human health than conventional food, but should be tested on a case-by-case basis.[141][142][143][144] No reports of ill effects have been proven in the human population from ingesting GM food.[141][145][146][147] Although labeling of GMO products in the marketplace is required in many countries, it is not required in the United States and no distinction between marketed GMO and non-GMO foods is recognized by the US FDA. In a May 2014 article in The Economist it was argued that, while GM foods could potentially help feed 842 million malnourished people globally, laws such as those being considered by Vermont's governor, Peter Shumlin, to require labeling of foods containing genetically modified ingredients, could have the unintended consequence of interrupting the process of spreading GM technologies to impoverished countries that suffer with food security problems.[141]

The Organic Consumers Association, and the Union of Concerned Scientists,[148][149][150][151][152] and Greenpeace stated that risks have not been adequately identified and managed, and they have questioned the objectivity of regulatory authorities. Some health groups say there are unanswered questions regarding the potential long-term impact on human health from food derived from GMOs, and propose mandatory labeling[153][154] or a moratorium on such products.[155][156][157] Concerns include contamination of the non-genetically modified food supply,[158][159] effects of GMOs on the environment and nature,[155][157] the rigor of the regulatory process,[156][160] and consolidation of control of the food supply in companies that make and sell GMOs,[155] or concerns over the use of herbicides with glyphosate.[161]

5.6 See also

- BioSteel
- Chimera (genetics)
- Council for Responsible Genetics
- Detection of genetically modified organisms
- Gene flow
- Gene pool
- Genetic erosion
- Horizontal gene transfer
- Non-GMO Project
- Organic farming

- SMART breeding

- Sperm-mediated gene transfer

- Timeline of genetically modified organisms

5.7   References

[1] "History of Genetically Modified Foods". *umich.edu*.

[2] *Cornell Chronicle*. 14 May 1987, page 3. Biologists invent gun for shooting cells with DNA

[3] Sanford, JC; et al. (1987). "Delivery of substances into cells and tissues using a particle bombardment process". *Journal of Particulate Science and Technology* **5**: 27–37. doi:10.1080/02726358708904533.

[4] Klein, TM; et al. (1987). "High-velocity microprojectiles for delivering nucleic acids into living cells". *Nature* **327** (6117): 70–73. Bibcode:1987Natur.327...70K. doi:10.1038/327070a0.

[5] Lee LY, Gelvin SB (February 2008). "T-DNA binary vectors and systems". *Plant Physiol.* **146** (2): 325–332. doi:10.1104/pp.107.113001. OCLC 1642351. PMC 2245830. PMID 18250230.

[6] Park F (October 2007). "Lentiviral vectors: are they the future of animal transgenesis?". *Physiol. Genomics* **31** (2): 159–173. doi:10.1152/physiolgenomics.00069.2007. OCLC 37367250. PMID 17684037.

[7] Noel Kingsbury. Hybrid: The History and Science of Plant Breeding University of Chicago Press, Oct 15, 2009

[8] Clive Root (2007). *Domestication*. Greenwood Publishing Groups.

[9] Daniel Zohary, Maria Hopf, Ehud Weiss (2012). *Domestication of Plants in the Old World: The Origin and Spread of Plants in the Old World*. Oxford University Press.

[10] Jackson, DA; Symons, RH; Berg, P (1 October 1972). "Biochemical Method for Inserting New Genetic Information into DNA of Simian Virus 40: Circular SV40 DNA Molecules Containing Lambda Phage Genes and the Galactose Operon of *Escherichia coli*". *PNAS* **69** (10): 2904–2909. Bibcode:1972PNAS...69.2904J. doi:10.1073/pnas.69.10.2904. PMC 389671. PMID 4342968.

[11] M. K. Sateesh (25 August 2008). *Bioethics And Biosafety*. I. K. International Pvt Ltd. pp. 456–. ISBN 978-81-906757-0-3. Retrieved 27 March 2013.

[12] "Genome and genetics timeline – 1973". Genome news network.

[13] Morrow, J. F.; Cohen, S. N.; Chang, A. C.; Boyer, H. W.; Goodman, H. M.; Helling, R. B. (1974-05-01). "Replication and transcription of eukaryotic DNA in Escherichia coli". *Proceedings of the National Academy of Sciences of the United States of America* **71** (5): 1743–1747. doi:10.1073/pnas.71.5.1743. ISSN 0027-8424. PMC 388315. PMID 4600264.

[14] Jaenisch, R. and Mintz, B. (1974) Simian virus 40 DNA sequences in DNA of healthy adult mice derived from preimplantation blastocysts injected with viral DNA. Proc. Natl. Acad. 71(4):1250–1254

[15] Gordon, J.; Ruddle, F. (1981). "Integration and stable germ line transmission of genes injected into mouse pronuclei". *Science* **214** (4526): 1244–6. Bibcode:1981Sci...214.1244G. doi:10.1126/science.6272397. PMID 6272397.

[16] Costantini, F.; Lacy, E. (1981). "Introduction of a rabbit β-globin gene into the mouse germ line". *Nature* **294** (5836): 92–4. Bibcode:1981Natur.294...92C. doi:10.1038/294092a0. PMID 6945481.

[17] Hanahan, D.; Wagner, E. F.; Palmiter, R. D. (2007). "The origins of oncomice: A history of the first transgenic mice genetically engineered to develop cancer". *Genes & Development* **21** (18): 2258–2270. doi:10.1101/gad.1583307. PMID 17875663.

[18] Brophy, B.; Smolenski, G.; Wheeler, T.; Wells, D.; l'Huillier, P.; Laible, G. T. (2003). "Cloned transgenic cattle produce milk with higher levels of β-casein and κ-casein". *Nature Biotechnology* **21** (2): 157–162. doi:10.1038/nbt783. PMID 12548290.

[19] A. John Clark. "The Mammary Gland as a Bioreactor: Expression, Processing, and Production of Recombinant Proteins". *Journal of Mammary Gland Biology and Neoplasia* **3** (3): 337–350. doi:10.1023/a:1018723712996.

[20] K. Gordon, E. Lee, J. Vitale, A. Smith, H. Westphal, and L. Hennighausen (1987). "Production of human tissue plasminogen activator in transgenic mouse milk". *Biotechnology* **5**: 1183±1187. doi:10.1038/nbt1187-1183.

[21] Bevan, M. W.; Flavell, R. B.; Chilton, M. D. (1983). "A chimaeric antibiotic resistance gene as a selectable marker for plant cell transformation". *Nature* **304** (5922): 184–187. doi:10.1038/304184a0.

[22] Roger Segelken for the Cornell Chronicle. Mary 14, 1987. Biologists Invent Gun for Shooting Cells with DNA Issue available as pdf download here. page 3

[23] Ye, Xudong; Al-Babili, Salim; Klöti, Andreas; Zhang, Jing; Lucca, Paola; Beyer, Peter; Potrykus, Ingo (2000-01-14). "Engineering the Provitamin A (β-Carotene) Biosynthetic Pathway into (Carotenoid-Free) Rice Endosperm". *Science* **287** (5451): 303–305. doi:10.1126/science.287.5451.303. ISSN 0036-8075. PMID 10634784.

[24] Goeddel, D. V.; Kleid, D. G.; Bolivar, F.; Heyneker, H. L.; Yansura, D. G.; Crea, R.; Hirose, T.; Kraszewski, A.; Itakura, K.; Riggs, A. D. (1979). "Expression in Escherichia coli of chemically synthesized genes for human insulin". *Proceedings of the National Academy of Sciences* **76** (1): 106–10. doi:10.1073/pnas.76.1.106. PMC 382885. PMID 85300.

[25] "Artificial Genes". TIME. 15 November 1982. Retrieved 17 July 2010.

[26] Woodard, S. L.; Woodard, J. A.; Howard, M. E. (2004). "Plant molecular farming: Systems and products". *Plant Cell Reports* **22** (10): 711–720. doi:10.1007/s00299-004-0767-1. PMID 14997337.

[27] BBC News 14 June 2002 GM crops: A bitter harvest?

[28] Thomas H. Maugh II for the Los Angeles Times. June 09, 1987. Altered Bacterium Does Its Job : Frost Failed to Damage Sprayed Test Crop, Company Says

[29] Fraley, RT; et al. (1983). "Expression of bacterial genes in plant cells" (PDF). *Proc. Natl. Acad. Sci. USA* **80**: 4803–4807.

[30] James, Clive (1997). "Global Status of Transgenic Crops in 1997" (PDF). *ISAAA Briefs No. 5.*: 31.

[31] Bruening, G.; Lyons, J. M. (2000). "The case of the FLAVR SAVR tomato". *California Agriculture* **54** (4): 6–7. doi:10.3733/ca.v054n04p6.

[32] Debora MacKenzie (18 June 1994). "Transgenic tobacco is European first". New Scientist.

[33] Genetically Altered Potato Ok'd For Crops Lawrence Journal-World - 6 May 1995

[34] James, Clive (1996). "Global Review of the Field Testing and Commercialization of Transgenic Plants: 1986 to 1995" (PDF). The International Service for the Acquisition of Agri-biotech Applications. Retrieved 17 July 2010.

[35] Gibson, D. G.; Glass, J. I.; Lartigue, C.; Noskov, V. N.; Chuang, R.-Y.; Algire, M. A.; Benders, G. A.; Montague, M. G.; Ma, L.; Moodie, M. M.; Merryman, C.; Vashee, S.; Krishnakumar, R.; Assad-Garcia, N.; Andrews-Pfannkoch, C.; Denisova, E. A.; Young, L.; Qi, Z.-Q.; Segall-Shapiro, T. H.; Calvey, C. H.; Parmar, P. P.; Hutchison Ca, C. A.; Smith, H. O.; Venter, J. C. (2010). "Creation of a Bacterial Cell Controlled by a Chemically Synthesized Genome". *Science* **329** (5987): 52–6. doi:10.1126/science.1190719. PMID 20488990.

[36] Sample, Ian (20 May 2010). "Craig Venter creates synthetic life form". London: guardian.co.uk.

[37] Vàzquez-Salat, Núria; Salter, Brian; Smets, Greet; Houdebine, Louis-Marie (2012-11-01). "The current state of GMO governance: Are we ready for GM animals?". *Biotechnology Advances*. Special issue on ACB 2011 **30** (6): 1336–1343. doi:10.1016/j.biotechadv.2012.02.006.

[38] "AQUABOUNTY CLEARED TO SELL SALMON IN USA FOR COMMERCIAL PURPOSES".

[39] Bodnar, Anastasia (October 2010). "Risk Assessment and Mitigation of AquAdvantage Salmon" (PDF). ISB News Report.

[40] http://www.fda.gov/AboutFDA/WhatWeDo/History/ProductRegulation/SelectionsFromFDLIUpdateSeriesonFDAHistory/ucm081964.htm

[41] Melo, Eduardo O.; Canavessi, Aurea M. O.; Franco, Mauricio M.; Rumpf, Rodolpho (2007). "Animal transgenesis: state of the art and applications". *J. Appl. Genet.* **48** (1): 47–61. doi:10.1007/BF03194657. PMID 17272861. Archived from the original (PDF) on 26 September 2009.

[42] Arpino, JA; et al. (Jul 2013). "Tuning the dials of Synthetic Biology". *Microbiology* **159** (7): 1236–53. doi:10.1099/mic.0.067975-0. PMC 3749727. PMID 23704788.

[43] Pollack, Andrew (7 May 2014). "Researchers Report Breakthrough in Creating Artificial Genetic Code". *New York Times*. Retrieved 7 May 2014.

[44] Malyshev, Denis A.; Dhami, Kirandeep; Lavergne, Thomas; Chen, Tingjian; Dai, Nan; Foster, Jeremy M.; Corrêa, Ivan R.; Romesberg, Floyd E. (7 May 2014). "A semi-synthetic organism with an expanded genetic alphabet". *Nature* **509**: 385–388. Bibcode:2014Natur.509..385M. doi:10.1038/nature13314. PMC 4058825. PMID 24805238. Retrieved 7 May 2014.

[45] Leader, Benjamin; Baca, Qentin J.; Golan, David E. (January 2008). "Protein therapeutics: a summary and pharmacological classification". *Nature Reviews Drug Discovery*. A guide to drug discovery **7** (1): 21–39. doi:10.1038/nrd2399. PMID 18097458.
Leader 2008 — Fee required for access to full text.

[46] Walsh, Gary (April 2005). "Therapeutic insulins and their large-scale manufacture". *Appl. Microbiol. Biotechnol.* **67** (2): 151–159. doi:10.1007/s00253-004-1809-x. PMID 15580495.
Walsh 2005 — Fee required for access to full text.

[47] Summers, Rebecca (24 April 2013) "Bacteria churn out first ever petrol-like biofuel" *New Scientist*. Retrieved 27 April 2013

[48] Pipe, Steven W. (May 2008). "Recombinant clotting factors". *Thromb. Haemost.* **99** (5): 840–850. doi:10.1160/TH07-10-0593. PMID 18449413.

[49] Bryant, Jackie; Baxter, Louise; Cave, Carolyn B.; Milne, Ruairidh; Bryant, Jackie (2007). Bryant, Jackie, ed. "Recombinant growth hormone for idiopathic short stature in children and adolescents". *Cochrane Database Syst Rev* (3): CD004440. doi:10.1002/14651858.CD004440.pub2. PMID 17636758.
Bryant 2007 — Fee required for access to full text.

[50] Baxter L, Bryant J, Cave CB, Milne R (2007). Bryant, Jackie, ed. "Recombinant growth hormone for children and adolescents with Turner syndrome". *Cochrane Database Syst Rev* (1): CD003887. doi:10.1002/14651858.CD003887.pub2. PMID 17253498.

[51] Panesar, Pamit *et al.* (2010) *Enzymes in Food Processing: Fundamentals and Potential Applications*, Chapter 10, I K International Publishing House, ISBN 978-93-80026-33-6

[52] Jefferson R. A. Kavanagh T. A. Bevan M. W. (1987). "GUS fusions: beta-glucuronidase as a sensitive and versatile gene fusion marker in higher plants". *The EMBO Journal* 6 (13): 3901–3907. ISSN 0261-4189. PMC 553867. PMID 3327686.

[53] Nosowitz, Dan (15 September 2011) "Suntory Creates Mythical Blue (Or, Um, Lavender-ish) Rose" *Popular Science*, Retrieved 30 August 2012

[54] Phys.Org website. 4 April 2005 Plant gene replacement results in the world's only blue rose

[55] Kyodo (11 September 2011 "Suntory to sell blue roses overseas" *The Japan Times*, Retrieved 30 August 2012

[56] "World's First 'Blue' Rose Soon Available in U.S.". *WIRED*. 14 September 2011.

[57] Gasdaska JR *et al.* (2003) "Advantages of Therapeutic Protein Production in the Aquatic Plant *Lemna*". *BioProcessing Journal* Mar/Apr 2003 pp 49–56

[58] (10 December 2012) "Engineering algae to make complex anti-cancer 'designer' drug" *PhysOrg*. Retrieved 15 April 2013

[59] Büttner-Mainik, A., *et al.* (2011): "Production of biologically active recombinant human factor H in *Physcomitrella*". *Plant Biotechnology Journal* 9, 373–383.

[60] Baur, A.; Reski, R.; Gorr, G. (2005). "Enhanced recovery of a secreted recombinant human growth factor using stabilizing additives and by co-expression of human serum albumin in the moss *Physcomitrella patens*". *Plant Biotech. J.* 3: 331–340. doi:10.1111/j.1467-7652.2005.00127.x. PMID 17129315.

[61] Protalix technology platform

[62] Gali Weinreb and Koby Yeshayahou for Globes 2 May 2012. "FDA approves Protalix Gaucher treatment"

[63] ISAAA 2013 Annual Report Executive Summary, Global Status of Commercialized Biotech/GM Crops: 2013 ISAAA Brief 46-2013, Retrieved 6 August 2014

[64] James, C (2011). "ISAAA Brief 43, Global Status of Commercialized Biotech/GM Crops: 2011". *ISAAA Briefs*. Ithaca, New York: International Service for the Acquisition of Agri-biotech Applications (ISAAA). Retrieved 2012-06-02.

[65] Jorge Fernandez-Cornejo, Seth James Wechsler. "USDA ERS - Adoption of Genetically Engineered Crops in the U.S.". *usda.gov*.

[66] Nielsen, K. M. (2003). "Transgenic organisms—time for conceptual diversification?". *Nature Biotechnology* 21 (3): 227–228. doi:10.1038/nbt0303-227. PMID 12610561.

[67] Schouten, H.; Krens, F.; Jacobsen, E. (2006). "Cisgenic plants are similar to traditionally bred plants: international regulations for genetically modified organisms should be altered to exempt cisgenesis". *EMBO Reports* 7 (8): 750–753. doi:10.1038/sj.embor.7400769. PMC 1525145. PMID 16880817.

[68] Prins, T. W. and Kok, E. J. (2010) *Food and feed safety aspects of cisgenic crop plant varieties* Report 2010.001, Project number: 120.72.667.01, RIKILT – Institute of Food Safety, Netherlands. Retrieved 6 September 2010.

[69] EFSA (2012). Genetically modified animals Europe: EFSA

[70] Murray, Joo (20). Genetically modified animals. Canada: Brainwaving

[71] Jaenisch, R. and Mintz, B. (1974). "Simian virus 40 DNA sequences in DNA of healthy adult mice derived from preimplantation blastocysts injected with viral DNA.". *Proc. Natl. Acad. Sci.* 71 (4): 1250–1254. Bibcode:1974PNAS...71.1250J. doi:10.1073/pnas.71.4.1250. PMC 388203. PMID 4364530.

[72] Rudinko, Larisa (20). Guidance for industry. USA: Center for veterinary medicine Link.

[73] Sathasivam K, Hobbs C, Mangiarini L, et al. (June 1999). "Transgenic models of Huntington's disease". *Philosophical Transactions of the Royal Society B* 354 (1386): 963–9. doi:10.1098/rstb.1999.0447. PMC 1692600. PMID 10434294.

[74] Spencer, L; Humphries, J; Brantly, M. (12 May 2005). "Antibody Response to Aerosolized Transgenic Human Alpha1-Antitrypsin". *New England Journal of Medicine* 352: 19. doi:10.1056/nejm200505123521923. Retrieved 28 April 2011.

[75] "Green fluorescent protein takes Nobel prize". Lewis Brindley. Retrieved 2015-05-31.

[76] Randall S. *et al.* (2008) "Genetically Modified Pigs for Medicine and Agriculture" *Biotechnology and Genetic Engineering Reviews* – Vol. 25, 245–266. Retrieved 31 August 2012

[77] Wongsrikeao P, Saenz D, Rinkoski T, Otoi T, Poeschla E (2011). "Antiviral restriction factor transgenesis in the domestic cat". *Nature Methods* 8 (10): 853–9. doi:10.1038/nmeth.1703. PMC 4006694. PMID 21909101.

[78] Staff (3 April 2012) Biology of HIV National Institute of Allergy and Infectious Diseases, Retrieved 31 August 2012.

[79] Sasaki, E.; Suemizu, H.; Shimada, A.; Hanazawa, K.; Oiwa, R.; Kamioka, M.; Tomioka, I.; Sotomaru, Y.; Hirakawa, R.; Eto, T.; Shiozawa, S.; Maeda, T.; Ito, M.; Ito, R.; Kito, C.; Yagihashi, C.; Kawai, K.; Miyoshi, H.; Tanioka, Y.; Tamaoki, N.; Habu, S.; Okano, H.; Nomura, T. (2009). "Generation of transgenic non-human primates with germline transmission". *Nature* **459** (7246): 523–527. Bibcode:2009Natur.459..523S. doi:10.1038/nature08090. PMID 19478777.

[80] Schatten, G.; Mitalipov, S. (2009). "Developmental biology: Transgenic primate offspring". *Nature* **459** (7246): 515–516. Bibcode:2009Natur.459..515S. doi:10.1038/459515a. PMC 2777739. PMID 19478771.

[81] Cyranoski, D. (2009). "Marmoset model takes centre stage". *Nature* **459** (7246): 492–492. doi:10.1038/459492a. PMID 19478751.

[82] Louis-Marie Houdebine (2009) "Production of Pharmaceutical by transgenic animals". *Comparative Immunology, Microbiology & Infectious Diseases* 32(2): 107–121

[83] Britt Erickson, 10 February 2009, for *Chemical & Engineering News*. FDA Approves Drug From Transgenic Goat Milk Accessed 6 October 2012

[84] Lai L., et al. (2006). "Generation of cloned transgenic pigs rich in omega-3 fatty acids" (PDF). *Nature Biotechnology* **24** (4): 435–436. doi:10.1038/nbt1198. PMC 2976610. PMID 16565727. Retrieved 2009-03-29.

[85] Guelph(2010). Enviropig. Canada:

[86] Schimdt, Sarah. "Genetically engineered pigs killed after funding ends", *Postmedia News*, 22 June 2012. Accessed 31 July 2012.

[87] Canada. "Enviropig — Environmental Benefits | University of Guelph". Uoguelph.ca. Retrieved 8 March 2010.

[88] Gray,Richard(2011). "Genetically modified cows produce 'human' milk".

[89] Classical Medicine Journal (14 April 2010). "Genetically modified cows producing human milk.".

[90] Yapp, Robin (11 June 2011). "Scientists create cow that produces 'human' milk". *The Daily Telegraph* (London). Retrieved 15 June 2012.

[91] Jabed, A.; Wagner, S.; McCracken, J.; Wells, D. N.; Laible, G. (2012). "Targeted microRNA expression in dairy cattle directs production of -lactoglobulin-free, high-casein milk". *Proceedings of the National Academy of Sciences* **109** (42): 16811–16816. Bibcode:2012PNAS..10916811J. doi:10.1073/pnas.1210057109.

[92] Zyga, Lisa(2010). "Scientist bred goats that produce spider silk".

[93] Selkirk SM (October 2004). "Gene therapy in clinical medicine". *Postgrad Med J* **80** (948): 560–70. doi:10.1136/pgmj.2003.017764. PMC 1743106. PMID 15466989.

[94] Cavazzana-Calvo M, Fischer A (June 2007). "Gene therapy for severe combined immunodeficiency: are we there yet?". *J. Clin. Invest.* **117** (6): 1456–65. doi:10.1172/JCI30953. PMC 1878528. PMID 17549248.

[95] Richards, Sabrina (6 November 2012) "Gene therapy arrives in Europe" *The Scientist*. Retrieved 15 April 2013

[96] Rosenecker J, Huth S, Rudolph C (October 2006). "Gene therapy for cystic fibrosis lung disease: current status and future perspectives". *Current Opinion in Molecular Therapeutics* **8** (5): 439–45. PMID 17078386.

[97] Persons DA, Nienhuis AW (July 2003). "Gene therapy for the hemoglobin disorders". *Curr. Hematol. Rep.* **2** (4): 348–55. PMID 12901333.

[98] Lewitt, P. A.; Rezai, A. R.; Leehey, M. A.; Ojemann, S. G.; Flaherty, A. W.; Eskandar, E. N.; Kostyk, S. K.; Thomas, K.; Sarkar, A.; Siddiqui, M. S.; Tatter, S. B.; Schwalb, J. M.; Poston, K. L.; Henderson, J. M.; Kurlan, R. M.; Richard, I. H.; Van Meter, L.; Sapan, C. V.; During, M. J.; Kaplitt, M. G.; Feigin, A. (2011). "AAV2-GAD gene therapy for advanced Parkinson's disease: A double-blind, sham-surgery controlled, randomised trial". *The Lancet Neurology* **10** (4): 309–319. doi:10.1016/S1474-4422(11)70039-4. PMID 21419704.

[99] Gallaher, James "Gene therapy 'treats' Parkinson's disease" BBC News Health, 17 March 2011. Retrieved 24 April 2011

[100] Urbina, Zachary (12 February 2013) "Genetically Engineered Virus Fights Liver Cancer" United Academics. Retrieved 15 February 2013

[101] "Treatment for Leukemia Is Showing Early Promise". *The New York Times*. Associated Press. 11 August 2011. p. A15. Retrieved 21 January 2013.

[102] Coghlan, Andy (26 March 2013) "Gene therapy cures leukaemia in eight days" *The New Scientist*. Retrieved 15 April 2013

[103] Staff (13 February 2013) "Gene therapy cures diabetic dogs" *New Scientist*. Retrieved 15 February 2013

[104] (30 April 2013) "New gene therapy trial gives hope to people with heart failure" British Heart Foundation. Retrieved 5 May 2013

[105] Foster K, Foster H, Dickson JG (December 2006). "Gene therapy progress and prospects: Duchenne muscular dystrophy". *Gene Ther.* **13** (24): 1677–85. doi:10.1038/sj.gt.3302877. PMID 17066097.

[106] Hackett, P. B., Ekker, S. E. and Essner, J. J. (2004) *Applications of transposable elements in fish for transgenesis and functional genomics. Fish Development and Genetics* (Z. Gong and V. Korzh, eds.) World Scientific, Inc., Chapter 16, 532–580.

[107] Published PCT Application WO2000049150 "Chimeric Gene Constructs for Generation of Fluorescent Transgenic Ornamental Fish". National University of Singapore

[108] Eric Hallerman "Glofish, The First GM Animal Commercialized: Profits amid Controversy". June, 2004. Accessed 3 September 2012.

[109] Schuchat, S. (17 December 2003). "Why GloFish won't glow in California". *San Francisco Chronicle*.

[110] Shao Jun Du *et al.* (1992) "Growth Enhancement in Transgenic Atlantic Salmon by the Use of an 'All Fish' Chimeric Growth Hormone Gene Construct". *Nature Biotechnology* 10, 176–181

[111] Devlin RF *et al.* (2001) "Growth of domesticated transgenic fish". *Nature* 409, 781–782

[112] Rahman MA *et al.* (2001) "Growth and nutritional trials on transgenic Nile tilapia containing an exogenous fish growth hormone gene". *Journal of Fish Biology* 59(1):62–78

[113] Pollack, Andrew (December 21, 2012). "Engineered Fish Moves a Step Closer to Approval". *The New York Times*.

[114] http://www.fda.gov/ForConsumers/ConsumerUpdates/ucm472487.htm. Missing or empty |title= (help)

[115] Goldenberg, Suzanne (25 November 2013). "Canada approves production of GM salmon eggs on commercial scale". *The Guardian*. Retrieved 26 November 2013.

[116] National University of Singapore Enterprise webpage Archived July 5, 2014, at the Wayback Machine.

[117] "Zebra Fish as Pollution Indicators" Page last modified on 31 July 2001. Accessed October 2012

[118] Carvan, MJ; et al. (2000). "Transgenic zebrafish as sentinels for aquatic pollution". *Ann N Y Acad Sci* 919: 133–47. PMID 11083105.

[119] Nebert, DW; et al. (2002). "Use of Reporter Genes and Vertebrate DNA Motifs in Transgenic Zebrafish as Sentinels for Assessing Aquatic Pollution". *Environmental Health Perspectives* 110 (1): A15. doi:10.1289/ehp.110-a15. PMC 1240712. PMID 11813700.

[120] Mattingly, CJ; et al. (Aug 2001). "Green fluorescent protein (GFP) as a marker of aryl hydrocarbon receptor (AhR) function in developing zebrafish (*Danio rerio*)". *Environ Health Perspect* 109 (8): 845–9. doi:10.1289/ehp.01109845. PMC 1240414. PMID 11564622.

[121] Huntingford, F.A., Adams, C., Braithwaite, V.A., Kadri, S., Pottinger, T.G., Sandøe, P. and Turnbull, J.F. (2006). "Review paper: Current issues in fish welfare" (PDF). *Journal of Fish Biology* 68 (2): 332–372. doi:10.1111/j.0022-1112.2006.001046.x.

[122] Fini, Jean-Baptiste; Le Mevel, Sebastien; Turque, Nathalie; Palmier, Karima; Zalko, Daniel; Cravedi, Jean-Pierre; Demeneix, Barbara A. (2007-08-15). "An in vivo multiwell-based fluorescent screen for monitoring vertebrate thyroid hormone disruption". *Environmental Science & Technology* 41 (16): 5908–5914. doi:10.1021/es0704129. ISSN 0013-936X. PMID 17874805.

[123] "Online Education Kit: 1981-82: First Transgenic Mice and Fruit Flies". genome.gov.

[124] Gallagher, James "GM mosquitoes offer malaria hope" BBC News, Health, 20 April 2011. Retrieved 22 April 2011

[125] Corby-Harris, V.; Drexler, A.; Watkins De Jong, L.; Antonova, Y.; Pakpour, N.; Ziegler, R.; Ramberg, F.; Lewis, E. E.; Brown, J. M.; Luckhart, S.; Riehle, M. A. (2010). Vernick, Kenneth D., ed. "Activation of Akt Signaling Reduces the Prevalence and Intensity of Malaria Parasite Infection and Lifespan in Anopheles stephensi Mosquitoes". *PLoS Pathogens* 6 (7): e1001003. doi:10.1371/journal.ppat.1001003. PMC 2904800. PMID 20664791.

[126] Windbichler, N.; Menichelli, M.; Papathanos, P. A.; Thyme, S. B.; Li, H.; Ulge, U. Y.; Hovde, B. T.; Baker, D.; Monnat Jr, R. J.; Burt, A.; Crisanti, A. (2011). "A synthetic homing endonuclease-based gene drive system in the human malaria mosquito". *Nature* 473 (7346): 212–215. Bibcode:2011Natur.473..212W. doi:10.1038/nature09937. PMC 3093433. PMID 21508956.

[127] World Health Organization, Malaria, Key Facts Retrieved 22 April 2011

[128] Wise De Valdez, M. R.; Nimmo, D.; Betz, J.; Gong, H. -F.; James, A. A.; Alphey, L.; Black, W. C. (2011). "Genetic elimination of dengue vector mosquitoes". *Proceedings of the National Academy of Sciences* 108 (12): 4772–4775. Bibcode:2011PNAS..108.4772W. doi:10.1073/pnas.1019295108.

[129] Knapton, Sarah (6 February 2016). "Releasing millions of GM mosquitoes 'could solve zika crisis'". The Telegraph. Retrieved 14 March 2016.

[130] Harris, A. F.; Nimmo, D.; McKemey, A. R.; Kelly, N.; Scaife, S.; Donnelly, C. A.; Beech, C.; Petrie, W. D.; Alphey, L. (2011). "Field performance of engineered male mosquitoes". *Nature Biotechnology* 29 (11): 1034–1037. doi:10.1038/nbt.2019. PMID 22037376.

[131] Staff (March 2011) "Cayman demonstrates RIDL potential" Oxitec Newsletter, March 2011. Retrieved 20 September 2011

[132] Nicholls, Henry (14 September 2011) "Swarm troopers: Mutant armies waging war in the wild" *The New Scientist*. Retrieved 20 September 2011

[133] Staff Pink Bollworm Oxitec, Retrieved 17 August 2014

[134] Walters, M.; et al. (2012). "Field longevity of a fluorescent protein marker in an engineered strain of the pink bollworm, Pectinophora gossypiella (Saunders)". *PLoS ONE* **7** (6): e38547. Bibcode:2012PLoSO...738547W. doi:10.1371/journal.pone.0038547. PMID 22693645.

[135] Wittlieb J, Khalturin K, Lohmann JU, Anton-Erxleben F and Bosch TCG (2006). "Transgenic Hydra allow in vivo tracking of individual stem cells during morphogenesis". *Proc. Natl. Acad. Sci. U.S.A.* **103** (16): 6208–6211. Bibcode:2006PNAS..103.6208W. doi:10.1073/pnas.0510163103. PMC 1458856. PMID 16556723.

[136] Gaskell, G.; Bauer, M. W.; Durant, J.; Allum, N. C. (1999). "Worlds Apart? The Reception of Genetically Modified Foods in Europe and the U.S". *Science* **285** (5426): 384–387. doi:10.1126/science.285.5426.384. PMID 10411496.

[137] "The History and Future of GM Potatoes". *PotatoPro.com*.

[138] Wesseler, J. and N. Kalaitzandonakes (2011): "Present and Future EU GMO policy". In Arie Oskam, Gerrit Meesters and Huib Silvis (eds.), *EU Policy for Agriculture, Food and Rural Areas*. Second Edition, pp. 23–323 – 23-332. Wageningen: Wageningen Academic Publishers

[139] Beckmann, V., C. Soregaroli, J. Wesseler (2011): "Coexistence of genetically modified (GM) and non-modified (non GM) crops: Are the two main property rights regimes equivalent with respect to the coexistence value?" In *Genetically modified food and global welfare* edited by Colin Carter, GianCarlo Moschini and Ian Sheldon, pp 201–224. Volume 10 in Frontiers of Economics and Globalization Series. Bingley, UK: Emerald Group Publishing

[140] Smithsonian (2015). "Some Brands Are Labeling Products "GMO-free" Even if They Don't Have Genes".

[141] "Vermont v science", *The Economist* (Montpelier) **411** (8886), 10 May 2014, pp. 25–26

[142] American Association for the Advancement of Science (AAAS), Board of Directors (2012). Statement by the AAAS Board of Directors On Labeling of Genetically Modified Foods, and associated Press release: Legally Mandating GM Food Labels Could Mislead and Falsely Alarm Consumers

[143] *A decade of EU-funded GMO research (2001–2010)* (PDF). Directorate-General for Research and Innovation. Biotechnologies, Agriculture, Food. European Union. 2010. doi:10.2777/97784. ISBN 978-92-79-16344-9. "The main conclusion to be drawn from the efforts of more than 130 research projects, covering a period of more than 25 years of research, and involving more than 500 independent research groups, is that biotechnology, and in particular GMOs, are not per se more risky than e.g. conventional plant breeding technologies." (p. 16)

[144] Ronald, Pamela (2011). "Plant genetics, sustainable agriculture and global food security". *Genetics* **188** (1): 11–20. doi:10.1534/genetics.111.128553. PMC 3120150. PMID 21546547.

[145] American Medical Association (2012). "Report 2 of the Council on Science and Public Health: Labeling of Bioengineered Foods" "Bioengineered foods have been consumed for close to 20 years, and during that time, no overt consequences on human health have been reported and/or substantiated in the peer-reviewed literature." (first page)

[146] United States Institute of Medicine and National Research Council (2004). "Safety of Genetically Engineered Foods: Approaches to Assessing Unintended Health Effects". National Academies Press. Free full-text. National Academies Press. pp R9-10: "In contrast to adverse health effects that have been associated with some traditional food production methods, similar serious health effects have not been identified as a result of genetic engineering techniques used in food production. This may be because developers of bioengineered organisms perform extensive compositional analyses to determine that each phenotype is desirable and to ensure that unintended changes have not occurred in key components of food."

[147] Key S, Ma JK, Drake PM (June 2008). "Genetically modified plants and human health". *J R Soc Med* **101** (6): 290–8. doi:10.1258/jrsm.2008.070372. PMC 2408621. PMID 18515776. pp 292-293. "Foods derived from GM crops have been consumed by hundreds of millions of people across the world for more than 15 years, with no reported ill effects (or legal cases related to human health), despite many of the consumers coming from that most litigious of countries, the USA."

[148] Nathanael Johnson for Grist. Jul 8, 2013 The genetically modified food debate: Where do we begin?

[149] JoAnna Wendel for the Genetic Literacy Project. 10 September 2013 Scientists, journalists and farmers join lively GMO forum

[150] Keith Kloor for Discover Magazine's CollideAScape 22 August 2014 On Double Standards and the Union of Concerned Scientists

[151] Union of Concerned Scientists. Alternatives to Genetic Engineering. Page source description: "Biotechnology companies produce genetically engineered crops to control insects and weeds and to manufacture pharmaceuticals and other chemicals. The Union of Concerned Scientists works to strengthen the federal oversight needed to prevent such products from contaminating our food supply."

[152] Emily Marden, Risk and Regulation: U.S. Regulatory Policy on Genetically Modified Food and Agriculture 44 B.C.L.

Rev. 733 (2003). Quote: "By the late 1990s, public awareness of GM foods reached a critical level and a number of
public interest groups emerged to focus on the issue. One
of the early groups to focus on the issue was Mothers for
Natural Law ("MFNL"), an Iowa based organization that
aimed to ban GM foods from the market....The Union of
Concerned Scientists ("UCS"), an alliance of 50,000 citizens and scientists, has been another prominent voice on the
issue.... As the pace of GM products entering the market
increased in the 1990s, UCS became a vocal critic of what
it saw as the agency's collusion with industry and failure to
fully take account of allergenicity and other safety issues."

[153] British Medical Association Board of Science and Education
(2004). "Genetically modified food and health: A second
interim statement". March.

[154] Public Health Association of Australia (2007) "Genetically
Modified Foods" *PHAA AGM* 2007 Archived 20 January
2014 at the Wayback Machine.

[155] Canadian Association of Physicians for the Environment
(2013) "Statement on Genetically Modified Organisms in
the Environment and the Marketplace". October 2013

[156] Irish Doctors' Environmental Association "IDEA Position
on Genetically Modified Foods". Retrieved 3/25/14

[157] PR Newswire "Genetically Modified Maize: Doctors'
Chamber Warns of 'Unpredictable Results' to Humans". 11
November 2013

[158] Chartered Institute of Environmental Health (2006)
"Proposals for managing the coexistence of GM, conven-
tional and organic crops Response to the Department for
Environment, Food and Rural Affairs consultation paper".
October 2006

[159] Paull, John (2015) GMOs and organic agriculture: Six
lessons from Australia, Agriculture & Forestry, 61(1): 7-14.

[160] American Medical Association (2012). "Report 2 of the
Council on Science and Public Health: Labeling of Bioengineered Foods". "To better detect potential harms of bioengineered foods, the Council believes that pre-market safety assessment should shift from a voluntary notification process
to a mandatory requirement." page 7

[161] "GMOs, Herbicides, and Public Health". New England
Journal of Medicine. 2015.

5.8 External links

- Everything you wanted to know about GM organisms — Provided by *New Scientist*.
- Transgenic Organism Research
- International Society for Transgenic Technologies (ISTT)
- GMO-Compass: Information on genetically modified organisms
- Co-Extra: Research on co-existence and traceability of GM and non-GM supply chains
- ISAAA Knowledge Center: Information on genetically modified organisms

Chapter 6

Genetic engineering techniques

Genetic engineering techniques enable modification of the DNA of living organisms. A variety of editing techniques have been developed since DNA's structure was first discovered.

6.1 Targets

Bacteria are commonly engineered for research purposes. Typically this is through transformation to add a plasmid containing a gene of interest, but editing of the chromosome is also used. Plants and animals have both been genetically modified for research, agricultural and medical applications. In plants, the most widely inserted genes provide herbicide resistance or insecticidal properties.[1] In animals, the most widely used are growth hormone genes. Finally, genetically modified viruses are also used as viral vectors to transfer target genes to another organism in gene therapy.

6.2 Procedure

The first step involves choosing and isolating the gene that will be inserted into/removed from the genetically modified organism.

The gene must generally be combined with a promoter and terminator region as well as a selectable marker gene.

Then the genes must be spliced into the target's DNA. For animals, the gene must be inserted into embryonic stem cells.

The resulting organism must have the presence of the target gene confirmed.

First generation offspring are heterozygous, requiring them to be inbred to create the homozygous pattern necessary for stable inheritance. Homozygosity must be confirmed in second generation specimens, which then become the final product.

6.3 History

See also: History of genetic engineering

Human directed genetic manipulation began with the domestication of plants and animals through artificial selection in about 12,000 BC.[2]:1 Various techniques were developed to aid in breeding and selection. Hybridization was one way rapid changes in an organisms makeup could be introduced. Hybridization most likely first occurred when humans first grew similar, yet slightly different plants in close proximity.[3]:32 Some plants were able to be propagated by vegetative cloning.[3]:31 X-rays were first used to deliberately mutate plants in 1927. Between 1927 and 2007, more than 2,540 genetically mutated plant varieties had been produced using x-rays.[4]

It wasn't until the mid 1800s that DNA and genes were discovered, which would form the basis of modern genetic manipulation. Genetic inheritance was first discovered by Gregor Mendel in 1865 following experiments crossing peas.[5] In 1928 Frederick Griffith proved the existence of a "transforming principle" involved in inheritance, which was identified as DNA in 19. Frederick Sanger developed a method for sequencing DNA in 1977, greatly increasing the genetic information available to researchers.

As well as discovering how DNA works, tools had to be developed that allowed it to be manipulated. In 1970 Hamilton Smiths lab discovered restriction enzymes, enabling scientists to isolate genes from an organism's genome.[6] DNA ligases, that join broken DNA together, had been discovered earlier in 1967[7] and by combining the two enzymes it was possible to "cut and paste" DNA sequences to create recombinant DNA. Plasmids, discovered in 1952,[8] became important tools for transferring information between cells and replicating DNA sequences. Polymerase chain reaction (PCR), developed by Kary Mullis in 1983, allowed small sections of DNA to be amplified and aided identification and isolation of genetic material.

As well as manipulating the DNA, techniques had to be developed for its insertion (known as transformation) into an organism's genome. Griffiths experiment had already shown that some bacteria had the ability to naturally uptake and express foreign DNA. Artificial competence was induced in *Escherichia coli* in 1970 by treating them with calcium chloride solution ($CaCl_2$).[9] Transformation using electroporation was developed in the late 1980s, increasing the efficiency and bacterial range.[10] In 1907 a bacterium that caused plant tumors, *Agrobacterium tumefaciens*, had been discovered and in the early 1970s it was found that the bacteria inserted its DNA into plants using a Ti plasmid.[11] By removing the genes in the plasmid that caused the tumor and adding in novel genes researchers were able to infect plants with *A. tumefaciens* and let the bacteria insert their chosen DNA into the genomes of the plants.[12]

An important part of genetic engineering is to identify useful genes to transform into the genetically modified organism. The bacteria *Bacillus thuringiensis* was first discovered in 1901 as the causative agent in the death of silkworms. Due to these insecticidal properties the bacteria was used as an biological insecticide, commercially developed in 1938. The cry proteins were discovered to provide the insecticidal activity in 1956 and by the 1980s scientists had successfully cloned the gene coding for this protein and expressed it in plants.[13] The gene that provides resistance to the glyphosate herbicide was found, after seven years searching, in the outflow pipe of a Monsanto roundup manufacturing facility.[14] In animals the majority of genes used are growth hormone genes.[15]

6.4 Libraries

Target genes can be cloned from a DNA segment after the creation of a DNA library. The libraries generally cover the organism's genome multiple times and its size will depend on how large the genome is.

6.5 Techniques

6.5.1 Gene isolation

The DNA is first digested with a random digestion method, commonly by cutting the DNA with restriction enzymes (enzymes that cut DNA). A partial restriction digest cuts only some of the restriction sites, resulting in overlapping DNA fragment lengths. The DNA fragments are put into individual plasmid vectors and grown inside bacteria. Once in the bacteria the plasmid is copied as the bacteria divides. To determine if a useful gene is present on a particular fragment the bacterial library is screened for the desired phenotype. If the phenotype is detected then it is possible that the bacteria contains the target gene. If the gene does not have a detectable phenotype or a DNA library does not contain the correct gene, other methods can be used to isolate it. If the position of the gene can be determined using molecular markers then chromosome walking is one way to isolate the correct DNA fragment. If the gene expresses close homology to a known gene in another species, then it could be isolated by searching for genes in the library that closely match the known gene.[16]

If the DNA sequence of the gene and the organism is known, restriction enzymes can cut the DNA on either side of the gene and gel electrophoresis can sort the fragments according to length.[17] The DNA band at the correct size should contain the gene, where it can be excised from the gel. Polymerase chain reaction (PCR) can be used to amplify the gene, which can then be isolated through gel electrophoresis.[18] It is also possible to synthesize the gene.[19]

6.5.2 Additional material

The gene to be inserted into the genetically modified organism must be combined with other genetic elements in order for it to work properly. The gene can also be modified at this stage for better expression or effectiveness. As well as the gene to be inserted most constructs contain a promoter and terminator region as well as a selectable marker gene. The promoter region initiates transcription of the gene and can be used to control the location and level of gene expression, while the terminator region ends transcription. The selectable marker, which in most cases confers antibiotic resistance to the organism it is expressed in, is needed to determine which cells are transformed with the new gene. The constructs are made using recombinant DNA techniques, such as restriction digests, ligations and molecular cloning.[20]

6.5.3 Gene targeting

Main article: Gene targeting

Gene targeting uses homologous recombination to target desired changes to a specific endogenous gene. This tends to occur at a relatively low frequency in plants and animals and generally requires the use of selectable markers. The success of gene targeting can be enhanced with the use of engineered nucleases such as zinc finger nucleases,[21][22] engineered homing endonucleases,[23][24] transcription activator-like effector nuclease,[25][26] or CRISPR. Engineered nucleases can also introduce mutations at endogenous genes that generate a gene knockout.[27][28]

6.5.4   Transformation

Main article: Transformation (genetics)
 About 1% of bacteria are naturally able to take up foreign

A. tumefaciens *attaching itself to a carrot cell*

DNA but it can be induced in other bacteria.[29] Stressing the bacteria with a heat shock or an electric shock can make the cell membrane permeable to DNA that may then incorporate into the genome or exist as extrachromosomal DNA. DNA is generally inserted into animal cells using microinjection, where it can be injected through the cells nuclear envelope directly into the nucleus or through the use of viral vectors. In plants the DNA is generally inserted using *Agrobacterium*-mediated recombination or biolistics.[30]

In *Agrobacterium*-mediated recombination the plasmid construct must also contain T-DNA. *Agrobacterium* naturally inserts DNA from a tumor inducing plasmid into any susceptible plant that it infects, causing crown gall disease. The T-DNA region of this plasmid is responsible for insertion of the DNA. The DNA to be inserted is cloned into a binary vector that contains T-DNA and can be grown in both *E. coli* and *Agrobacterium*. Once the binary vector is constructed the plasmid is transformed into *Agrobacterium* containing no plasmids and plant cells are infected. The

Agrobacterium naturally inserts the genetic material into the plant cells.[31]

In biolistic particles of gold or tungsten are coated with DNA and then shot into young plant cells or plant embryos. Some genetic material enters the cells and transforms them. This method can be used on plants that are not susceptible to *Agrobacterium* infection and also allows transformation of plant plastids.

Another transformation method for plant and animal cells is electroporation, which involves subjecting cells to an electric shock, which can make the cell membrane permeable to plasmid DNA. In some cases the electroporated cells will incorporate the DNA. Due to the associated cell and DNA damage, the transformation efficiency of biolistics and electroporation is lower than with agrobacteria and microinjection.[32]

6.5.5   Selection

Not all the organism's cells will be transformed with the new genetic material; typically a selectable marker is used to differentiate transformed from untransformed cells. Cells that have been successfully transformed with the DNA it will also contain the marker gene. By growing the cells in the presence of an antibiotic or chemical that selects or marks the cells expressing that gene, it is possible to separate modified from unmodified cells. Another screening method involves a DNA probe that sticks only to the inserted gene. Multiple strategies can remove the marker from the mature plant.[33]

6.5.6   Regeneration

As often only a single cell is transformed with genetic material the modified organism must be grown from that single cell. Bacteria consist of a single cell and reproduce clonally, so regeneration is not necessary for them. In plants this is accomplished through the use of tissue culture. Each plant species has different requirements for successful regeneration. If successful, the technique produces an adult plant that contains the transgene in every cell.

In animals it is necessary to ensure that the inserted DNA is present in embryonic stem cells. Offspring can be screened for the gene. All offspring from the first generation will be heterozygous for the inserted gene and must be inbred to produce a homozygous specimen.

6.5.7   Confirmation

The finding that a recombinant organism contains the inserted genes is not usually sufficient to ensure that they will be appropriately expressed in the intended tissues. To confirm the presence of the gene, PCR, Southern hybridization and DNA sequencing are employed to determine the chromosomal location and number of gene copies.

To assess gene expression, transcription, RNA processing patterns and expression and localization of protein product(s) must usually be assessed, using methods including northern hybridization, quantitative RT-PCR, Western blot, immunofluorescence and phenotypic analysis. When appropriate, the organism's offspring are studied to confirm that the trans-gene and associated phenotype are stably inherited.

In some cases further generations must be produced and confirmed, to ensure the absence of undesirable traits in the modified organism. For hybrid products such as maize, the modified organism is crossbred with other cultivars that possess required traits.

6.6 References

[1] James, Clive (2008). "Global Status of Commercialized Biotech/GM Crops:2008". *ISSA Brief No. 39*.

[2] Clive Root (2007). *Domestication*. Greenwood Publishing Groups.

[3] Noel Kingsbury. Hybrid: The History and Science of Plant Breeding University of Chicago Press, Oct 15, 2009

[4] Schouten, H. J.; Jacobsen, E. (2007). "Are Mutations in Genetically Modified Plants Dangerous?". *Journal of Biomedicine and Biotechnology* **2007**: 1. doi:10.1155/2007/82612.

[5] D. L. Hartl and V. Orel (1992). "What Did Gregor Mendel Think He Discovered?". *Genetics* **131** (2): 245–25.

[6] Roberts, R. J. (2005). "Classic Perspective: How restriction enzymes became the workhorses of molecular biology". *Proceedings of the National Academy of Sciences* **102** (17): 5905–5908. doi:10.1073/pnas.0500923102. PMC 1087929. PMID 15840723.

[7] Weiss, B.; Richardson, C. C. (1967). "Enzymatic breakage and joining of deoxyribonucleic acid. 1. Repair of single-strand breaks in DNA by an enzyme system from Escherichia coli infected with T4 bacteriophage". *Proceedings of the National Academy of Sciences* **57** (4): 1021–8. doi:10.1073/pnas.57.4.1021. PMC 224649. PMID 5340583.

[8] Lederberg, J (1952). "Cell genetics and hereditary symbiosis". *Physiological reviews* **32** (4): 403–30. PMID 13003535.

[9] Mandel, Morton; Higa, Akiko (1970). "Calcium-dependent bacteriophage DNA infection". *Journal of Molecular Biology* **53** (1): 159–162. doi:10.1016/0022-2836(70)90051-3. PMID 4922220.

[10] Wirth, Reinhard; Friesenegger, Anita; Fiedlerand, Stefan (1989). "Transformation of various species of gram-negative bacteria belonging to 11 different genera by electroporation". *Molecular and General Genetics MGG*.

[11] Nester, Eugene. "Agrobacterium: The Natural Genetic Engineer (100 Years Later)". Retrieved 14 January 2011.

[12] Zambryski, P.; Joos, H.; Genetello, C.; Leemans, J.; Montagu, M. V.; Schell, J. (1983). "Ti plasmid vector for the introduction of DNA into plant cells without alteration of their normal regeneration capacity". *The EMBO Journal* **2** (12): 2143–2150. PMC 555426. PMID 16453482.

[13] Roh JY, Choi JY, Li MS, Jin BR, Je YH (2007). "Bacillus thuringiensis as a specific, safe, and effective tool for insect pest control". *J Microbiol Biotechnol* **17** (4): 547–59.

[14] Jerry Adler (May 2011). "The Growing Menace From Superweeds". Scientific American.

[15] Food and Agricultural Organisation of the United Nations. "The process of genetic modification".

[16] Corinne A. Michels (2002). "7". *Genetic Techniques for Biological Research: A Case Study Approach*. John Wiley & Sons. pp. 85–88. ISBN 0-471-89919-4.

[17] Alberts B, Johnson A, Lewis J, et al. (2002). "8". *Isolating, Cloning, and Sequencing DNA*. (4th ed.). New York: Garland Science.

[18] R I Kaufman and B T Nixon (1996). "Use of PCR to isolate genes encoding sigma54-dependent activators from diverse bacteria.". *J Bacteriol* **178** (13): 3967–3970.

[19] Liang, J.; Luo, Y.; Zhao, H. (2011). "Synthetic biology: Putting synthesis into biology". *Wiley Interdisciplinary Reviews: Systems Biology and Medicine* **3**: 7. doi:10.1002/wsbm.104.

[20] Berg, P.; Mertz, J. (2010). "Personal reflections on the origins and emergence of recombinant DNA technology". *Genetics* **184** (1): 9–17. doi:10.1534/genetics.109.112144. PMC 2815933. PMID 20061565.

[21] Townsend JA, Wright DA, Wintrey RJ, et al. (May 2009). "High-frequency modification of plant genes using engineered zinc-finger nucleases". *Nature* **459** (7245): 442–5. Bibcode:2009Natur.459..442T. doi:10.1038/nature07845. PMC 2743854. PMID 19404258.

[22] Shukla VK, Doyon Y, Miller JC, et al. (May 2009). "Precise genome modification in the crop species Zea mays using zinc-finger nucleases". *Nature* **459** (7245): 437–41. Bibcode:2009Natur.459..437S. doi:10.1038/nature07992. PMID 19404259.

[23] Grizot S, Smith J, Daboussi F, et al. (September 2009). "Efficient targeting of a SCID gene by an engineered single-chain homing endonuclease". *Nucleic Acids Res.* **37** (16): 5405–19. doi:10.1093/nar/gkp548. PMC 2760784. PMID 19584299.

[24] Gao H, Smith J, Yang M, et al. (January 2010). "Heritable targeted mutagenesis in maize using a designed endonuclease". *Plant J.* **61** (1): 176–87. doi:10.1111/j.1365-313X.2009.04041.x. PMID 19811621.

[25] Christian M, Cermak T, Doyle EL, et al. (July 2010). "TAL Effector Nucleases Create Targeted DNA Double-strand Breaks". *Genetics* **186** (2): 757–61. doi:10.1534/genetics.110.120717. PMC 2942870. PMID 20660643.

[26] Li T, Huang S, Jiang WZ, et al. (August 2010). "TAL nucleases (TALNs): hybrid proteins composed of TAL effectors and FokI DNA-cleavage domain". *Nucleic Acids Res* **39** (1): 359–72. doi:10.1093/nar/gkq704. PMC 3017587. PMID 20699274.

[27] S.C. Ekker (2008). "Zinc finger-based knockout punches for zebrafish genes". *Zebrafish* **5** (2): 1121–3. doi:10.1089/zeb.2008.9988. PMC 2849655. PMID 18554175.

[28] Geurts AM, Cost GJ, Freyvert Y, et al. (July 2009). "Knockout rats via embryo microinjection of zinc-finger nucleases". *Science* **325** (5939): 433. Bibcode:2009Sci...325..433G. doi:10.1126/science.1172447. PMC 2831805. PMID 19628861.

[29] Chen I, Dubnau D (2004). "DNA uptake during bacterial transformation". *Nature Reviews Microbiology* **2** (3): 241–9. doi:10.1038/nrmicro844. PMID 15083159.

[30] Graham Head; Hull, Roger H; Tzotzos, George T. (2009). *Genetically Modified Plants: Assessing Safety and Managing Risk*. London: Academic Pr. p. 244. ISBN 0-12-374106-8.

[31] Gelvin, S. B. (2003). "Agrobacterium-Mediated Plant Transformation: the Biology behind the "Gene-Jockeying" Tool". *Microbiology and Molecular Biology Reviews* **67** (1): 16–37, table of contents. doi:10.1128/MMBR.67.1.16-37.2003. PMC 150518. PMID 12626681.

[32] Behrooz Darbani, Safar Farajnia, Mahmoud Toorchi, Saeed Zakerbostanabad, Shahin Noeparvar and C. Neal Stewart Jr. (2010). "DNA-Delivery Methods to Produce Transgenic Plants". Science Alert.

[33] Barbara Hohn, Avraham A Levy and Holger Puchta (2001). "Elimination of selection markers from transgenic plants". *Current Opinion in Biotechnology* **12** (2): 139–143. doi:10.1016/S0958-1669(00)00188-9. PMID 11287227.

Chapter 7

Genetically modified food

For related content, see Genetic engineering, Genetically modified organism, Genetically modified crops, and Genetically modified food controversies.

Genetically modified foods or **GM foods**, also **genetically engineered foods**, are foods produced from organisms that have had changes introduced into their DNA using the methods of genetic engineering. Genetic engineering techniques allow for the introduction of new traits as well as greater control over traits than previous methods such as selective breeding and mutation breeding.[1]

Commercial sale of genetically modified foods began in 1994, when Calgene first marketed its unsuccessful Flavr Savr delayed-ripening tomato.[2][3] Most food modifications have primarily focused on cash crops in high demand by farmers such as soybean, corn, canola, and cotton seed oil. Genetically modified crops have been engineered for resistance to pathogens and herbicides and for better nutrient profiles. GM livestock have been developed, although as of November 2013 none were on the market.[4]

There is general scientific agreement that food from genetically modified crops is not inherently riskier to human health than conventional food, but should be tested on a case-by-case basis.[5][6][7][8][9][10] However, there are ongoing public concerns related to food safety, regulation, labelling, environmental impact, research methods, and the fact that some GM seeds are subject to intellectual property rights owned by corporations.[11]

7.1 Definition

Genetically modified foods, GM foods or genetically engineered foods, are foods produced from organisms that have had changes introduced into their DNA using the methods of genetic engineering as opposed to traditional cross breeding.[7][12] In the US, the Department of Agriculture (USDA) and the Food and Drug Administration (FDA) favor the use of "genetic engineering" over "genetic modifi-

cation" as the more precise term; the USDA defines genetic modification to include "genetic engineering or other more traditional methods."[13][14]

According to the World Health Organization, "Genetically modified organisms (GMOs) can be defined as organisms (i.e. plants, animals or microorganisms) in which the genetic material (DNA) has been altered in a way that does not occur naturally by mating and/or natural recombination. The technology is often called 'modern biotechnology' or 'gene technology', sometimes also 'recombinant DNA technology' or 'genetic engineering'. ... Foods produced from or using GM organisms are often referred to as GM foods."[7]

7.2 History

Main article: History of genetic engineering

Human directed genetic manipulation of food began with the domestication of plants and animals through artificial selection at about 10,500 to 10,100 BC.[15]:1 The process of selective breeding, in which organisms with desired traits (and thus with the desired genes) are used to breed the next generation and organisms lacking the trait are not bred, is a precursor to the modern concept of genetic modification (GM).[15]:1[16]:1 With the discovery of DNA in the early 1900s and various advancements in genetic techniques through the 1970s[17] it became possible to directly alter the DNA and genes within food.

Genetically modified microbial enzymes were the first application of genetically modified organisms in food production and were approved in 1988 by the US Food and Drug Administration.[18] In the early 1990s, recombinant chymosin was approved for use in several countries.[18][19] Cheese had typically been made using the enzyme complex rennet that had been extracted from cows' stomach lining. Scientists modified bacteria to produce chymosin, which was also able to clot milk, resulting in cheese curds.[20]

The first genetically modified food approved for release was the Flavr Savr tomato in 1994.[2] Developed by Calgene, it was engineered to have a longer shelf life by inserting an antisense gene that delayed ripening.[21] In 1995, *Bacillus thuringiensis* (Bt) Potato was approved for cultivation, making it the first pesticide producing crop to be approved in the USA.[22] Other genetically modified crops receiving marketing approval in 1995 were: canola with modified oil composition, Bt maize, cotton resistant to the herbicide bromoxynil, Bt cotton, glyphosate-tolerant soybeans, virus-resistant squash, and another delayed ripening tomato.[2]

By 2010, 29 countries had planted commercialized biotech crops and a further 31 countries had granted regulatory approval for transgenic crops to be imported.[23] The US was the leading country in the production of GM foods in 2011, with twenty-five GM crops having received regulatory approval.[24] In 2015, 92% of corn, 94% of soybeans, and 94% of cotton produced in the US were genetically modified strains.[25]

With the creation of golden rice in 2000, scientists had genetically modified food to increase its nutrient value for the first time.[26]

The first genetically modified animal to be approved for food use was AquAdvantage salmon in 2015.[27] The salmon were transformed with a growth hormone-regulating gene from a Pacific Chinook salmon and a promoter from an ocean pout enabling it to grow year-round instead of only during spring and summer.[28]

The most widely planted GMOs are designed to tolerate herbicides. By 2006 some weed populations had evolved to tolerate some of the same herbicides. Palmer amaranth is a weed that competes with cotton. A native of the southwestern US, it traveled east and was first found resistant to glyphosate in 2006, less than 10 years after GM cotton was introduced.[29][30][31]

7.3   Process

Main article: Genetic engineering

Genetically engineered organisms are generated and tested in the laboratory for desired qualities. The most common modification is to add one or more genes to an organism's genome. Less commonly, genes are removed or their expression is increased or silenced or the number of copies of a gene is increased or decreased.

Once satisfactory strains are produced, the producer applies for regulatory approval to field-test them, called a "field release." Field-testing involves cultivating the plants on farm fields or growing animals in a controlled environment. If these field tests are successful, the producer applies for regulatory approval to grow and market the crop. Once approved, specimens (seeds, cuttings, breeding pairs, etc.) are cultivated and sold to farmers. The farmers cultivate and market the new strain. In some cases, the approval covers marketing but not cultivation.

According to the USDA, the number of field releases for genetically engineered organisms has grown from four in 1985 to an average of about 800 per year. Cumulatively, more than 17,000 releases had been approved through September 2013.[32]

7.4   Crops

Main article: Genetically modified crops

7.4.1   Fruits and vegetables

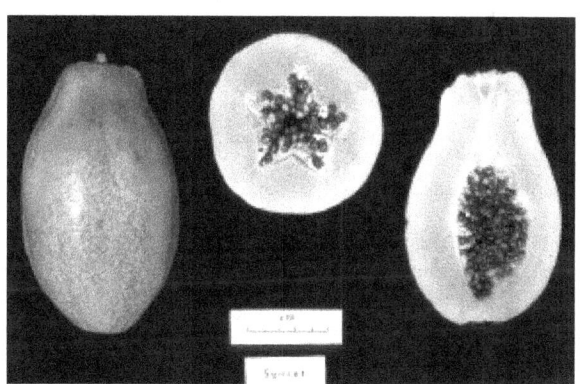

3 views of the Sunset papaya cultivar, which was genetically modified to create the SunUp cultivar, resistant to PRSV.[]

Papaya was genetically modified to resist the ringspot virus. 'SunUp' is a transgenic red-fleshed Sunset papaya cultivar that is homozygous for the coat protein gene PRSV; 'Rainbow' is a yellow-fleshed F1 hybrid developed by crossing 'SunUp' and nontransgenic yellow-fleshed 'Kapoho'.[33] The New York Times stated, "in the early 1990s, Hawaii's papaya industry was facing disaster because of the deadly papaya ringspot virus. Its single-handed savior was a breed engineered to be resistant to the virus. Without it, the state's papaya industry would have collapsed. Today, 80% of Hawaiian papaya is genetically engineered, and there is still no conventional or organic method to control ringspot virus."[34] The GM cultivar was approved in 1998.[35] In China, a transgenic PRSV-resistant papaya was developed by South China Agricultural University and was first approved for commercial planting in 2006; as of 2012 95% of the papaya grown in Guangdong province and 40%

of the papaya grown in Hainan province was genetically modified.[36]

The New Leaf potato, brought to market by Monsanto in the late 1990s, was developed for the fast food market. It was withdrawn in 2001 after retailers rejected it and food processors ran into export problems.[37]

As of 2005, about 13% of the Zucchini (a form of squash) grown in the US was genetically modified to resist three viruses; that strain is also grown in Canada.[38][39]

Plums genetically engineered for resistance to plum pox, a disease carried by aphids.

In 2011, BASF requested the European Food Safety Authority's approval for cultivation and marketing of its Fortuna potato as feed and food. The potato was made resistant to late blight by adding resistant genes blb1 and blb2 that originate from the Mexican wild potato Solanum bulbocastanum.[40][41] In February 2013, BASF withdrew its application.[42]

In 2013, the USDA approved the import of a GM pineapple that is pink in color and that "overexpresses" a gene derived from tangerines and suppress other genes, increasing production of lycopene. The plant's flowering cycle was changed to provide for more uniform growth and quality. The fruit "does not have the ability to propagate and persist in the environment once they have been harvested," according to USDA APHIS. According to Del Monte's submission, the pineapples are commercially grown in a "monoculture" that prevents seed production, as the plant's flowers aren't exposed to compatible pollen sources. Importation into Hawaii is banned for "plant sanitation" reasons.[43]

In 2014, the USDA approved a genetically modified potato developed by J.R. Simplot Company that contained ten genetic modifications that prevent bruising and produce less acrylamide when fried. The modifications eliminate specific proteins from the potatoes, via RNA interference, rather than introducing novel proteins.[44][45]

In February 2015 Arctic Apples were approved by the USDA,[46] becoming the first genetically modified apple approved for sale in the US.[47] Gene silencing is used to reduce the expression of polyphenol oxidase (PPO), thus preventing the fruit from browning.[48]

7.4.2 Corn

Corn used for food and ethanol has been genetically modified to tolerate various herbicides and to express a protein from Bacillus thuringiensis (Bt) that kills certain insects.[49] About 90% of the corn grown in the U.S. was genetically modified in 2010.[50] In the US in 2015, 81% of corn acreage contained the Bt trait and 89% of corn acreage contained the glyphosate-tolerant trait.[25] Corn can be processed into grits, meal and flour as an ingredient in pancakes, muffins, doughnuts, breadings and batters, as well as baby foods, meat products, cereals and some fermented products. Corn-based masa flour and masa dough are used in the production of taco shells, corn chips and tortillas.[51]

7.4.3 Soy

Genetically modified soybean has been modified to tolerate herbicides, express Bt and produce healthier oils.[52] In 2015, 94% of soybean acreage in the U.S. was genetically modified to be glyphosate-tolerant.[25] Soybeans contain about 20% oil. In the most common method used to extract the oil, the soybeans are cracked, adjusted for moisture content, rolled into flakes and solvent-extracted with commercial hexane. The remaining soy meal has a 50% soy protein content. The meal is 'toasted' (actually heated with moist steam) and ground in a hammer mill. Part of the balance is processed further into high protein soy products that are used in a variety of foods, such as salad dressings, soups, meat analogues, beverage powders, cheeses, nondairy creamer, frozen desserts, whipped topping, infant formulas, breads, breakfast cereals, pasta and pet foods.[53][54] Processed soy protein appears in foods mainly in three forms: soy flour, soy protein isolates and soy protein concentrates.[54][55]

Food-grade soy protein isolate first became available on October 2, 1959.[56]:227-28 Soy protein isolate is a highly refined form of soy protein with a minimum protein content of 90% on a moisture-free basis. It is made from soy meal that has had most of the fats and carbohydrates removed. Soy isolates are mainly used to improve the texture of processed meat products and to increase protein content, enhance moisture retention and as an emulsifier.[57][58]

Soy protein concentrate is about 70% soy protein and is basically soybean meal without carbohydrates. Soy protein concentrate retains most of the bean fiber. It is used as a functional or nutritional ingredient in food products, mainly in baked foods, breakfast cereals and in some meat products. Soy protein concentrate is used in meat and poultry products to increase water and fat retention and to improve nutritional values (more protein, less fat).[57][59]

Soy flour is made by grinding soybeans into a fine powder. It comes in three forms: natural or full-fat (contains natural oils); defatted (oils removed) with 50% protein content and with either high water solubility or low water solubility; and lecithinated (lecithin added). As soy flour is gluten-free, yeast-raised breads made with soy flour are dense in texture. Soy grits are similar to soy flour except the soybeans have been toasted and cracked into coarse pieces. *Kinako* is a soy flour used in Japanese cuisine.[57][60]

Textured soy protein (TSP) is a fibrous, spongy matrix similar in texture to meat. TSP is used as a low-cost substitute in meat and poultry products.[57][61]

7.5 Derivative products

7.5.1 Corn starch and starch sugars, including syrups

Starch or amylum is a polysaccharide produced by all green plants as an energy store. Pure starch is a white, tasteless and odourless powder. It consists of two types of molecules: the linear and helical amylose and the branched amylopectin. Depending on the plant, starch generally contains 20 to 25% amylose and 75 to 80% amylopectin by weight.[62]

Starch can be further modified to create modified starch for specific purposes,[63] including creation of many of the sugars in processed foods. They include:

- Maltodextrin, a lightly hydrolyzed starch product used as a bland-tasting filler and thickener.

- Various glucose syrups, also called corn syrups in the US, viscous solutions used as sweeteners and thickeners in many kinds of processed foods.

- Dextrose, commercial glucose, prepared by the complete hydrolysis of starch.

- High fructose syrup, made by treating dextrose solutions with the enzyme glucose isomerase, until a substantial fraction of the glucose has been converted to fructose.

- Sugar alcohols, such as maltitol, erythritol, sorbitol, mannitol and hydrogenated starch hydrolysate, are sweeteners made by reducing sugars.

7.5.2 Lecithin

Lecithin is a naturally occurring lipid. It can be found in egg yolks and oil-producing plants. it is an emulsifier and thus is used in many foods. Corn, soy and safflower oil are sources of lecithin, though the majority of lecithin commercially available is derived from soy.[64][65][66][67] Sufficiently processed lecithin is often undetectable with standard testing practices.[62] According to the FDA, no evidence shows or suggests hazard to the public when lecithin is used at common levels. Lecithin added to foods amounts to only 2 to 10 percent of the 1 to 5 g of phosphoglycerides consumed daily on average.[64][65] Nonetheless, consumer concerns about GM food extend to such products.[68] This concern led to policy and regulatory changes in Europe in 2000, when Regulation (EC) 50/2000 was passed[69] which required labelling of food containing additives derived from GMOs, including lecithin. Because of the difficulty of detecting the origin of derivatives like lecithin with current testing practices, European regulations require those who wish to sell lecithin in Europe to employ a comprehensive system of Identity preservation (IP).[70][71]

7.5.3 Sugar

The US imports 10% of its sugar, while the remaining 90% is extracted from sugar beet and sugarcane. After deregulation in 2005, glyphosate-resistant sugar beet was extensively adopted in the United States. 95% of beet acres in the US were planted with glyphosate-resistant seed in 2011.[72] GM sugar beets are approved for cultivation in the US, Canada and Japan; the vast majority are grown in the US. GM beets are approved for import and consumption in Australia, Canada, Colombia, EU, Japan, Korea, Mexico, New Zealand, Philippines, Russian Federation and Singapore.[73] Pulp from the refining process is used as animal feed. The sugar produced from GM sugarbeets contains no DNA or protein—it is just sucrose that is chemically indistinguishable from sugar produced from non-GM sugarbeets.[62][74]

Independent analyses conducted by internationally recognized laboratories found that sugar from Roundup Ready sugar beets is identical to the sugar from comparably grown conventional (non-Roundup Ready) sugar beets. And, like all sugar, sugar from Roundup Ready sugar beets contains no genetic material or detectable protein (including the protein that provides glyphosate tolerance).[75]

7.5.4 Vegetable oil

Most vegetable oil used in the US is produced from GM crops canola,[76] corn,[66][77] cotton[78] and soybeans.[79]

Vegetable oil is sold directly to consumers as cooking oil, shortening and margarine[80] and is used in prepared foods. There is a vanishingly small amount of protein or DNA from the original crop in vegetable oil.[62][81] Vegetable oil is made of triglycerides extracted from plants or seeds and then refined and may be further processed via hydrogenation to turn liquid oils into solids. The refining process[82] removes all, or nearly all non-triglyceride ingredients.[83]

7.6 Other uses

7.6.1 Animal feed

Livestock and poultry are raised on animal feed, much of which is composed of the leftovers from processing crops, including GM crops. For example, approximately 43% of a canola seed is oil. What remains after oil extraction is a meal that becomes an ingredient in animal feed and contains canola protein.[84] Likewise, the bulk of the soybean crop is grown for oil and meal. The high-protein defatted and toasted soy meal becomes livestock feed and dog food. 98% of the US soybean crop goes for livestock feed.[85][86] In 2011, 49% of the US maize harvest was used for livestock feed (including the percentage of waste from distillers grains).[87] "Despite methods that are becoming more and more sensitive, tests have not yet been able to establish a difference in the meat, milk, or eggs of animals depending on the type of feed they are fed. It is impossible to tell if an animal was fed GM soy just by looking at the resulting meat, dairy, or egg products. The only way to verify the presence of GMOs in animal feed is to analyze the origin of the feed itself."[88]

A 2012 literature review of studies evaluating the effect of GM feed on the health of animals did not find evidence that animals were adversely affected, although small biological differences were occasionally found. The studies included in the review ranged from 90 days to two years, with several of the longer studies considering reproductive and intergenerational effects.[89]

7.6.2 Proteins

Rennet is a mixture of enzymes used to coagulate milk into cheese. Originally it was available only from the fourth stomach of calves, and was scarce and expensive, or was available from microbial sources, which often produced unpleasant tastes. Genetic engineering made it possible to extract rennet-producing genes from animal stomachs and insert them into bacteria, fungi or yeasts to make them produce chymosin, the key enzyme.[90][91] The mod-

ified microorganism is killed after fermentation. Chymosin is isolated from the fermentation broth, so that the Fermentation-Produced Chymosin (FPC) used by cheese producers has an amino acid sequence that is identical to bovine rennet.[92] The majority of the applied chymosin is retained in the whey. Trace quantities of chymosin may remain in cheese.[92]

FPC was the first artificially produced enzyme to be approved by the US Food and Drug Administration.[18][19] FPC products have been on the market since 1990 and as of 2015 had yet to be surpassed in commercial markets.[93] In 1999, about 60% of US hard cheese was made with FPC.[94] Its global market share approached 80%.[95] By 2008, approximately 80% to 90% of commercially made cheeses in the US and Britain were made using FPC.[92] The most widely used FPC is produced either by the fungus *Aspergillus niger* (CHY-MAX®)[96][97]

In some countries, recombinant (GM) bovine somatotropin (also called rBST, or bovine growth hormone or BGH) is approved for administration to increase milk production. rBST may be present in milk from rBST treated cows, but it is destroyed in the digestive system and even if directly injected into the human bloodstream, has no observable effect on humans.[98][99][100] The FDA, World Health Organization, American Medical Association, American Dietetic Association and the National Institutes of Health have independently stated that dairy products and meat from rBST-treated cows are safe for human consumption.[101] However, on 30 September 2010, the United States Court of Appeals, Sixth Circuit, analyzing submitted evidence, found a "compositional difference" between milk from rBGH-treated cows and milk from untreated cows.[102][103] The court stated that milk from rBGH-treated cows has: increased levels of the hormone Insulin-like growth factor 1 (IGF-1); higher fat content and lower protein content when produced at certain points in the cow's lactation cycle; and more somatic cell counts, which may "make the milk turn sour more quickly."[103]

7.6.3 Livestock

Main article: Genetically modified livestock

Genetically modified livestock are organisms from the group of cattle, sheep, pigs, goats, birds, horses and fish kept for human consumption, whose genetic material (DNA) has been altered using genetic engineering techniques. In some cases, the aim is to introduce a new trait to the animals which does not occur naturally in the species, i.e. transgenesis.

A 2003 review published on behalf of Food Standards Australia New Zealand examined transgenic experimentation on terrestrial livestock species as well as aquatic species such as fish and shellfish. The review examined the molecular techniques used for experimentation as well as techniques for tracing the transgenes in animals and products as well as issues regarding transgene stability.[104]

Some mammals typically used for food production have been modified to produce non-food products, a practice sometimes called Pharming.

Salmon

See also: Genetically modified fish § AquAdvantage salmon and Genetically modified fish § AquAdvantage salmon 2

A GM salmon, awaiting regulatory approval[105][106][107] since 1997,[108] was approved for human consumption by the American FDA in November 2015, to be raised in specific land-based hatcheries in Canada and Panama.[109]

7.6.4 Recombinant food-grade organisms for healthcare

The use of genetically modified food-grade organisms as recombinant vaccine expression hosts and delivery vehicles can open new avenues for vaccinology. Considering that oral immunization is a beneficial approach in terms of costs, patient comfort, and protection of mucosal tissues, the use of food-grade organisms can lead to highly advantageous vaccines in terms of costs, easy administration, and safety. The organisms currently used for this purpose are bacteria (Lactobacillus and Bacillus), yeasts, algae, plants, and insect species. Several such organisms are under clinical evaluation, and the current adoption of this technology by the industry indicates a potential to benefit global healthcare systems.[110]

7.7 Health and safety

See also: Genetically modified food controversies § Health

There is general scientific agreement that food on the market from genetically modified crops is not inherently riskier to human health than conventional food.[5][111][112] A 2004 report by the Institute of Medicine and National Research Council found that "genetic engineering is not an inherently hazardous process".[113] The report also stated "Adverse health effects from genetic engineering have not been documented in the human population, but the technique is new

and concerns about its safety remain". The report stated that any method of producing new foods could lead to unwanted changes so that singling out genetic engineering is "scientifically unjustified," and called for case-by-case assessment for all novel foods.[113]

Opponents claim that long-term health risks have not been adequately assessed and propose various combinations of additional testing, labeling[114] or removal from the market.[115][116][117][118] The advocacy group European Network of Scientists for Social and Environmental Responsibility (ENSSER), disputes the claim that "science" supports the safety of current GM foods, proposing that each GM food must be judged on case-by-case basis.[119] The Canadian Association of Physicians for the Environment called for removing GM foods from the market pending long term health studies.[115] Multiple disputed studies have claimed health effects relating to GM foods or to the pesticides used with them.[120]

7.7.1 Testing

The requirements for safety testing of GMO food varies substantially between countries.[121] Countries such as the United States, Canada, Lebanon and Egypt use *substantial equivalence* to determine if further testing is required, while many countries such as those in the European Union, Brazil and China only authorize GMO cultivation on a case-by-case basis. In the U.S. the FDA determined that GMO's are "Generally Recognized as Safe" (GRAS) and therefore do not require additional testing if the GMO product is substantially equivalent to the non-modified product.[122] If new substances are found, further testing may be required to satisfy concerns over potential toxicity, allergenicity, possible gene transfer to humans or genetic outcrossing to other organisms.[7]

7.8 Regulation

See also: Regulation of the release of genetic modified organisms and Regulation of genetic engineering
Government regulation of GMO development and release varies widely between countries. Marked differences separate GMO regulation in the U.S. and GMO regulation in the European Union.[123] Regulation also varies depending on the intended product's use. For example, a crop not intended for food use is generally not reviewed by authorities responsible for food safety.[124]

Green: Mandatory labeling required; Red:Ban on import and cultivation of genetically engineered food.

7.8.1 United States Regulations

Main article: Genetic engineering in the United States §
Regulation

In the U.S., three government organizations regulate
GMOs. The FDA checks the chemical composition of
organisms for potential allergens. The United States Department of Agriculture (USDA) supervises field testing
and monitors the distribution of GM seeds. The United
States Environmental Protection Agency (EPA) is responsible for monitoring pesticide usage, including plants modified to contain proteins toxic to insects. Like USDA, EPA
also oversees field testing and the distribution of crops that
have had contact with pesticides to ensure environmental
safety.[125] In 2015 the Obama administration announced
that it would update the way the government regulated GM
crops.[126]

In 1992 FDA published "Statement of Policy: Foods derived from New Plant Varieties." This statement is a clarification of FDA's interpretation of the Food, Drug, and
Cosmetic Act with respect to foods produced from new
plant varieties developed using recombinant deoxyribonucleic acid (rDNA) technology. FDA encouraged developers to consult with the FDA regarding any bioengineered
foods in development. The FDA says developers routinely
do reach out for consultations. In 1996 FDA updated consultation procedures.[127][128]

7.8.2 Labeling

As of 2015, 64 countries require labeling of GMO products
in the marketplace.[129]

US and Canadian national policy is to require a label
only given significant composition differences or documented health impacts, although some individual US states
(Vermont, Connecticut and Maine) enacted laws requiring
them.[130][131][132][133]

In some jurisdictions, the labeling requirement depends on

the relative quantity of GMO in the product. A study that
investigated voluntary labeling in South Africa found that
31% of products labeled as GMO-free had a GM content
above 1.0%.[134]

In Europe all food (including processed food) or feed that
contains greater than 0.9% GMOs must be labelled.[135]

7.8.3 Detection

Main article: Detection of genetically modified organisms

Testing on GMOs in food and feed is routinely done using
molecular techniques such as PCR and bioinformatics.[136]

In a January 2010 paper, the extraction and detection of
DNA along a complete industrial soybean oil processing
chain was described to monitor the presence of Roundup
Ready (RR) soybean: "The amplification of soybean lectin
gene by end-point polymerase chain reaction (PCR) was
successfully achieved in all the steps of extraction and refining processes, until the fully refined soybean oil. The
amplification of RR soybean by PCR assays using event-specific primers was also achieved for all the extraction and
refining steps, except for the intermediate steps of refining (neutralisation, washing and bleaching) possibly due to
sample instability. The real-time PCR assays using specific
probes confirmed all the results and proved that it is possible
to detect and quantify genetically modified organisms in the
fully refined soybean oil. To our knowledge, this has never
been reported before and represents an important accomplishment regarding the traceability of genetically modified
organisms in refined oils."[137]

According to Thomas Redick, detection and prevention of
cross-pollination is possible through the suggestions offered
by the Farm Service Agency (FSA) and Natural Resources
Conservation Service (NRCS). Suggestions include educating farmers on the importance of coexistence, providing
farmers with tools and incentives to promote coexistence,
conduct research to understand and monitor gene flow, provide assurance of quality and diversity in crops, provide
compensation for actual economic losses for farmers.[138]

7.9 Controversies

Main article: Genetically modified food controversies

The genetically modified foods controversy consists of a
set of disputes over the use of food made from genetically
modified crops. The disputes involve consumers, farmers,
biotechnology companies, governmental regulators, non-governmental organizations, environmental and political

activists and scientists. The major disagreements include whether GM foods can be safely consumed, harm the environment and/or are adequately tested and regulated.[116][139] The objectivity of scientific research and publications has been challenged.[115] Farming-related disputes include the use and impact of pesticides, seed production and use, side effects on non-GMO crops/farms,[140] and potential control of the GM food supply by seed companies.[115]

The conflicts have continued since GM foods were invented. They have occupied the media, the courts, local, regional and national governments and international organizations.

7.10   See also

- California Proposition 37 (2012)

- Chemophobia

- Genetic engineering

- Genetically modified crops

- Genetically modified food controversies

- Genetically modified organisms

- List of genetically modified crops

- Pharming (genetics) – use of genetically modified mammals to produce drugs

- Regulation of the release of genetic modified organisms

- Starlink corn recall

7.11   References

[1] GM Science Review First Report, Prepared by the UK GM Science Review panel (July 2003). Chairman Professor Sir David King, Chief Scientific Advisor to the UK Government, P 9

[2] James, Clive (1996). "Global Review of the Field Testing and Commercialization of Transgenic Plants: 1986 to 1995" (PDF). The International Service for the Acquisition of Agri-biotech Applications. Retrieved 17 July 2010.

[3] Weasel, Lisa H. 2009. *Food Fray*. Amacom Publishing

[4] "Consumer Q&A". Fda.gov. 2009-03-06. Retrieved 2012-12-29.

[5] American Association for the Advancement of Science (AAAS), Board of Directors (2012). Statement by the AAAS Board of Directors On Labeling of Genetically Modified Foods, and associated Press release: Legally Mandating GM Food Labels Could Mislead and Falsely Alarm Consumers

[6] American Medical Association (2012). Report 2 of the Council on Science and Public Health: Labeling of Bioengineered Foods

[7] World Health Organization. "Frequently asked questions on genetically modified foods". Retrieved 29 March 2016.

[8] United States Institute of Medicine and National Research Council (2004). Safety of Genetically Engineered Foods: Approaches to Assessing Unintended Health Effects. National Academies Press. Free full-text. See pp11ff on need for better standards and tools to evaluate GM food.

[9] *A decade of EU-funded GMO research (2001-2010)* (PDF). Directorate-General for Research and Innovation. Biotechnologies, Agriculture, Food. European Union. 2010. p. 16. doi:10.2777/97784. ISBN 978-92-79-16344-9.

[10] Other sources:

- Tamar Haspel for the Washington Post. October 15, 2013. Genetically modified foods: What is and isn't true

- Winter CK and Gallegos LK (2006). Safety of Genetically Engineered Food. University of California Agriculture and Natural Resources Communications, Publication 8180.

- Ronald, Pamela (2011). "Plant Genetics, Sustainable Agriculture and Global Food Security". *Genetics* **188** (1): 11–20. doi:10.1534/genetics.111.128553. PMC 3120150. PMID 21546547.

- Miller, Henry (2009). "A golden opportunity, squandered" (PDF). *Trends in Biotechnology* **27** (3): 129–130. doi:10.1016/j.tibtech.2008.11.004. PMID 19185375.

- Dr. Christopher Preston, AgBioWorld 2011. Peer Reviewed Publications on the Safety of GM Foods.

[11] Cowan, Tadlock (18 Jun 2011). "Agricultural Biotechnology: Background and Recent Issues" (PDF). Congressional Research Service (Library of Congress). pp. 33–38. Retrieved 27 September 2015.

[12] "Genetically engineered foods". University of Maryland Medical Center. Retrieved 29 September 2015.

[13] "Glossary of Agricultural Biotechnology Terms". United States Department of Agriculture. 27 Feb 2013. Retrieved 29 September 2015.

[14] "Questions & Answers on Food from Genetically Engineered Plants". US Food and Drug Administration. 22 Jun 2015. Retrieved 29 September 2015.

[15] Daniel Zohary, Maria Hopf, Ehud Weiss (2012). *Domestication of Plants in the Old World: The Origin and Spread of Plants in the Old World*. Oxford University Press.

[16] Clive Root (2007). *Domestication*. Greenwood Publishing Groups.

[17] Jackson, DA; Symons, RH; Berg, P (1 October 1972). "Biochemical Method for Inserting New Genetic Information into DNA of Simian Virus 40: Circular SV40 DNA Molecules Containing Lambda Phage Genes and the Galactose Operon of Escherichia coli". *PNAS* **69** (10): 2904–2909. Bibcode:1972PNAS...69.2904J. doi:10.1073/pnas.69.10.2904. PMC 389671. PMID 4342968.

[18] "FDA Approves 1st Genetically Engineered Product for Food". *Los Angeles Times*. 24 March 1990. Retrieved 1 May 2014.

[19] Staff, National Centre for Biotechnology Education. 2006. Case Study: Chymosin

[20] Campbell-Platt, Geoffrey (26 August 2011). *Food Science and Technology*. Ames, IA: John Wiley & Sons. ISBN 978-1-4443-5782-0.

[21] Bruening, G.; Lyons, J. M. (2000). "The case of the FLAVR SAVR tomato". *California Agriculture* **54** (4): 6–7. doi:10.3733/ca.v054n04p6.

[22] Genetically Altered Potato Ok'd For Crops Lawrence Journal-World - 6 May 1995

[23] Global Status of Commercialized Biotech/GM Crops: 2011 ISAAA Brief ISAAA Brief 43-2011. Retrieved 14 October 2012

[24] James, C (2011). "ISAAA Brief 43, Global Status of Commercialized Biotech/GM Crops: 2011". *ISAAA Briefs*. Ithaca, New York: International Service for the Acquisition of Agri-biotech Applications (ISAAA). Retrieved 2012-06-02.

[25] "Adoption of Genetically Engineered Crops in the U.S.". *Economic Research Service*. USDA. Retrieved 26 August 2015.

[26] Ye, Xudong; Al-Babili, Salim; Klöti, Andreas; Zhang, Jing; Lucca, Paola; Beyer, Peter; Potrykus, Ingo (2000-01-14). "Engineering the Provitamin A (β-Carotene) Biosynthetic Pathway into (Carotenoid-Free) Rice Endosperm". *Science* **287** (5451): 303–305. doi:10.1126/science.287.5451.303. ISSN 0036-8075. PMID 10634784.

[27] "AQUABOUNTY CLEARED TO SELL SALMON IN USA FOR COMMERCIAL PURPOSES".

[28] Bodnar, Anastasia (October 2010). "Risk Assessment and Mitigation of AquAdvantage Salmon" (PDF). ISB News Report.

[29] Culpepper, Stanley A; et al. (2006). "Glyphosate-resistant Palmer amaranth (Amaranthus palmeri) confirmed in Georgia.". *Weed Science* **54** (4): 620–626. doi:10.1614/ws-06-001r.1.

[30] Gallant, Andre. "Pigweed in the Cotton: A superweed invades Georgia". *Modern Farmer*.

[31] Webster, TM; Grey, TL. (2015). "Glyphosate-Resistant Palmer Amaranth (Amaranthus palmeri) Morphology, Growth, and Seed Production in Georgia.". *Weed Science* **63** (1): 264–272. doi:10.1614/ws-d-14-00051.1.

[32] Fernandez-Cornejo J, Wechsler S, Livingston M, Mitchell L. (Feb 2014). "Genetically engineered crops in the United States". *Economic Research Service*.

[33] Gonsalves, D. (2004). "Transgenic papaya in Hawaii and beyond". *AgBioForum* **7** (1&2): 36–40.

[34] Ronald, Pamela; McWilliams, James (May 14, 2010). "Genetically Engineered Distortions". The New York Times. Retrieved July 26, 2010.

[35] "The Rainbow Papaya Story". Hawaii Papaya Industry Association. Retrieved April 2015.

[36] Li, Y; et al. (April 2014). "Biosafety management and commercial use of genetically modified crops in China". *Plant Cell Reports* **33** (4): 565–73. doi:10.1007/s00299-014-1567-x. PMID 24493253.

[37] "The History and Future of GM Potatoes". Potatopro.com. 2010-03-10. Retrieved 2012-12-29.

[38] Johnson, Stanley R. (February 2008). "Quantification of the Impacts on US Agriculture of Biotechnology-Derived Crops Planted in 2006" (PDF). Washington DC: National Center for Food and Agricultural Policy. Retrieved August 12, 2010.

[39] "GMO Database: Zucchini (courgette)". GMO Compass. November 7, 2007. Retrieved February 28, 2015.

[40] "Business BASF applies for approval for another biotech potato". Research in Germany. November 17, 2011.

[41] Burger, Ludwig (October 31, 2011). "BASF applies for EU approval for Fortuna GM potato". Frankfurt: Reuters. Retrieved December 29, 2011.

[42] Turley, Andrew (February 7, 2013). "BASF drops GM potato projects". Royal Society of Chemistry News.

[43] PERKOWSKI, MATEUSZ (April 16, 2013). "Del Monte Gets Approval to Import GMO Pineapple". *Food Democracy Now*.

[44] Pollack, Andrew (November 7, 2014). "U.S.D.A. Approves Modified Potato. Next Up: French Fry Fans". The New York Times.

[45] "Availability of Petition for Determination of Nonregulated Status of Potato Genetically Engineered for Low Acrylamide Potential and Reduced Black Spot Bruise". Federal Register. May 3, 2013.

[46] Pollack, A. (February 13, 2015). "Gene-Altered Apples Get U.S. Approval". The New York Times.

[47] Tennille, Tracy (Feb 13, 2015). "First Genetically Modified Apple Approved for Sale in U.S.". *Wall Street Journal*. Retrieved Feb 2015.

[48] "Apple-to-apple transformation". Okanagan Specialty Fruits. Retrieved August 3, 2012.

[49] For a list of all traits, see table As of September 2012 that site listed 13 traits in nearly 30 different products.

[50] "Acreage NASS" (PDF). *National Agricultural Statistics Board annual report*. June 2010. Retrieved July 23, 2010.

[51] "Corn-Based Food Production in South Dakota: A Preliminary Feasibility Study" (PDF). South Dakota State University, College of Agriculture and Biological Sciences, Agricultural Experiment Station. June 2004.

[52] "GMO Compass - GM Soy".

[53] Lusas, Edmund W.; Riaz, Mian N (1995). "Soy Protein Products: Processing and Use" (PDF). 125 (3_Suppl). Journal of Nutrition: 573S–580S.

[54] Sipos, E.S. "Edible Uses of Soybean Protein" (PDF).

[55] Singh, Preeti; Kumar, R.; Sahapathy, S. N.; Bawa, A. S. (2008). "Functional and Edible Uses of". *Comprehensive Reviews in Food Science and Food Safety* **7**: 14–28. doi:10.1111/j.1541-4337.2007.00025.x.

[56] Shurtleff, William; Aoyagi, Akiko (2008). "History of Cooperative Soybean Processing in the United States: Extensively Annotated Bibliography and Sourcebook" (PDF). Soyinfo Center.

[57] Weingartner, Karl; Owen, Bridget (March 2009). "Soy Protein Applications in Nutrition & Food Technology" (PDF). National Soybean Research Laboratory, University of Illinois at Urbana-Champaign.

[58] Isolated Soy Proteins

[59] Staff, World Initiative for Soy in Human Health (WISHH) Soy Protein Concentrate Reference Guide

[60] Soy Flours

[61] Textured Soy Proteins

[62] Jaffe,Greg (Director of Biotechnology at the Center for Science in the Public Interest) (February 7, 2013). "What You Need to Know About Genetically Engineered Food". *Atlantic*.

[63] "International Starch: Production of corn starch". Starch.dk. Retrieved 2011-06-12.

[64] "Lecithin". Oct 2015. Retrieved 18 October 2015.

[65] "Select Committee on GRAS Substances (SCOGS) Opinion: Lecithin". Aug 10, 2015. Retrieved 18 October 2015.

[66] "Poster of corn products" (PDF). Retrieved 2012-12-29.

[67] "Corn Oil, 5th Edition" (PDF). *Corn Refiners Association*. 2006.

[68] Staff (July 1, 2005). "Danisco emulsifier to substitute non-GM soy lecithin as demand outstrips supply". *FoodNavigator.com*.

[69] "Regulation (EC) 50/2000". *Eur-lex.europa.eu*.

[70] Marx,Gertruida M. (December 2010). "Dissertation submitted in fulfilment of requirements for the degree Doctor of Philosophy in the Faculty of Health Sciences" (PDF). *MONITORING OF GENETICALLY MODIFIED FOOD PRODUCTS IN SOUTH AFRICA]* (South Africa: University of the Free State).

[71] Davison, John; Bertheau, Yves Bertheau (2007). "EU regulations on the traceability and detection of GMOs: difficulties in interpretation, implementation and compliance". *CAB Reviews: Perspectives in Agriculture, Veterinary Science, Nutrition and Natural Resources* **2** (77). doi:10.1079/pavsnnr20072077.

[72] "ISAAA Brief 43-2011. Executive Summary: Global Status of Commercialized Biotech/GM Crops: 2011". Isaaa.org. Retrieved 2012-12-29.

[73] http://www.gmo-compass.org/eng/database/plants/13. sugar_beet.html

[74] Food and Agriculture Organization of the United Nations (2009). *Sugar Beet: White Sugar* (PDF). p. 9.

[75] Klein, Joachim; Altenbuchner, Josef; Mattes, Ralf (1998-02-26). "Nucleic acid and protein elimination during the sugar manufacturing process of conventional and transgenic sugar beets". *Journal of Biotechnology* **60** (3): 145–153. doi:10.1016/S0168-1656(98)00006-6. PMID 9608751.

[76] "Soyatech.com". Soyatech.com. Retrieved 2012-12-29.

[77] "Food Fats and Oils" (PDF). Institute of Shortening and Edible Oils. 2006. Retrieved 2011-11-19.

[78] "Twenty Facts about Cottonseed Oil". National Cottonseed Producers Association. Archived from the original on January 7, 2016.

[79] Simon, Michelle (August 24, 2011). "ConAgra Sued Over GMO '100% Natural' Cooking Oils". Food Safety News.

[80] "ingredients of margarine". Imace.org. Retrieved 2012-12-29.

[81] "USDA Protein(g) in Fats and Oils". Retrieved 2015-05-31.

[82] "How Cooking Oil is Made". Madehow.com. 1991-04-27. Retrieved 2012-12-29.

[83] Crevel, R.W.R; Kerkhoff, M.A.T; Koning, M.M.G (2000). "Allergenicity of refined vegetable oils". *Food and Chemical Toxicology* **38** (4): 385–93. doi:10.1016/S0278-6915(99)00158-1. PMID 10722892.

[84] "What is Canola Oil?". CanolaInfo. Retrieved 2012-12-29.

[85] David Bennett for Southeast Farm Press, February 5, 2003 World soybean consumption quickens

[86] "Soybean". Encyclopedia Britannica Online. Retrieved February 18, 2012.

[87] "2012 World of Corn, National Corn Growers Association" (PDF). Retrieved 2012-12-29.

[88] Staff, GMO Compass. December 7, 2006. Genetic Engineering: Feeding the EU's Livestock

[89] Snell C; Bernheim A; Berge JB; Kuntz M; Pascal G; paris A; Ricroch AE (2012). "Assessment of the health impact of GM plant diets in long-term and multigenerational animal feeding trials: A literature review". *Food and Chemical Toxicology* **50** (3–4): 1134–1148. doi:10.1016/j.fct.2011.11.048. PMID 22155268.

[90] Emtage, JS; Angal, S; Doel, MT; Harris, TJ; Jenkins, B; Lilley, G; Lowe, PA (1983). "Synthesis of calf prochymosin (prorennin) in *Escherichia coli*". *Proceedings of the National Academy of Sciences of the United States of America* **80** (12): 3671–5. Bibcode:1983PNAS...80.3671E. doi:10.1073/pnas.80.12.3671. PMC 394112. PMID 6304731.

[91] Harris TJ, Lowe PA, Lyons A, Thomas PG, Eaton MA, Millican TA, Patel TP, Bose CC, Carey NH, Doel MT (April 1982). "Molecular cloning and nucleotide sequence of cDNA coding for calf preprochymosin". *Nucleic Acids Res.* **10** (7): 2177–87. doi:10.1093/nar/10.7.2177. PMC 320601. PMID 6283469.

[92] "Chymosin". GMO Compass. Retrieved 2011-03-03.

[93] Law, Barry A. (2010). *Technology of Cheesemaking*. UK: WILEY-BLACKWELL. pp. 100–101. ISBN 978-1-4051-8298-0.

[94] "Food Biotechnology in the United States: Science, Regulation, and Issues". U.S. Department of State. Retrieved 2006-08-14.

[95] Johnson, M.E.; Lucey, J.A. (2006). "Major Technological Advances and Trends in Cheese". *Journal of Dairy Science* **89** (4): 1174–8. doi:10.3168/jds.S0022-0302(06)72186-5. PMID 16537950.

[96] Hansen, C. "Enzymes". Improving Food & Health. Retrieved 2014-01-14.

[97] "DMS cheese enzymes page".

[98] Baumana, Dale E.; Collier, Robert J (September 15, 2010). "Use of Bovine Somatotropin in Dairy Production" (PDF).

[99] Staff (2011-02-18). *Last Medical Review*. American Cancer Society. Missing or empty |title= (help);

[100] "Recombinant Bovine Growth Hormone".

[101] Brennand, Charlotte P. "Bovine Somatotropin in Milk" (PDF). Retrieved 2011-03-06.

[102] Cima, Greg (November 18, 2010). "Appellate court gives mixed ruling on Ohio rBST labeling rules". JAVMA News.

[103] leafcom. "INTERNATIONAL DAIRY FOODS ASS'N v. BOGGS – Argued: June 10, 2010". Leagle.com.

[104] Harper, G.S., Brownlee, A., Hall, T.E., Seymour, R., Lyons, R. and Ledwith, P. (2003). "Global progress toward transgenic food animals: A survey of publicly available information." (PDF). Food Standards Australia and New Zealand. Retrieved August 27, 2015.

[105] Rick MacInnes-Rae, Rick (November 27, 2013). "GMO salmon firm clears one hurdle but still waits for key OKs AquaBounty began seeking American approval in 1995". CBC News.

[106] Pollack, Andrew (May 21, 2012). "An Entrepreneur Bankrolls a Genetically Engineered Salmon". The New York Times. Retrieved September 3, 2012.

[107] Staff (December 26, 2012). "Draft Environmental Assessment and Preliminary Finding of No Significant Impact Concerning a Genetically Engineered Atlantic Salmon" (PDF). Federal Register. Retrieved January 2, 2013.

[108] Naik, Gautam (September 21, 2010). "Gene-Altered Fish Closer to Approval". *Wall Street Journal*.

[109] Commissioner, Office of the. "Press Announcements - FDA takes several actions involving genetically engineered plants and animals for food". www.fda.gov. Retrieved 2015-12-03.

[110] Rosales-Mendoza, S.; Angulo, C.; Meza, B. (2015). "Food-Grade Organisms as Vaccine Biofactories and Oral Delivery Vehicles". *Trends in Biotechnology*. doi:10.1016/j.tibtech.2015.11.007.

[111] Ronald, Pamela (2011). "Plant Genetics, Sustainable Agriculture and Global Food Security". *Genetics* **188** (1): 11–20. doi:10.1534/genetics.111.128553. PMC 3120150. PMID 21546547.

[112] Bett, Charles; Ouma, James Okuro; Groote, Hugo De (August 2010). "Perspectives of gatekeepers in the Kenyan food industry towards genetically modified food". *Food Policy* **35** (4): 332–340. doi:10.1016/j.foodpol.2010.01.003.

[113] "Composition of Altered Food Products, Not Method Used to Create Them, Should Be Basis for Federal Safety Assessment". National Academies of Sciences. Retrieved 2 January 2016.

[114] "Genetically modified foods" (PDF). Public Health Association of Australia. 2007.

[115] "CAPE's Position Statement on GMOs". Canadian Association of Physicians for the Environment. November 11, 2013.

[116] "IDEA Position on Genetically Modified Foods". Irish Doctors' Environmental Association. Retrieved 2014-03-25.

[117] "American Academy of Environmental Medicine Calls for Immediate Moratorium on Genetically Modified Foods, position paper". American Academy of Environmental Medicine. Retrieved 18 October 2015.

[118] "Press Advisory". American Academy of Environmental Medicine. Retrieved 18 October 2015.

[119] Hilbeck; et al. (2015). "No scientific consensus on GMO safety" (PDF). *Environmental Sciences Europe* **27**. doi:10.1186/s12302-014-0034-1.

[120] Martinelli, L; et al. (2013). "Science, safety, and trust: the case of transgenic food". *Croat Med J* **54** (1): 91–6. doi:10.3325/cmj.2013.54.91. PMC 3584506. PMID 23444254.

[121] http://www.loc.gov/law/help/restrictions-on-gmos/

[122] Emily Marden, Risk and Regulation: U.S. Regulatory Policy on Genetically Modified Food and Agriculture 44 B.C.L. Rev. 733 (2003).

[123] http://www.cfr.org/agricultural-policy/regulation-gmos-europe-united-states-case-study-contemporary-european-regulatory-politics/p8688

[124] "The History and Future of GM Potatoes". *PotatoPro.com*.

[125] APPDMZ\ccvivr. "Commonly Asked Questions about the Food Safety of GMOs". *monsanto.com*.

[126] Pollack, Andrew (2015-07-02). "White House Orders Review of Rules for Genetically Modified Crops". *The New York Times*. ISSN 0362-4331. Retrieved 2015-07-03.

[127] "Food from Genetically Engineered Plants". FDA. Retrieved 18 October 2015.

[128] "Statement of Policy - Foods Derived from New Plant Varieties". Retrieved 18 October 2015.

[129] "International Labeling Laws". Center for Food Safety.|

[130] Chokshi, Niraj (9 May 2014). "Vermont just passed the nation's first GMO food labeling law. Now it prepares to get sued". *The Washington Post*. Retrieved 19 January 2016.

[131] "The Regulation of Genetically Modified Food".

[132] Van Eenennaam, Alison; Chassy, Bruce; Kalaitzandonakes, Nicholas; Redick, Thomas (2014). "The Potential Impacts of Mandatory Labeling for Genetically Engineered Food in the United States" (PDF). *Council for Agricultural Science and Technology (CAST)* **54** (April 2014). ISSN 1070-0021. Retrieved 2014-05-28. To date, no material differences in composition or safety of commercialized GE crops have been identified that would justify a label based on the GE nature of the product.

[133] Hallenbeck, Terri (2014-04-27). "How GMO labeling came to pass in Vermont". *Burlington Free Press*. Retrieved 2014-05-28.

[134] Botha, Gerda M.; Viljoen, Christopher D. (2009). "South Africa: A case study for voluntary GM labelling". *Food Chemistry* **112** (4): 1060–4. doi:10.1016/j.foodchem.2008.06.050.

[135] Davison, John (2010). "GM plants: Science, politics and EC regulations". *Plant Science* **178** (2): 94–8. doi:10.1016/j.plantsci.2009.12.005.

[136] "EU GMO testing homepage". European Commission Join Research Centre. Retrieved May 31, 2015.

[137] Costa, Joana; Mafra, Isabel; Amaral, Joana S.; Oliveira, M.B.P.P. (2010). "Monitoring genetically modified soybean along the industrial soybean oil extraction and refining processes by polymerase chain reaction techniques". *Food Research International* **43**: 301–306. doi:10.1016/j.foodres.2009.10.003.

[138] "Mason Central Authentication Service". *www.heinonline.org.mutex.gmu.edu*. Archived from the original on December 22, 2015. Retrieved 2015-12-15.

[139] American Medical Association (2012). Report 2 of the Council on Science and Public Health: Labeling of Bioengineered Foods. "To better detect potential harms of bioengineered foods, the Council believes that pre-market safety assessment should shift from a voluntary notification process to a mandatory requirement." page 7

[140] Chartered Institute of Environmental Health (2006) Proposals for managing the coexistence of GM, conventional and organic crops Response to the Department for Environment, Food and Rural Affairs consultation paper. October 2006

7.12 External links

- Documentary on YouTube

- Library resources in your library and in other libraries about Genetically modified food

- Media related to Genetically modified organisms at Wikimedia Commons

Chapter 8

Gene expression

For vocabulary, see Glossary of gene expression terms. For a non-technical introduction to the topic, see Introduction to genetics.

Gene expression is the process by which information from

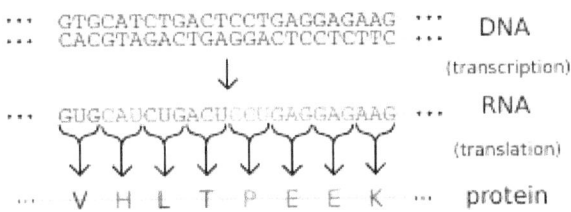

Genes are expressed by being transcribed into RNA, and this transcript may then be translated into protein.

a gene is used in the synthesis of a functional gene product. These products are often proteins, but in non-protein coding genes such as transfer RNA (tRNA) or small nuclear RNA (snRNA) genes, the product is a functional RNA. The process of gene expression is used by all known life—eukaryotes (including multicellular organisms), prokaryotes (bacteria and archaea), and utilized by viruses—to generate the macromolecular machinery for life.

Several steps in the gene expression process may be modulated, including the transcription, RNA splicing, translation, and post-translational modification of a protein. Gene regulation gives the cell control over structure and function, and is the basis for cellular differentiation, morphogenesis and the versatility and adaptability of any organism. Gene regulation may also serve as a substrate for evolutionary change, since control of the timing, location, and amount of gene expression can have a profound effect on the functions (actions) of the gene in a cell or in a multicellular organism.

In genetics, gene expression is the most fundamental level at which the genotype gives rise to the phenotype, i.e. observable trait. The genetic code stored in DNA is "interpreted" by gene expression, and the properties of the expression give rise to the organism's phenotype. Such phenotypes are of-

ten expressed by the synthesis of proteins that control the organism's shape, or that act as enzymes catalysing specific metabolic pathways characterising the organism.

8.1 Mechanism

8.1.1 Transcription

The process of transcription is carried out by RNA polymerase (RNAP), which uses DNA (black) as a template and produces RNA (blue).

Main article: Transcription (genetics)

A gene is a stretch of DNA that encodes information. Genomic DNA consists of two antiparallel and reverse complementary strands, each having 5' and 3' ends. With respect to a gene, the two strands may be labeled the "template strand," which serves as a blueprint for the production of an RNA transcript, and the "coding strand," which includes the DNA version of the transcript sequence. The production of RNA copies of the DNA is called transcription, and is performed in the nucleus by RNA polymerase, which adds one RNA nucleotide at a time to a growing RNA strand. This RNA is complementary to the template $3' \rightarrow 5'$ DNA strand,[1] which is itself complementary to the coding $5' \rightarrow 3'$ DNA strand. Therefore, the resulting $5' \rightarrow 3'$ RNA strand is identical to the coding DNA strand with the exception that thymines (T) are replaced with uracils (U) in the RNA. A coding DNA strand reading "ATG" is indirectly transcribed through the non-coding strand as "AUG" in RNA.

Transcription in prokaryotes is carried out by a single type of RNA polymerase, which needs a DNA sequence called a

Pribnow box as well as a sigma factor (σ factor) to start transcription. In eukaryotes, transcription is performed by three types of RNA polymerases, each of which needs a special DNA sequence called the promoter and a set of DNA-binding proteins—transcription factors—to initiate the process. RNA polymerase I is responsible for transcription of ribosomal RNA (rRNA) genes. RNA polymerase II (Pol II) transcribes all protein-coding genes but also some non-coding RNAs (e.g., snRNAs, snoRNAs or long non-coding RNAs). Pol II includes a C-terminal domain (CTD) that is rich in serine residues. When these residues are phosphorylated, the CTD binds to various protein factors that promote transcript maturation and modification. RNA polymerase III transcribes 5S rRNA, transfer RNA (tRNA) genes, and some small non-coding RNAs (e.g., 7SK). Transcription ends when the polymerase encounters a sequence called the terminator.

8.1.2 RNA processing

Main article: Post-transcriptional modification

While transcription of prokaryotic protein-coding genes creates messenger RNA (mRNA) that is ready for translation into protein, transcription of eukaryotic genes leaves a primary transcript of RNA (pre-mRNA), which first has to undergo a series of modifications to become a mature mRNA.

These include 5' *capping*, which is set of enzymatic reactions that add 7-methylguanosine (m⁷G) to the 5' end of pre-mRNA and thus protect the RNA from degradation by exonucleases. The m⁷G cap is then bound by cap binding complex heterodimer (CBC20/CBC80), which aids in mRNA export to cytoplasm and also protect the RNA from decapping.

Another modification is 3' *cleavage and polyadenylation*. They occur if polyadenylation signal sequence (5'-AAUAAA-3') is present in pre-mRNA, which is usually between protein-coding sequence and terminator. The pre-mRNA is first cleaved and then a series of ~200 adenines (A) are added to form poly(A) tail, which protects the RNA from degradation. Poly(A) tail is bound by multiple poly(A)-binding proteins (PABP) necessary for mRNA export and translation re-initiation.

A very important modification of eukaryotic pre-mRNA is *RNA splicing*. The majority of eukaryotic pre-mRNAs consist of alternating segments called exons and introns. During the process of splicing, an RNA-protein catalytical complex known as spliceosome catalyzes two transesterification reactions, which remove an intron and release it in form of lariat structure, and then splice neighbouring ex-

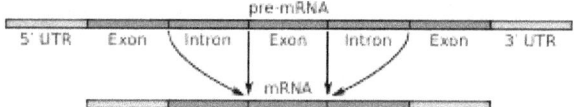

Simple illustration of exons and introns in pre-mRNA and the formation of mature mRNA by splicing. The UTRs are non-coding parts of exons at the ends of the mRNA.

ons together. In certain cases, some introns or exons can be either removed or retained in mature mRNA. This so-called alternative splicing creates series of different transcripts originating from a single gene. Because these transcripts can be potentially translated into different proteins, splicing extends the complexity of eukaryotic gene expression.

Extensive RNA processing may be an evolutionary advantage made possible by the nucleus of eukaryotes. In prokaryotes, transcription and translation happen together, whilst in eukaryotes, the nuclear membrane separates the two processes, giving time for RNA processing to occur.

8.1.3 Non-coding RNA maturation

Main articles: tRNA maturation, rRNA maturation and miRNA maturation

In most organisms non-coding genes (ncRNA) are transcribed as precursors that undergo further processing. In the case of ribosomal RNAs (rRNA), they are often transcribed as a pre-rRNA that contains one or more rRNAs. The pre-rRNA is cleaved and modified (2'-O-methylation and pseudouridine formation) at specific sites by approximately 150 different small nucleolus-restricted RNA species, called snoRNAs. SnoRNAs associate with proteins, forming snoRNPs. While snoRNA part basepair with the target RNA and thus position the modification at a precise site, the protein part performs the catalytical reaction. In eukaryotes, in particular a snoRNP called RNase MRP cleaves the 45S pre-rRNA into the 28S, 5.8S, and 18S rRNAs. The rRNA and RNA processing factors form large aggregates called the nucleolus.[2]

In the case of transfer RNA (tRNA), for example, the 5' sequence is removed by RNase P,[3] whereas the 3' end is removed by the tRNase Z enzyme[4] and the non-templated 3' CCA tail is added by a nucleotidyl transferase.[5] In the case of micro RNA (miRNA), miRNAs are first transcribed as primary transcripts or pri-miRNA with a cap and poly-A tail and processed to short, 70-nucleotide stem-loop structures known as pre-miRNA in the cell nucleus by the enzymes Drosha and Pasha. After being exported, it is then processed to mature miRNAs in the cytoplasm by interaction with the endonuclease Dicer, which also initiates the formation of the RNA-induced silencing complex (RISC), composed of the Argonaute protein.

Even snRNAs and snoRNAs themselves undergo series of modification before they become part of functional RNP complex. This is done either in the nucleoplasm or in the specialized compartments called Cajal bodies. Their bases are methylated or pseudouridinilated by a group of small Cajal body-specific RNAs (scaRNAs), which are structurally similar to snoRNAs.

8.1.4   RNA export

In eukaryotes most mature RNA must be exported to the cytoplasm from the nucleus. While some RNAs function in the nucleus, many RNAs are transported through the nuclear pores and into the cytosol. Notably this includes all RNA types involved in protein synthesis.[6] In some cases RNAs are additionally transported to a specific part of the cytoplasm, such as a synapse; they are then towed by motor proteins that bind through linker proteins to specific sequences (called "zipcodes") on the RNA.[7]

8.1.5   Translation

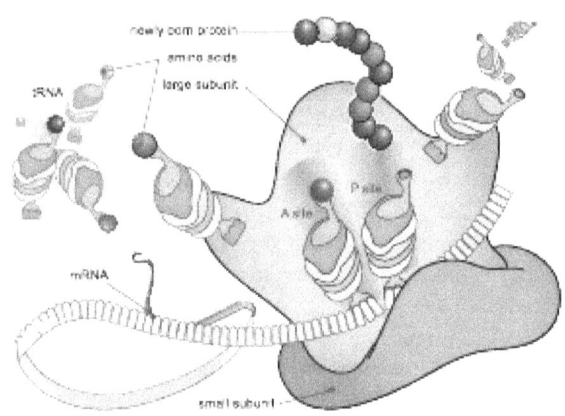

During the translation, tRNA charged with amino acid enters the ribosome and aligns with the correct mRNA triplet. Ribosome then adds amino acid to growing protein chain.

Main article: Translation (genetics)

For some RNA (non-coding RNA) the mature RNA is the final gene product.[8] In the case of messenger RNA (mRNA) the RNA is an information carrier coding for the synthesis of one or more proteins. mRNA carrying a single protein sequence (common in eukaryotes) is monocistronic whilst mRNA carrying multiple protein sequences (common in prokaryotes) is known as polycistronic.

Every mRNA consists of three parts: a 5' untranslated region (5'UTR), a protein-coding region or open reading frame (ORF), and a 3' untranslated region (3'UTR). The coding region carries information for protein synthesis encoded by the genetic code to form triplets. Each triplet of nucleotides of the coding region is called a codon and corresponds to a binding site complementary to an anti-codon triplet in transfer RNA. Transfer RNAs with the same anticodon sequence always carry an identical type of amino acid. Amino acids are then chained together by the ribosome according to the order of triplets in the coding region. The ribosome helps transfer RNA to bind to messenger RNA and takes the amino acid from each transfer RNA and makes a structure-less protein out of it.[9][10] Each mRNA molecule is translated into many protein molecules, on average ~2800 in mammals.[11][12]

In prokaryotes translation generally occurs at the point of transcription (co-transcriptionally), often using a messenger RNA that is still in the process of being created. In eukaryotes translation can occur in a variety of regions of the cell depending on where the protein being written is supposed to be. Major locations are the cytoplasm for soluble cytoplasmic proteins and the membrane of the endoplasmic reticulum for proteins that are for export from the cell or insertion into a cell membrane. Proteins that are supposed to be expressed at the endoplasmic reticulum are recognised part-way through the translation process. This is governed by the signal recognition particle—a protein that binds to the ribosome and directs it to the endoplasmic reticulum when it finds a signal peptide on the growing (nascent) amino acid chain.[13] Translation is the communication of the meaning of a source-language text by means of an equivalent target-language text

8.1.6   Folding

Main article: Protein folding
The polypeptide folds into its characteristic and functional

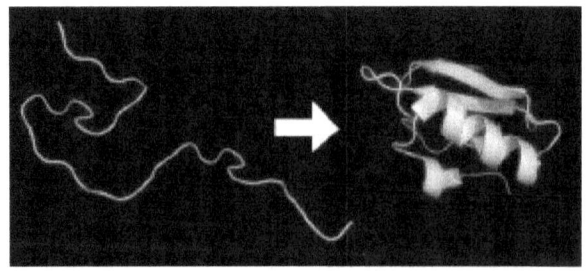

Protein before (left) and after (right) folding.

three-dimensional structure from a random coil.[14] Each protein exists as an unfolded polypeptide or random coil when translated from a sequence of mRNA into a linear chain of amino acids. This polypeptide lacks any developed three-dimensional structure (the left hand side of the neighboring figure). Amino acids interact with each other to produce a well-defined three-dimensional structure, the folded protein (the right hand side of the figure) known as the native state. The resulting three-dimensional structure is determined by the amino acid sequence (Anfinsen's dogma).[15]

The correct three-dimensional structure is essential to function, although some parts of functional proteins may remain unfolded.[16] Failure to fold into the intended shape usually produces inactive proteins with different properties including toxic prions. Several neurodegenerative and other diseases are believed to result from the accumulation of *misfolded* (incorrectly folded) proteins.[17] Many allergies are caused by the folding of the proteins, for the immune system does not produce antibodies for certain protein structures.[18]

Enzymes called chaperones assist the newly formed protein to attain (fold into) the 3-dimensional structure it needs to function.[19] Similarly, RNA chaperones help RNAs attain their functional shapes.[20] Assisting protein folding is one of the main roles of the endoplasmic reticulum in eukaryotes.

8.1.7 Translocation

Secretory proteins of eukaryotes or prokaryotes must be translocated to enter the secretory pathway. Newly synthesized proteins are directed to the eukaryotic Sec61 or prokaryotic SecYEG translocation channel by signal peptides. The efficiency of protein secretion in eukaryotes is very dependent on the signal peptide which has been used.[21]

8.1.8 Protein transport

Many proteins are destined for other parts of the cell than the cytosol and a wide range of signalling sequences or (signal peptides) are used to direct proteins to where they are supposed to be. In prokaryotes this is normally a simple process due to limited compartmentalisation of the cell. However, in eukaryotes there is a great variety of different targeting processes to ensure the protein arrives at the correct organelle.

Not all proteins remain within the cell and many are exported, for example, digestive enzymes, hormones and extracellular matrix proteins. In eukaryotes the export pathway is well developed and the main mechanism for the export of these proteins is translocation to the endoplasmic reticulum, followed by transport via the Golgi apparatus.[22][23]

8.2 Regulation of gene expression

The patchy colours of a tortoiseshell cat are the result of different levels of expression of pigmentation genes in different areas of the skin.

Main article: Regulation of gene expression

Regulation of gene expression refers to the control of the amount and timing of appearance of the functional product of a gene. Control of expression is vital to allow a cell to produce the gene products it needs when it needs them; in turn, this gives cells the flexibility to adapt to a variable environment, external signals, damage to the cell, etc. Some simple examples of where gene expression is important are:

- Control of insulin expression so it gives a signal for blood glucose regulation.

- X chromosome inactivation in female mammals to prevent an "overdose" of the genes it contains.

- Cyclin expression levels control progression through the eukaryotic cell cycle.

More generally, gene regulation gives the cell control over all structure and function, and is the basis for cellular differentiation, morphogenesis and the versatility and adaptability of any organism.

Any step of gene expression may be modulated, from the DNA-RNA transcription step to post-translational modification of a protein. The stability of the final gene product, whether it is RNA or protein, also contributes to the expression level of the gene—an unstable product results in a low expression level. In general gene expression is regulated through changes[24] in the number and type of interactions between molecules[25] that collectively influence transcription of DNA[26] and translation of RNA.[27]

Numerous terms are used to describe types of genes depending on how they are regulated; these include:

- A **constitutive gene** is a gene that is transcribed continually as opposed to a facultative gene, which is only transcribed when needed.

- A *housekeeping gene* is typically a constitutive gene that is transcribed at a relatively constant level. The housekeeping gene's products are typically needed for maintenance of the cell. It is generally assumed that their expression is unaffected by experimental conditions. Examples include actin, GAPDH and ubiquitin.

- A **facultative gene** is a gene only transcribed when needed as opposed to a constitutive gene.

- An **inducible gene** is a gene whose expression is either responsive to environmental change or dependent on the position in the cell cycle.

8.2.1   Transcriptional regulation

Main article: Transcriptional regulation

Regulation of transcription can be broken down into three main routes of influence; genetic (direct interaction of a control factor with the gene), modulation interaction of a control factor with the transcription machinery and epigenetic (non-sequence changes in DNA structure that influence transcription).

Direct interaction with DNA is the simplest and the most direct method by which a protein changes transcription levels. Genes often have several protein binding sites around the coding region with the specific function of regulating transcription. There are many classes of regulatory DNA binding sites known as enhancers, insulators and silencers. The mechanisms for regulating transcription are very varied, from blocking key binding sites on the DNA for RNA polymerase to acting as an activator and promoting transcription by assisting RNA polymerase binding.

The activity of transcription factors is further modulated by intracellular signals causing protein post-translational modification including phosphorylated, acetylated, or glycosylated. These changes influence a transcription factor's ability to bind, directly or indirectly, to promoter DNA, to recruit RNA polymerase, or to favor elongation of a newly synthesized RNA molecule.

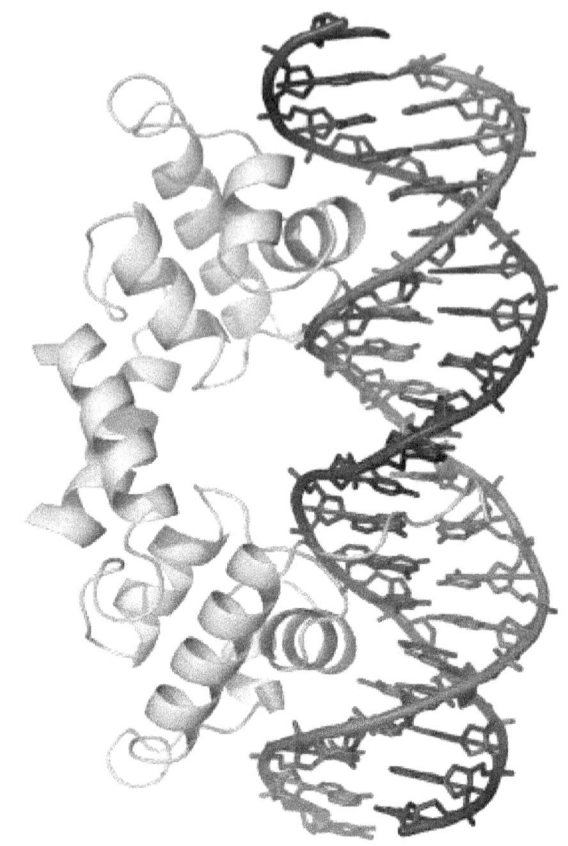

The lambda repressor transcription factor (green) binds as a dimer to major groove of DNA target (red and blue) and disables initiation of transcription. From PDB: 1LMB.

The nuclear membrane in eukaryotes allows further regulation of transcription factors by the duration of their presence in the nucleus, which is regulated by reversible changes in their structure and by binding of other proteins.[28] Environmental stimuli or endocrine signals[29] may cause modification of regulatory proteins[30] eliciting cascades of intracellular signals,[31] which result in regulation of gene expression.

More recently it has become apparent that there is a significant influence of non-DNA-sequence specific effects on translation. These effects are referred to as epigenetic and involve the higher order structure of DNA, non-sequence specific DNA binding proteins and chemical modification of DNA. In general epigenetic effects alter the accessibility of DNA to proteins and so modulate transcription.

DNA methylation is a widespread mechanism for epigenetic influence on gene expression and is seen in bacteria and eukaryotes and has roles in heritable transcription silencing and transcription regulation. In eukaryotes the structure of chromatin, controlled by the histone code, regulates access to DNA with significant impacts on the ex-

In eukaryotes, DNA is organized in form of nucleosomes. Note how the DNA (blue and green) is tightly wrapped around the protein core made of histone octamer (ribbon coils), restricting access to the DNA. From PDB: 1KX5.

pression of genes in euchromatin and heterochromatin areas.

8.2.2 Post-transcriptional regulation

Main article: Post-transcriptional regulation

In eukaryotes, where export of RNA is required before translation is possible, nuclear export is thought to provide additional control over gene expression. All transport in and out of the nucleus is via the nuclear pore and transport is controlled by a wide range of importin and exportin proteins.

Expression of a gene coding for a protein is only possible if the messenger RNA carrying the code survives long enough to be translated. In a typical cell, an RNA molecule is only stable if specifically protected from degradation. RNA degradation has particular importance in regulation of expression in eukaryotic cells where mRNA has to travel significant distances before being translated. In eukaryotes, RNA is stabilised by certain post-transcriptional modifications, particularly the 5' cap and poly-adenylated tail.

Intentional degradation of mRNA is used not just as a defence mechanism from foreign RNA (normally from viruses) but also as a route of mRNA *destabilisation*. If an mRNA molecule has a complementary sequence to a small interfering RNA then it is targeted for destruction via the RNA interference pathway.

8.2.3 Three prime untranslated regions and microRNAs

Main article: Three prime untranslated region
Main article: MicroRNA

Three prime untranslated regions (3'UTRs) of messenger RNAs (mRNAs) often contain regulatory sequences that post-transcriptionally influence gene expression. Such 3'-UTRs often contain both binding sites for microRNAs (miRNAs) as well as for regulatory proteins. By binding to specific sites within the 3'-UTR, miRNAs can decrease gene expression of various mRNAs by either inhibiting translation or directly causing degradation of the transcript. The 3'-UTR also may have silencer regions that bind repressor proteins that inhibit the expression of a mRNA.

The 3'-UTR often contains microRNA response elements (MREs). MREs are sequences to which miRNAs bind. These are prevalent motifs within 3'-UTRs. Among all regulatory motifs within the 3'-UTRs (e.g. including silencer regions), MREs make up about half of the motifs.

As of 2014, the miRBase web site,[32] an archive of miRNA sequences and annotations, listed 28,645 entries in 233 biologic species. Of these, 1,881 miRNAs were in annotated human miRNA loci. miRNAs were predicted to have an average of about four hundred target mRNAs (affecting expression of several hundred genes).[33] Freidman et al.[33] estimate that >45,000 miRNA target sites within human mRNA 3'UTRs are conserved above background levels, and >60% of human protein-coding genes have been under selective pressure to maintain pairing to miRNAs.

Direct experiments show that a single miRNA can reduce the stability of hundreds of unique mRNAs.[34] Other experiments show that a single miRNA may repress the production of hundreds of proteins, but that this repression often is relatively mild (less than 2-fold).[35][36]

The effects of miRNA dysregulation of gene expression seem to be important in cancer.[37] For instance, in gastrointestinal cancers, nine miRNAs have been identified as epigenetically altered and effective in down regulating DNA repair enzymes.[38]

The effects of miRNA dysregulation of gene expression also seem to be important in neuropsychiatric disorders, such as schizophrenia, bipolar disorder, major depression, Parkinson's disease, Alzheimer's disease and autism spectrum disorders.[39][40][41]

8.2.4 Translational regulation

Main article: Translation (genetics)

Neomycin is an example of a small molecule that reduces expression of all protein genes inevitably leading to cell death; it thus acts as an antibiotic.

Direct regulation of translation is less prevalent than control of transcription or mRNA stability but is occasionally used. Inhibition of protein translation is a major target for toxins and antibiotics, so they can kill a cell by overriding its normal gene expression control. Protein synthesis inhibitors include the antibiotic neomycin and the toxin ricin.

8.2.5   Protein degradation

Main article: Proteasome

Once protein synthesis is complete, the level of expression of that protein can be reduced by protein degradation. There are major protein degradation pathways in all prokaryotes and eukaryotes, of which the proteasome is a common component. An unneeded or damaged protein is often labeled for degradation by addition of ubiquitin.

8.3   Measurement

Measuring gene expression is an important part of many life sciences, as the ability to quantify the level at which a particular gene is expressed within a cell, tissue or organism can provide a lot of valuable information. For example, measuring gene expression can:

- Identify viral infection of a cell (viral protein expression).

- Determine an individual's susceptibility to cancer (oncogene expression).

- Find if a bacterium is resistant to penicillin (beta-lactamase expression).

Similarly, the analysis of the location of protein expression is a powerful tool, and this can be done on an organismal or cellular scale. Investigation of localization is particularly important for the study of development in multicellular organisms and as an indicator of protein function in single cells. Ideally, measurement of expression is done by detecting the final gene product (for many genes, this is the protein); however, it is often easier to detect one of the precursors, typically mRNA and to infer gene-expression levels from these measurements.

8.3.1   mRNA quantification

Levels of mRNA can be quantitatively measured by northern blotting, which provides size and sequence information about the mRNA molecules. A sample of RNA is separated on an agarose gel and hybridized to a radioactively labeled RNA probe that is complementary to the target sequence. The radiolabeled RNA is then detected by an autoradiograph. Because the use of radioactive reagents makes the procedure time consuming and potentially dangerous, alternative labeling and detection methods, such as digoxigenin and biotin chemistries, have been developed. Perceived disadvantages of Northern blotting are that large quantities of RNA are required and that quantification may not be completely accurate, as it involves measuring band strength in an image of a gel. On the other hand, the additional mRNA size information from the Northern blot allows the discrimination of alternately spliced transcripts.

Another approach for measuring mRNA abundance is RT-qPCR. In this technique, reverse transcription is followed by quantitative PCR. Reverse transcription first generates a DNA template from the mRNA; this single-stranded template is called cDNA. The cDNA template is then amplified in the quantitative step, during which the fluorescence emitted by labeled hybridization probes or intercalating dyes changes as the DNA amplification process progresses. With a carefully constructed standard curve, qPCR can produce an absolute measurement of the number of copies of original mRNA, typically in units of copies per nanolitre of homogenized tissue or copies per cell. qPCR is very sensitive (detection of a single mRNA molecule is theoretically possible), but can be expensive depending on the type of reporter used; fluorescently labeled oligonucleotide probes are more expensive than non-specific intercalating fluorescent dyes.

For expression profiling, or high-throughput analysis of many genes within a sample, quantitative PCR may be performed for hundreds of genes simultaneously in the case of low-density arrays. A second approach is the hybridization microarray. A single array or "chip" may contain probes to determine transcript levels for every known gene in the genome of one or more organisms. Alternatively, "tag based" technologies like Serial analysis of gene expression (SAGE) and RNA-Seq, which can provide a relative measure of the cellular concentration of different mRNAs, can be used. An advantage of tag-based methods is the "open architecture", allowing for the exact measurement of any transcript, with a known or unknown sequence. Next-generation sequencing (NGS) such as RNA-Seq is another approach, producing vast quantities of sequence data that can be matched to a reference genome. Although NGS is comparatively time-consuming, expensive, and resource-intensive, it can identify single-nucleotide polymorphisms, splice-variants, and novel genes, and can also be used to profile expression in organisms for which little or no sequence information is available.

8.3.2 RNA Profiles in Wikipedia

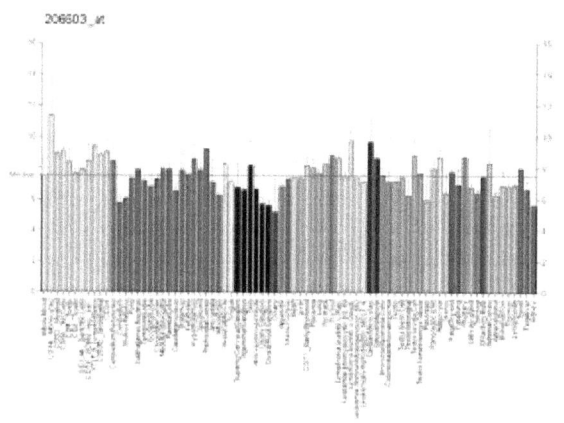

The RNA Expression profile of the GLUT4 Transporter (one of the main glucose transporters found in the human body.

Profiles like these are found for almost all proteins listed in Wikipedia. They are generated by organizations such as the Genomics Institute of the Novartis Research Foundation and the European Bioinformatics Institute. Additional information can be found by searching their databases (for an example of the GLUT4 transporter pictured here, see citation).[42] These profiles indicate the level of DNA expression (and hence RNA produced) of a certain protein in a certain tissue, and are color-coded accordingly in the images located in the Protein Box on the right side of each Wikipedia page.

8.3.3 Protein quantification

For genes encoding proteins, the expression level can be directly assessed by a number of methods with some clear analogies to the techniques for mRNA quantification.

The most commonly used method is to perform a Western blot against the protein of interest—this gives information on the size of the protein in addition to its identity. A sample (often cellular lysate) is separated on a polyacrylamide gel, transferred to a membrane and then probed with an antibody to the protein of interest. The antibody can either be conjugated to a fluorophore or to horseradish peroxidase for imaging and/or quantification. The gel-based nature of this assay makes quantification less accurate, but it has the advantage of being able to identify later modifications to the protein, for example proteolysis or ubiquitination, from changes in size.

8.3.4 Localisation

Main articles: In situ hybridization and Immunofluorescence

Analysis of expression is not limited to quantification:

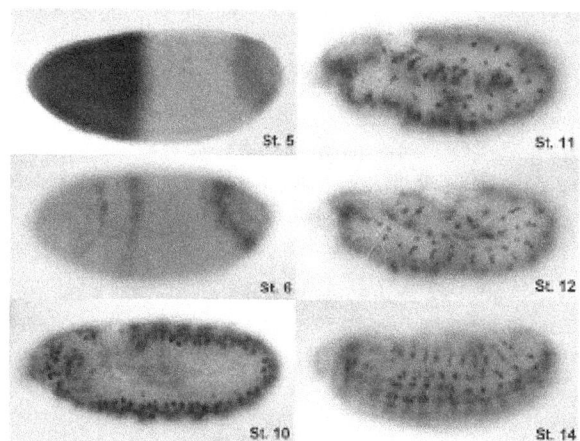

In situ-hybridization of Drosophila embryos at different developmental stages for the mRNA responsible for the expression of hunchback. High intensity of blue color marks places with high hunchback mRNA quantity.

localisation can also be determined. mRNA can be detected with a suitably labelled complementary mRNA strand and protein can be detected via labelled antibodies. The probed sample is then observed by microscopy to identify where the mRNA or protein is.

By replacing the gene with a new version fused to a green fluorescent protein (or similar) marker, expression may be directly quantified in live cells. This is done by imaging using a fluorescence microscope. It is very difficult to clone a GFP-fused protein into its native location in the genome without affecting expression levels so this method often cannot be used to measure endogenous gene expression. It is, however, widely used to measure the expression of a gene artificially introduced into the cell, for example via an expression vector. It is important to note that by fusing a

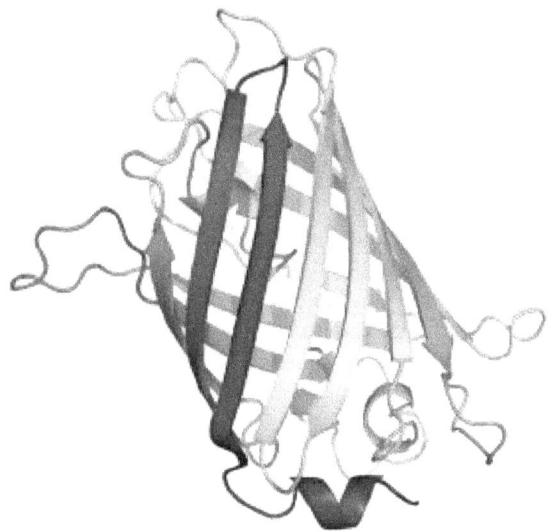

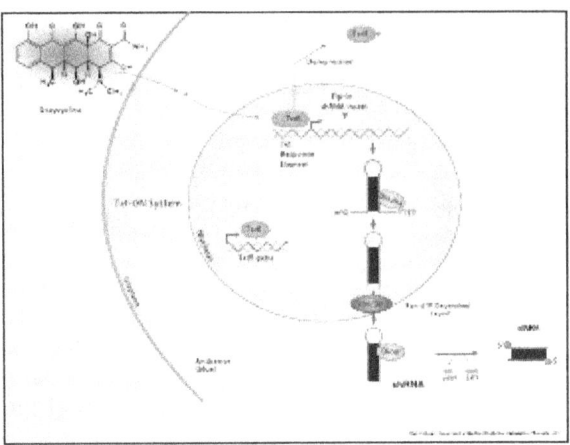

Tet-ON inducible shRNA system

The three-dimensional structure of green fluorescent protein. The residues in the centre of the "barrel" are responsible for production of green light after exposing to higher energetic blue light. From PDB: 1EMA.

target protein to a fluorescent reporter the protein's behavior, including its cellular localization and expression level, can be significantly changed.

The enzyme-linked immunosorbent assay works by using antibodies immobilised on a microtiter plate to capture proteins of interest from samples added to the well. Using a detection antibody conjugated to an enzyme or fluorophore the quantity of bound protein can be accurately measured by fluorometric or colourimetric detection. The detection process is very similar to that of a Western blot, but by avoiding the gel steps more accurate quantification can be achieved.

8.4   Expression system

Main article: Protein production (biotechnology)

An expression system is a system specifically designed for the production of a gene product of choice. This is normally a protein although may also be RNA, such as tRNA or a ribozyme. An expression system consists of a gene, normally encoded by DNA, and the molecular machinery required to transcribe the DNA into mRNA and translate the mRNA into protein using the reagents provided. In the broadest sense this includes every living cell but the term is more normally used to refer to expression as a laboratory tool. An expression system is therefore often artificial in some manner. Expression systems are, however, a funda-

mentally natural process. Viruses are an excellent example where they replicate by using the host cell as an expression system for the viral proteins and genome.

8.4.1   Inducible expression

Doxycycline is also used in "Tet-on" and "Tet-off" tetracycline controlled transcriptional activation to regulate transgene expression in organisms and cell cultures.

8.4.2   In nature

In addition to these biological tools, certain naturally observed configurations of DNA (genes, promoters, enhancers, repressors) and the associated machinery itself are referred to as an expression system. This term is normally used in the case where a gene or set of genes is switched on under well defined conditions, for example, the simple repressor switch expression system in Lambda phage and the lac operator system in bacteria. Several natural expression systems are directly used or modified and used for artificial expression systems such as the Tet-on and Tet-off expression system.

8.5   Gene networks

Main article: Gene regulatory network

Genes have sometimes been regarded as nodes in a network, with inputs being proteins such as transcription factors, and outputs being the level of gene expression. The node itself performs a function, and the operation of these functions have been interpreted as performing a kind of information processing within cells and determines cellular behavior.

Gene networks can also be constructed without formulating an explicit causal model. This is often the case when assembling networks from large expression data sets. Covariation and correlation of expression is computed across a large sample of cases and measurements (often transcriptome or proteome data). The source of variation can be either experimental or natural (observational). There are several ways to construct gene expression networks, but one common approach is to compute a matrix of all pair-wise correlations of expression across conditions, time points, or individuals and convert the matrix (after thresholding at some cut-off value) into a graphical representation in which nodes represent genes, transcripts, or proteins and edges connecting these nodes represent the strength of association (see).[43] Weighted correlation network analysis involves weighted networks defined by soft-thresholding the pairwise correlations among variables (e.g. measures of transcript abundance).

8.6 Techniques and tools

The following experimental techniques are used to measure gene expression and are listed in roughly chronological order, starting with the older, more established technologies. They are divided into two groups based on their degree of multiplexity.

- Low-to-mid-plex techniques:
  - Reporter gene
  - Northern blot
  - Western blot[44]
  - Fluorescent in situ hybridization
  - Reverse transcription PCR
- Higher-plex techniques:
  - SAGE[45]
  - DNA microarray[46]
  - Tiling array[47]
  - RNA-Seq[48]

8.7 See also

- AlloMap molecular expression testing
- Bookmarking

- Expressed sequence tag
- Expression profiling
- Genetic engineering
- Genetically modified organism
- List of human genes
- Oscillating gene
- Paramutation
- Protein production
- Protein_purification
- Ribonomics
- Ridges
- Sequence profiling tool
- Transcriptional bursting
- Transcriptional noise

8.8 References

[1] Brueckner F, Armache KJ, Cheung A, et al. (February 2009). "Structure–function studies of the RNA polymerase II elongation complex". Acta Crystallogr. D Biol. Crystallogr. 65 (Pt 2): 112–20. doi:10.1107/S0907444908039875. PMC 2631633. PMID 19171965.

[2] Sirri V, Urcuqui-Inchima S, Roussel P, Hernandez-Verdun D (January 2008). "Nucleolus: the fascinating nuclear body". Histochem. Cell Biol. 129 (1): 13–31. doi:10.1007/s00418-007-0359-6. PMC 2137947. PMID 18046571.

[3] Frank DN, Pace NR (1998). "Ribonuclease P: unity and diversity in a tRNA processing ribozyme". Annu. Rev. Biochem. 67: 153–80. doi:10.1146/annurev.biochem.67.1.153. PMID 9759486.

[4] Ceballos M, Vioque A (2007). "tRNase Z". Protein Pept. Lett. 14 (2): 137–45. doi:10.2174/092986607779816050. PMID 17305600.

[5] Weiner AM (October 2004). "tRNA maturation: RNA polymerization without a nucleic acid template". Curr. Biol. 14 (20): R883–5. doi:10.1016/j.cub.2004.09.069. PMID 15498478.

[6] Köhler A, Hurt E (October 2007). "Exporting RNA from the nucleus to the cytoplasm". Nat. Rev. Mol. Cell Biol. 8 (10): 761–73. doi:10.1038/nrm2255. PMID 17786152.

[7] Jambhekar A. Derisi JL (May 2007). "Cis-acting determinants of asymmetric, cytoplasmic RNA transport". *RNA* **13** (5): 625–42. doi:10.1261/rna.262607. PMC 1852811. PMID 17449729.

[8] Amaral PP, Dinger ME, Mercer TR, Mattick JS (March 2008). "The eukaryotic genome as an RNA machine". *Science* **319** (5871): 1787–9. Bibcode:2008Sci...319.1787A. doi:10.1126/science.1155472. PMID 18369136.

[9] Hansen TM, Baranov PV, Ivanov IP, Gesteland RF, Atkins JF (May 2003). "Maintenance of the correct open reading frame by the ribosome". *EMBO Rep.* **4** (5): 499–504. doi:10.1038/sj.embor.embor825. PMC 1319180. PMID 12717454.

[10] Berk V, Cate JH (June 2007). "Insights into protein biosynthesis from structures of bacterial ribosomes". *Curr. Opin. Struct. Biol.* **17** (3): 302–9. doi:10.1016/j.sbi.2007.05.009. PMID 17574829.

[11] Schwanhäusser B, Busse D, Dittmar G, Schuchhardt J, Wolf J, Chen W, Selbach M (2011). "Global quantification of mammalian gene expression control". *Nature* **473** (7347): 337–42. Bibcode:2011Natur.473..337S. doi:10.1038/nature10098. PMID 21593866.

[12] Schwanhäusser B, Busse D, Dittmar G, Schuchhardt J, Wolf J, Chen W, Selbach M (2013). "Corrigendum: Global quantification of mammalian gene expression control". *Nature* **495** (7439): 126–7. Bibcode:2013Natur.495..126S. doi:10.1038/nature11848. PMID 23407496.

[13] Hegde RS, Kang SW (July 2008). "The concept of translocational regulation". *J. Cell Biol.* **182** (2): 225–32. doi:10.1083/jcb.200804157. PMC 2483521. PMID 18644895.

[14] Alberts, Bruce; Alexander Johnson; Julian Lewis; Martin Raff; Keith Roberts; Peter Walters (2002). "The Shape and Structure of Proteins". *Molecular Biology of the Cell; Fourth Edition*. New York and London: Garland Science. ISBN 0-8153-3218-1.

[15] Anfinsen, C. (1972). "The formation and stabilization of protein structure". *Biochem. J.* **128** (4): 737–49. PMC 1173893. PMID 4565129.

[16] Jeremy M. Berg, John L. Tymoczko, Lubert Stryer; Web content by Neil D. Clarke (2002). "3. Protein Structure and Function". *Biochemistry*. San Francisco: W. H. Freeman. ISBN 0-7167-4684-0.

[17] Dennis J. Selkoe (2003). "Folding proteins in fatal ways". *Nature* **426** (6968): 900–904. Bibcode:2003Natur.426..900S. doi:10.1038/nature02264. PMID 14685251.

[18] Alberts, Bruce, Dennis Bray, Karen Hopkin, Alexander Johnson, Julian Lewis, Martin Raff, Keith Roberts, and Peter Walter. "Protein Structure and Function." Essential Cell Biology. Edition 3. New York: Garland Science, Taylor and Francis Group, LLC, 2010. Pg 120-170.

[19] Hebert DN, Molinari M (October 2007). "In and out of the ER: protein folding, quality control, degradation, and related human diseases". *Physiol. Rev.* **87** (4): 1377–408. doi:10.1152/physrev.00050.2006. PMID 17928587.

[20] Russell R (2008). "RNA misfolding and the action of chaperones". *Front. Biosci.* **13** (13): 1–20. doi:10.2741/2557. PMC 2610265. PMID 17981525.

[21] Kober L, Zehe C, Bode J (April 2013). "Optimized signal peptides for the development of high expressing CHO cell lines". *Biotechnol. Bioeng.* **110** (4): 1164–73. doi:10.1002/bit.24776. PMID 23124363.

[22] Moreau P, Brandizzi F, Hanton S, et al. (2007). "The plant ER-Golgi interface: a highly structured and dynamic membrane complex". *J. Exp. Bot.* **58** (1): 49–64. doi:10.1093/jxb/erl135. PMID 16990376.

[23] Prudovsky I, Tarantini F, Landriscina M, et al. (April 2008). "Secretion Without Golgi". *J. Cell. Biochem.* **103** (5): 1327–43. doi:10.1002/jcb.21513. PMC 2613191. PMID 17786931.

[24] Zaidi SK, Young DW, Choi JY, Pratap J, Javed A, Montecino M, Stein JL, Lian JB, van Wijnen AJ, Stein GS (October 2004). "Intranuclear trafficking: organization and assembly of regulatory machinery for combinatorial biological control". *J. Biol. Chem.* **279** (42): 43363–6. doi:10.1074/jbc.R400020200. PMID 15277516.

[25] Mattick JS, Amaral PP, Dinger ME, Mercer TR, Mehler MF (January 2009). "RNA regulation of epigenetic processes". *BioEssays* **31** (1): 51–9. doi:10.1002/bies.080099. PMID 19154003.

[26] Martinez NJ, Walhout AJ (April 2009). "The interplay between transcription factors and microRNAs in genome-scale regulatory networks". *BioEssays* **31** (4): 435–45. doi:10.1002/bies.200800212. PMC 3118512. PMID 19274664.

[27] Tomilin NV (April 2008). "Regulation of mammalian gene expression by retroelements and non-coding tandem repeats". *BioEssays* **30** (4): 338–48. doi:10.1002/bies.20741. PMID 18348251.

[28] Veitia RA (November 2008). "One thousand and one ways of making functionally similar transcriptional enhancers". *BioEssays* **30** (11–12): 1052–7. doi:10.1002/bies.20849. PMID 18937349.

[29] Nguyen T, Nioi P, Pickett CB (May 2009). "The Nrf2-Antioxidant Response Element Signaling Pathway and Its Activation by Oxidative Stress". *J. Biol. Chem.* **284** (20): 13291–5. doi:10.1074/jbc.R900010200. PMC 2679427. PMID 19182219.

[30] Paul S (November 2008). "Dysfunction of the ubiquitin-proteasome system in multiple disease conditions: therapeutic approaches". *BioEssays* **30** (11–12): 1172–84. doi:10.1002/bies.20852. PMID 18937370.

[31] Los M, Maddika S, Erb B, Schulze-Osthoff K (May 2009). "Switching Akt: from survival signaling to deadly response". *BioEssays* **31** (5): 492–5. doi:10.1002/bies.200900005. PMC 2954189. PMID 19319914.

[32] miRBase.org

[33] Friedman RC, Farh KK, Burge CB, Bartel DP (2009). "Most mammalian mRNAs are conserved targets of microRNAs". *Genome Res.* **19** (1): 92–105. doi:10.1101/gr.082701.108. PMC 2612969. PMID 18955434.

[34] Lim LP, Lau NC, Garrett-Engele P, Grimson A, Schelter JM, Castle J, Bartel DP, Linsley PS, Johnson JM; Lau; Garrett-Engele; Grimson; Schelter; Castle; Bartel; Linsley; Johnson (February 2005). "Microarray analysis shows that some microRNAs downregulate large numbers of target mRNAs". *Nature* **433** (7027): 769–73. Bibcode:2005Natur.433..769L. doi:10.1038/nature03315. PMID 15685193.

[35] Selbach M, Schwanhäusser B, Thierfelder N, Fang Z, Khanin R, Rajewsky N; Schwanhäusser; Thierfelder; Fang; Khanin; Rajewsky (September 2008). "Widespread changes in protein synthesis induced by microRNAs". *Nature* **455** (7209): 58–63. Bibcode:2008Natur.455...58S. doi:10.1038/nature07228. PMID 18668040.

[36] Baek D, Villén J, Shin C, Camargo FD, Gygi SP, Bartel DP; Villén; Shin; Camargo; Gygi; Bartel (September 2008). "The impact of microRNAs on protein output". *Nature* **455** (7209): 64–71. Bibcode:2008Natur.455...64B. doi:10.1038/nature07242. PMC 2745094. PMID 18668037.

[37] Palmero EI, de Campos SG, Campos M, de Souza NC, Guerreiro ID, Carvalho AL, Marques MM (2011). "Mechanisms and role of microRNA deregulation in cancer onset and progression". *Genet. Mol. Biol.* **34** (3): 363–70. doi:10.1590/S1415-47572011000300001. PMC 3168173. PMID 21931505.

[38] Bernstein C, Bernstein H (2015). "Epigenetic reduction of DNA repair in progression to gastrointestinal cancer". *World J Gastrointest Oncol* **7** (5): 30–46. doi:10.4251/wjgo.v7.i5.30. PMC 4434036. PMID 25987950.

[39] Maffioletti E, Tardito D, Gennarelli M, Bocchio-Chiavetto L (2014). "Micro spies from the brain to the periphery: new clues from studies on microRNAs in neuropsychiatric disorders". *Front Cell Neurosci* **8**: 75. doi:10.3389/fncel.2014.00075. PMC 3949217. PMID 24653674.

[40] Mellios N, Sur M (2012). "The Emerging Role of microRNAs in Schizophrenia and Autism Spectrum Disorders". *Front Psychiatry* **3**: 39. doi:10.3389/fpsyt.2012.00039. PMC 3336189. PMID 22539927.

[41] Geaghan M, Cairns MJ (2015). "MicroRNA and Posttranscriptional Dysregulation in Psychiatry". *Biol. Psychiatry* **78** (4): 231–9. doi:10.1016/j.biopsych.2014.12.009. PMID 25636176.

[42] "GLUT4 RNA Expression Profile".

[43] Chesler EJ, Lu L, Wang J, Williams RW, Manly KF (2004). "WebQTL: rapid exploratory analysis of gene expression and genetic networks for brain and behavior". *Nat Neurosci* **7** (5): 485–86. doi:10.1038/nn0504-485. PMID 15114364.

[44] Song Y, Wang W, Qu X, Sun S (February 2009). "Effects of hypoxia inducible factor-1alpha (HIF-1alpha) on the growth & adhesion in tongue squamous cell carcinoma cells". *Indian J. Med. Res.* **129** (2): 154–63. PMID 19293442.

[45] Hanriot L, Keime C, Gay N, et al. (2008). "A combination of LongSAGE with Solexa sequencing is well suited to explore the depth and the complexity of transcriptome". *BMC Genomics* **9**: 418. doi:10.1186/1471-2164-9-418. PMC 2562395. PMID 18796152.

[46] Wheelan SJ, Martinez Murillo F, Boeke JD (July 2008). "The incredible shrinking world of DNA microarrays". *Mol Biosyst* **4** (7): 726–32. doi:10.1039/b706237k. PMC 2535915. PMID 18563246.

[47] Miyakoshi M, Nishida H, Shintani M, Yamane H, Nojiri H (2009). "High-resolution mapping of plasmid transcriptomes in different host bacteria". *BMC Genomics* **10**: 12. doi:10.1186/1471-2164-10-12. PMC 2642839. PMID 19134166.

[48] Denoeud F, Aury JM, Da Silva C, F; Artiguenave; et al. (2008). "Annotating genomes with massive-scale RNA sequencing". *Genome Biol.* **9** (12): R175. doi:10.1186/gb-2008-9-12-r175. PMC 2646279. PMID 19087247.

8.9 External links

- "Genes & Gene Expression". *The Virtual Library of Biochemistry and Cell Biology*. BioChemWeb.org. 2005-12-04. Retrieved 2008-06-10.

- John Kryk (2008-05-28). "DNA makes RNA". Retrieved 2008-06-10.

- "Advancing Gene Expression Studies". *Genetic Engineering & Biotechnology News*. Mary Ann Liebert, Inc. 2008-08-01.

- "Optimizing Transient Gene Expression". *Genetic Engineering & Biotechnology News*. Mary Ann Liebert, Inc. 2008-03-01.

-

-

Chapter 9

Genome

For a non-technical introduction to the topic, see Introduction to genetics. For other uses, see Genome (disambiguation).

In modern molecular biology and genetics, the **genome**

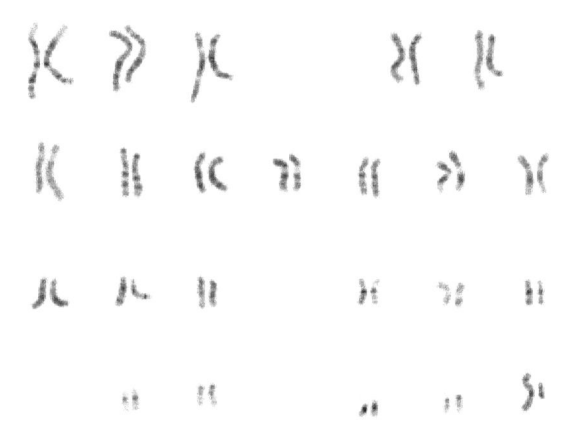

An image of the 46 chromosomes making up the diploid genome of a human male. (The mitochondrial chromosome is not shown.)

is the genetic material of an organism. It consists of DNA (or RNA in RNA viruses). The genome includes both the genes and the non-coding sequences of the DNA/RNA.[1]

9.1   Origin of term

The term was created in 1920 by Hans Winkler,[2] professor of botany at the University of Hamburg, Germany. The Oxford Dictionary suggests the name to be a blend of the words *gene* and *chromosome*.[3] However, see omics for a more thorough discussion. A few related *-ome* words already existed—such as *biome*, *rhizome*, forming a vocabulary into which *genome* fits systematically.[4]

9.2   Overview

Some organisms have multiple copies of chromosomes: diploid, triploid, tetraploid and so on. In classical genetics, in a sexually reproducing organism (typically eukarya) the gamete has half the number of chromosomes of the somatic cell and the genome is a full set of chromosomes in a diploid cell. The halving of the genetic material in gametes is accomplished by the segregation of homologous chromosomes during meiosis.[5] In haploid organisms, including cells of bacteria, archaea, and in organelles including mitochondria and chloroplasts, or viruses, that similarly contain genes, the single or set of circular or linear chains of DNA (or RNA for some viruses), likewise constitute the genome. The term *genome* can be applied specifically to mean what is stored on a complete set of nuclear DNA (i.e., the "nuclear genome") but can also be applied to what is stored within organelles that contain their own DNA, as with the "mitochondrial genome" or the "chloroplast genome". Additionally, the genome can comprise non-chromosomal genetic elements such as viruses, plasmids, and transposable elements.[6]

When people say that the genome of a sexually reproducing species has been "sequenced", typically they are referring to a determination of the sequences of one set of autosomes and one of each type of sex chromosome, which together represent both of the possible sexes. Even in species that exist in only one sex, what is described as a "genome sequence" may be a composite read from the chromosomes of various individuals. Colloquially, the phrase "genetic makeup" is sometimes used to signify the genome of a particular individual or organism. The study of the global properties of genomes of related organisms is usually referred to as genomics, which distinguishes it from genetics which generally studies the properties of single genes or groups of genes.

Both the number of base pairs and the number of genes vary widely from one species to another, and there is only a rough correlation between the two (an observation known as the C-value paradox). At present, the highest known number of genes is around 60,000, for the protozoan causing trichomoniasis (see List of sequenced eukaryotic genomes), almost three times as many as in the human genome.

An analogy to the human genome stored on DNA is that of instructions stored in a book:

- The book (genome) would contain 23 chapters (chromosomes);

- Each chapter contains 48 to 250 million letters (A,C,G,T) without spaces;

- Hence, the book contains over 3.2 billion letters total;

- The book fits into a cell nucleus the size of a pinpoint;

- At least one copy of the book (all 23 chapters) is contained in most cells of our body. The only exception in humans is found in mature red blood cells which become enucleated during development and therefore lack a genome.

9.3 Sequencing and mapping

For more details on this topic, see Genome project.

In 1976, Walter Fiers at the University of Ghent (Bel-

Part of DNA sequence - prototypification of complete genome of virus

gium) was the first to establish the complete nucleotide sequence of a viral RNA-genome (Bacteriophage MS2). The next year Fred Sanger completed the first DNA-genome sequence: Phage Φ-X174, of 5386 base pairs.[7] The first complete genome sequences among all three domains of

life were released within a short period during the mid-1990s: The first bacterial genome to be sequenced was that of Haemophilus influenzae, completed by a team at The Institute for Genomic Research in 1995. A few months later, the first eukaryotic genome was completed, with sequences of the 16 chromosomes of budding yeast *Saccharomyces cerevisiae* published as the result of a European-led effort begun in the mid-1980s. The first genome sequence for an archaeon, *Methanococcus jannaschii*, was completed in 1996, again by The Institute for Genomic Research.

The development of new technologies has made it dramatically easier and cheaper to do sequencing, and the number of complete genome sequences is growing rapidly. The US National Institutes of Health maintains one of several comprehensive databases of genomic information.[8] Among the thousands of completed genome sequencing projects include those for rice, a mouse, the plant *Arabidopsis thaliana*, the puffer fish, and the bacteria E. coli. In December 2013, scientists first sequenced the entire *genome* of a Neanderthal, an extinct species of humans. The genome was extracted from the toe bone of a 130,000-year-old Neanderthal found in a Siberian cave.[9][10]

New sequencing technologies, such as massive parallel sequencing have also opened up the prospect of personal genome sequencing as a diagnostic tool, as pioneered by Manteia Predictive Medicine. A major step toward that goal was the completion in 2007 of the full genome of James D. Watson, one of the co-discoverers of the structure of DNA.[11]

Whereas a genome sequence lists the order of every DNA base in a genome, a genome map identifies the landmarks. A genome map is less detailed than a genome sequence and aids in navigating around the genome. The Human Genome Project was organized to map and to sequence the human genome. A fundamental step in the project was the release of a detailed genomic map by Jean Weissenbach and his team at the Genoscope in Paris.[12][13]

9.4 Genome compositions

Genome composition is used to describe the make up of contents of a haploid genome, which should include **genome size**, proportions of **non-repetitive DNA** and **repetitive DNA** in details. By comparing the genome compositions between genomes, scientists can better understand the evolutionary history of a given genome.

When talking about genome composition, one should distinguish between prokaryotes and eukaryotes as the big differences on contents structure they have. In prokaryotes, most of the genome (85–90%) is non-repetitive DNA, which means coding DNA mainly forms it, while non-coding regions only take a small part.[14] On the contrary, eukaryotes have the feature of exon-intron organization of protein coding genes; the variation of repetitive DNA content in eukaryotes is also extremely high. In mammals and plants, the major part of the genome is composed of repetitive DNA.[15]

Most biological entities that are more complex than a virus sometimes or always carry additional genetic material besides that which resides in their chromosomes. In some contexts, such as sequencing the genome of a pathogenic microbe, "genome" is meant to include information stored on this auxiliary material, which is carried in plasmids. In such circumstances then, "genome" describes all of the genes and information on non-coding DNA that have the potential to be present.

In eukaryotes such as plants, protozoa and animals, however, "genome" carries the typical connotation of only information on chromosomal DNA. So although these organisms contain chloroplasts or mitochondria that have their own DNA, the genetic information contained by DNA within these organelles is not considered part of the genome. In fact, mitochondria are sometimes said to have their own genome often referred to as the "mitochondrial genome". The DNA found within the chloroplast may be referred to as the "plastome".

9.4.1   Genome size

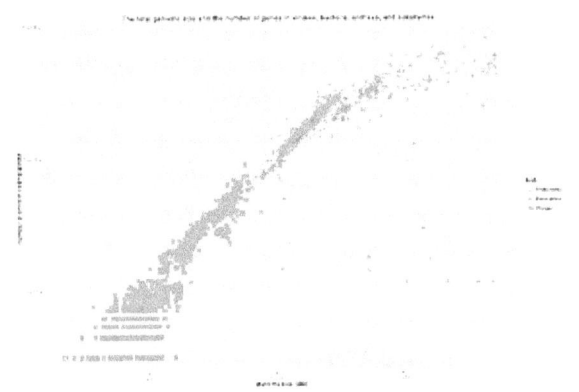

Log-log plot of the total number of annotated proteins in genomes submitted to GenBank as a function of genome size.[16]

Genome size is the total number of DNA base pairs in one copy of a haploid genome. The genome size is positively correlated with the morphological complexity among prokaryotes and lower eukaryotes; however, after mollusks and all the other higher eukaryotes above, this correlation is no longer effective.[15][17] This phenomenon also indicates the mighty influence coming from repetitive DNA act on the genomes.

Since genomes are very complex, one research strategy is to reduce the number of genes in a genome to the bare minimum and still have the organism in question survive. There is experimental work being done on minimal genomes for single cell organisms as well as minimal genomes for multicellular organisms (see Developmental biology). The work is both *in vivo* and *in silico*.[18][19]

Here is a table of some significant or representative genomes. See #See also for lists of sequenced genomes.

9.4.2   Proportion of non-repetitive DNA

The **proportion of non-repetitive DNA** is calculated by using the length of non-repetitive DNA divided by genome size. Protein-coding genes and RNA-coding genes are generally non-repetitive DNA.[60] A bigger genome does not mean more genes, and the proportion of non-repetitive DNA decreases along with increasing genome size in higher eukaryotes.[15]

It had been found that the proportion of non-repetitive DNA can vary a lot between species. Some *E. coli* as prokaryotes only have non-repetitive DNA, lower eukaryotes such as *C. elegans* and fruit fly, still possess more non-repetitive DNA than repetitive DNA.[15][61] Higher eukaryotes tend to have more repetitive DNA than non-repetitive ones. In some plants and amphibians, the proportion of non-repetitive DNA is no more than 20%, becoming a minority component.[15]

9.4.3   Proportion of repetitive DNA

The **proportion of repetitive DNA** is calculated by using length of repetitive DNA divide by genome size. There are two categories of repetitive DNA in genome: tandem repeats and interspersed repeats.[62]

Tandem repeats

Tandem repeats are usually caused by slippage during replication, unequal crossing-over and gene conversion.[63] satellite DNA and microsatellites are forms of tandem repeats in the genome.[64] Although tandem repeats count for a significant proportion in genome, the largest proportion in mammalian is the other type, interspersed repeats.

Interspersed repeats

Interspersed repeats mainly come from transposable elements (TEs), but they also include some protein coding gene families and pseudogenes. Transposable elements are able to integrate into the genome at another site within the cell.[114][165] It is believed that TEs are an important driving force on genome evolution of higher eukaryotes.[166] TEs can be classified into two categories, Class 1 (retrotransposons) and Class 2 (DNA transposons).[165]

Retrotransposons　Retrotransposons can be transcribed into RNA, which are then duplicated at another site into the genome.[67] Retrotransposons can be divided into Long terminal repeats (LTRs) and Non-Long Terminal Repeats (Non-LTR).[166]

Long Terminal Repeats (LTRs)　similar to retroviruses, which have both gag and pol genes to make cDNA from RNA and proteins to insert into genome, but LTRs can only act within the cell as they lack the env gene in retroviruses.[165] It has been reported that LTRs consist of the largest fraction in most plant genome and might account for the huge variation in genome size.[168]

Non-Long Terminal Repeats (Non-LTRs)　can be divided into long interspersed elements (LINEs), short interspersed elements (SINEs) and Penelope-like elements. In *Dictyostelium discoideum*, there is another DIRS-like elements belong to Non-LTRs. Non-LTRs are widely spread in eukaryotic genomes.[169]

Long interspersed elements (LINEs)　are able to encode two Open Reading Frames (ORFs) to generate transcriptase and endonuclease, which are essential in retrotransposition. The human genome has around 500,000 LINEs, taking around 17% of the genome.[170]

Short interspersed elements (SINEs)　are usually less than 500 base pairs and need to co-opt with the LINEs machinery to function as nonautonomous retrotransposons.[171] The Alu element is the most common SINEs found in primates, it has a length of about 350 base pairs and takes about 11% of the human genome with around 1,500,000 copies.[166]

DNA transposons　DNA transposons generally move by "cut and paste" in the genome, but duplication has also been observed. Class 2 TEs do not use RNA as intermediate and are popular in bacteria, in metazoan it has also been found.[166]

9.5　Genome evolution

Genomes are more than the sum of an organism's genes and have traits that may be measured and studied without reference to the details of any particular genes and their products. Researchers compare traits such as *chromosome number* (karyotype), genome size, gene order, codon usage bias, and GC-content to determine what mechanisms could have produced the great variety of genomes that exist today (for recent overviews, see Brown 2002; Saccone and Pesole 2003; Benfey and Protopapas 2004; Gibson and Muse 2004; Reese 2004; Gregory 2005).

Duplications play a major role in shaping the genome. Duplication may range from extension of short tandem repeats, to duplication of a cluster of genes, and all the way to duplication of entire chromosomes or even entire genomes. Such duplications are probably fundamental to the creation of genetic novelty.

Horizontal gene transfer is invoked to explain how there is often extreme similarity between small portions of the genomes of two organisms that are otherwise very distantly related. Horizontal gene transfer seems to be common among many microbes. Also, eukaryotic cells seem to have experienced a transfer of some genetic material from their chloroplast and mitochondrial genomes to their nuclear chromosomes.

9.6　See also

- Bacterial genome size

- Genome Browser

- Genome project

- Genome-wide association study

- Genomics

  - Genome Compiler

- List of sequenced eukaryotic genomes

- List of sequenced animal genomes

- List of sequenced archaeal genomes

- List of sequenced bacterial genomes

- List of sequenced fungi genomes

- List of sequenced plastomes

- List of sequenced protist genomes

- Metagenomics

- Microbiome

- Molecular epidemiology

- Molecular pathological epidemiology

- Molecular pathology

- Nucleic acid sequence

- Pan-genome

- Precision medicine

- Sequenceome

- Whole genome sequencing

- Genome topology

9.7   References

[1] Ridley. M. (2006). *Genome*. New York. NY: Harper Perennial. ISBN 0-06-019497-9

[2] Winkler. HL (1920). *Verbreitung und Ursache der Parthenogenesis im Pflanzen- und Tierreiche*. Jena: Verlag Fischer.

[3] "definition of Genome in Oxford dictionary". Retrieved 25 March 2014.

[4] Lederberg, Joshua: McCray. Alexa T. (2001). "Ome Sweet 'Omics – A Genealogical Treasury of Words" (PDF). *The Scientist* **15** (7).

[5] Griffiths JF; Gelbart WM; Lewontin RC; Wessler SR; Suzuki DT; Miller JH (2005). *Introduction to Genetic Analysis*. New York: W.H. Freeman and Co. pp. 34–40, 473–476, 626–629. ISBN 0-7167-4939-4.

[6] Madigan M; Martinko J. eds. (2006). *Brock Biology of Microorganisms* (11th ed.). Prentice Hall. ISBN 0-13-144329-1.

[7]

[8] "Genome Home". 2010-12-08. Retrieved 27 January 2011.

[9] Zimmer. Carl (December 18. 2013). "Toe Fossil Provides Complete Neanderthal Genome". *New York Times*. Retrieved 18 December 2013.

[10] Prüfer. Kay: Racimo. Fernando: Patterson. Nick: Jay. Flora: Sankararaman, Sriram: Sawyer, Susanna: Heinze. Anja: Renaud, Gabriel: Sudmant. Peter H.: De Filippo. Cesare: Li. Heng: Mallick, Swapan: Dannemann. Michael: Fu. Qiaomei: Kircher. Martin: Kuhlwilm. Martin: Lachmann. Michael: Meyer, Matthias; Ongyerth. Matthias: Siebauer. Michael: Theunert, Christoph: Tandon. Arti: Moorjani. Priya: Pickrell, Joseph: Mullikin, James C.: Vohr. Samuel H.: Green, Richard E.: Hellmann, Ines: Johnson, Philip L. F.: et al. (December 18. 2013). "The complete genome sequence of a Neanderthal from the Altai Mountains". *Nature* **505** (7481): 43–49. Bibcode:2014Natur.505...43P. doi:10.1038/nature12886. Retrieved 18 December 2013.

[11] Wade. Nicholas (2007-05-31). "Genome of DNA Pioneer Is Deciphered". *The New York Times*. Retrieved 2 April 2010.

[12] "What's a Genome?". Genomenewsnetwork.org. 2003-01-15. Retrieved 27 January 2011.

[13] NCBI_user_services (2004-03-29). "Mapping Factsheet". Retrieved 27 January 2011.

[14] Koonin. Eugene V.: Wolf, Yuri I. (2010). "Constraints and plasticity in genome and molecular-phenome evolution". *Nature Reviews Genetics* **11** (7): 487–498. doi:10.1038/nrg2810. PMC 3273317. PMID 20548290.

[15] Lewin, Benjamin (2004). *Genes VIII* (8th ed.). Upper Saddle River, NJ: Pearson/Prentice Hall. ISBN 0-13-143981-2.

[16] Koonin, Eugene V. (2011-08-31). *The Logic of Chance: The Nature and Origin of Biological Evolution*. FT Press. ISBN 9780132542494.

[17] Gregory TR: Nicol JA: Tamm H: Kullman B: Kullman K: Leitch IJ: Murray BG: Kapraun DF: Greilhuber J: Bennett MD (3 January 2007). "Eukaryotic genome size databases". *Nucleic Acids Research* **35** (Database): D332–D338. doi:10.1093/nar/gkl828.

[18] Glass JI: Assad-Garcia N: Alperovich N: Yooseph S: Lewis MR: Maruf M: Hutchison CA 3rd: Smith HO: Venter JC (2006). "Essential genes of a minimal bacterium". *Proc Natl Acad Sci USA* **103** (2): 425–30. Bibcode:2006PNAS..103..425G. doi:10.1073/pnas.0510013103. PMC 1324956. PMID 16407165.

[19] Forster AC: Church GM (2006). "Towards synthesis of a minimal cell". *Mol Syst Biol*. **2** (1): 45. doi:10.1038/msb4100090. PMC 1681520. PMID 16924266.

[20] Mankertz P (2008). "Molecular Biology of Porcine Circoviruses". *Animal Viruses: Molecular Biology*. Caister Academic Press. ISBN 978-1-904455-22-6.

[21] Fiers W: Contreras. R.: Duerinck. F.: Haegeman. G.: Iserentant. D.: Merregaert. J.: Min Jou. W.: Molemans. F.: Raeymaekers. A.: Van Den Berghe. A.: Volckaert. G.: Ysebaert. M. (1976). "Complete nucleotide-sequence of bacteriophage MS2-RNA – primary and secondary structure of replicase gene". *Nature* **260** (5551): 500–507. Bibcode:1976Natur.260..500F. doi:10.1038/260500a0. PMID 1264203.

[22] Fiers. W.: Contreras. R.: Haegeman. G.: Rogiers. R.: Van De Voorde. A.: Van Heuverswyn. H.: Van Herreweghe. J.: Volckaert. G.: Ysebaert. M. (1978). "Complete nucleotide sequence of SV40 DNA". *Nature* **273** (5658): 113–120. Bibcode:1978Natur.273..113F. doi:10.1038/273113a0. PMID 205802.

[23] Sanger, F.; Air, G.M.; Barrell, B.G.; Brown, N.L.; Coulson, A.R.; Fiddes, J.C.; Hutchison, C.A.; Slocombe, P. M.; Smith, M. (1977). "Nucleotide sequence of bacteriophage phi X174 DNA". *Nature* **265** (5596): 687–695. Bibcode:1977Natur.265..687S. doi:10.1038/265687a0. PMID 870828.

[24] "Virology – Human Immunodeficiency Virus And Aids. Structure: The Genome And Proteins Of HIV". Pathmicro.med.sc.edu. 2010-07-01. Retrieved 27 January 2011.

[25] Thomason, Lynn; Court, Donald L.; Bubunenko, Mikail; Costantino, Nina; Wilson, Helen; Datta, Simanti; Oppenheim, Amos (2007). "Recombineering: genetic engineering in bacteria using homologous recombination". *Current Protocols in Molecular Biology*, Chapter 1: Unit 1.16. doi:10.1002/0471142727.mb0116s78. ISBN 0471142727. PMID 18265390.

[26] Court, D. L.; Oppenheim, A. B.; Adhya, S. L. (2007). "A new look at bacteriophage lambda genetic networks". *Journal of Bacteriology* **189** (2): 298–304. doi:10.1128/JB.01215-06. PMC 1797383. PMID 17085553.

[27] Sanger, F.; Coulson, A.R.; Hong, G.F.; Hill, D.F.; Petersen, G.B. (1982). "Nucleotide sequence of bacteriophage lambda DNA". *Journal of Molecular Biology* **162** (4): 729–73. doi:10.1016/0022-2836(82)90546-0. PMID 6221115.

[28] Legendre, M; Arslan, D; Abergel, C; Claverie, JM (2012). "Genomics of Megavirus and the elusive fourth domain of life[journal". *Communicative & Integrative Biology* **5** (1): 102–106. doi:10.4161/cib.18624. PMC 3291303. PMID 22482024.

[29] Philippe, N.; Legendre, M.; Doutre, G.; Coute, Y.; Poirot, O.; Lescot, M.; Arslan, D.; Seltzer, V.; Bertaux, L.; Bruley, C.; Garin, J.; Claverie, J.-M.; Abergel, C. (2013). "Pandoraviruses: Amoeba Viruses with Genomes Up to 2.5 Mb Reaching That of Parasitic Eukaryotes". *Science* **341** (6143): 281–6. Bibcode:2013Sci...341..281P. doi:10.1126/science.1239181. PMID 23869018.

[30] Bennett, G. M.; Moran, N. A. (5 August 2013). "Small, Smaller, Smallest: The Origins and Evolution of Ancient Dual Symbioses in a Phloem-Feeding Insect". *Genome Biology and Evolution* **5** (9): 1675–1688. doi:10.1093/gbe/evt118. PMID 23918810.

[31] Shigenobu, S; Watanabe, H; Hattori, M; Sakaki, Y; Ishikawa, H (Sep 7, 2000). "Genome sequence of the endocellular bacterial symbiont of aphids Buchnera sp. APS". *Nature* **407** (6800): 81–6. doi:10.1038/35024074. PMID 10993077.

[32] Fleischmann R; Adams M; White O; Clayton R; Kirkness E; Kerlavage A; Bult C; Tomb J; Dougherty B; Merrick J; McKenney; Sutton; Fitzhugh; Fields; Gocyne; Scott; Shirley; Liu; Glodek; Kelley; Weidman; Phillips; Spriggs; Hedblom; Cotton; Utterback; Hanna; Nguyen;

Saudek; et al. (1995). "Whole-genome random sequencing and assembly of Haemophilus influenzae Rd". *Science* **269** (5223): 496–512. Bibcode:1995Sci...269..496F. doi:10.1126/science.7542800. PMID 7542800.

[33] Frederick R. Blattner; Guy Plunkett III; et al. (1997). "The Complete Genome Sequence of Escherichia coli K-12". *Science* **277** (5331): 1453–1462. doi:10.1126/science.277.5331.1453. PMID 9278503.

[34] Challacombe, Jean F.; Eichorst, Stephanie A.; Hauser, Loren; Land, Miriam; Xie, Gary; Kuske, Cheryl R.; Steinke, Dirk (15 September 2011). Steinke, Dirk. ed. "Biological Consequences of Ancient Gene Acquisition and Duplication in the Large Genome of Candidatus Solibacter usitatus Ellin6076". *PLoS ONE* **6** (9): e24882. Bibcode:2011PLoSO...624882C. doi:10.1371/journal.pone.0024882. PMC 3174227. PMID 21949776.

[35] Rocap, G.; Larimer, F. W.; Lamerdin, J.; Malfatti, S.; Chain, P.; Ahlgren, N. A.; Arellano, A.; Coleman, M.; Hauser, L.; Hess, W. R.; Johnson, Z. I.; Land, M.; Lindell, D.; Post, A. F.; Regala, W.; Shah, M.; Shaw, S. L.; Steglich, C.; Sullivan, M. B.; Ting, C. S.; Tolonen, A.; Webb, E. A.; Zinser, E. R.; Chisholm, S. W. (2003). "Genome divergence in two Prochlorococcus ecotypes reflects oceanic niche differentiation". *Nature* **424** (6952): 1042–7. Bibcode:2003Natur.424.1042R. doi:10.1038/nature01947. PMID 12917642.

[36] Dufresne, A.; Salanoubat, M.; Partensky, F.; Artiguenave, F.; Axmann, I. M.; Barbe, V.; Duprat, S.; Galperin, M. Y.; Koonin, E. V.; Le Gall, F.; Makarova, K. S.; Ostrowski, M.; Oztas, S.; Robert, C.; Rogozin, I. B.; Scanlan, D. J.; De Marsac, N. T.; Weissenbach, J.; Wincker, P.; Wolf, Y. I.; Hess, W. R. (2003). "Genome sequence of the cyanobacterium Prochlorococcus marinus SS120, a nearly minimal oxyphototrophic genome". *Proceedings of the National Academy of Sciences* **100** (17): 10020–5. Bibcode:2003PNAS..10010020D. doi:10.1073/pnas.1733211100. PMC 187748. PMID 12917486.

[37] Meeks, J. C.; Elhai, J; Thiel, T; Potts, M; Larimer, F; Lamerdin, J; Predki, P; Atlas, R (2001). "An overview of the genome of Nostoc punctiforme, a multicellular, symbiotic cyanobacterium". *Photosynthesis Research* **70** (1): 85–106. doi:10.1023/A:1013840025518. PMID 16228364.

[38] Parfrey LW; Lahr DJG; Katz LA (2008). "The Dynamic Nature of Eukaryotic Genomes". *Molecular Biology and Evolution* **25** (4): 787–94. doi:10.1093/molbev/msn032. PMC 2933061. PMID 18258610.

[39] ScienceShot: Biggest Genome Ever, comments: "The measurement for Amoeba dubia and other protozoa which have been reported to have very large genomes were made in the 1960s using a rough biochemical approach which is now considered to be an unreliable method for accurate genome size determinations."

[40] Fleischmann A; Michael TP; Rivadavia F; Sousa A; Wang W; Temsch EM; Greilhuber J; Müller KF & Heubl G (2014). "Evolution of genome size and chromosome number in the carnivorous plant genus *Genlisea* (Lentibulariaceae), with a new estimate of the minimum genome size in angiosperms". *Annals of Botany* **114** (8): 1651–1663. doi:10.1093/aob/mcu189. PMID 25274549.

[41] Greilhuber J; Borsch T; Müller K; Worberg A; Porembski S & Barthlott W (2006). "Smallest angiosperm genomes found in Lentibulariaceae, with chromosomes of bacterial size". *Plant Biology* **8** (6): 770–777. doi:10.1055/s-2006-924101. PMID 17203433.

[42] Tuskan, GA; Difazio, S; Jansson, S; Bohlmann, J; Grigoriev, I; Hellsten, U; Putnam, N; Ralph, S; Rombauts, S; Salamov, A; Schein, J; Sterck, L; Aerts, A; Bhalerao, RR; Bhalerao, RP; Blaudez, D; Boerjan, W; Brun, A; Brunner, A; Busov, V; Campbell, M; Carlson, J; Chalot, M; Chapman, J; Chen, GL; Cooper, D; Coutinho, PM; Couturier, J; Covert, S; Cronk, Q; Cunningham, R; Davis, J; Degroeve, S; Déjardin, A; Depamphilis, C; Detter, J; Dirks, B; Dubchak, I; Duplessis, S; Ehlting, J; Ellis, B; Gendler, K; Goodstein, D; Gribskov, M; Grimwood, J; Groover, A; Gunter, L; Hamberger, B; Heinze, B; Helariutta, Y; Henrissat, B; Holligan, D; Holt, R; Huang, W; Islam-Faridi, N; Jones, S; Jones-Rhoades, M; Jorgensen, R; Joshi, C; Kangasjärvi, J; Karlsson, J; Kelleher, C; Kirkpatrick, R; Kirst, M; Kohler, A; Kalluri, U; Larimer, F; Leebens-Mack, J; Leplé, JC; Locascio, P; Lou, Y; Lucas, S; Martin, F; Montanini, B; Napoli, C; Nelson, DR; Nelson, C; Nieminen, K; Nilsson, O; Pereda, V; Peter, G; Philippe, R; Pilate, G; Poliakov, A; Razumovskaya, J; Richardson, P; Rinaldi, C; Ritland, K; Rouzé, P; Ryaboy, D; Schmutz, J; Schrader, J; Segerman, B; Shin, H; Siddiqui, A; Sterky, F; Terry, A; Tsai, CJ; Uberbacher, E; Unneberg, P; Vahala, J; Wall, K; Wessler, S; Yang, G; Yin, T; Douglas, C; Marra, M; Sandberg, G; Van de Peer, Y; Rokhsar, D (Sep 15, 2006). "The genome of black cottonwood, Populus trichocarpa (Torr. & Gray)". *Science* **313** (5793): 1596–604. Bibcode:2006Sci...313.1596T. doi:10.1126/science.1128691. PMID 16973872.

[43] PELLICER, JAUME; FAY, MICHAEL F.; LEITCH, ILIA J. (15 September 2010). "The largest eukaryotic genome of them all?". *Botanical Journal of the Linnean Society* **164** (1): 10–15. doi:10.1111/j.1095-8339.2010.01072.x.

[44] Lang D; Zimmer AD; Rensing SA; Reski R (October 2008). "Exploring plant biodiversity: the Physcomitrella genome and beyond". *Trends Plant Sci* **13** (10): 542–549. doi:10.1016/j.tplants.2008.07.002. PMID 18762443.

[45] "Saccharomyces Genome Database". Yeastgenome.org. Retrieved 27 January 2011.

[46] Galagan JE, Calvo SE, Cuomo C, Ma LJ, Wortman JR, Batzoglou S, Lee SI, Baştürkmen M, Spevak CC, Clutterbuck J, Kapitonov V, Jurka J, Scazzocchio C, Farman M, Butler J, Purcell S, Harris S, Braus GH, Draht O, Busch S, D'Enfert C, Bouchier C, Goldman GH, Bell-Pedersen D, Griffiths-Jones S, Doonan JH, Yu J, Vienken K, Pain A, Freitag M,

Selker EU, Archer DB, Peñalva MA, Oakley BR, Momany M, Tanaka T, Kumagai T, Asai K, Machida M, Nierman WC, Denning DW, Caddick M, Hynes M, Paoletti M, Fischer R, Miller B, Dyer P, Sachs MS, Osmani SA, Birren BW (2005). "Sequencing of Aspergillus nidulans and comparative analysis with A. fumigatus and A. oryzae". *Nature* **438** (7071): 1105–15. Bibcode:2005Natur.438.1105G. doi:10.1038/nature04341. PMID 16372000.

[47] Leroy, S., S. Bouamer, S. Morand, and M. Fargette (2007). Genome size of plant-parasitic nematodes. Nematology 9: 449-450.

[48] Gregory TR (2005). "Animal Genome Size Database". http://www.genomesize.com. External link in |publisher= (help)

[49] The *C. elegans* Sequencing Consortium (1998). "Genome sequence of the nematode *C. elegans*: a platform for investigating biology". *Science* **282** (5396): 2012–2018. doi:10.1126/science.282.5396.2012. PMID 9851916.

[50] Ellis LL; Huang W; Quinn AM; et al. (2014). "Intrapopulation Genome Size Variation in "Drosophila melanogaster" Reflects Life History Variation and Plasticity". *PLoS Genetics* **10** (7): e1004522. doi:10.1371/journal.pgen.1004522. Retrieved 17 March 2016.

[51] Honeybee Genome Sequencing Consortium; Weinstock; Robinson; Gibbs; Weinstock; Weinstock; Robinson; Worley; Evans; Maleszka; Robertson; Weaver; Beye; Bork; Elsik; Evans; Hartfelder; Hunt; Robertson; Robinson; Maleszka; Weinstock; Worley; Zdobnov; Hartfelder; Amdam; Bitondi; Collins; Cristino; Evans (October 2006). "Insights into social insects from the genome of the honeybee Apis mellifera". *Nature* **443** (7114): 931–49. Bibcode:2006Natur.443..931T. doi:10.1038/nature05260. PMC 2048586. PMID 17073008.

[52] The International Silkworm Genome (2008). "The genome of a lepidopteran model insect, the silkworm Bombyx mori". *Insect Biochemistry and Molecular Biology* **38** (12): 1036–1045. doi:10.1016/j.ibmb.2008.11.004. PMID 19121390.

[53] Wurm Y; Wang, J.; Riba-Grognuz, O.; Corona, M.; Nygaard, S.; Hunt, B. G.; Ingram, K. K.; Falquet, L.; Nipitwattanaphon, M.; Gotzek, D.; Dijkstra, M. B.; Oettler, J.; Comtesse, F.; Shih, C.-J.; Wu, W.-J.; Yang, C.-C.; Thomas, J.; Beaudoing, E.; Pradervand, S.; Flegel, V.; Cook, E. D.; Fabbretti, R.; Stockinger, H.; Long, L.; Farmerie, W. G.; Oakey, J.; Boomsma, J. J.; Pamilo, P.; Yi, S. V.; et al. (2011). "The genome of the fire ant *Solenopsis invicta*". *PNAS* **108** (14): 5679–5684. Bibcode:2011PNAS..108.5679W. doi:10.1073/pnas.1009690108. PMC 3078418. PMID 21282665. Retrieved 1 February 2011.

[54] Church, DM; Goodstadt, L; Hillier, LW; Zody, MC; Goldstein, S; She, X; Bult, CJ; Agarwala, R; Cherry, JL; DiCuccio, M; Hlavina, W; Kapustin, Y; Meric, P; Maglott, D; Birtle, Z; Marques, AC; Graves, T; Zhou, S; Teague, B; Potamousis, K; Churas, C; Place, M; Herschleb, J; Runnheim,

R; Forrest, D; Amos-Landgraf, J; Schwartz, DC; Cheng, Z; Lindblad-Toh, K; Eichler, EE; Ponting, CP; Mouse Genome Sequencing, Consortium (May 5, 2009). Roberts, Richard J. ed. "Lineage-specific biology revealed by a finished genome assembly of the mouse". *PLoS Biology* **7** (5): e1000112. doi:10.1371/journal.pbio.1000112. PMC 2680341. PMID 19468303.

[55] "Human Genome Project Information Site Has Been Updated". Ornl.gov. 2013-07-23. Retrieved 6 February 2014.

[56] Venter, J. C.; Adams, M.; Myers, E.; Li, P.; Mural, R.; Sutton, G.; Smith, H.; Yandell, M.; Evans, C.; Holt, R. A.; Gocayne, J. D.; Amanatides, P.; Ballew, R. M.; Huson, D. H.; Wortman, J. R.; Zhang, Q.; Kodira, C. D.; Zheng, X. H.; Chen, L.; Skupski, M.; Subramanian, G.; Thomas, P. D.; Zhang, J.; Gabor Miklos, G. L.; Nelson, C.; Broder, S.; Clark, A. G.; Nadeau, J.; McKusick, V. A.; Zinder, N. (2001). "The Sequence of the Human Genome". *Science* **291** (5507): 1304–1351. Bibcode:2001Sci...291.1304V. doi:10.1126/science.1058040. PMID 11181995.

[57] Crollius, HR; Jaillon, O; Dasilva, C; Ozouf-Costaz, C; Fizames, C; Fischer, C; Bouneau, L; Billault, A; Quetier, F; Saurin, W; Bernot, A; Weissenbach, J (2000). "Characterization and Repeat Analysis of the Compact Genome of the Freshwater Pufferfish Tetraodon nigroviridis". *Genome Research* **10** (7): 939–949. doi:10.1101/gr.10.7.939. PMC 310905. PMID 10899143.

[58] Olivier Jaillon; et al. (21 October 2004). "Genome duplication in the teleost fish Tetraodon nigroviridis reveals the early vertebrate proto-karyotype". *Nature* **431** (7011): 946–957. Bibcode:2004Natur.431..946J. doi:10.1038/nature03025. PMID 15496914.

[59] "Tetraodon Project Information". Retrieved 17 October 2012.

[60] Britten, RJ; Davidson, EH (June 1971). "Repetitive and non-repetitive DNA sequences and a speculation on the origins of evolutionary novelty". *The Quarterly Review of Biology* **46** (2): 111–38. doi:10.1086/406830. PMID 5160087.

[61] Naclerio, G; Cangiano, G; Coulson, A; Levitt, A; Ruvolo, V; La Volpe, A (1992-07-05). "Molecular and genomic organization of clusters of repetitive DNA sequences in Caenorhabditis elegans". *Journal of Molecular Biology* **226** (1): 159–68. doi:10.1016/0022-2836(92)90131-3. PMID 1619649.

[62] Stojanovic, edited by Nikola (2007). *Computational genomics : current methods*. Wymondham: Horizon Bioscience. ISBN 1-904933-30-0.

[63] Li, YC; Korol, AB; Fahima, T; Beiles, A; Nevo, E (December 2002). "Microsatellites: genomic distribution, putative functions and mutational mechanisms: a review". *Molecular Ecology* **11** (12): 2453–65. doi:10.1046/j.1365-294X.2002.01643.x. PMID 12453231.

[64] Schlötterer, C (December 2000). "Microsatellite analysis indicates genetic differentiation of the neo-sex chromosomes in Drosophila americana americana". *Heredity* **85** (Pt 6): 610–6. doi:10.1046/j.1365-2540.2000.00797.x. PMID 11240628.

[65] Wessler, S. R. (13 November 2006). "Eukaryotic Transposable Elements and Genome Evolution Special Feature: Transposable elements and the evolution of eukaryotic genomes". *Proceedings of the National Academy of Sciences* **103** (47): 17600–17601. Bibcode:2006PNAS..10317600W. doi:10.1073/pnas.0607612103.

[66] Kazazian, H. H. (12 March 2004). "Mobile Elements: Drivers of Genome Evolution". *Science* **303** (5664): 1626–1632. Bibcode:2004Sci...303.1626K. doi:10.1126/science.1089670. PMID 15016989.

[67] Deininger PL; Moran JV; Batzer MA; Kazazian, HH Jr. (December 2003). "Mobile elements and mammalian genome evolution". *Current opinion in genetics & development* **13** (6): 651–8. doi:10.1016/j.gde.2003.10.013. PMID 14638329.

[68] Kidwell MG; Lisch DR (March 2000). "Transposable elements and host genome evolution". *Trends in Ecology & Evolution* **15** (3): 95–99. doi:10.1016/S0169-5347(99)01817-0. PMID 10675923.

[69] Richard G.-F.; Kerrest A; Dujon B (3 December 2008). "Comparative Genomics and Molecular Dynamics of DNA Repeats in Eukaryotes". *Microbiology and Molecular Biology Reviews* **72** (4): 686–727. doi:10.1128/MMBR.00011-08. PMC 2593564. PMID 19052325.

[70] Cordaux R; Batzer MA (1 October 2009). "The impact of retrotransposons on human genome evolution". *Nature Reviews Genetics* **10** (10): 691–703. doi:10.1038/nrg2640. PMC 2884099. PMID 19763152.

[71] Han, Jeffrey S.; Boeke, Jef D. (1 August 2005). "LINE-1 retrotransposons: Modulators of quantity and quality of mammalian gene expression?". *BioEssays* **27** (8): 775–784. doi:10.1002/bies.20257. PMID 16015595.

9.8 Further reading

- Benfey, P.; Protopapas, A.D. (2004). *Essentials of Genomics*. Prentice Hall.

- Brown, Terence A. (2002). *Genomes 2*. Oxford: Bios Scientific Publishers. ISBN 978-1-85996-029-5.

- Gibson, Greg; Muse, Spencer V. (2004). *A Primer of Genome Science* (Second ed.). Sunderland, Mass: Sinauer Assoc. ISBN 0-87893-234-8.

- Gregory (2005). T. Ryan. ed. *The Evolution of the Genome*. Elsevier. ISBN 0-12-301463-8.

- Reece, Richard J. (2004). *Analysis of Genes and Genomes*. Chichester: John Wiley & Sons. ISBN 0-470-84379-9.

- Saccone, Cecilia; Pesole, Graziano (2003). *Handbook of Comparative Genomics*. Chichester: John Wiley & Sons. ISBN 0-471-39128-X.

- Werner, E. (2003). "In silico multicellular systems biology and minimal genomes". *Drug Discov Today* **8** (24): 1121–1127. doi:10.1016/S1359-6446(03)02918-0. PMID 14678738.

9.9   External links

- UCSC Genome Browser – view the genome and annotations for more than 80 organisms.

- genomecenter.howard.edu

- Build a DNA Molecule

- Some comparative genome sizes

- DNA Interactive: The History of DNA Science

- DNA From The Beginning

- All About The Human Genome Project—from Genome.gov

- Animal genome size database

- Plant genome size database

- GOLD:Genomes OnLine Database

- The Genome News Network

- NCBI Entrez Genome Project database

- NCBI Genome Primer

- GeneCards—an integrated database of human genes

- Visualization of nucleotide sequence - prototypification of complete genome of virus, sequence of 5418 nucleotides

- BBC News – Final genome 'chapter' published

- IMG (The Integrated Microbial Genomes system)—for genome analysis by the DOE-JGI

- GeKnome Technologies Next-Gen Sequencing Data Analysis—next-generation sequencing data analysis for Illumina and 454 Service from GeKnome Technologies.

Chapter 10

Gene

This article is about the heritable unit for transmission of biological traits. For other uses, see Gene (disambiguation).

A **gene** is a locus (or region) of DNA that encodes a functional RNA or protein product, and is the molecular unit of heredity.[1][2]:Glossary The transmission of genes to an organism's offspring is the basis of the inheritance of phenotypic traits. Most biological traits are under the influence of polygenes (many different genes) as well as the gene–environment interactions. Some genetic traits are instantly visible, such as eye colour or number of limbs, and some are not, such as blood type, risk for specific diseases, or the thousands of basic biochemical processes that comprise life.

Genes can acquire mutations in their sequence, leading to different variants, known as alleles, in the population. These alleles encode slightly different versions of a protein, which cause different phenotype traits. Colloquial usage of the term "having a gene" (e.g., "good genes," "hair colour gene") typically refers to having a different allele of the gene. Genes evolve due to natural selection or survival of the fittest of the alleles.

The concept of a gene continues to be refined as new phenomena are discovered.[3] For example, regulatory regions of a gene can be far removed from its coding regions, and coding regions can be split into several exons. Some viruses store their genome in RNA instead of DNA and some gene products are functional non-coding RNAs. Therefore, a broad, modern working definition of a gene is any discrete locus of heritable, genomic sequence which affect an organism's traits by being expressed as a functional product or by regulation of gene expression.[4][5]

10.1 History

Main article: History of genetics

Gregor Mendel

10.1.1 Discovery of discrete inherited units

The existence of discrete inheritable units was first suggested by Gregor Mendel (1822–1884).[6] From 1857 to 1864, he studied inheritance patterns in 8000 common edible pea plants, tracking distinct traits from parent to offspring. He described these mathematically as 2^n combinations where n is the number of differing characteristics in the original peas. Although he did not use the term *gene*, he explained his results in terms of discrete inherited units that give rise to observable physical characteristics. This description prefigured the distinction between genotype (the genetic material of an organism) and phenotype (the visible traits of that organism). Mendel was also the first to demonstrate independent assortment, the distinction between dominant and recessive traits, the distinction between a heterozygote and homozygote, and the phenomenon of discontinuous inheritance.

Prior to Mendel's work, the dominant theory of heredity was one of blending inheritance, which suggested that each parent contributed fluids to the fertilisation process and that the traits of the parents blended and mixed to produce the offspring. Charles Darwin developed a theory of inheritance he termed pangenesis,[7] which used the term *gemmule* to describe hypothetical particles that would mix during reproduction. Although Mendel's work was largely unrecognized after its first publication in 1866, it was 'rediscovered' in 1900 by three European scientists, Hugo de Vries, Carl Correns, and Erich von Tschermak, who claimed to have reached similar conclusions in their own research.[8]

The word *gene* is derived (via *pangene*) from the Ancient Greek word γένος (*génos*) meaning "race, offspring".[9] *Gene* was coined in 1909 by Danish botanist Wilhelm Johannsen to describe the fundamental physical and functional unit of heredity,[10] while the related word *genetics* was first used by William Bateson in 1905.[11]

10.1.2   Discovery of DNA

Advances in understanding genes and inheritance continued throughout the 20th century. Deoxyribonucleic acid (DNA) was shown to be the molecular repository of genetic information by experiments in the 1940s to 1950s.[12][13] The structure of DNA was studied by Rosalind Franklin using X-ray crystallography, which led James D. Watson and Francis Crick to publish a model of the double-stranded DNA molecule whose paired nucleotide bases indicated a compelling hypothesis for the mechanism of genetic replication.[14][15] Collectively, this body of research established the central dogma of molecular biology, which states that proteins are translated from RNA, which is transcribed from DNA. This dogma has since been shown to have exceptions, such as reverse transcription in retroviruses. The modern study of genetics at the level of DNA is known as molecular genetics.

In 1972, Walter Fiers and his team at the University of Ghent were the first to determine the sequence of a gene: the gene for Bacteriophage MS2 coat protein.[16] The subsequent development of chain-termination DNA sequencing in 1977 by Frederick Sanger improved the efficiency of sequencing and turned it into a routine laboratory tool.[17] An automated version of the Sanger method was used in early phases of the Human Genome Project.[18]

10.1.3   Modern evolutionary synthesis

Main article: Modern evolutionary synthesis

The theories developed in the 1930s and 1940s to integrate molecular genetics with Darwinian evolution are called the modern evolutionary synthesis, a term introduced by Julian Huxley.[19] Evolutionary biologists subsequently refined this concept, such as George C. Williams' gene-centric view of evolution. He proposed an evolutionary concept of the gene as a unit of natural selection with the definition: "that which segregates and recombines with appreciable frequency."[20]:24 In this view, the molecular gene *transcribes* as a unit, and the evolutionary gene *inherits* as a unit. Related ideas emphasizing the centrality of genes in evolution were popularized by Richard Dawkins.[21][22]

10.2   Molecular basis

Main article: DNA

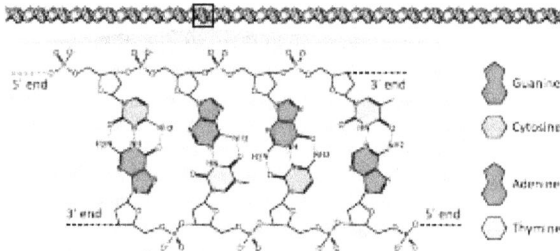

The chemical structure of a four base pair fragment of a DNA double helix. The sugar-phosphate backbone chains run in opposite directions with the bases pointing inwards, base-pairing A to T and C to G with hydrogen bonds.

10.2.1   DNA

The vast majority of living organisms encode their genes in long strands of DNA (deoxyribonucleic acid). DNA consists of a chain made from four types of nucleotide subunits, each composed of: a five-carbon sugar (2'-deoxyribose), a phosphate group, and one of the four bases adenine, cytosine, guanine, and thymine.[2]:2.1

Two chains of DNA twist around each other to form a DNA double helix with the phosphate-sugar backbone spiralling around the outside, and the bases pointing inwards with adenine base pairing to thymine and guanine to cytosine. The specificity of base pairing occurs because adenine and thymine align form two hydrogen bonds, whereas cytosine and guanine form three hydrogen bonds. The two strands in a double helix must therefore be complementary, with their sequence of bases matching such that the adenines of one strand are paired with the thymines of the other strand, and so on.[2]:4.1

Due to the chemical composition of the pentose residues of the bases, DNA strands have directionality. One end of a DNA polymer contains an exposed hydroxyl group on the deoxyribose; this is known as the 3' end of the molecule. The other end contains an exposed phosphate group; this is the 5' end. The two strands of a double-helix run in opposite directions. Nucleic acid synthesis, including DNA replication and transcription occurs in the 5'→3' direction, because new nucleotides are added via a dehydration reaction that uses the exposed 3' hydroxyl as a nucleophile.[23]:27.2

The expression of genes encoded in DNA begins by transcribing the gene into RNA, a second type of nucleic acid that is very similar to DNA, but whose monomers contain the sugar ribose rather than deoxyribose. RNA also contains the base uracil in place of thymine. RNA molecules are less stable than DNA and are typically single-stranded. Genes that encode proteins are composed of a series of three-nucleotide sequences called codons, which serve as the "words" in the genetic "language". The genetic code specifies the correspondence during protein translation between codons and amino acids. The genetic code is nearly the same for all known organisms.[2]:4.1

10.2.2 Chromosomes

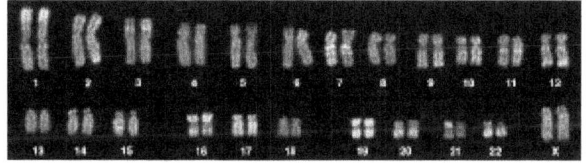

Fluorescent microscopy image of a human female karyotype, showing 23 pairs of chromosomes. The DNA is stained red, with regions rich in housekeeping genes further stained in green. The largest chromosomes are around 10 times the size of the smallest.[34]

The total complement of genes in an organism or cell is known as its genome, which may be stored on one or more chromosomes. A chromosome consists of a single, very long DNA helix on which thousands of genes are encoded.[2]:4.2 The region of the chromosome at which a particular gene is located is called its locus. Each locus contains one allele of a gene; however, members of a population may have different alleles at the locus, each with a slightly different gene sequence.

The majority of eukaryotic genes are stored on a set of large, linear chromosomes. The chromosomes are packed within the nucleus in complex with storage proteins called histones to form a unit called a nucleosome. DNA packaged and condensed in this way is called chromatin.[2]:4.2 The manner in which DNA is stored on the histones, as well as chemical modifications of the histone itself, regu-

late whether a particular region of DNA is accessible for gene expression. In addition to genes, eukaryotic chromosomes contain sequences involved in ensuring that the DNA is copied without degradation of end regions and sorted into daughter cells during cell division: replication origins, telomeres and the centromere.[2]:4.2 Replication origins are the sequence regions where DNA replication is initiated to make two copies of the chromosome. Telomeres are long stretches of repetitive sequence that cap the ends of the linear chromosomes and prevent degradation of coding and regulatory regions during DNA replication. The length of the telomeres decreases each time the genome is replicated and has been implicated in the aging process.[25] The centromere is required for binding spindle fibres to separate sister chromatids into daughter cells during cell division.[2]:18.2

Prokaryotes (bacteria and archaea) typically store their genomes on a single large, circular chromosome. Similarly, some eukaryotic organelles contain a remnant circular chromosome with a small number of genes.[2]:14.4 Prokaryotes sometimes supplement their chromosome with additional small circles of DNA called plasmids, which usually encode only a few genes and are transferable between individuals. For example, the genes for antibiotic resistance are usually encoded on bacterial plasmids and can be passed between individual cells, even those of different species, via horizontal gene transfer.[26]

Whereas the chromosomes of prokaryotes are relatively gene-dense, those of eukaryotes often contain regions of DNA that serve no obvious function. Simple single-celled eukaryotes have relatively small amounts of such DNA, whereas the genomes of complex multicellular organisms, including humans, contain an absolute majority of DNA without an identified function.[27] This DNA has often been referred to as "junk DNA". However, more recent analyses suggest that, although protein-coding DNA makes up barely 2% of the human genome, about 80% of the bases in the genome may be expressed, so the term "junk DNA" may be a misnomer.[5]

10.3 Structure and function

The structure of a gene consists of many elements of which the actual protein coding sequence is often only a small part. These include DNA regions that are not transcribed as well as untranslated regions of the RNA.

Firstly, flanking the open reading frame, all genes contain a regulatory sequence that is required for their expression. In order to be expressed, genes require a promoter sequence. The promoter is recognized and bound by transcription factors and RNA polymerase to initiate transcription.[2]:7.1 A gene can have more than one promoter, resulting in messenger RNAs (mRNA) that differ in how far they extend in the 5' end.[28] Promoter regions have a consensus sequence, however highly transcribed genes have "strong" promoter sequences that bind the transcription machinery well, whereas others have "weak" promoters that bind poorly and initiate transcription less frequently.[2]:7.2 Eukaryotic promoter regions are much more complex and difficult to identify than prokaryotic promoters.[2]:7.3

Additionally, genes can have regulatory regions many kilobases upstream or downstream of the open reading frame. These act by binding to transcription factors which then cause the DNA to loop so that the regulatory sequence (and bound transcription factor) become close to the RNA polymerase binding site.[29] For example, enhancers increase transcription by binding an activator protein which then helps to recruit the RNA polymerase to the promoter; conversely silencers bind repressor proteins and make the DNA less available for RNA polymerase.[30]

The transcribed pre-mRNA contains untranslated regions at both ends which contain a ribosome binding site, terminator and start and stop codons.[31] In addition, most eukaryotic open reading frames contain untranslated introns which are removed before the exons are translated. The sequences at the ends of the introns, dictate the splice sites to generate the final mature mRNA which encodes the protein or RNA product.[32]

Many prokaryotic genes are organized into operons, with multiple protein-coding sequences that are transcribed as a unit.[33][34] The products of operon genes typically have related functions and are involved in the same regulatory network.[2]:7.3

10.3.1 Functional definitions

Defining exactly what section of a DNA sequence comprises a gene is difficult.[3] Regulatory regions of a gene such as enhancers do not necessarily have to be close to the coding sequence on the linear molecule because the intervening DNA can be looped out to bring the gene and its regulatory region into proximity. Similarly, a gene's introns can be much larger than its exons. Regulatory regions can even be on entirely different chromosomes and operate in trans to allow regulatory regions on one chromosome to come in contact with target genes on another chromosome.[35][36]

Early work in molecular genetics suggested the model that one gene makes one protein. This model has been refined since the discovery of genes that can encode multiple proteins by alternative splicing and coding sequences split in short section across the genome whose mRNAs are concatenated by trans-splicing.[5][37][38]

A broad operational definition is sometimes used to encompass the complexity of these diverse phenomena, where a gene is defined as a union of genomic sequences encoding a coherent set of potentially overlapping functional products.[11] This definition categorizes genes by their functional products (proteins or RNA) rather than their specific DNA loci, with regulatory elements classified as gene-associated regions.[11]

10.4 Gene expression

Main article: Gene expression

In all organisms, two steps are required to read the information encoded in a gene's DNA and produce the protein it specifies. First, the gene's DNA is transcribed to messenger RNA (mRNA).[2]:6.1 Second, that mRNA is translated to protein.[2]:6.2 RNA-coding genes must still go through the first step, but are not translated into protein.[39] The process of producing a biologically functional molecule of either RNA or protein is called gene expression, and the resulting molecule is called a gene product.

10.4.1 Genetic code

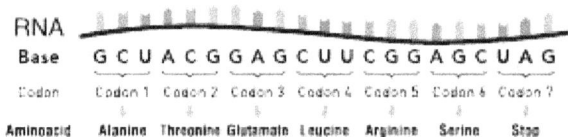

Schematic of a single-stranded RNA molecule illustrating a series of three-base codons. Each three-nucleotide codon corresponds to an amino acid when translated to protein

The nucleotide sequence of a gene's DNA specifies the amino acid sequence of a protein through the genetic code. Sets of three nucleotides, known as codons, each correspond to a specific amino acid.[2]:6 Additionally, a "start codon", and three "stop codons" indicate the beginning and end of the protein coding region. There are 64 possible codons (four possible nucleotides at each of three positions, hence 4^3 possible codons) and only 20 standard amino acids; hence the code is redundant and multiple codons can specify the same amino acid. The correspondence between codons and amino acids is nearly universal among all known living organisms.[40]

10.4.2 Transcription

Transcription produces a single-stranded RNA molecule

known as messenger RNA. whose nucleotide sequence is complementary to the DNA from which it was transcribed.[2]:6.1 The mRNA acts as an intermediate between the DNA gene and its final protein product. The gene's DNA is used as a template to generate a complementary mRNA. The mRNA matches the sequence of the gene's DNA coding strand because it is synthesised as the complement of the template strand. Transcription is performed by an enzyme called an RNA polymerase, which reads the template strand in the 3' to 5' direction and synthesizes the RNA from 5' to 3'. To initiate transcription, the polymerase first recognizes and binds a promoter region of the gene. Thus, a major mechanism of gene regulation is the blocking or sequestering the promoter region, either by tight binding by repressor molecules that physically block the polymerase, or by organizing the DNA so that the promoter region is not accessible.[2]:7

In prokaryotes, transcription occurs in the cytoplasm; for very long transcripts, translation may begin at the 5' end of the RNA while the 3' end is still being transcribed. In eukaryotes, transcription occurs in the nucleus, where the cell's DNA is stored. The RNA molecule produced by the polymerase is known as the primary transcript and undergoes post-transcriptional modifications before being exported to the cytoplasm for translation. One of the modifications performed is the splicing of introns which are sequences in the transcribed region that do not encode protein. Alternative splicing mechanisms can result in mature transcripts from the same gene having different sequences and thus coding for different proteins. This is a major form of regulation in eukaryotic cells and also occurs in some prokaryotes.[2]:7.5[41]

10.4.3 Translation

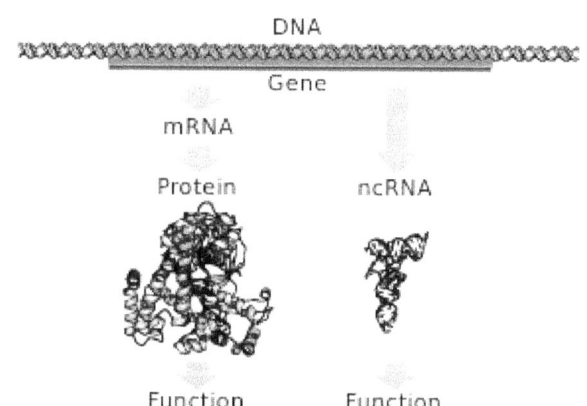

Protein coding genes are transcribed to an mRNA intermediate, then translated to a functional protein. RNA-coding genes are transcribed to a functional non-coding RNA. (PDB: 3BSE, 1OBB, 3TRA)

Translation is the process by which a mature mRNA molecule is used as a template for synthesizing a new protein.[2]:6.2 Translation is carried out by ribosomes, large complexes of RNA and protein responsible for carrying out the chemical reactions to add new amino acids to a growing polypeptide chain by the formation of peptide bonds. The genetic code is read three nucleotides at a time, in units called codons, via interactions with specialized RNA molecules called transfer RNA (tRNA). Each tRNA has three unpaired bases known as the anticodon that are complementary to the codon it reads on the mRNA. The tRNA is also covalently attached to the amino acid specified by the complementary codon. When the tRNA binds to its complementary codon in an mRNA strand, the ribosome attaches its amino acid cargo to the new polypeptide chain, which is synthesized from amino terminus to carboxyl terminus. During and after synthesis, most new proteins must folds to their active three-dimensional structure before they can carry out their cellular functions.[2]:3

10.4.4 Regulation

Genes are regulated so that they are expressed only when the product is needed, since expression draws on limited resources.[2]:7 A cell regulates its gene expression depending on its external environment (e.g. available nutrients, temperature and other stresses), its internal environment (e.g. cell division cycle, metabolism, infection status), and its specific role if in a multicellular organism. Gene expression can be regulated at any step: from transcriptional initiation, to RNA processing, to post-translational modification of the protein. The regulation of lactose metabolism genes in *E. coli* (*lac* operon) was the first such mechanism to be described in 1961.[42]

10.4.5 RNA genes

A typical protein-coding gene is first copied into RNA as an intermediate in the manufacture of the final protein product.[2]:6.1 In other cases, the RNA molecules are the actual functional products, as in the synthesis of ribosomal RNA and transfer RNA. Some RNAs known as ribozymes are capable of enzymatic function, and microRNA has a regulatory role. The DNA sequences from which such RNAs are transcribed are known as non-coding RNA genes.[39]

Some viruses store their entire genomes in the form of RNA, and contain no DNA at all.[43][44] Because they use RNA to store genes, their cellular hosts may synthesize their proteins as soon as they are infected and without the delay in waiting for transcription.[45] On the other hand, RNA retroviruses, such as HIV, require the reverse transcription of their genome from RNA into DNA before their proteins can be synthesized. RNA-mediated epigenetic inheritance has also been observed in plants and very rarely in animals.[46]

10.5 Inheritance

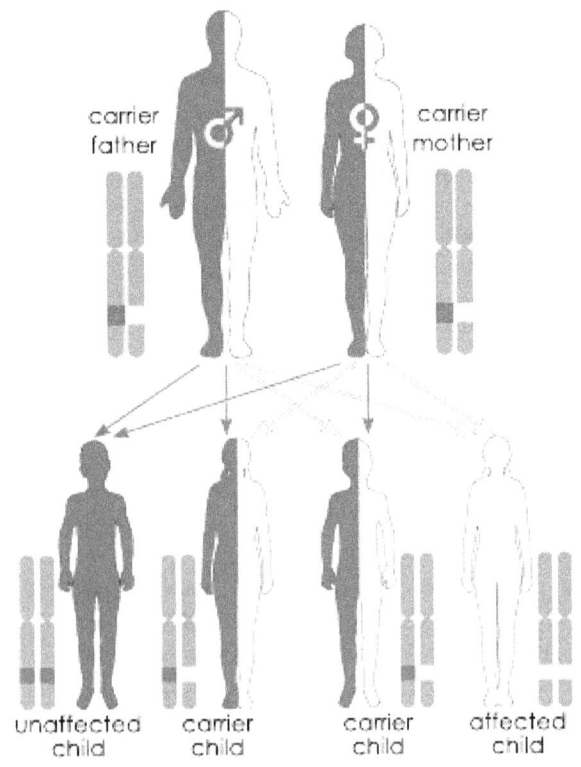

Inheritance of a gene that has two different alleles (blue and white). The gene is located on an autosomal chromosome. The blue allele is recessive to the white allele. The probability of each outcome in the children's generation is one quarter, or 25 percent.

Main articles: Mendelian inheritance and Heredity

Organisms inherit their genes from their parents. Asexual organisms simply inherit a complete copy of their parent's genome. Sexual organisms have two copies of each chromosome because they inherit one complete set from each parent.[2]:1

10.5.1 Mendelian inheritance

According to Mendelian inheritance, variations in an organism's phenotype (observable physical and behavioral characteristics) are due in part to variations in its genotype (particular set of genes). Each gene specifies a particular trait

with different sequence of a gene (alleles) giving rise to different phenotypes. Most eukaryotic organisms (such as the pea plants Mendel worked on) have two alleles for each trait, one inherited from each parent.[2]:20

Alleles at a locus may be dominant or recessive; dominant alleles give rise to their corresponding phenotypes when paired with any other allele for the same trait, whereas recessive alleles give rise to their corresponding phenotype only when paired with another copy of the same allele. For example, if the allele specifying tall stems in pea plants is dominant over the allele specifying short stems, then pea plants that inherit one tall allele from one parent and one short allele from the other parent will also have tall stems. Mendel's work demonstrated that alleles assort independently in the production of gametes, or germ cells, ensuring variation in the next generation. Although Mendelian inheritance remains a good model for many traits determined by single genes (including a number of well-known genetic disorders) it does not include the physical processes of DNA replication and cell division.[47][48]

10.5.2 DNA replication and cell division

The growth, development, and reproduction of organisms relies on cell division, or the process by which a single cell divides into two usually identical daughter cells. This requires first making a duplicate copy of every gene in the genome in a process called DNA replication.[2]:5.2 The copies are made by specialized enzymes known as DNA polymerases, which "read" one strand of the double-helical DNA, known as the template strand, and synthesize a new complementary strand. Because the DNA double helix is held together by base pairing, the sequence of one strand completely specifies the sequence of its complement; hence only one strand needs to be read by the enzyme to produce a faithful copy. The process of DNA replication is semiconservative; that is, the copy of the genome inherited by each daughter cell contains one original and one newly synthesized strand of DNA.[2]:5.2

After DNA replication is complete, the cell must physically separate the two copies of the genome and divide into two distinct membrane-bound cells.[2]:18.2 In prokaryotes (bacteria and archaea) this usually occurs via a relatively simple process called binary fission, in which each circular genome attaches to the cell membrane and is separated into the daughter cells as the membrane invaginates to split the cytoplasm into two membrane-bound portions. Binary fission is extremely fast compared to the rates of cell division in eukaryotes. Eukaryotic cell division is a more complex process known as the cell cycle; DNA replication occurs during a phase of this cycle known as S phase, whereas the process of segregating chromosomes and splitting the cytoplasm occurs during M phase.[2]:18.1

10.5.3 Molecular inheritance

The duplication and transmission of genetic material from one generation of cells to the next is the basis for molecular inheritance, and the link between the classical and molecular pictures of genes. Organisms inherit the characteristics of their parents because the cells of the offspring contain copies of the genes in their parents' cells. In asexually reproducing organisms, the offspring will be a genetic copy or clone of the parent organism. In sexually reproducing organisms, a specialized form of cell division called meiosis produces cells called gametes or germ cells that are haploid, or contain only one copy of each gene.[2]:20.2 The gametes produced by females are called eggs or ova, and those produced by males are called sperm. Two gametes fuse to form a diploid fertilized egg, a single cell that has two sets of genes, with one copy of each gene from the mother and one from the father.[2]:20

During the process of meiotic cell division, an event called genetic recombination or *crossing-over* can sometimes occur, in which a length of DNA on one chromatid is swapped with a length of DNA on the corresponding sister chromatid. This has no effect if the alleles on the chromatids are the same, but results in reassortment of otherwise linked alleles if they are different.[2]:5.5 The Mendelian principle of independent assortment asserts that each of a parent's two genes for each trait will sort independently into gametes; which allele an organism inherits for one trait is unrelated to which allele it inherits for another trait. This is in fact only true for genes that do not reside on the same chromosome, or are located very far from one another on the same chromosome. The closer two genes lie on the same chromosome, the more closely they will be associated in gametes and the more often they will appear together; genes that are very close are essentially never separated because it is extremely unlikely that a crossover point will occur between them. This is known as genetic linkage.[49]

10.6 Molecular evolution

Main article: Molecular evolution

10.6.1 Mutation

DNA replication is for the most part extremely accurate, however errors (mutations) do occur.[2]:7.6 The error rate in eukaryotic cells can be as low as 10^{-8} per nucleotide per replication,[50][51] whereas for some RNA viruses it can be

as high as 10^{-3}.[52] This means that each generation, each human genome accumulates 1–2 new mutations.[52] Small mutations can be caused by DNA replication and the aftermath of DNA damage and include point mutations in which a single base is altered and frameshift mutations in which a single base is inserted or deleted. Either of these mutations can change the gene by missense (change a codon to encode a different amino acid) or nonsense (a premature stop codon).[53] Larger mutations can be caused by errors in recombination to cause chromosomal abnormalities including the duplication, deletion, rearrangement or inversion of large sections of a chromosome. Additionally, the DNA repair mechanisms that normally revert mutations can introduce errors when repairing the physical damage to the molecule is more important than restoring an exact copy, for example when repairing double-strand breaks.[2]:5.4

When multiple different alleles for a gene are present in a species's population it is called polymorphic. Most different alleles are functionally equivalent, however some alleles can give rise to different phenotypic traits. A gene's most common allele is called the wild type, and rare alleles are called mutants. The genetic variation in relative frequencies of different alleles in a population is due to both natural selection and genetic drift.[54] The wild-type allele is not necessarily the ancestor of less common alleles, nor is it necessarily fitter.

Most mutations within genes are neutral, having no effect on the organism's phenotype (silent mutations). Some mutations do not change the amino acid sequence because multiple codons encode the same amino acid (synonymous mutations). Other mutations can be neutral if they lead to amino acid sequence changes, but the protein still functions similarly with the new amino acid (e.g. conservative mutations). Many mutations, however, are deleterious or even lethal, and are removed from populations by natural selection. Genetic disorders are the result of deleterious mutations and can be due to spontaneous mutation in the affected individual, or can be inherited. Finally, a small fraction of mutations are beneficial, improving the organism's fitness and are extremely important for evolution, since their directional selection leads to adaptive evolution.[2]:7.6

10.6.2 Sequence homology

Genes with a most recent common ancestor, and thus a shared evolutionary ancestry, are known as homologs.[55] These genes appear either from gene duplication within an organism's genome, where they are known as paralogous genes, or are the result of divergence of the genes after a speciation event, where they are known as orthologous genes,[2]:7.6 and often perform the same or similar functions in related organisms. It is often assumed that the functions

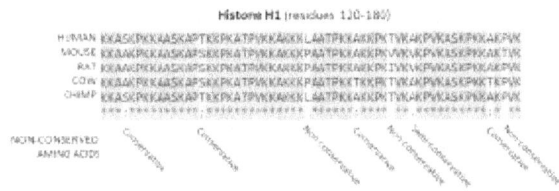

A sequence alignment, produced by ClustalO, of mammalian histone proteins

of orthologous genes are more similar than those of paralogous genes, although the difference is minimal.[56][57]

The relationship between genes can be measured by comparing the sequence alignment of their DNA.[2]:7.6 The degree of sequence similarity between homologous genes is called conserved sequence. Most changes to a gene's sequence do not affect its function and so genes accumulate mutations over time by neutral molecular evolution. Additionally, any selection on a gene will cause its sequence to diverge at a different rate. Genes under stabilizing selection are constrained and so change more slowly whereas genes under directional selection change sequence more rapidly.[58] The sequence differences between genes can be used for phylogenetic analyses to study how those genes have evolved and how the organisms they come from are related.[59][60]

10.6.3   Origins of new genes

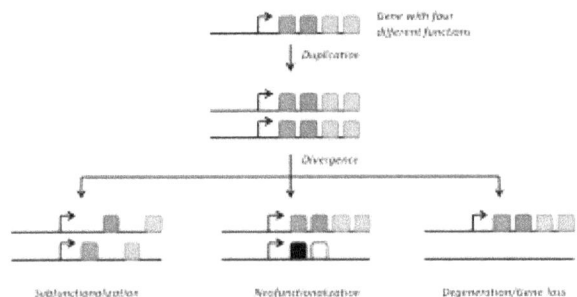

Evolutionary fate of duplicate genes

The most common source of new genes in eukaryotic lineages is gene duplication, which creates copy number variation of an existing gene in the genome.[61][62] The resulting genes (paralogs) may then diverge in sequence and in function. Sets of genes formed in this way comprise a gene family. Gene duplications and losses within a family are common and represent a major source of evolutionary biodiversity.[63] Sometimes, gene duplication may result in a nonfunctional copy of a gene, or a functional copy may be subject to mutations that result in loss of function; such nonfunctional genes are called pseudogenes.[2]:7.6

De novo or "orphan" genes, whose sequence shows no similarity to existing genes, are extremely rare. Estimates of the number of de novo genes in the human genome range from 18[64] to 60.[65] Such genes are typically shorter and simpler in structure than most eukaryotic genes, with few if any introns.[61] Two primary sources of orphan protein-coding genes are gene duplication followed by extremely rapid sequence change, such that the original relationship is undetectable by sequence comparisons, and formation through mutation of "cryptic" transcription start sites that introduce a new open reading frame in a region of the genome that did not previously code for a protein.[66][67]

Horizontal gene transfer refers to the transfer of genetic material through a mechanism other than reproduction. This mechanism is a common source of new genes in prokaryotes, sometimes thought to contribute more to genetic variation than gene duplication.[68] It is a common means of spreading antibiotic resistance, virulence, and adaptive metabolic functions.[26][69] Although horizontal gene transfer is rare in eukaryotes, likely examples have been identified of protist and alga genomes containing genes of bacterial origin.[70][71]

10.7   Genome

The genome is the total genetic material of an organism and includes both the genes and non-coding sequences.[72]

10.7.1   Number of genes

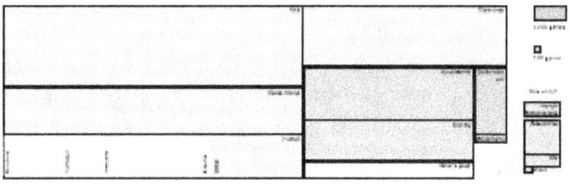

Representative genome sizes for plants (green), vertebrates (blue), invertebrates (red), fungus (yellow), bacteria (purple), and viruses (grey). An inset on the right shows the smaller genomes expanded 100-fold.[73][74][75][76][77][78][79][80]

The genome size, and the number of genes it encodes varies widely between organisms. The smallest genomes occur in viruses (which can have as few as 2 protein-coding genes),[81] and viroids (which act as a single non-coding RNA gene).[82] Conversely, plants can have extremely large genomes,[83] with rice containing >46,000 protein-coding genes.[84] The total number of protein-coding genes (the Earth's proteome) is estimated to be 5 million sequences.[85]

Although the number of base-pairs of DNA in the human

genome has been known since the 1960s, the estimated number of genes has changed over time as definitions of genes, and methods of detecting them have been refined. Initial theoretical predictions of the number of human genes were as high as 2,000,000.[86] Early experimental measures indicated there to be 50,000–100,000 *transcribed* genes (expressed sequence tags).[87] Subsequently, the sequencing in the Human Genome Project indicated that many of these transcripts were alternative variants of the same genes, and the total number of protein-coding genes was revised down to ~20,000[80] with 13 genes encoded on the mitochondrial genome.[78] Of the human genome, only 1–2% consists of protein-coding genes,[88] with the remainder being 'noncoding' DNA such as introns, retrotransposons, and noncoding RNAs.[88][89]

10.7.2 Essential genes

Main article: Essential gene

Essential genes are the set of genes thought to be critical for

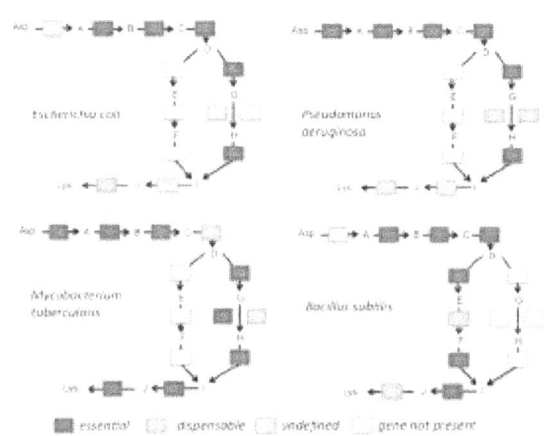

A schematic view of essential genes in lysine biosynthesis of different bacteria. The same protein may be essential in one species but not another.

an organism's survival.[90] This definition assumes the abundant availability of all relevant nutrients and the absence of environmental stress. Only a small portion of an organism's genes are essential. In bacteria, an estimated 250–400 genes are essential for *Escherichia coli* and *Bacillus subtilis*, which is less than 10% of their genes.[91][92][93] Half of these genes are orthologs in both organisms and are largely involved in protein synthesis.[93] In the budding yeast *Saccharomyces cerevisiae* the number of essential genes is slightly higher, at 1000 genes (~20% of their genes).[94] Although the number is more difficult to measure in higher eukaryotes, mice and humans are estimated to have around 2000 essential genes (~10% of their genes).[95] The synthetic organism, Syn3, has a minimal genome of 473 es-

sential genes and quasi-essential genes (necessary for fast growth), although 149 have unknown function.[96]

Essential genes include Housekeeping genes (critical for basic cell functions)[97] as well as genes that are expressed at different times in the organisms development or life cycle.[98] Housekeeping genes are used as experimental controls when analysing gene expression, since they are constitutively expressed at a relatively constant level.

10.7.3 Genetic and genomic nomenclature

Gene nomenclature has been established by the HUGO Gene Nomenclature Committee (HGNC) for each known human gene in the form of an approved gene name and symbol (short-form abbreviation), which can be accessed through a database maintained by HGNC. Symbols are chosen to be unique, and each gene has only one symbol (although approved symbols sometimes change). Symbols are preferably kept consistent with other members of a gene family and with homologs in other species, particularly the mouse due to its role as a common model organism.[99]

10.8 Genetic engineering

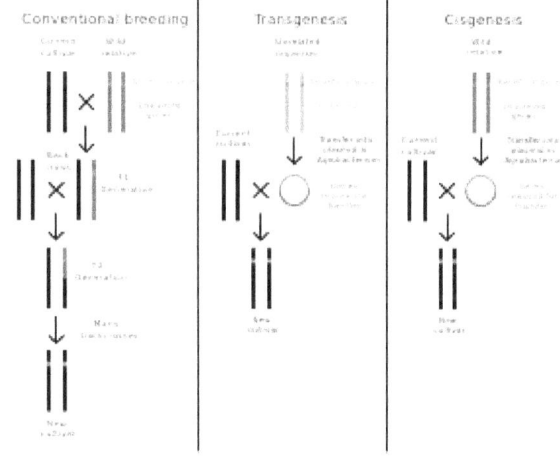

Comparison of conventional plant breeding with transgenic and cisgenic genetic modification.

Main article: Genetic engineering

Genetic engineering is the modification of an organism's genome through biotechnology. Since the 1970s, a variety of techniques have been developed to specifically add, remove and edit genes in an organism.[100] Recently developed genome engineering techniques use engineered nuclease enzymes to create targeted DNA repair in a chromosome to either disrupt or edit a gene when the break is repaired.[101][102][103][104] The related term synthetic biology is sometimes used to refer to extensive genetic engineering of an organism.[105]

Genetic engineering is now a routine research tool with model organisms. For example, genes are easily added to bacteria[106] and lineages of knockout mice with a specific gene's function disrupted are used to investigate that gene's function.[107][108] Many organisms have been genetically modified for applications in agriculture, industrial biotechnology, and medicine.

For multicellular organisms, typically the embryo is engineered which grows into the adult genetically modified organism.[109] However, the genomes of cells in an adult organism can be edited using gene therapy techniques to treat genetic diseases.

10.9   See also

- Copy number variation

- Epigenetics

- Full genome sequencing

- Gene-centric view of evolution

- Gene dosage

- Gene expression

- Gene family

- Gene nomenclature

- Gene patent

- Gene pool

- Gene redundancy

- Genetic algorithm

- List of gene prediction software

- List of notable genes

- Predictive medicine

- Pseudogene

- Quantitative trait locus

10.10   References

10.10.1   Main textbook

Alberts B, Johnson A, Lewis J, Raff M, Roberts K, Walter P (2002). *Molecular Biology of the Cell* (Fourth ed.). New York: Garland Science. ISBN 978-0-8153-3218-3. – A molecular biology textbook available free online through NCBI Bookshelf.

Referenced chapters of *Molecular Biology of the Cell*

Glossary

Ch 1: Cells and genomes

   1.1: The Universal Features of Cells on Earth

Ch 2: Cell Chemistry and Biosynthesis

   2.1: The Chemical Components of a Cell

Ch 3: Proteins

Ch 4: DNA and Chromosomes

   4.1: The Structure and Function of DNA
   4.2: Chromosomal DNA and Its Packaging in the Chromatin Fiber

Ch 5: DNA Replication, Repair, and Recombination

   5.2: DNA Replication Mechanisms
   5.4: DNA Repair
   5.5: General Recombination

Ch 6: How Cells Read the Genome: From DNA to Protein

   6.1: DNA to RNA
   6.2: RNA to Protein

Ch 7: Control of Gene Expression

   7.1: An Overview of Gene Control
   7.2: DNA-Binding Motifs in Gene Regulatory Proteins
   7.3: How Genetic Switches Work
   7.5: Posttranscriptional Controls
   7.6: How Genomes Evolve

10.10.2 References

[1] Slack, J.M.W. Genes-A Very Short Introduction. Oxford University Press 2014

[2] Alberts B, Johnson A, Lewis J, Raff M, Roberts K, Walter P (2002). *Molecular Biology of the Cell* (Fourth ed.). New York: Garland Science. ISBN 978-0-8153-3218-3.

[3] Gericke, Niklas Markus; Hagberg, Mariana (5 December 2006). "Definition of historical models of gene function and their relation to students' understanding of genetics". *Science & Education* **16** (7-8): 849–881. Bibcode:2007Sc&Ed..16..849G. doi:10.1007/s11191-006-9064-4.

[4] Pearson H (May 2006). "Genetics: what is a gene?". *Nature* **441** (7092): 398–401. Bibcode:2006Natur.441..398P. doi:10.1038/441398a. PMID 16724031.

[5] Pennisi E (June 2007). "Genomics. DNA study forces rethink of what it means to be a gene". *Science* **316** (5831): 1556–1557. doi:10.1126/science.316.5831.1556. PMID 17569836.

[6] Noble D (September 2008). "Genes and causation" (Free full text). *Philosophical Transactions. Series A, Mathematical, Physical, and Engineering Sciences* **366** (1878): 3001–3015. Bibcode:2008RSPTA.366.3001N. doi:10.1098/rsta.2008.0086. PMID 18559318.

[7] Magner, Lois N. (2002). *A History of the Life Sciences* (Third ed.). Marcel Dekker, CRC Press. p. 371. ISBN 978-0-203-91100-6.

[8] Henig, Robin Marantz (2000). *The Monk in the Garden: The Lost and Found Genius of Gregor Mendel, the Father of Genetics*. Boston: Houghton Mifflin. pp. 1–9. ISBN 978-0395-97765-1.

[9] "gene". *Oxford English Dictionary* (3rd ed.). Oxford University Press. September 2005. (Subscription or UK public library membership required.)

[10] "The Human Genome Project Timeline". Retrieved 13 September 2006.

[11] Gerstein MB, Bruce C, Rozowsky JS, Zheng D, Du J, Korbel JO, Emanuelsson O, Zhang ZD, Weissman S, Snyder M (June 2007). "What is a gene, post-ENCODE? History and updated definition". *Genome Research* **17** (6): 669–681. doi:10.1101/gr.6339607. PMID 17567988.

[12] Avery, OT; MacLeod, CM; McCarty, M (1944). "Studies on the Chemical Nature of the Substance Inducing Transformation of Pneumococcal Types: Induction of Transformation by a Desoxyribonucleic Acid Fraction Isolated from Pneumococcus Type III". *The Journal of Experimental Medicine* **79** (2): 137–58. doi:10.1084/jem.79.2.137. PMC 2135445. PMID 19871359. Reprint: Avery, OT; MacLeod, CM; McCarty, M (1979). "Studies on the chemical nature of the substance inducing transformation of pneumococcal types. Inductions of transformation by a desoxyribonucleic acid fraction isolated from pneumococcus type III". *The Journal of Experimental Medicine* **149** (2): 297–326. doi:10.1084/jem.149.2.297. PMC 2184805. PMID 33226.

[13] Hershey, AD; Chase, M (1952). "Independent functions of viral protein and nucleic acid in growth of bacteriophage". *The Journal of General Physiology* **36** (1): 39–56. doi:10.1085/jgp.36.1.39. PMC 2147348. PMID 12981234.

[14] Judson, Horace (1979). *The Eighth Day of Creation: Makers of the Revolution in Biology*. Cold Spring Harbor Laboratory Press. pp. 51–169. ISBN 0-87969-477-7.

[15] Watson, J. D.; Crick, FH (1953). "Molecular Structure of Nucleic Acids: A Structure for Deoxyribose Nucleic Acid" (PDF). *Nature* **171** (4356): 737–8. Bibcode:1953Natur.171..737W. doi:10.1038/171737a0. PMID 13054692.

[16] Min Jou W, Haegeman G, Ysebaert M, Fiers W (May 1972). "Nucleotide sequence of the gene coding for the bacteriophage MS2 coat protein". *Nature* **237** (5350): 82–8. Bibcode:1972Natur.237...82J. doi:10.1038/237082a0. PMID 4555447.

[17] Sanger, F; Nicklen, S; Coulson, AR (1977). "DNA sequencing with chain-terminating inhibitors". *Proceedings of the National Academy of Sciences of the United States of America* **74** (12): 5463–7. Bibcode:1977PNAS...74.5463S. doi:10.1073/pnas.74.12.5463. PMC 431765. PMID 271968.

[18] Adams, Jill U. (2008). "DNA Sequencing Technologies". *Nature Education Knowledge*. SciTable (Nature Publishing Group) **1** (1): 193.

[19] Huxley, Julian (1942). *Evolution: the modern synthesis* (Definitive ed.). Cambridge, Mass.: MIT Press. ISBN 978-0262513661.

[20] Williams, George C. (2001). *Adaptation and Natural Selection a Critique of Some Current Evolutionary Thought*. ([Online-Ausg.]. ed.). Princeton: Princeton University Press. ISBN 9781400820108.

[21] Dawkins, Richard (1977). *The selfish gene* (Repr. (with corr.) ed.). London: Oxford Univ. Press. ISBN 0-19-857519-X.

[22] Dawkins, Richard (1989). *The extended phenotype*. (Pbk. ed.). Oxford: Oxford University Press. ISBN 0-19-286088-7.

[23] Stryer L, Berg JM, Tymoczko JL (2002). *Biochemistry* (5th ed.). San Francisco: W.H. Freeman. ISBN 0-7167-4955-6.

[24] Bolzer, Andreas; Kreth, Gregor; Solovei, Irina; Koehler, Daniela; Saracoglu, Kaan; Fauth, Christine; Müller, Stefan; Eils, Roland; Cremer, Christoph; Speicher, Michael R.; Cremer, Thomas (2005). "Three-Dimensional Maps of All Chromosomes in Human Male Fibroblast Nuclei and Prometaphase Rosettes". *PLoS Biology* **3** (5): e157. doi:10.1371/journal.pbio.0030157. PMID 15839726.

[25] Braig M, Schmitt CA (March 2006). "Oncogene-induced senescence: putting the brakes on tumor development". *Cancer Research* **66** (6): 2881–4. doi:10.1158/0008-5472.CAN-05-4006. PMID 16540631.

[26] Bennett, PM (March 2008). "Plasmid encoded antibiotic resistance: acquisition and transfer of antibiotic resistance genes in bacteria.". *British Journal of Pharmacology*. 153 Suppl 1: S347–57. doi:10.1038/sj.bjp.0707607. PMC 2268074. PMID 18193080.

[27] International Human Genome Sequencing Consortium (October 2004). "Finishing the euchromatic sequence of the human genome". *Nature* **431** (7011): 931–45. Bibcode:2004Natur.431..931H. doi:10.1038/nature03001. PMID 15496913.

[28] Mortazavi A, Williams BA, McCue K, Schaeffer L, Wold B (July 2008). "Mapping and quantifying mammalian transcriptomes by RNA-Seq". *Nature Methods* **5** (7): 621–8. doi:10.1038/nmeth.1226. PMID 18516045.

[29] Pennacchio, L. A.; Bickmore, W.; Dean, A.; Nobrega, M. A.; Bejerano, G. (2013). "Enhancers: Five essential questions". *Nature Reviews Genetics* **14** (4): 288–95. doi:10.1038/nrg3458. PMID 23503198.

[30] Maston, G. A.; Evans, S. K.; Green, M. R. (2006). "Transcriptional Regulatory Elements in the Human Genome". *Annual Review of Genomics and Human Genetics* 7: 29–59. doi:10.1146/annurev.genom.7.080505.115623.

[31] Mignone, Flavio; Gissi, Carmela; Liuni, Sabino; Pesole, Graziano (2002-02-28). "Untranslated regions of mRNAs". *Genome Biology* 3 (3): reviews0004. doi:10.1186/gb-2002-3-3-reviews0004. ISSN 1465-6906. PMID 11897027.

[32] Bicknell AA, Cenik C, Chua HN, Roth FP, Moore MJ (December 2012). "Introns in UTRs: why we should stop ignoring them.". *BioEssays* **34** (12): 1025–34. doi:10.1002/bies.201200073. PMID 23108796.

[33] Salgado, H.; Moreno-Hagelsieb, G.; Smith, T.; Collado-Vides, J. (2000). "Operons in Escherichia coli: Genomic analyses and predictions". *Proceedings of the National Academy of Sciences* **97** (12): 6652–6657. Bibcode:2000PNAS...97.6652S. doi:10.1073/pnas.110147297. PMC 18690. PMID 10823905.

[34] Blumenthal, Thomas (November 2004). "Operons in eukaryotes". *Briefings in Functional Genomics & Proteomics* **3** (3): 199–211. doi:10.1093/bfgp/3.3.199. ISSN 2041-2649. PMID 15642184.

[35] Spilianakis CG, Lalioti MD, Town T, Lee GR, Flavell RA (June 2005). "Interchromosomal associations between alternatively expressed loci". *Nature* **435** (7042): 637–45. Bibcode:2005Natur.435..637S. doi:10.1038/nature03574. PMID 15880101.

[36] Williams, A; Spilianakis, CG; Flavell, RA (April 2010). "Interchromosomal association and gene regulation in trans.". *Trends in genetics : TIG* **26** (4): 188–97. doi:10.1016/j.tig.2010.01.007. PMID 20236724.

[37] Marande W, Burger G (October 2007). "Mitochondrial DNA as a genomic jigsaw puzzle". *Science* (AAAS) **318** (5849): 415. Bibcode:2007Sci...318..415M. doi:10.1126/science.1148033. PMID 17947575.

[38] Parra G, Reymond A, Dabbouseh N, Dermitzakis ET, Castelo R, Thomson TM, Antonarakis SE, Guigó R (January 2006). "Tandem chimerism as a means to increase protein complexity in the human genome". *Genome Research* **16** (1): 37–44. doi:10.1101/gr.4145906. PMC 1356127. PMID 16344564.

[39] Eddy SR (December 2001). "Non-coding RNA genes and the modern RNA world". *Nat. Rev. Genet.* **2** (12): 919–29. doi:10.1038/35103511. PMID 11733745.

[40] Crick, Francis (1962). *The genetic code*. WH Freeman and Company. PMID 13882204.

[41] Woodson SA (May 1998). "Ironing out the kinks: splicing and translation in bacteria". *Genes & Development* **12** (9): 1243–7. doi:10.1101/gad.12.9.1243. PMID 9573040.

[42] Jacob F, Monod J (June 1961). "Genetic regulatory mechanisms in the synthesis of proteins". *J Mol Biol.* **3** (3): 318–56. doi:10.1016/S0022-2836(61)80072-7. PMID 13718526.

[43] Koonin, Eugene V.; Dolja, Valerian V.; Morris, T. Jack (January 1993). "Evolution and Taxonomy of Positive-Strand RNA Viruses: Implications of Comparative Analysis of Amino Acid Sequences". *Critical Reviews in Biochemistry and Molecular Biology* **28** (5): 375–430. doi:10.3109/10409239309078440. PMID 8269709.

[44] Domingo, Esteban (2001). "RNA Virus Genomes". *eLS*. doi:10.1002/9780470015902.a0001488.pub2.

[45] Domingo, E; Escarmís, C; Sevilla, N; Moya, A; Elena, SF; Quer, J; Novella, IS; Holland, JJ (June 1996). "Basic concepts in RNA virus evolution.". *FASEB Journal* **10** (8): 859–64. PMID 8666162.

[46] Morris, KV; Mattick, JS (June 2014). "The rise of regulatory RNA.". *Nature reviews. Genetics* **15** (6): 423–37. doi:10.1038/nrg3722. PMID 24776770.

[47] Miko, Ilona (2008). "Gregor Mendel and the Principles of Inheritance". *Nature Education Knowledge*. SciTable (Nature Publishing Group) **1** (1): 134.

[48] Chial, Heidi (2008). "Mendelian Genetics: Patterns of Inheritance and Single-Gene Disorders". *Nature Education Knowledge*. SciTable (Nature Publishing Group) **1** (1): 63.

[49] Lobo, Ingrid; Shaw, Kelly (2008). "Discovery and Types of Genetic Linkage". *Nature Education Knowledge*. SciTable (Nature Publishing Group) **1** (1): 139.

[50] Nachman MW, Crowell SL (September 2000). "Estimate of the mutation rate per nucleotide in humans". *Genetics* **156** (1): 297–304. PMC 1461236. PMID 10978293.

[51] Roach JC, Glusman G, Smit AF, et al. (April 2010). "Analysis of genetic inheritance in a family quartet by whole-genome sequencing". *Science* **328** (5978): 636–9. Bibcode:2010Sci...328..636R. doi:10.1126/science.1186802. PMC 3037280. PMID 20220176.

[52] Drake JW, Charlesworth B, Charlesworth D, Crow JF (April 1998). "Rates of spontaneous mutation". *Genetics* **148** (4): 1667–86. PMC 1460098. PMID 9560386.

[53] "What kinds of gene mutations are possible?". *Genetics Home Reference*. United States National Library of Medicine. 11 May 2015. Retrieved 19 May 2015.

[54] Andrews, Christine A. (2010). "Natural Selection, Genetic Drift, and Gene Flow Do Not Act in Isolation in Natural Populations". *Nature Education Knowledge*. SciTable (Nature Publishing Group) **3** (10): 5.

[55] Patterson, C (November 1988). "Homology in classical and molecular biology.". *Molecular Biology and Evolution* **5** (6): 603–25. PMID 3065587.

[56] Studer, RA; Robinson-Rechavi, M (May 2009). "How confident can we be that orthologs are similar, but paralogs differ?". *Trends in genetics : TIG* **25** (5): 210–6. doi:10.1016/j.tig.2009.03.004. PMID 19368988.

[57] Altenhoff, AM; Studer, RA; Robinson-Rechavi, M; Dessimoz, C (2012). "Resolving the ortholog conjecture: orthologs tend to be weakly, but significantly, more similar in function than paralogs.". *PLOS Computational Biology* **8** (5): e1002514. doi:10.1371/journal.pcbi.1002514. PMID 22615551.

[58] NOSIL, PATRIK; FUNK, DANIEL J.; ORTIZ-BARRIENTOS, DANIEL (February 2009). "Divergent selection and heterogeneous genomic divergence". *Molecular Ecology* **18** (3): 375–402. doi:10.1111/j.1365-294X.2008.03946.x.

[59] Emery, Laura. "Introduction to Phylogenetics". EMBL-EBI. Retrieved 19 May 2015.

[60] Mitchell, Matthew W.; Gonder, Mary Katherine (2013). "Primate Speciation: A Case Study of African Apes". *Nature Education Knowledge*. SciTable (Nature Publishing Group) **4** (2): 1.

[61] Guerzoni, D; McLysaght, A (November 2011). "De novo origins of human genes.". *PLOS Genetics* **7** (11): e1002381. doi:10.1371/journal.pgen.1002381. PMID 22102832.

[62] Reams, AB; Roth, JR (2 February 2015). "Mechanisms of gene duplication and amplification.". *Cold Spring Harbor perspectives in biology* **7** (2): a016592. doi:10.1101/cshperspect.a016592. PMID 25646380.

[63] Demuth, JP; De Bie, T; Stajich, JE; Cristianini, N; Hahn, MW (20 December 2006). "The evolution of mammalian gene families.". *PLOS ONE* **1**: e85. Bibcode:2006PLoSO...1...85D. doi:10.1371/journal.pone.0000085. PMC 1762380. PMID 17183716.

[64] Knowles, DG; McLysaght, A (October 2009). "Recent de novo origin of human protein-coding genes.". *Genome Research* **19** (10): 1752–9. doi:10.1101/gr.095026.109. PMC 2765279. PMID 19726446.

[65] Wu, DD; Irwin, DM; Zhang, YP (November 2011). "De novo origin of human protein-coding genes.". *PLOS Genetics* **7** (11): e1002379. doi:10.1371/journal.pgen.1002379. PMC 3213175. PMID 22102831.

[66] Tautz, D; Domazet-Lošo, T (31 August 2011). "The evolutionary origin of orphan genes.". *Nature reviews. Genetics* **12** (10): 692–702. doi:10.1038/nrg3053. PMID 21878963.

[67] Carvunis, AR; Rolland, T; Wapinski, I; Calderwood, MA; Yildirim, MA; Simonis, N; Charloteaux, B; Hidalgo, CA; Barbette, J; Santhanam, B; Brar, GA; Weissman, JS; Regev, A; Thierry-Mieg, N; Cusick, ME; Vidal, M (19 July 2012). "Proto-genes and de novo gene birth.". *Nature* **487** (7407): 370–4. Bibcode:2012Natur.487..370C. doi:10.1038/nature11184. PMID 22722833.

[68] Treangen, TJ; Rocha, EP (27 January 2011). "Horizontal transfer, not duplication, drives the expansion of protein families in prokaryotes.". *PLOS Genetics* **7** (1): e1001284. doi:10.1371/journal.pgen.1001284. PMID 21298028.

[69] Ochman, H; Lawrence, JG; Groisman, EA (18 May 2000). "Lateral gene transfer and the nature of bacterial innovation.". *Nature* **405** (6784): 299–304. Bibcode:2000Natur.405..299O. doi:10.1038/35012500. PMID 10830951.

[70] Keeling, PJ; Palmer, JD (August 2008). "Horizontal gene transfer in eukaryotic evolution.". *Nature reviews. Genetics* **9** (8): 605–18. doi:10.1038/nrg2386. PMID 18591983.

[71] Schönknecht, G; Chen, WH; Ternes, CM; Barbier, GG; Shrestha, RP; Stanke, M; Bräutigam, A; Baker, BJ; Banfield, JF; Garavito, RM; Carr, K; Wilkerson, C; Rensing, SA; Gagneul, D; Dickenson, NE; Oesterhelt, C; Lercher, MJ; Weber, AP (8 March 2013). "Gene transfer from bacteria and archaea facilitated evolution of an extremophilic eukaryote.". *Science* **339** (6124): 1207–10. Bibcode:2013Sci...339.1207S. doi:10.1126/science.1231707. PMID 23471408.

[72] Ridley, M. (2006). *Genome*. New York, NY: Harper Perennial. ISBN 0-06-019497-9

[73] Watson, JD, Baker TA, Bell SP, Gann A, Levine M, Losick R. (2004). "Ch9-10", Molecular Biology of the Gene, 5th ed., Peason Benjamin Cummings; CSHL Press.

[74] "Integr8 – A.thaliana Genome Statistics:".

[75] "Understanding the Basics". *The Human Genome Project*. Retrieved 26 April 2015.

[76] "WS227 Release Letter". WormBase. 10 August 2011. Retrieved 19 November 2013.

[77] Yu, J. (5 April 2002). "A Draft Sequence of the Rice Genome (Oryza sativa L. ssp. indica)". *Science* **296** (5565): 79–92. Bibcode:2002Sci...296...79Y. doi:10.1126/science.1068037. PMID 11935017.

[78] Anderson, S.; Bankier, A. T.; Barrell, B. G.; de Bruijn, M. H. L.; Coulson, A. R.; Drouin, J.; Eperon, I. C.; Nierlich, D. P.; Roe, B. A.; Sanger, F.; Schreier, P. H.; Smith, A. J. H.; Staden, R.; Young, I. G. (9 April 1981). "Sequence and organization of the human mitochondrial genome". *Nature* **290** (5806): 457–465. Bibcode:1981Natur.290..457A. doi:10.1038/290457a0. PMID 7219534.

[79] Adams, M. D. (24 March 2000). "The Genome Sequence of Drosophila melanogaster". *Science* **287** (5461): 2185–2195. Bibcode:2000Sci...287.2185.. doi:10.1126/science.287.5461.2185. PMID 10731132.

[80] Pertea, Mihaela; Salzberg, Steven L. (2010). "Between a chicken and a grape: estimating the number of human genes". *Genome Biology* **11** (5): 206. doi:10.1186/gb-2010-11-5-206. PMC 2898077. PMID 20441615.

[81] Belyi, V. A.; Levine, A. J.; Skalka, A. M. (22 September 2010). "Sequences from Ancestral Single-Stranded DNA Viruses in Vertebrate Genomes: the Parvoviridae and Circoviridae Are More than 40 to 50 Million Years Old". *Journal of Virology* **84** (23): 12458–12462. doi:10.1128/JVI.01789-10. PMC 2976387. PMID 20861255.

[82] Flores, Ricardo; Di Serio, Francesco; Hernández, Carmen (February 1997). "Viroids: The Noncoding Genomes". *Seminars in Virology* **8** (1): 65–73. doi:10.1006/smvy.1997.0107.

[83] Zonneveld, B. J. M. (2010). "New Record Holders for Maximum Genome Size in Eudicots and Monocots". *Journal of Botany* **2010**: 1–4. doi:10.1155/2010/527357.

[84] Yu J, Hu S, Wang J, Wong GK, Li S, Liu B, Deng Y, Dai L, Zhou Y, Zhang X, Cao M, Liu J, Sun J, Tang J, Chen Y, Huang X, Lin W, Ye C, Tong W, Cong L, Geng J, Han Y, Li L, Li W, Hu G, Huang X, Li W, Li J, Liu Z, Li L, Liu J, Qi Q, Liu J, Li L, Li T, Wang X, Lu H, Wu T, Zhu M, Ni P, Han H, Dong W, Ren X, Feng X, Cui P, Li X, Wang H, Xu X, Zhai W, Xu Z, Zhang J, He S, Zhang J, Xu J, Zhang K, Zheng X, Dong J, Zeng W, Tao L, Ye J, Tan J, Ren X, Chen X, He J, Liu D, Tian W, Tian C, Xia H, Bao Q, Li G, Gao H, Cao T, Wang J, Zhao W, Li P, Chen W, Wang X, Zhang Y, Hu J, Wang J, Liu S, Yang J, Zhang G, Xiong Y, Li Z, Mao L, Zhou C, Zhu Z, Chen R, Hao B, Zheng W, Chen S, Guo W, Li G, Liu S, Tao M, Wang J, Zhu L, Yuan L, Yang H (April 2002). "A draft sequence of the rice genome (Oryza sativa L. ssp. indica)". *Science* **296** (5565): 79–92. Bibcode:2002Sci...296...79Y. doi:10.1126/science.1068037. PMID 11935017.

[85] Perez-Iratxeta C, Palidwor G, Andrade-Navarro MA (December 2007). "Towards completion of the Earth's proteome". *EMBO Reports* **8** (12): 1135–1141. doi:10.1038/sj.embor.7401117. PMC 2267224. PMID 18059312.

[86] Kauffman SA (1969). "Metabolic stability and epigenesis in randomly constructed genetic nets". *Journal of Theoretical Biology* (Elsevier) **22** (3): 437–467. doi:10.1016/0022-5193(69)90015-0. PMID 5803332.

[87] Schuler GD, Boguski MS, Stewart EA, Stein LD, Gyapay G, Rice K, White RE, Rodriguez-Tomé P, Aggarwal A, Bajorek E, Bentolila S, Birren BB, Butler A, Castle AB, Chiannilkulchai N, Chu A, Clee C, Cowles S, Day PJ, Dibling T, Drouot N, Dunham I, Duprat S, East C, Edwards C, Fan JB, Fang N, Fizames C, Garrett C, Green L, Hadley D, Harris M, Harrison P, Brady S, Hicks A, Holloway E, Hui L, Hussain S, Louis-Dit-Sully C, Ma J, MacGilvery A, Mader C, Maratukulam A, Matise TC, McKusick KB, Morissette J, Mungall A, Muselet D, Nusbaum HC, Page DC, Peck A, Perkins S, Piercy M, Qin F, Quackenbush J, Ranby S, Reif T, Rozen S, Sanders C, She X, Silva J, Slonim DK, Soderlund C, Sun WL, Tabar P, Thangarajah T, Vega-Czarny N, Vollrath D, Voyticky S, Wilmer T, Wu X, Adams MD, Auffray C, Walter NA, Brandon R, Dehejia A, Goodfellow PN, Houlgatte R, Hudson JR, Ide SE, Iorio KR, Lee WY, Seki N, Nagase T, Ishikawa K, Nomura N, Phillips C, Polymeropoulos MH, Sandusky M, Schmitt K, Berry R, Swanson K, Torres R, Venter JC, Sikela JM, Beckmann JS, Weissenbach J, Myers RM, Cox DR, James MR, Bentley D, Deloukas P, Lander ES, Hudson TJ (October 1996). "A gene map of the human genome". *Science* **274** (5287): 540–6. Bibcode:1996Sci...274..540S. doi:10.1126/science.274.5287.540. PMID 8849440.

[88] Claverie JM (September 2005). "Fewer genes, more noncoding RNA". *Science* **309**

(5740): 1529–30. Bibcode:2005Sci...309.1529C. doi:10.1126/science.1116800. PMID 16141064.

[89] Carninci P, Hayashizaki Y (April 2007). "Noncoding RNA transcription beyond annotated genes". *Current Opinion in Genetics & Development* **17** (2): 139–44. doi:10.1016/j.gde.2007.02.008. PMID 17317145.

[90] Glass, J. I.; Assad-Garcia, N.; Alperovich, N.; Yooseph, S.; Lewis, M. R.; Maruf, M.; Hutchison, C. A.; Smith, H. O.; Venter, J. C. (3 January 2006). "Essential genes of a minimal bacterium". *Proceedings of the National Academy of Sciences* **103** (2): 425–430. Bibcode:2006PNAS..103..425G. doi:10.1073/pnas.0510013103. PMC 1324956. PMID 16407165.

[91] Gerdes, SY; Scholle, MD; Campbell, JW; Balázsi, G; Ravasz, E; Daugherty, MD; Somera, AL; Kyrpides, NC; Anderson, I; Gelfand, MS; Bhattacharya, A; Kapatral, V; D'Souza, M; Baev, MV; Grechkin, Y; Mseeh, F; Fonstein, MY; Overbeek, R; Barabási, AL; Oltvai, ZN; Osterman, AL (October 2003). "Experimental determination and system level analysis of essential genes in Escherichia coli MG1655.". *Journal of Bacteriology* **185** (19): 5673–84. doi:10.1128/jb.185.19.5673-5684.2003. PMC 193955. PMID 13129938.

[92] Baba, T; Ara, T; Hasegawa, M; Takai, Y; Okumura, Y; Baba, M; Datsenko, KA; Tomita, M; Wanner, BL; Mori, H (2006). "Construction of Escherichia coli K-12 in-frame, single-gene knockout mutants: the Keio collection.". *Molecular systems biology* **2**: 2006.0008. doi:10.1038/msb4100050. PMC 1681482. PMID 16738554.

[93] Juhas, M; Reuß, DR; Zhu, B; Commichau, FM (November 2014). "Bacillus subtilis and Escherichia coli essential genes and minimal cell factories after one decade of genome engineering.". *Microbiology (Reading, England)* **160** (Pt 11): 2341–51. doi:10.1099/mic.0.079376-0. PMID 25092907.

[94] Tu, Z; Wang, L; Xu, M; Zhou, X; Chen, T; Sun, F (21 February 2006). "Further understanding human disease genes by comparing with housekeeping genes and other genes.". *BMC Genomics* **7**: 31. doi:10.1186/1471-2164-7-31. PMID 16504025.

[95] Georgi, B; Voight, BF; Bućan, M (May 2013). "From mouse to human: evolutionary genomics analysis of human orthologs of essential genes.". *PLOS Genetics* **9** (5): e1003484. doi:10.1371/journal.pgen.1003484. PMC 3649967. PMID 23675308.

[96] Hutchison, Clyde A.; Chuang, Ray-Yuan; Noskov, Vladimir N.; Assad-Garcia, Nacyra; Deerinck, Thomas J.; Ellisman, Mark H.; Gill, John; Kannan, Krishna; Karas, Bogumil J. (2016-03-25). "Design and synthesis of a minimal bacterial genome". *Science* **351** (6280): aad6253. doi:10.1126/science.aad6253. ISSN 0036-8075. PMID 27013737.

[97] Eisenberg, E; Levanon, EY (October 2013). "Human housekeeping genes, revisited.". *Trends in genetics : TIG* **29** (10): 569–74. doi:10.1016/j.tig.2013.05.010. PMID 23810203.

[98] Amsterdam, A; Hopkins, N (September 2006). "Mutagenesis strategies in zebrafish for identifying genes involved in development and disease.". *Trends in genetics : TIG* **22** (9): 473–8. doi:10.1016/j.tig.2006.06.011. PMID 16844256.

[99] "About the HGNC". *HGNC Database of Human Gene Names*. HUGO Gene Nomenclature Committee. Retrieved 14 May 2015.

[100] Stanley N. Cohen and Annie C. Y. Chang (1 May 1973). "Recircularization and Autonomous Replication of a Sheared R-Factor DNA Segment in Escherichia coli Transformants — PNAS". Pnas.org. Retrieved 17 July 2010.

[101] Esvelt, KM.; Wang, HH. (2013). "Genome-scale engineering for systems and synthetic biology". *Mol Syst Biol* **9** (1): 641. doi:10.1038/msb.2012.66. PMC 3564264. PMID 23340847.

[102] Tan, WS.; Carlson, DF.; Walton, MW.; Fahrenkrug, SC.; Hackett, PB. (2012). "Precision editing of large animal genomes". *Adv Genet*. Advances in Genetics **80**: 37–97. doi:10.1016/B978-0-12-404742-6.00002-8. ISBN 9780124047426. PMC 3683964. PMID 23084873.

[103] Puchta, H.; Fauser, F. (2013). "Gene targeting in plants: 25 years later". *Int. J. Dev. Biol* **57** (6–7–8): 629–637. doi:10.1387/ijdb.130194hp.

[104] Ran FA, Hsu PD, Wright J, Agarwala V, Scott DA, Zhang F (2013). "Genome engineering using the CRISPR-Cas9 system". *Nat Protoc* **8** (11): 2281–308. doi:10.1038/nprot.2013.143. PMC 3969860. PMID 24157548.

[105] Kittleson, Joshua (2012). "Successes and failures in modular genetic engineering". *Current Opinion in Chemical Biology* **16**: 329–336. doi:10.1016/j.cbpa.2012.06.009.

[106] Berg, P.; Mertz, J. E. (2010). "Personal Reflections on the Origins and Emergence of Recombinant DNA Technology". *Genetics* **184** (1): 9–17. doi:10.1534/genetics.109.112144. PMC 2815933. PMID 20061565.

[107] Austin, Christopher P.; Battey, James F.; Bradley, Allan; Bucan, Maja; Capecchi, Mario; Collins, Francis S.; Dove, William F.; Duyk, Geoffrey; Dymecki, Susan (September 2004). "The Knockout Mouse Project". *Nature Genetics* **36** (9): 921–924. doi:10.1038/ng0904-921. ISSN 1061-4036. PMC 2716027. PMID 15340423.

[108] "A review of current large-scale mouse knockout efforts". *genesis*. doi:10.1002/dvg.20594.

[109] Deng C (2007). "In celebration of Dr. Mario R. Capecchi's Nobel Prize". *International Journal of Biological Sciences* **3** (7): 417–419. doi:10.7150/ijbs.3.417. PMID 17998949.

10.10.3 Further reading

- Watson JD, Baker TA, Bell SP, Gann A, Levine M, Losick R (2013). *Molecular Biology of the Gene* (7th ed.). Benjamin Cummings. ISBN 978-0-321-90537-6.

- Dawkins R (1990). *The Selfish Gene*. Oxford University Press. ISBN 0-19-286092-5. Google Book Search: first published 1976.

- Ridley M (1999). *Genome: The Autobiography of a Species in 23 Chapters*. Fourth Estate. ISBN 0-00-763573-7.

- Brown. T (2002). *Genomes* (2nd ed.). New York: Wiley-Liss. ISBN 0-471-25046-5.

10.11 External links

- Comparative Toxicogenomics Database

- DNA From The Beginning – a primer on genes and DNA

- Genes And DNA – Introduction to genes and DNA aimed at non-biologist

- Entrez Gene – a searchable database of genes

- IDconverter – converts gene IDs between public databases

- iHOP – Information Hyperlinked over Proteins

- TranscriptomeBrowser – Gene expression profile analysis

- The Protein Naming Utility, a database to identify and correct deficient gene names

- *Genes* – an Open Access journal

- IMPC (International Mouse Phenotyping Consortium) – Encyclopedia of mammalian gene function

- Global Genes Project – Leading non-profit organization supporting people living with genetic diseases

- ENCODE threads Explorer Characterization of intergenic regions and gene definition. *Nature*

Chapter 11

Gene therapy

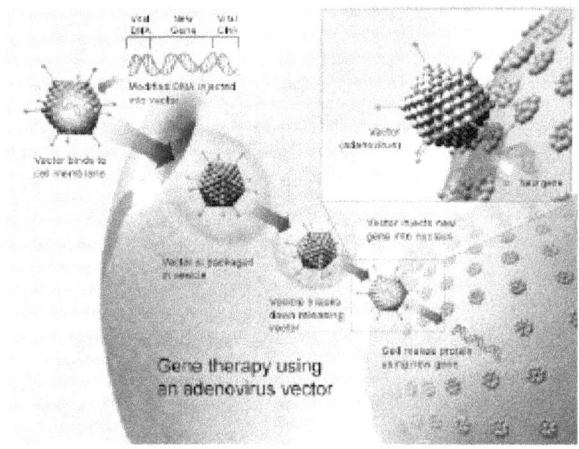

Gene therapy using an adenovirus vector. In some cases, the adenovirus will insert the new gene into a cell. If the treatment is successful, the new gene will make a functional protein to treat a disease.

Gene therapy is the therapeutic delivery of nucleic acid polymers into a patient's cells as a drug to treat disease.

The origins of gene therapy can be traced back to the first live attenuated vaccines in the 1950s.[1] Although attenuated vaccines do not alter extant human genes, viruses are RNA polymers with their own genetic code that acts upon human cells, thus live vaccines can be considered a primitive form of gene therapy, albeit not in the sense that is generally implied today.

The first attempt at modifying human DNA was performed in 1980 by Martin Cline, but the first successful and approved nuclear gene transfer in humans was performed in May 1989.[2] The first therapeutic use of gene transfer as well as the first direct insertion of human DNA into the nuclear genome was performed by French Anderson in a trial starting in September 1990.

Between 1989 and July 2015 over 2,200 clinical trials had been conducted.[3]

It should be noted that not all medical procedures that introduce alterations to a patient's genetic makeup can be considered gene therapy. Bone marrow transplantation, and organ transplants in general have been found to introduce foreign DNA into patients.[4] Gene therapy is defined by the precision of the procedure and the intention of direct therapeutic effects.

11.1 Background

Gene therapy was conceptualized in 1972, by authors who urged caution before commencing human gene therapy studies.

The first attempt, albeit an unsuccessful one, at gene therapy (as well as the first case of medical transfer of foreign genes into humans not counting organ transplantation) was performed by Martin Cline on 10 July 1980.[5][6] Cline claimed that one of the genes in his patients was active six months later, though he never published this data or had it verified [7] and even if he is correct, it's unlikely it produced any significant beneficial effects treating beta-thalassemia.[8]

After extensive research on animals throughout the 1980s and a 1989 bacterial gene tagging trial on humans, the first gene therapy widely accepted as a success was demonstrated in a trial that started on September 14, 1990, when Ashi DeSilva was treated for ADA-SCID.[9]

The first somatic treatment that produced a permanent genetic change was performed in 1993.[10]

The first germ line gene therapy consisted of producing a genetically engineered embryo in October 1996. The baby was born on July 21, 1997 and was produced by taking a donor's egg with healthy mitochondria, removing its nuclear DNA and filling it with the nuclear DNA of the biological mother - a procedure known as cytoplasmic transfer.[11]

This procedure was referred to sensationally and somewhat inaccurately in the media as a "three parent baby", though mtDNA is not the primary human genome and has little effect on an organism's individual characteristics beyond powering their cells.

Gene therapy is a way to fix a genetic problem at its source.

The polymers are either translated into proteins, interfere with target gene expression, or possibly correct genetic mutations.

The most common form uses DNA that encodes a functional, therapeutic gene to replace a mutated gene. The polymer molecule is packaged within a "vector", which carries the molecule inside cells.

Early clinical failures led to dismissals of gene therapy. Clinical successes since 2006 regained researchers' attention, although as of 2014, it was still largely an experimental technique.[12] These include treatment of retinal disease Leber's congenital amaurosis,[13][14][15][16] X-linked SCID,[17] ADA-SCID,[18][19] adrenoleukodystrophy,[20] chronic lymphocytic leukemia (CLL),[21] acute lymphocytic leukemia (ALL),[22] multiple myeloma,[23] haemophilia[19] and Parkinson's disease.[24] Between 2013 and April 2014, US companies invested over $600 million in the field.[25]

The first commercial gene therapy, Gendicine, was approved in China in 2003 for the treatment of certain cancers.[26] In 2011 Neovasculgen was registered in Russia as the first-in-class gene-therapy drug for treatment of peripheral artery disease, including critical limb ischemia.[27] In 2012 Glybera, a treatment for a rare inherited disorder, became the first treatment to be approved for clinical use in either Europe or the United States after its endorsement by the European Commission.[12][28]

11.2   Approaches

Following early advances in genetic engineering of bacteria, cells and small animals, scientists started considering how to apply it to medicine. Two main approaches were considered – replacing or disrupting defective genes.[29] Scientists focused on diseases caused by single-gene defects, such as cystic fibrosis, haemophilia, muscular dystrophy, thalassemia and sickle cell anemia. Glybera treats one such disease, caused by a defect in lipoprotein lipase.[28]

DNA must be administered, reach the damaged cells, enter the cell and express/disrupt a protein.[30] Multiple delivery techniques have been explored. The initial approach incorporated DNA into an engineered virus to deliver the DNA into a chromosome.[31][32] Naked DNA approaches have also been explored, especially in the context of vaccine development.[33]

Generally, efforts focused on administering a gene that causes a needed protein to be expressed. More recently, increased understanding of nuclease function has led to more direct DNA editing, using techniques such as zinc finger nucleases and CRISPR. The vector incorporates genes into chromosomes. The expressed nucleases then knock out and replace genes in the chromosome. As of 2014 these approaches involve removing cells from patients, editing a chromosome and returning the transformed cells to patients.[34]

11.2.1   Future of CRISPR-Cas 9

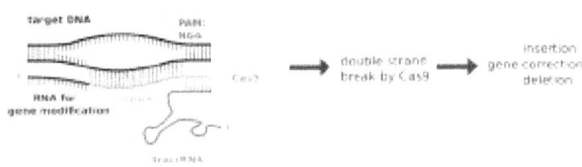

A duplex of crRNA and tracrRNA acts as guide RNA to introduce a specifically located gene modification based on the RNA 5' upstream of the crRNA. Cas9 binds the tracrRNA and needs a DNA binding sequence (5'NGG3'), which is called protospacer adjacent motif (PAM). After binding, Cas9 introduces a DNA double strand break, which is then followed by gene modification via homologous recombination (HDR) or non-homologous end joining (NHEJ).

Gene editing has been a potential therapy for many genetic diseases. Targeted genome editing using nucleases provides a general method for inducing deletions or insertion. An earlier method for targeting relies on protein-DNA interactions; however, the most recent one, using CRISPR – associated protein 9 (Cas9), provides better specificity, simplicity, speed and pricing. The CRISPR System was first identified in single cell archaea (Prokaryotes). It is now widely used in different cell types and organisms including human cells (HEK293T, HeLa, iPSC), mouse, fruit fly, rice, wheat etc. CRISPR –Cas9 genome editing has potential applications in human gene therapy, screening drug targets, synthetic biology, agriculture, programmable RNA targeting, and viral gene disruption.

Other technologies employ antisense, small interfering RNA and other DNA. To the extent that these technologies do not alter DNA, but instead directly interact with molecules such as RNA, they are not considered "gene therapy" per se.

11.3   Cell types

Gene therapy may be classified into two types:

11.3.1   Somatic cell

In somatic cell gene therapy (SCGT), the therapeutic genes are transferred into any cell other than a gamete, germ cell, gametocyte or undifferentiated stem cell. Any such modifications affect the individual patient only, and are not inherited by offspring. Somatic gene therapy represents mainstream basic and clinical research, in which therapeutic DNA (either integrated in the genome or as an external episome or plasmid) is used to treat disease.

Over 600 clinical trials utilizing SCGT are underway in the US. Most focus on severe genetic disorders, including immunodeficiencies, haemophilia, thalassaemia and cystic fibrosis. Such single gene disorders are good candidates for somatic cell therapy. The complete correction of a genetic disorder or the replacement of multiple genes is not yet possible. Only a few of the trials are in the advanced stages.[35]

11.3.2 Germline

In germline gene therapy (GGT), germ cells (sperm or eggs) are modified by the introduction of functional genes into their genomes. Modifying a germ cell causes all the organism's cells to contain the modified gene. The change is therefore heritable and passed on to later generations. Australia, Canada, Germany, Israel, Switzerland and the Netherlands[36] prohibit GGT for application in human beings, for technical and ethical reasons, including insufficient knowledge about possible risks to future generations[36] and higher risks versus SCGT.[37] The US has no federal controls specifically addressing human genetic modification (beyond FDA regulations for therapies in general).[36][38][39][40]

11.4 Vectors

Main article: Vectors in gene therapy

The delivery of DNA into cells can be accomplished by multiple methods. The two major classes are recombinant viruses (sometimes called biological nanoparticles or viral vectors) and naked DNA or DNA complexes (non-viral methods).

11.4.1 Viruses

Main article: Viral vector

In order to replicate, viruses introduce their genetic material into the host cell, tricking the host's cellular machinery into using it as blueprints for viral proteins. Scientists exploit this by substituting a virus's genetic material with therapeutic DNA. (The term 'DNA' may be an oversimplification, as some viruses contain RNA, and gene therapy could take this form as well.) A number of viruses have been used for human gene therapy, including retrovirus, adenovirus, lentivirus, herpes simplex, vaccinia and adeno-associated virus.[3] Like the genetic material (DNA or RNA) in viruses, therapeutic DNA can be designed to simply serve as a temporary blueprint that is degraded naturally or (at least theoretically) to enter the host's genome, becoming a permanent part of the host's DNA in infected cells.

11.4.2 Non-viral

Non-viral methods present certain advantages over viral methods, such as large scale production and low host immunogenicity. However, non-viral methods initially produced lower levels of transfection and gene expression, and thus lower therapeutic efficacy. Later technology remedied this deficiency.

Methods for non-viral gene therapy include the injection of naked DNA, electroporation, the gene gun, sonoporation, magnetofection, the use of oligonucleotides, lipoplexes, dendrimers, and inorganic nanoparticles.

11.5 Hurdles

Some of the unsolved problems include:

- Short-lived nature – Before gene therapy can become a permanent cure for a condition, the therapeutic DNA introduced into target cells must remain functional and the cells containing the therapeutic DNA must be stable. Problems with integrating therapeutic DNA into the genome and the rapidly dividing nature of many cells prevent it from achieving long-term benefits. Patients require multiple treatments.

- Immune response – Any time a foreign object is introduced into human tissues, the immune system is stimulated to attack the invader. Stimulating the immune system in a way that reduces gene therapy effectiveness is possible. The immune system's enhanced response to viruses that it has seen before reduces the effectiveness to repeated treatments.

- Problems with viral vectors – Viral vectors carry the risks of toxicity, inflammatory responses, and gene control and targeting issues.

- Multigene disorders – Some commonly occurring disorders, such as heart disease, high blood pressure, Alzheimer's disease, arthritis, and diabetes, are affected by variations in multiple genes, which complicate gene therapy.

- Some therapies may breach the Weismann barrier (between soma and germ-line) protecting the testes, potentially modifying the germline, falling afoul of regulations in countries that prohibit the latter practice.[41]

- Insertional mutagenesis – If the DNA is integrated in a sensitive spot in the genome, for example in a tumor suppressor gene, the therapy could induce a tumor. This has occurred in clinical trials for X-linked severe combined immunodeficiency (X-SCID) patients, in which hematopoietic stem cells were transduced with a corrective transgene using a retrovirus, and this led to the development of T cell leukemia in 3 of 20 patients.[42][43] One possible solution is to add a functional tumor suppressor gene to the DNA to be integrated. This may be problematic since the longer the DNA is, the harder it is to integrate into cell genomes. CRISPR technology allows researchers to make much more precise genome changes at exact locations.[44]

- Cost – Alipogene tiparvovec or Glybera, for example, at a cost of $1.6 million per patient, was reported in 2013 to be the world's most expensive drug.[45][46]

11.5.1   Deaths

Three patients' deaths have been reported in gene therapy trials, putting the field under close scrutiny. The first was that of Jesse Gelsinger in 1999.[47] One X-SCID patient died of leukemia in 2003.[9] In 2007, a rheumatoid arthritis patient died from an infection; the subsequent investigation concluded that the death was not related to gene therapy.[48]

11.6   History

11.6.1   1970s and earlier

In 1972 Friedmann and Roblin authored a paper in *Science* titled "Gene therapy for human genetic disease?"[49] Rogers (1970) was cited for proposing that *exogenous good DNA* be used to replace the defective DNA in those who suffer from genetic defects.[50]

11.6.2   1980s

In 1984 a retrovirus vector system was designed that could efficiently insert foreign genes into mammalian chromosomes.[51]

11.6.3   1990s

The first approved gene therapy in the US took place on 14 September 1990, at the National Institutes of Health (NIH), under the direction of William French Anderson.[52] Four-year-old Ashanti DeSilva received treatment for a genetic defect that left her with ADA-SCID, a severe immune system deficiency. The effects were temporary, but successful.[53]

Cancer gene therapy was introduced in 1992/93.[54] The treatment of glioblastoma multiforme, the malignant brain tumor whose outcome is always fatal, was done using a vector expressing antisense IGF-I RNA (clinical trial approved by NIH n° 1602, and FDA in 1994). The therapy proved to be effective due to the anti-tumor mechanism of IGF-I antisense, which is related to strong immune and apoptotic phenomena.

In 1992 Claudio Bordignon, working at the Vita-Salute San Raffaele University, performed the first gene therapy procedure using hematopoietic stem cells as vectors to deliver genes intended to correct hereditary diseases.[55] In 2002 this work led to the publication of the first successful gene therapy treatment for adenosine deaminase-deficiency (SCID). The success of a multi-center trial for treating children with SCID (severe combined immune deficiency or "bubble boy" disease) from 2000 and 2002, was questioned when two of the ten children treated at the trial's Paris center developed a leukemia-like condition. Clinical trials were halted temporarily in 2002, but resumed after regulatory review of the protocol in the US, the United Kingdom, France, Italy and Germany.[56]

In 1993 Andrew Gobea was born with SCID following prenatal genetic screening. Blood was removed from his mother's placenta and umbilical cord immediately after birth, to acquire stem cells. The allele that codes for adenosine deaminase (ADA) was obtained and inserted into a retrovirus. Retroviruses and stem cells were mixed, after which the viruses inserted the gene into the stem cell chromosomes. Stem cells containing the working ADA gene were injected into Andrew's blood. Injections of the ADA enzyme were also given weekly. For four years T cells (white blood cells), produced by stem cells, made ADA enzymes using the ADA gene. After four years more treatment was needed.

Jesse Gelsinger's death in 1999 impeded gene therapy research in the US.[57][58] As a result, the FDA suspended several clinical trials pending the reevaluation of ethical and procedural practices.[59]

11.6. HISTORY 123

11.6.4 2000s

The modified cancer gene therapy strategy of antisense IGF-I RNA (NIH n° 1602)[60] using antisense / triple helix anti IGF-I approach was registered in 2002 by Wiley gene therapy clinical trial - n° 635 and 636. The approach has shown promising results in the treatment of six different malignant tumors: glioblastoma, cancers of liver, colon, prostate, uterus and ovary (Collaborative NATO Science Programme on Gene Therapy USA, France, Poland n° LST 980517 conducted by J. Trojan) (Trojan et al., 2012). This anti–gene antisense/triple helix therapy has proven to be efficient, due to the mechanism stopping simultaneously IGF-I expression on translation and transcription levels, strengthening anti-tumor immune and apoptotic phenomenons (Trojan et al., 2013).[61]

2002

Sickle-cell disease can be treated in mice.[62] The mice – which have essentially the same defect that causes human cases – used a viral vector to induce production of fetal hemoglobin (HbF), which normally ceases to be produced shortly after birth. In humans, the use of hydroxyurea to stimulate the production of HbF temporarily alleviates sickle cell symptoms. The researchers demonstrated this treatment to be a more permanent means to increase therapeutic HbF production.[63]

A new gene therapy approach repaired errors in messenger RNA derived from defective genes. This technique has the potential to treat thalassaemia, cystic fibrosis and some cancers.[64]

Researchers created liposomes 25 nanometers across that can carry therapeutic DNA through pores in the nuclear membrane.[65]

2003

In 2003 a research team inserted genes into the brain for the first time. They used liposomes coated in a polymer called polyethylene glycol, which, unlike viral vectors, are small enough to cross the blood–brain barrier.[66]

Short pieces of double-stranded RNA (short, interfering RNAs or siRNAs) are used by cells to degrade RNA of a particular sequence. If a siRNA is designed to match the RNA copied from a faulty gene, then the abnormal protein product of that gene will not be produced.[67]

Gendicine is a cancer gene therapy that delivers the tumor suppressor gene p53 using an engineered adenovirus. In 2003, it was approved in China for the treatment of head and neck squamous cell carcinoma.[26]

2006

In March researchers announced the successful use of gene therapy to treat two adult patients for X-linked chronic granulomatous disease, a disease which affects myeloid cells and damages the immune system. The study is the first to show that gene therapy can treat the myeloid system.[68]

In May a team reported a way to prevent the immune system from rejecting a newly delivered gene.[69] Similar to organ transplantation, gene therapy has been plagued by this problem. The immune system normally recognizes the new gene as foreign and rejects the cells carrying it. The research utilized a newly uncovered network of genes regulated by molecules known as microRNAs. This natural function selectively obscured their therapeutic gene in immune system cells and protected it from discovery. Mice infected with the gene containing an immune-cell microRNA target sequence did not reject the gene.

In August scientists successfully treated metastatic melanoma in two patients using killer T cells genetically retargeted to attack the cancer cells.[70]

In November researchers reported on the use of VRX496, a gene-based immunotherapy for the treatment of HIV that uses a lentiviral vector to deliver an antisense gene against the HIV envelope. In a phase I clinical trial, five subjects with chronic HIV infection who had failed to respond to at least two antiretroviral regimens were treated. A single intravenous infusion of autologous CD4 T cells genetically modified with VRX496 was well tolerated. All patients had stable or decreased viral load; four of the five patients had stable or increased CD4 T cell counts. All five patients had stable or increased immune response to HIV antigens and other pathogens. This was the first evaluation of a lentiviral vector administered in a US human clinical trial.[71][72]

2007

In May researchers announced the first gene therapy trial for inherited retinal disease. The first operation was carried out on a 23-year-old British male, Robert Johnson, in early 2007.[73]

2008

Main article: Gene therapy of the human retina

Leber's congenital amaurosis is an inherited blinding disease caused by mutations in the RPE65 gene. The results of a small clinical trial in children were published in April.[13] Delivery of recombinant adeno-associated virus (AAV) carrying RPE65 yielded positive results. In May two more groups reported positive results in independent clinical trials using gene therapy to treat the condition. In all three clinical trials, patients recovered functional vision without apparent side-effects.[13][14][15][16]

2009

In September researchers were able to give trichromatic vision to squirrel monkeys.[74] In November 2009, researchers halted a fatal genetic disorder called adrenoleukodystrophy in two children using a lentivirus vector to deliver a functioning version of ABCD1, the gene that is mutated in the disorder.[75]

11.6.5   2010s

2010

An April paper reported that gene therapy addressed achromatopsia (color blindness) in dogs by targeting cone photoreceptors. Cone function and day vision were restored for at least 33 months in two young specimens. The therapy was less efficient for older dogs.[76]

In September it was announced that an 18-year-old male patient in France with beta-thalassemia major had been successfully treated.[77] Beta-thalassemia major is an inherited blood disease in which beta haemoglobin is missing and patients are dependent on regular lifelong blood transfusions.[78] The technique used a lentiviral vector to transduce the human β-globin gene into purified blood and marrow cells obtained from the patient in June 2007.[79] The patient's haemoglobin levels were stable at 9 to 10 g/dL. About a third of the hemoglobin contained the form introduced by the viral vector and blood transfusions were not needed.[79][80] Further clinical trials were planned.[81] Bone marrow transplants are the only cure for thalassemia, but 75% of patients do not find a matching donor.[80]

2011

In 2007 and 2008, a man was cured of HIV by repeated Hematopoietic stem cell transplantation (see also Allogeneic stem cell transplantation, Allogeneic bone marrow transplantation, Allotransplantation) with double-delta-32 mutation which disables the CCR5 receptor. This cure was accepted by the medical community in 2011.[82] It required complete ablation of existing bone marrow, which is very debilitating.

In August two of three subjects of a pilot study were confirmed to have been cured from chronic lymphocytic leukemia (CLL). The therapy used genetically modified T

cells to attack cells that expressed the CD19 protein to fight the disease.[21] In 2013, the researchers announced that 26 of 59 patients had achieved complete remission and the original patient had remained tumor-free.[83]

Human HGF plasmid DNA therapy of cardiomyocytes is being examined as a potential treatment for coronary artery disease as well as treatment for the damage that occurs to the heart after myocardial infarction.[84][85]

In 2011 Neovasculgen was registered in Russia as the first-in-class gene-therapy drug for treatment of peripheral artery disease, including critical limb ischemia.[27] Neovasculogen is a plasmid encoding the CMV promoter and the 165 amino acid form of VEGF.[86][87]

2012

The FDA approved Phase 1 clinical trials on thalassemia major patients in the US for 10 participants in July.[88] The study was expected to continue until 2015.[89]

In July 2012, the European Medicines Agency recommended approval of a gene therapy treatment for the first time in either Europe or the United States. The treatment used Alipogene tiparvovec (Glybera) to compensate for lipoprotein lipase deficiency, which can cause severe pancreatitis.[90] The recommendation was endorsed by the European Commission in November 2012[12][28][91][92] and commercial rollout began in late 2014.[93]

In December 2012, it was reported that 10 of 13 patients with multiple myeloma were in remission "or very close to it" three months after being injected with a treatment involving genetically engineered T cells to target proteins NY-ESO-1 and LAGE-1, which exist only on cancerous myeloma cells.[23]

2013

In March researchers reported that three of five subjects who had acute lymphocytic leukemia (ALL) had been in remission for five months to two years after being treated with genetically modified T cells which attacked cells with CD19 genes on their surface, i.e. all B-cells, cancerous or not. The researchers believed that the patients' immune systems would make normal T-cells and B-cells after a couple of months. They were also given bone marrow. One patient relapsed and died and one died of a blood clot unrelated to the disease.[22]

Following encouraging Phase 1 trials, in April, researchers announced they were starting Phase 2 clinical trials (called CUPID2 and SERCA-LVAD) on 250 patients[94] at several hospitals to combat heart disease. The therapy was designed to increase the levels of SERCA2a protein in heart muscles, improving muscle function.[95] The FDA granted this a Breakthrough Therapy Designation to accelerate the trial and approval process.[96]

In July researchers reported promising results for six children with two severe hereditary diseases had been treated with a partially deactivated lentivirus to replace a faulty gene and after 7–32 months. Three of the children had metachromatic leukodystrophy, which causes children to lose cognitive and motor skills.[97] The other children had Wiskott-Aldrich syndrome, which leaves them to open to infection, autoimmune diseases and cancer.[98] Follow up trials with gene therapy on another six children with Wiskott-Aldrich syndrome were also reported as promising.[99][100]

In October researchers reported that two children born with adenosine deaminase severe combined immunodeficiency disease (ADA-SCID) had been treated with genetically engineered stem cells 18 months previously and that their immune systems were showing signs of full recovery. Another three children were making progress.[19] In 2014 a further 18 children with ADA-SCID were cured by gene therapy.[101] ADA-SCID children have no functioning immune system and are sometimes known as "bubble children."[19]

Also in October researchers reported that they had treated six haemophilia sufferers in early 2011 using an adeno-associated virus. Over two years later all six were producing clotting factor.[19][102]

Data from three trials on Topical cystic fibrosis transmembrane conductance regulator gene therapy were reported to not support its clinical use as a mist inhaled into the lungs to treat cystic fibrosis patients with lung infections.[103]

2014

In January researchers reported that six choroideremia patients had been treated with adeno-associated virus with a copy of REP1. Over a six-month to two-year period all had improved their sight. Choroideremia is an inherited genetic eye disease with no approved treatment, leading to loss of sight.[104][105]

In March researchers reported that 12 HIV patients had been treated since 2009 in a trial with a genetically engineered virus with a rare mutation (CCR5 deficiency) known to protect against HIV with promising results.[106][107]

Clinical trials of gene therapy for sickle cell disease were started in 2014[108][109] although one review failed to find any such trials.[110]

2015

In February LentiGlobin BB305, a gene therapy treatment undergoing clinical trials for treatment of beta thalassemia gained FDA "breakthrough" status after several patients were able to forgo the frequent blood transfusions usually required to treat the disease.[111]

In March researchers delivered a recombinant gene encoding a broadly neutralizing antibody into monkeys infected with simian HIV; the monkey's cells produced the antibody, which cleared them of HIV. The technique is named immunoprophylaxis by gene transfer (IGT). Animal tests for antibodies to ebola, malaria, influenza and hepatitis are underway.[112][113]

In March scientists, including an inventor of CRISPR, urged a worldwide moratorium on germline gene therapy, writing "scientists should avoid even attempting, in lax jurisdictions, germline genome modification for clinical application in humans" until the full implications "are discussed among scientific and governmental organizations".[114][115][116][117]

Also in 2015 Glybera was approved for the German market.[118]

In October, researchers announced that they had treated a baby girl, Layla Richards, with an experimental treatment using donor T-cells genetically engineered to attack cancer cells. Two months after the treatment she was still free of her cancer (a highly aggressive form of acute lymphoblastic leukaemia [ALL]). Children with highly aggressive ALL normally have a very poor prognosis and Lalya's disease had been regarded as terminal before the treatment.[119]

In December, scientists of major world academies called for a moratorium on inheritable human genome edits, including those related to CRISPR-Cas9 technologies[120] but that basic research including embryo gene editing should continue.[121]

11.7 Speculative uses

Speculated uses for gene therapy include:

11.7.1 Fertility

Gene Therapy techniques have the potential to provide alternative treatments for those with infertility. Recently, successful experimentation on mice has proven that fertility can be restored by using the gene therapy method, CRISPR.[122] Spermatogenical stem cells from another organism were transplanted into the testes of an infertile male mouse. The stem cells re-established spermatogenesis and fertility.[123]

11.7.2   Gene doping

Main article: Gene doping

Athletes might adopt gene therapy technologies to improve their performance.[124] Gene doping is not known to occur, but multiple gene therapies may have such effects. Kayser et al. argue that gene doping could level the playing field if all athletes receive equal access. Critics claim that any therapeutic intervention for non-therapeutic/enhancement purposes compromises the ethical foundations of medicine and sports.[125]

11.7.3   Human genetic engineering

See also: germinal choice technology

Genetic engineering could be used to change physical appearance, metabolism, and even improve physical capabilities and mental faculties such as memory and intelligence. Ethical claims about germline engineering include beliefs that every fetus has a right to remain genetically unmodified, that parents hold the right to genetically modify their offspring, and that every child has the right to be born free of preventable diseases.[126][127][128] For adults, genetic engineering could be seen as another enhancement technique to add to diet, exercise, education, cosmetics and plastic surgery.[129][130] Another theorist claims that moral concerns limit but do not prohibit germline engineering.[131]

Possible regulatory schemes include a complete ban, provision to everyone, or professional self-regulation. The American Medical Association's Council on Ethical and Judicial Affairs stated that "genetic interventions to enhance traits should be considered permissible only in severely restricted situations: (1) clear and meaningful benefits to the fetus or child; (2) no trade-off with other characteristics or traits; and (3) equal access to the genetic technology, irrespective of income or other socioeconomic characteristics."[132]

As early in the history of biotechnology as 1990, there have been scientists opposed to attempts to modify the human germline using these new tools,[133] and such concerns have continued as technology progressed.[134] With the advent of new techniques like CRISPR, in March 2015 a group of scientists urged a worldwide moratorium on clinical use of gene editing technologies to edit the human genome in a way that can be inherited.[114][115][116][117] In April 2015, researchers sparked controversy when they reported results of basic research to edit the DNA of non-viable human embryos using CRISPR.[122][135]

11.8   Regulations

Regulations covering genetic modification are part of general guidelines about human-involved biomedical research.

The Helsinki Declaration (Ethical Principles for Medical Research Involving Human Subjects) was amended by the World Medical Association's General Assembly in 2008. This document provides principles physicians and researchers must consider when involving humans as research subjects. The Statement on Gene Therapy Research initiated by the Human Genome Organization (HUGO) in 2001 provides a legal baseline for all countries. HUGO's document emphasizes human freedom and adherence to human rights, and offers recommendations for somatic gene therapy, including the importance of recognizing public concerns about such research.[136]

11.8.1   United States

No federal legislation lays out protocols or restrictions about human genetic engineering. This subject is governed by overlapping regulations from local and federal agencies, including the Department of Health and Human Services, the FDA and NIH's Recombinant DNA Advisory Committee. Researchers seeking federal funds for an investigational new drug application, (commonly the case for somatic human genetic engineering), must obey international and federal guidelines for the protection of human subjects.[137]

NIH serves as the main gene therapy regulator for federally funded research. Privately funded research is advised to follow these regulations. NIH provides funding for research that develops or enhances genetic engineering techniques and to evaluate the ethics and quality in current research. The NIH maintains a mandatory registry of human genetic engineering research protocols that includes all federally funded projects.

An NIH advisory committee published a set of guidelines on gene manipulation.[138] The guidelines discuss lab safety as well as human test subjects and various experimental types that involve genetic changes. Several sections specifically pertain to human genetic engineering, including Section III-C-1. This section describes required review processes and other aspects when seeking approval to begin clinical research involving genetic transfer into a human patient.[139]

The FDA regulates the quality and safety of gene therapy products and supervises how these products are used clinically. Therapeutic alteration of the human genome falls under the same regulatory requirements as any other medical treatment. Research involving human subjects, such as clinical trials, must be reviewed and approved by the FDA and an Institutional Review Board.[140][141]

11.9 Popular culture

Gene therapy is the basis for the plotline of the film *I Am Legend*[142] and the TV show Will Gene Therapy Change the Human Race?[143]

11.10 See also

- Antisense therapy
- Bioethics
- Gene therapy for color blindness
- Gene therapy for osteoarthritis
- Genetic engineering
- Therapeutic gene modulation
- Synthetic rescue
- Synthetic lethality

11.11 References

[1] http://vaccine-safety-training.org/live-attenuated-vaccines.html

[2] Rosenberg SA, Aebersold P, Cornetta K; et al. (August 1990). "Gene transfer into humans--immunotherapy of patients with advanced melanoma, using tumor-infiltrating lymphocytes modified by retroviral gene transduction". *N. Engl. J. Med.* **323**: 570–8. doi:10.1056/NEJM199008303230904. PMID 2381442.

[3] Gene Therapy Clinical Trials Worldwide Database. *The Journal of Gene Medicine.* Wiley (January 2014)

[4] http://www.nytimes.com/2013/09/17/science/dna-double-take.html?pagewanted=all&_r=1

[5] https://books.google.com/books?id=FxGjBqEL-3kC&pg=PA45&lpg=PA45&dq=%22martin+cline%22+%22gene+therapy%22&source=bl&ots=Goj5oCAVBp&sig=KbB9EHE4ucGkS1C2AUVltD-A4-U&hl=en&sa=X&ved=0CD8Q6AEwBWoVChMIyczEkPrRyAIVTcNjCh1NBg3U#v=onepage&q=%22martin%20cline%22%20%22gene%20therapy%22&f=false

[6] "Martin Cline loses appeal on NIH grant". *Science* **218**: 37. 1982. doi:10.1126/science.7123214. PMID 7123214.

[7] https://books.google.com/books?id=FzuwRwb--X4C&pg=PA30&lpg=PA30&dq=%22gene+therapy%22+martin+cline&source=bl&ots=LOoOWFnD-i&sig=k751o-g-TgVgJnECRIRMcM4EcFU&hl=en&sa=X&ved=0CDQQ6AEwBjgKahUKEwja2rCXtvfIAhVP5G MKHVPWDWQ#v=onepage&q=%22gene%20therapy%22%20martin%20cline&f=false

[8] http://cshmonographs.org/index.php/monographs/article/viewFile/4773/3874

[9] Sheridan C (2011). "Gene therapy finds its niche". *Nature Biotechnology* **29** (2): 121–128. doi:10.1038/nbt.1769. PMID 21301435.

[10] http://www.asgct.org/UserFiles/kohnslides.pdf

[11] http://www.docguide.com/healthy-baby-born-after-worlds-first-successful-cytopla

[12] Richards, Sabrina (6 November 2012). "Gene Therapy Arrives in Europe". *The Scientist.*

[13] Maguire, A. M.; Simonelli, F.; Pierce, E. A.; Pugh Jr, E. N.; Mingozzi, F.; Bennicelli, J.; Banfi, S.; Marshall, K. A.; Testa, F.; Surace, E. M.; Rossi, S.; Lyubarsky, A.; Arruda, V. R.; Konkle, B.; Stone, E.; Sun, J.; Jacobs, J.; Dell'Osso, L.; Hertle, R.; Ma, J. X.; Redmond, T. M.; Zhu, X.; Hauck, B.; Zelenaia, O.; Shindler, K. S.; Maguire, M. G.; Wright, J. F.; Volpe, N. J.; McDonnell, J. W.; Auricchio, A. (2008). "Safety and Efficacy of Gene Transfer for Leber's Congenital Amaurosis". *New England Journal of Medicine* **358** (21): 2240–2248. doi:10.1056/NEJMoa0802315. PMC 2829748. PMID 18441370.

[14] Simonelli, F.; Maguire, A. M.; Testa, F.; Pierce, E. A.; Mingozzi, F.; Bennicelli, J. L.; Rossi, S.; Marshall, K.; Banfi, S.; Surace, E. M.; Sun, J.; Redmond, T. M.; Zhu, X.; Shindler, K. S.; Ying, G. S.; Ziviello, C.; Acerra, C.; Wright, J. F.; McDonnell, J. W.; High, K. A.; Bennett, J.; Auricchio, A. (2009). "Gene Therapy for Leber's Congenital Amaurosis is Safe and Effective Through 1.5 Years After Vector Administration". *Molecular Therapy* **18** (3): 643–650. doi:10.1038/mt.2009.277. PMC 2839440. PMID 19953081.

[15] Cideciyan, A. V.; Hauswirth, W. W.; Aleman, T. S.; Kaushal, S.; Schwartz, S. B.; Boye, S. L.; Windsor, E. A. M.; Conlon, T. J.; Sumaroka, A.; Roman, A. J.; Byrne, B. J.; Jacobson, S. G. (2009). "Vision 1 Year after Gene Therapy for Leber's Congenital Amaurosis". *New England Journal of Medicine* **361** (7): 725–727. doi:10.1056/NEJMc0903652. PMC 2847775. PMID 19675341.

[16] Bainbridge, J. W. B.; Smith, A. J.; Barker, S. S.; Robbie, S.; Henderson, R.; Balaggan, K.; Viswanathan, A.; Holder, G. E.; Stockman, A.; Tyler, N.; Petersen-Jones, S.; Bhattacharya, S. S.; Thrasher, A. J.; Fitzke, F. W.; Carter, B. J.; Rubin, G. S.; Moore, A. T.; Ali, R. R. (2008). "Effect of Gene Therapy on Visual Function in Leber's Congenital Amaurosis". *New England Journal of Medicine* **358**

(21): 2231–2239. doi:10.1056/NEJMoa0802268. PMID 18441371.

[17] Fischer, A.; Hacein-Bey-Abina, S.; Cavazzana-Calvo, M. (2010). "20 years of gene therapy for SCID". *Nature Immunology* **11** (6): 457–460. doi:10.1038/ni0610-457. PMID 20485269.

[18] Ferrua, F.; Brigida, I.; Aiuti, A. (2010). "Update on gene therapy for adenosine deaminase-deficient severe combined immunodeficiency". *Current Opinion in Allergy and Clinical Immunology* **10** (6): 551–556. doi:10.1097/ACI.0b013e32833fea85. PMID 20966749.

[19] Geddes, Linda (30 October 2013) 'Bubble kid' success puts gene therapy back on track' The New Scientist. Retrieved 2 November 2013

[20] Cartier N; Aubourg P (2009). "Hematopoietic Stem Cell Transplantation and Hematopoietic Stem Cell Gene Therapy in X-Linked Adrenoleukodystrophy". *Brain Pathology* **20** (4): 857–862. doi:10.1111/j.1750-3639.2010.00394.x. PMID 20626747.

[21] Ledford, H. (2011). "Cell therapy fights leukaemia". *Nature*. doi:10.1038/news.2011.472.

[22] Coghlan, Andy (26 March 2013) Gene therapy cures leukaemia in eight days. The New Scientist. Retrieved 15 April 2013

[23] Coghlan, Andy (11 December 2013) Souped-up immune cells force leukaemia into remission. New Scientist. Retrieved 15 April 2013

[24] Lewitt, P. A.; Rezai, A. R.; Leehey, M. A.; Ojemann, S. G.; Flaherty, A. W.; Eskandar, E. N.; Kostyk, S. K.; Thomas, K.; Sarkar, A.; Siddiqui, M. S.; Tatter, S. B.; Schwalb, J. M.; Poston, K. L.; Henderson, J. M.; Kurlan, R. M.; Richard, I. H.; Van Meter, L.; Sapan, C. V.; During, M. J.; Kaplitt, M. G.; Feigin, A. (2011). "AAV2-GAD gene therapy for advanced Parkinson's disease: A double-blind, sham-surgery controlled, randomised trial". *The Lancet Neurology* **10** (4): 309–319. doi:10.1016/S1474-4422(11)70039-4. PMID 21419704.

[25] Herper, Matthew (26 March 2014) Gene Therapy's Big Comeback *Forbes*. Retrieved 28 April 2014

[26] Pearson, Sue; Jia, Hepeng; Kandachi, Keiko (2004). "China approves first gene therapy". *Nature Biotechnology* **22** (1): 3–4. doi:10.1038/nbt0104-3. PMID 14704685.

[27] "Gene Therapy for PAD Approved". 6 December 2011. Retrieved 5 August 2015.

[28] Gallagher, James. (2 November 2012) BBC News – Gene therapy: Glybera approved by European Commission. BBC. Retrieved 15 December 2012.

[29] U.S. National Library of Medicine, Genomics Home Reference. What is gene therapy?

[30] U.S. National Library of Medicine, Genomics Home Reference. How does gene therapy work?

[31] Pezzoli, D.; Chiesa, R.; De Nardo, L.; Candiani, G. (2012). "We still have a long way to go to effectively deliver genes!". *Journal of Applied Biomaterials & Functional Materials* **2** (10): 82–91. doi:10.5301/JABFM.2012.9707. PMID 23015375.

[32] Vannucci, L; Lai, M; Chiuppesi, F; Ceccherini-Nelli, L; Pistello, M (2013). "Viral vectors: A look back and ahead on gene transfer technology". *The new microbiologica* **36** (1): 1–22. PMID 23435812.

[33] Gothelf A; Gehl J (2012). "What you always needed to know about electroporation based DNA vaccines". *Hum Vaccin Immunother* **8** (11): 1694–702. doi:10.4161/hv.22062. PMC 3601144. PMID 23111168.

[34] Urnov, Fyodor D.; Rebar, Edward J.; Holmes, Michael C.; Zhang, H. Steve; Gregory, Philip D. (2010). "Genome editing with engineered zinc finger nucleases". *Nature Reviews Genetics* **11** (9): 636–646. doi:10.1038/nrg2842. PMID 20717154.

[35] Mavilio F; Ferrari G (2008). "Genetic modification of somatic stem cells. The progress, problems and prospects of a new therapeutic technology". *EMBO Rep.* 9 Suppl 1: S64–9. doi:10.1038/embor.2008.81. PMC 3327547. PMID 18578029.

[36] "International Law". The Genetics and Public Policy Center, Johns Hopkins University Berman Institute of Bioethics. 2010.

[37] Strachnan, T. and Read, A. P. (2004) *Human Molecular Genetics*, 3rd Edition. Garland Publishing. p. 616, ISBN 0815341849.

[38] Hanna, K., 2006, Germline Gene Transfer, National Human Genome Research Institute.

[39] 2013, Human Cloning and Genetic Modification, Association of Reproductive Health Officials.

[40] "Gene Therapy". *ama-assn.org*. 4 April 2014. Retrieved 22 March 2015.

[41] Korthof G. "The implications of Steele's soma-to-germline feedback for human gene therapy".

[42] Woods, N. B.; Bottero, V.; Schmidt, M.; Von Kalle, C.; Verma, I. M. (2006). "Gene therapy: Therapeutic gene causing lymphoma". *Nature* **440** (7088): 1123. Bibcode:2006Natur.440.1123W. doi:10.1038/4401123a. PMID 16641981.

[43] Thrasher, A. J.; Gaspar, H. B.; Baum, C.; Modlich, U.; Schambach, A.; Candotti, F.; Otsu, M.; Sorrentino, B.; Scobie, L.; Cameron, E.; Blyth, K.; Neil, J.; Abina, S. H. B.; Cavazzana-Calvo, M.; Fischer, A. (2006). "Gene therapy: X-SCID transgene leukaemogenicity". *Nature* **443** (7109): E5–E6; discussion E6–7. Bibcode:2006Natur.443E...5T. doi:10.1038/nature05219. PMID 16988659.

[44] Young, Susan (11 February 2014) Genome Surgery MIT Technology Review. Retrieved 17 February 2014

[45] (31 October 2013) Gene therapy needs a hero to live up to the hype The New Scientist. Retrieved 2 November 2012

[46] Crasto, Anthony Melvin (2013) Glybera – The Most Expensive Drug in the world & First Approved Gene Therapy in the West All About Drug. Retrieved 2 November 2013

[47] ORNL.gov. ORNL.gov. Retrieved 15 December 2012.

[48] Frank, K. M.; Hogarth, D. K.; Miller, J. L.; Mandal, S.; Mease, P. J.; Samulski, R. J.; Weisgerber, G. A.; Hart, J. (2009). "Investigation of the Cause of Death in a Gene-Therapy Trial". *New England Journal of Medicine* **361** (2): 161–169. doi:10.1056/NEJMoa0801066. PMID 19587341.

[49] Friedmann T; Roblin R (1972). "Gene Therapy for Human Genetic Disease?". *Science* **175** (4025): 949–955. Bibcode:1972Sci...175..949F. doi:10.1126/science.175.4025.949. PMID 5061866.

[50] Rogers S, New Scientist 1970, p. 194

[51] Cepko CL; Roberts BE; Mulligan RC (1984). "Construction and applications of a highly transmissible murine retrovirus shuttle vector". *Cell* **37** (3): 1053–62. doi:10.1016/0092-8674(84)90440-9. PMID 6331674.

[52] "The first gene therapy". Life Sciences Foundation. 21 June 2011. Archived from the original on 28 November 2012. Retrieved 7 January 2014.

[53] Blaese, R. M.; Culver, K. W.; Miller, A. D.; Carter, C. S.; Fleisher, T.; Clerici, M.; Shearer, G.; Chang, L.; Chiang, Y.; Tolstoshev, P.; Greenblatt, J. J.; Rosenberg, S. A.; Klein, H.; Berger, M.; Mullen, C. A.; Ramsey, W. J.; Muul, L.; Morgan, R. A.; Anderson, W. F. (1995). "T Lymphocyte-Directed Gene Therapy for ADA- SCID: Initial Trial Results After 4 Years". *Science* **270** (5235): 475–480. Bibcode:1995Sci...270..475B. doi:10.1126/science.270.5235.475.

[54] Trojan J., Johnson T., Rudin S., Ilan Ju., Tykocinski M., Ilan J. (1993). "Treatment and prevention of rat glioblastoma by immugenic C6 cells expressing antisense insulin-like growth factor I RNA". *Science* **259**: 94–97. Bibcode:1993Sci...259...94T. doi:10.1126/science.8418502.

[55] Abbott A (1992). "Gene therapy. Italians first to use stem cells". *Nature* **356** (6369): 465–199. Bibcode:1992Natur.356..465A. doi:10.1038/356465a0. PMID 1560817.

[56] Cavazzana-Calvo, M.; Thrasher, A.; Mavilio, F. (2004). "The future of gene therapy". *Nature* **427** (6977): 779–781. Bibcode:2004Natur.427..779C. doi:10.1038/427779a. PMID 14985734.

[57] Stein, Rob (11 October 2010). "First patient treated in stem cell study". *The Washington Post*. Retrieved 10 November 2010.

[58] "Death Prompts FDA to Suspend Arthritis Gene Therapy Trial". Medpage Today. 27 July 2007. Retrieved 10 November 2010.

[59] Stolberg, Sheryl Gay (22 January 2000). "Gene Therapy Ordered Halted At University". *The New York Times*. Retrieved 10 November 2010.

[60] Trojan J, Pan YX, Wei MX, Ly A, Shevelev A, Bierwagen M, Ardourel M-Y, Trojan L, Alvarez A, Andres C, Noguera MC, Briceño I, Aristizabal BH, Kasprzak H, Duc HT, Anthony DD (2012). "Methodology for anti - gene anti - IGF-I therapy of malignant tumours". *Chemother. Res. Pract* **2012**: 1–12. doi:10.1155/2012/721873.

[61] Trojan J. and Briceno I. IGF-I Antisense and Triple-Helix Gene Therapy of Glioblastoma In: A. Pantar "Evolution of the Molecular Biology of Brain Tumors and the Therapeutic Implications"

[62] Wilson, Jennifer Fisher (18 March 2002). "Murine Gene Therapy Corrects Symptoms of Sickle Cell Disease". *The Scientist – Magazine of the Life Sciences*. Retrieved 17 August 2010.

[63] St. Jude Children's Research Hospital (4 December 2008). "Gene Therapy Corrects Sickle Cell Disease In Laboratory Study". ScienceDaily. Retrieved 29 December 2012.

[64] Penman, Danny (11 October 2002). "Subtle gene therapy tackles blood disorder". *New Scientist*. Retrieved 17 August 2010.

[65] "DNA nanoballs boost gene therapy". *New Scientist*. 12 May 2002. Retrieved 17 August 2010.

[66] Ananthaswamy, Anil (20 March 2003). "Undercover genes slip into the brain". *New Scientist*. Retrieved 17 August 2010.

[67] Holmes, Bob (13 March 2003). "Gene therapy may switch off Huntington's". *New Scientist*. Retrieved 17 August 2010.

[68] Ott, M. G.; Schmidt, M.; Schwarzwaelder, K.; Stein, S.; Siler, U.; Koehl, U.; Glimm, H.; Kühlcke, K.; Schilz, A.; Kunkel, H.; Naundorf, S.; Brinkmann, A.; Deichmann, A.; Fischer, M.; Ball, C.; Pilz, I.; Dunbar, C.; Du, Y.; Jenkins, N. A.; Copeland, N. G.; Lüthi, U.; Hassan, M.; Thrasher, A. J.; Hoelzer, D.; Von Kalle, C.; Seger, R.; Grez, M. (2006). "Correction of X-linked chronic granulomatous disease by gene therapy, augmented by insertional activation of MDS1-EVI1, PRDM16 or SETBP1". *Nature Medicine* **12** (4): 401–409. doi:10.1038/nm1393. PMID 16582916.

[69] Brown, B. D.; Venneri, M. A.; Zingale, A.; Sergi, L. S.; Naldini, L. (2006). "Endogenous microRNA regulation suppresses transgene expression in hematopoietic lineages and enables stable gene transfer". *Nature Medicine* **12** (5): 585–591. doi:10.1038/nm1398. PMID 16633348.

[70] Morgan, R. A.; Dudley, M. E.; Wunderlich, J. R.; Hughes, M. S.; Yang, J. C.; Sherry, R. M.; Royal, R. E.; Topalian, S. L.; Kammula, U. S.; Restifo, N. P.; Zheng, Z.; Nahvi, A.; De Vries, C. R.; Rogers-Freezer, L. J.; Mavroukakis, S. A.; Rosenberg, S. A. (2006). "Cancer Regression in Patients After Transfer of Genetically Engineered Lymphocytes". *Science* **314** (5796): 126–129. Bibcode:2006Sci...314..126M. doi:10.1126/science.1129003. PMC 2267026. PMID 16946036.

[71] Levine, B. L.; Humeau, L. M.; Boyer, J.; MacGregor, R. -R.; Rebello, T.; Lu, X.; Binder, G. K.; Slepushkin, V.; Lemiale, F.; Mascola, J. R.; Bushman, F. D.; Dropulic, B.; June, C. H. (2006). "Gene transfer in humans using a conditionally replicating lentiviral vector". *Proceedings of the National Academy of Sciences* **103** (46): 17372–17377. Bibcode:2006PNAS..10317372L. doi:10.1073/pnas.0608138103. PMC 1635018. PMID 17090675.

[72] "Penn Medicine presents HIV gene therapy trial data at CROI 2009". EurekAlert!. 10 February 2009. Retrieved 19 November 2009.

[73] "Gene therapy first for poor sight". BBC News. 1 May 2007. Retrieved 3 May 2010.

[74] Dolgin, E. (2009). "Colour blindness corrected by gene therapy". *Nature*. doi:10.1038/news.2009.921.

[75] Cartier, N.; Hacein-Bey-Abina, S.; Bartholomae, C. C.; Veres, G.; Schmidt, M.; Kutschera, I.; Vidaud, M.; Abel, U.; Dal-Cortivo, L.; Caccavelli, L.; Mahlaoui, N.; Kiermer, V.; Mittelstaedt, D.; Bellesme, C.; Lahlou, N.; Lefrere, F.; Blanche, S.; Audit, M.; Payen, E.; Leboulch, P.; l'Homme, B.; Bougneres, P.; von Kalle, C.; Fischer, A.; Cavazzana-Calvo, M.; Aubourg, P. (2009). "Hematopoietic Stem Cell Gene Therapy with a Lentiviral Vector in X-Linked Adrenoleukodystrophy". *Science* **326** (5954): 818–823. Bibcode:2009Sci...326..818C. doi:10.1126/science.1171242. PMID 19892975.

[76] Komáromy, A.; Alexander, J.; Rowlan, J.; Garcia, M.; Chiodo, V.; Kaya, A.; Tanaka, J.; Acland, G.; Hauswirth, W.; Aguirre, G. D. (2010). "Gene therapy rescues cone function in congenital achromatopsia". *Human Molecular Genetics* **19** (13): 2581–2593. doi:10.1093/hmg/ddq136. PMC 2883338. PMID 20378608.

[77] Cavazzana-Calvo, M.; Payen, E.; Negre, O.; Wang, G.; Hehir, K.; Fusil, F.; Down, J.; Denaro, M.; Brady, T.; Westerman, K.; Cavallesco, R.; Gillet-Legrand, B.; Caccavelli, L.; Sgarra, R.; Maouche-Chrétien, L.; Bernaudin, F. O.; Girot, R.; Dorazio, R.; Mulder, G. J.; Polack, A.; Bank, A.; Soulier, J.; Larghero, J. R. M.; Kabbara, N.; Dalle, B.; Gourmel, B.; Socie, G. R.; Chrétien, S.; Cartier, N.; Aubourg, P. (2010). "Transfusion independence and HMGA2 activation after gene therapy of human β-thalassaemia". *Nature* **467** (7313): 318–22. Bibcode:2010Natur.467..318C. doi:10.1038/nature09328. PMC 3355472. PMID 20844535.

[78] Galanello, R.; Origa, R. (2010). "Beta-thalassemia". *Orphanet Journal of Rare Diseases* **5**: 11. doi:10.1186/1750-1172-5-11. PMC 2893117. PMID 20492708.

[79] Beals, Jacquelyn K. (16 September 2010). Gene Therapy Frees Beta-Thalassemia Patient From Transfusions for 2+ Years. Medscape.com (16 September 2010). Retrieved 15 December 2012.

[80] Leboulch P (20 March 2013). "Five year outcome of lentiviral gene therapy for human beta-thalassemia, lessons and prospects". *Thalassemia Reports* **3** (1s): 108.

[81] (11 July 2012) ß-Thalassemia Major With Autologous CD34+ Hematopoietic Progenitor Cells Transduced With TNS9.3.55 a Lentiviral Vector Encoding the Normal Human ß-Globin Gene ClinicalTrials.gov. Clinical trial NCT01639690 at the Memorial Sloan-Kettering Cancer Center. Retrieved 12 February 2014

[82] Rosenberg, Tina (29 May 2011) The Man Who Had HIV and Now Does Not. *New York*.

[83] "Gene Therapy Turns Several Leukemia Patients Cancer Free. Will It Work for Other Cancers, Too?". Singularity Hub. Retrieved 7 January 2014.

[84] Yang, Z. J.; Zhang, Y. R.; Chen, B.; Zhang, S. L.; Jia, E. Z.; Wang, L. S.; Zhu, T. B.; Li, C. J.; Wang, H.; Huang, J.; Cao, K. J.; Ma, W. Z.; Wu, B.; Wang, L. S.; Wu, C. T. (2008). "Phase I clinical trial on intracoronary administration of Ad-hHGF treating severe coronary artery disease". *Molecular Biology Reports* **36** (6): 1323–1329. doi:10.1007/s11033-008-9315-3. PMID 18649012.

[85] Hahn, W.; Pyun, W. B.; Kim, D. S.; Yoo, W. S.; Lee, S. D.; Won, J. H.; Shin, G. J.; Kim, J. M.; Kim, S. (2011). "Enhanced cardioprotective effects by coexpression of two isoforms of hepatocyte growth factor from naked plasmid DNA in a rat ischemic heart disease model". *The Journal of Gene Medicine* **13** (10): 549–555. doi:10.1002/jgm.1603. PMID 21898720.

[86] Eurolab. Neovasculogen listing in Eurolab Page accessed August 4, 2015

[87] Deev, R.; Bozo, I.; Mzhavanadze, N.; Voronov, D.; Gavrilenko, A.; Chervyakov, Yu.; Staroverov, I.; Kalinin, R.; Shvalb, P.; Isaev, A. (13 March 2015). "pCMV-vegf165 Intramuscular Gene Transfer is an Effective Method of Treatment for Patients With Chronic Lower Limb Ischemia". *Journal of cardiovascular pharmacology and therapeutics* **20**: 473–82. doi:10.1177/1074248415574336. PMID 25770117.

[88] On Cancer: Launch of Stem Cell Therapy Trial Offers Hope for Patients with Inherited Blood Disorder | Memorial Sloan-Kettering Cancer Center. Mskcc.org (16 July 2012). Retrieved 15 December 2012.

[89] (4 September 2014) ß-Thalassemia Major With Autologous CD34+ Hematopoietic Progenitor Cells Transduced With

TNS9.3.55 a Lentiviral Vector Encoding the Normal Human ß-Globin Gene ClinicalTrials.gov, US National Institutes of Health. Retrieved 17 December 2014.

[90] Pollack, Andrew (20 July 2012) European Agency Backs Approval of a Gene Therapy, *The New York Times*.

[91] First Gene Therapy Approved by European Commission. UniQure (2 November 2012). Retrieved 15 December 2012.

[92] "Chiesi and uniQure delay Glybera launch to add data". *Biotechnology*. The Pharma Letter. 4 August 2014. Retrieved 28 August 2014.

[93] BURGER, LUDWIG; HIRSCHLER, BEN (November 26, 2014). "First gene therapy drug sets million-euro price record". *Reuters*. Retrieved March 2015.

[94] Bosely, Sarah (30 April 2013) Pioneering gene therapy trials offer hope for heart patients *The Guardian*. Retrieved 28 April 2014

[95] First gene therapy trial for heart failure begins in UK. The Physicians Clinic (8 September 2013) Archived April 29, 2014, at the Wayback Machine.

[96] Celladon Receives Breakthrough Therapy Designation From FDA for MYDICAR(R), Novel, First-in-Class Therapy in Development to Treat Heart Failure. New York Times (10 April 2014)

[97] Biffi, A.; Montini, E.; Lorioli, L.; Cesani, M.; Fumagalli, F.; Plati, T.; Baldoli, C.; Martino, S.; Calabria, A.; Canale, S.; Benedicenti, F.; Vallanti, G.; Biasco, L.; Leo, S.; Kabbara, N.; Zanetti, G.; Rizzo, W. B.; Mehta, N. A. L.; Cicalese, M. P.; Casiraghi, M.; Boelens, J. J.; Del Carro, U.; Dow, D. J.; Schmidt, M.; Assanelli, A.; Neduva, V.; Di Serio, C.; Stupka, E.; Gardner, J.; Von Kalle, C. (2013). "Lentiviral Hematopoietic Stem Cell Gene Therapy Benefits Metachromatic Leukodystrophy". *Science* **341** (6148): 1233158. doi:10.1126/science.1233158. PMID 23845948.

[98] Aiuti, A.; Biasco, L.; Scaramuzza, S.; Ferrua, F.; Cicalese, M. P.; Baricordi, C.; Dionisio, F.; Calabria, A.; Giannelli, S.; Castiello, M. C.; Bosticardo, M.; Evangelio, C.; Assanelli, A.; Casiraghi, M.; Di Nunzio, S.; Callegaro, L.; Benati, C.; Rizzardi, P.; Pellin, D.; Di Serio, C.; Schmidt, M.; Von Kalle, C.; Gardner, J.; Mehta, N.; Neduva, V.; Dow, D. J.; Galy, A.; Miniero, R.; Finocchi, A.; Metin, A. (2013). "Lentiviral Hematopoietic Stem Cell Gene Therapy in Patients with Wiskott-Aldrich Syndrome". *Science* **341** (6148): 1233151. doi:10.1126/science.1233151. PMID 23845947.

[99] Gallagher, James (21 April 2015) Gene therapy: 'Tame HIV' used to cure disease BBC News, Health. Retrieved 21 April 2015

[100] Malech, H. L.; Ochs, H. D. (2015). "An Emerging Era of Clinical Benefit from Gene Therapy". *JAMA (Journal of the American Medical Association)* **313** (15): 1522. doi:10.1001/jama.2015.2055.

[101] Gene therapy cure for children with 'bubble baby' disease Science Daily (18 November 2014)

[102] Gene therapy provides safe, long-term relief for patients with severe hemophilia B Science Daily (20 November 2014)

[103] Lee, Tim WR; Southern, K. W. (26 November 2013). "Topical cystic fibrosis transmembrane conductance regulator gene replacement for cystic fibrosis-related lung disease". *Cochrane Database Syst Rev.* **11** (11): CD005599. doi:10.1002/14651858.CD005599.pub4. PMID 24282073.

[104] MacLaren, R. E.; Groppe, M.; Barnard, A. R.; Cottriall, C. L.; Tolmachova, T.; Seymour, L.; Clark, K. R.; During, M. J.; Cremers, F. P. M.; Black, G. C. M.; Lotery, A. J.; Downes, S. M.; Webster, A. R.; Seabra, M. C. (2014). "Retinal gene therapy in patients with choroideremia: Initial findings from a phase 1/2 clinical trial". *The Lancet* **383** (9923): 1129–37. doi:10.1016/S0140-6736(13)62117-0. PMID 24439297.

[105] Beali, Abigail (25 January 2014) Gene therapy restores sight in people with eye disease The New Scientist. Retrieved 25 January 2014

[106] Tebas, P.; Stein, D.; Tang, W. W.; Frank, I.; Wang, S. Q.; Lee, G.; Spratt, S. K.; Surosky, R. T.; Giedlin, M. A.; Nichol, G.; Holmes, M. C.; Gregory, P. D.; Ando, D. G.; Kalos, M.; Collman, R. G.; Binder-Scholl, G.; Plesa, G.; Hwang, W. T.; Levine, B. L.; June, C. H. (2014). "Gene Editing of CCR5 in Autologous CD4 T Cells of Persons Infected with HIV". *New England Journal of Medicine* **370** (10): 901–10. doi:10.1056/NEJMoa1300662. PMID 24597865.

[107] Dvorsky, George (6 March 2014) Scientists Create Genetically Modified Cells That Protect Against HIV io9, Biotechnology. Retrieved 6 March 2014

[108] (15 December 2014) Stem Cell Gene Therapy for Sickle Cell Disease. ClinicalTrials.gov Identifier: NCT02247843 ClinicalTrials.gov. U.S. National Institutes of Health. Retrieved 17 December 2014

[109] Collection and Storage of Umbilical Cord Stem Cells for Treatment of Sickle Cell Disease; ClinicalTrials.gov Identifier: NCT00012545 ClinicalTrials.gov. U.S. National Institutes of Health (15 December 2014)

[110] Olowoyeye, A; Okwundu. C. I. (October 2015). "Gene therapy for sickle cell disease". *Cochrane Database of Systematic Reviews* **11** (10): CD007652. doi:10.1002/14651858.CD007652.pub3. PMID 23152248.

[111] "Ten things you might have missed Monday from the world of business". *Boston Globe*. 3 February 2015. Retrieved 13 February 2015.

[112] Zimmer, Carl (9 March 2015). "Protection Without a Vaccine". *The New York Times*. Retrieved March 2015.

[113] Gardner, M. R.; Kattenhorn, L. M.; Kondur, H. R.; von Schaewen, M.; Dorfman, T.; Chiang, J. J.; Haworth, K. G.; Decker, J. M.; Alpert, M. D.; Bailey, C. C.; Neale, E. S.; Fellinger, C. H.; Joshi, V. R.; Fuchs, S. P.; Martinez-Navio, J. M.; Quinlan, B. D.; Yao, A. Y.; Mouquet, H.; Gorman, J.; Zhang, B.; Poignard, P.; Nussenzweig, M. C.; Burton, D. R.; Kwong, P. D.; Piatak, M.; Lifson, J. D.; Gao, G.; Desrosiers, R. C.; Evans, D. T.; et al. (2015). "AAV-expressed eCD4-Ig provides durable protection from multiple SHIV challenges". Nature 519 (7541): 87–91. Bibcode:2015Natur.519...87G. doi:10.1038/nature14264.

[114] Wade, Nicholas (19 March 2015). "Scientists Seek Ban on Method of Editing the Human Genome". New York Times. Retrieved 20 March 2015.

[115] Pollack, Andrew (3 March 2015). "A Powerful New Way to Edit DNA". New York Times. Retrieved 20 March 2015.

[116] Baltimore, David; Berg, Paul; Botchan, Dana; Charo, R. Alta; Church, George; Corn, Jacob E.; Daley, George Q.; Doudna, Jennifer A.; Fenner, Marsha; Greely, Henry T.; Jinek, Martin; Martin, G. Steven; Penhoet, Edward; Puck, Jennifer; Sternberg, Samuel H.; Weissman, Jonathan S.; Yamamoto, Keith R. (19 March 2015). "A prudent path forward for genomic engineering and germline gene modification". Science 348: 36–8. Bibcode:2015Sci...348...36B. doi:10.1126/science.aab1028. PMID 25791083. Retrieved 20 March 2015.

[117] Lanphier, Edward; Urnov, Fyodor; Haecker, Sarah Ehlen; Werner, Michael; Smolenski, Joanna (26 March 2015). "Don't edit the human germ line". Nature 519: 410–411. Bibcode:2015Natur.519..410L. doi:10.1038/519410a. PMID 25810189. Retrieved 20 March 2015.

[118] Die 1-Million-Euro-Spritze (4 April 2015)

[119] Sample, Ian (5 November 2015). "Baby girl is first in the world to be treated with 'designer immune cells'". The Guardian. Retrieved 6 November 2015.

[120] Wade, Nicholas (3 December 2015). "Scientists Place Moratorium on Edits to Human Genome That Could Be Inherited". New York Times. Retrieved 3 December 2015.

[121] Walsh, Fergus (3 December 2015). "Gene editing: Is era of designer humans getting closer?". BBC News Health. Retrieved 31 December 2015.

[122] Liang, Puping; et al. (18 April 2015). "CRISPR/Cas9-mediated gene editing in human tripronuclear zygotes". Protein & Cell 6: 363–372. doi:10.1007/s13238-015-0153-5. PMC 4417674. PMID 25894090. Retrieved 24 April 2015.

[123] Wu, Yuxuan; Zhou, Hai; Fan, Xiaoying; Zhang, Ying; Zhang, Man; Wang, Yinghua; Xie, Zhenfei; Bai, Meizhu; Yin, Qi (2015-01-01). "Correction of a genetic disease by CRISPR-Cas9-mediated gene editing in mouse spermatogonial stem cells". Cell Research 25 (1): 67–79. doi:10.1038/cr.2014.160. ISSN 1001-0602.

[124] "WADA Gene Doping". WADA. Archived from the original on 5 July 2013. Retrieved 27 September 2013.

[125] Kayser, B.; Mauron, A.; Miah, A. (2007). "Current anti-doping policy: A critical appraisal". BMC Medical Ethics 8: 2. doi:10.1186/1472-6939-8-2. PMC 1851967. PMID 17394662.

[126] Powell, R.; Buchanan, A. (2011). "Breaking Evolution's Chains: The Prospect of Deliberate Genetic Modification in Humans". Journal of Medicine and Philosophy 36: 6–27. doi:10.1093/jmp/jhq057. PMID 21228084.

[127] Baylis, F.; Robert, J. S. (2004). "The Inevitability of Genetic Enhancement Technologies". Bioethics 18: 1–26. doi:10.1111/j.1467-8519.2004.00376.x.

[128] Evans, John (2002). Playing God?: Human Genetic Engineering and the Rationalization of Public Bioethical Debate. University of Chicago Press.

[129] Gene Therapy and Genetic Engineering. The Center for Health Ethics. University of Missouri School of Medicine. 25 April 2013.

[130] Roco MC; Bainbridge WS (2002). "Converging Technologies for Improving Human Performance: Integrating From the Nanoscale". Journal of Nanoparticle Research 4 (4): 281–295. doi:10.1023/A:1021152023349.

[131] Allhoff, F. (2005). "Germ-Line Genetic Enhancement and Rawlsian Primary Goods". Kennedy Institute of Ethics Journal 15: 39–56. doi:10.1353/ken.2005.0007.

[132] AMA Council on Ethical and Judicial Affairs. Report on Ethical Issues Related to Prenatal Genetic Tests. 3 Archives Fam. Med. 633, 637–39 (1994). Archived September 28, 2011, at the Wayback Machine.

[133] The Declaration of Inuyama: Human Genome Mapping, Genetic Screening and Gene Therapy

[134] Smith KR, Chan S, Harris J. Human germline genetic modification: scientific and bioethical perspectives. Arch Med Res. 2012 Oct;43(7):491-513. doi: 10.1016/j.arcmed.2012.09.003. PMID 23072719

[135] Kolata, Gina (23 April 2015). "Chinese Scientists Edit Genes of Human Embryos, Raising Concerns". New York Times. Retrieved 24 April 2015.

[136] Human Genome Organization. HUGO Ethics Committee. Statement on Gene Therapy Research. April 2001.

[137] Isasi, R. M. et al. (October 2006) National Regulatory Frameworks Regarding Human Genetic Modification Technologies (Somatic and Germline Modification). Genetics & Public Policy Center.

[138] National Institutes of Health. NIH Guidelines for Research Involving Recombinant or Synthetic Nucleic Acid Molecules. Revised March 2013.

[139] U.S. Department of Health & Human Services. The National Commission for the Protection of Human Subjects of Biomedical and Behavioral Research. The Belmont Report: Ethical Principles and Guidelines for the Protection of Human Subjects of Research. 18 April 1979.

[140] U.S. Food and Drug Administration (14 October 1993). "Application of Current Statutory Authorities to Human Somatic Cell Therapy Products and Gene Therapy Products" (PDF). *Federal Register* **58** (197).

[141] U.S. Department of Health and Human Services. Food and Drug Administration. Center for Biologics Evaluation and Research. Guidance for Industry: Guidance for Human Somatic Cell Therapy and Gene Therapy. March 1998.

[142] "A Real-life 'I Am Legend?' Researcher Champions Development Of 'Reovirus' As Potential Treatment For Cancer". Sciencedaily.com. 9 May 2008. Retrieved 17 August 2010.

[143] Gene Therapy Change the Human Race?// *Gene therapy* at the Internet Movie Database

11.12 Further reading

- Tinkov S, Bekeredjian R, Winter G, Coester C (20 November 2000). "Polyplex-conjugated microbubbles for enhanced ultrasound targeted gene therapy" (PDF). Georgia World Congress Center, Atlanta, GA, USA: 2008 AAPS Annual Meeting and Exposition.

- Gardlík R, Pálffy R, Hodosy J, Lukács J, Turna J, Celec P; Pálffy; Hodosy; Lukács; Turna; Celec (Apr 2005). "Vectors and delivery systems in gene therapy". *Med Sci Monit.* **11** (4): RA110–21. PMID 15795707.

- Staff (18 November 2005). "Gene Therapy" (FAQ). *Human Genome Project Information.* Oak Ridge National Laboratory. Retrieved 28 May 2006.

- Salmons B, Günzburg WH; Günzburg (Apr 1993). "Targeting of retroviral vectors for gene therapy". *Hum Gene Ther.* **4** (2): 129–41. doi:10.1089/hum.1993.4.2-129. PMID 8494923.

- Baum C, Düllmann J, Li Z, Fehse B, Meyer J, Williams DA, von Kalle C; Düllmann; Li; Fehse; Meyer; Williams; von Kalle (Mar 2003). "Side effects of retroviral gene transfer into hematopoietic stem cells". *Blood* **101** (6): 2099–114. doi:10.1182/blood-2002-07-2314. PMID 12511419.

- Horn PA, Morris JC, Neff T, Kiem HP; Morris; Neff; Kiem (Sep 2004). "Stem cell gene transfer—efficacy and safety in large animal studies". *Mol. Ther.* **10** (3): 417–31. doi:10.1016/j.ymthe.2004.05.017. PMID 15336643.

- Wang H, Shayakhmetov DM, Leege T, Harkey M, Li Q, Papayannopoulou T, Stamatoyannopolous G, Lieber A; Shayakhmetov; Leege; Harkey; Li; Papayannopoulou; Stamatoyannopolous; Lieber (September 2005). "A Capsid-Modified Helper-Dependent Adenovirus Vector Containing the β-Globin Locus Control Region Displays a Nonrandom Integration Pattern and Allows Stable, Erythroid-Specific Gene Expression". *Journal of Virology* **79** (17): 10999–1013. doi:10.1128/JVI.79.17.10999-11013.2005. PMC 1193620. PMID 16103151.

11.13 External links

- Gene Therapy: Molecular Bandage? University of Utah's Genetic Science Learning Center

- The American Society of Gene & Cell Therapy

- The European Society of Gene & Cell Therapy

- Research Group at Cambridge, UK working on overcoming current hurdles to successful gene therapy

- Council for Responsible Genetics

- Molecular Medicine and Gene Therapy at Lund University

- Gene Therapy Frees β-Thalassemia Patient From Transfusions

- Clinical Trial at Sloan Kettering

- Stem Cell Therapy Trial Offers Hope

Chapter 12

Heritability

Studies of heritability ask questions such as how much genetic factors play a role in differences in height between people. This is not the same as asking how much genetic factors influence height in any one person.

Heritability is a statistic used in breeding and genetics works that estimates how much variation in a phenotypic trait in a population is due to genetic variation among individuals in that population.[1] It is calculated with the following equation (for broad-sense heritability): $H^2 = VG/VP$.[2] Other causes of measured variation in a trait are characterized as environmental factors, including measurement error. In human studies of heritability these are often apportioned into factors from "shared environment" and

"non-shared environment" based on whether they tend to result in persons brought up in the same household more or less similar to persons who were not.

Some humans in a population are taller than others; heritability attempts to identify how much genetics play a role in part of the population being taller. Heritability is estimated by comparing individual phenotypic variation among differently related individuals in a population. Heritability is an important concept in quantitative genetics, particularly in selective breeding and behavior genetics (for instance, twin studies), but is less widely used in population genetics.

Geoffrey Miller, an evolutionary psychologist, has said, writing about sexual selection and biological fitness, "The concept of heritability applies only to traits that differ between individuals. If a trait exists in precisely the same form across all individuals, it may be inherited, but it cannot be heritable."[3]

12.1   Overview

Heritability measures the fraction of phenotype variability that can be attributed to genetic variation. This is not the same as saying that this fraction of an individual phenotype is caused by genetics. In addition, heritability can change without any genetic change occurring, such as when the environment starts contributing to more variation. A case in point, consider that both genes and environment have the potential to influence intelligence. Heritability could increase if genetic variation increases, causing individuals to show more phenotypic variation, like showing different levels of intelligence. On the other hand, heritability might also increase if the environmental variation decreases, causing individuals to show less phenotypic variation, like showing more similar levels of intelligence. Heritability increases when genetics are contributing more variation or because non-genetic factors are contributing less variation; what matters is the relative contribution. Heritability is specific to a particular population in a particular environment.

The extent of dependence of phenotype on environment can also be a function of the genes involved. Matters of heritability are complicated because genes may canalize a phenotype, making its expression almost inevitable in all occurring environments. Individuals with the same genotype can also exhibit different phenotypes through a mechanism called phenotypic plasticity, which makes heritability difficult to measure in some cases. Recent insights in molecular biology have identified changes in transcriptional activity of individual genes associated with environmental changes. However, there are a large number of genes whose transcription is not affected by the environment.[4]

Estimates of heritability use statistical analyses to help to identify the causes of differences between individuals. Since heritability is concerned with variance, it is necessarily an account of the differences between individuals in a population. Heritability can be univariate – examining a single trait – or multivariate – examining the genetic and environmental associations between multiple traits at once. This allows a test of the genetic overlap between different phenotypes: for instance hair color and eye color. Environment and genetics may also interact, and heritability analyses can test for and examine these interactions (GxE models).

A prerequisite for heritability analyses is that there is some population variation to account for. This last point highlights the fact that heritability cannot take into account the effect of factors which are invariant in the population. Factors may be invariant if they are absent and do not exist in the population, such as no one having access to a particular antibiotic, or because they are omni-present, like if everyone is drinking coffee. In practice, all human behavioral traits vary and almost all traits show some heritability.[5]

12.2 Definition

Any particular phenotype can be modeled as the sum of genetic and environmental effects:[6]

Phenotype (P) = Genotype (G) + Environment (E).

Likewise the variance in the trait – Var (P) – is the sum of effects as follows:

$$\mathrm{Var}(P) = \mathrm{Var}(G) + \mathrm{Var}(E) + 2\,\mathrm{Cov}(G,E).$$

In a planned experiment Cov(G,E) can be controlled and held at 0. In this case, heritability is defined as:

$$H^2 = \frac{\mathrm{Var}(G)}{\mathrm{Var}(P)}$$

H^2 is the broad-sense heritability. This reflects all the genetic contributions to a population's phenotypic variance including additive, dominant, and epistatic (multi-genic interactions), as well as maternal and paternal effects, where individuals are directly affected by their parents' phenotype, such as with milk production in mammals.

A particularly important component of the genetic variance is the additive variance, Var(A), which is the variance due to the average effects (additive effects) of the alleles. Since each parent passes a single allele per locus to each offspring, parent-offspring resemblance depends upon the average effect of single alleles. Additive variance represents, therefore, the genetic component of variance responsible for parent-offspring resemblance. The additive genetic portion of the phenotypic variance is known as Narrow-sense heritability and is defined as

$$h^2 = \frac{\mathrm{Var}(A)}{\mathrm{Var}(P)}$$

An upper case H^2 is used to denote broad sense, and lower case h^2 for narrow sense.

Additive variance is important for selection. If a selective pressure such as improving livestock is exerted, the response of the trait is directly related to narrow-sense heritability. The mean of the trait will increase in the next generation as a function of how much the mean of the selected parents differs from the mean of the population from which the selected parents were chosen. The observed *response to selection* leads to an estimate of the narrow-sense heritability (called **realized heritability**). This is the principle underlying artificial selection or breeding.

12.2.1 Example

The simplest genetic model involves a single locus with two alleles (b and B) affecting one quantitative phenotype.

The number of **B** alleles can vary from 0, 1, or 2. For any genotype, B_iB_j, the expected phenotype can then be written as the sum of the overall mean, a linear effect, and a dominance deviation:

$$P_{ij} = \mu + \alpha_i + \alpha_j + d_{ij} = \text{Population mean} +$$
Additive Effect $(\alpha_{ij} = \alpha_i + \alpha_j)$ + Dominance Deviation (d_{ij}).

The additive genetic variance at this locus is the weighted average of the squares of the additive effects:

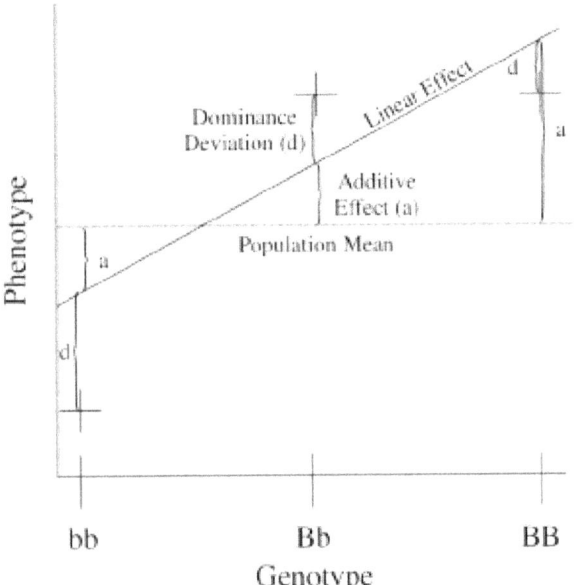

Figure 1. Relationship of phenotypic values to additive and dominance effects using a completely dominant locus.

$$\text{Var}(A) = f(bb)a_{bb}^2 + f(Bb)a_{Bb}^2 + f(BB)a_{BB}^2.$$

where $f(bb)a_{bb} + f(Bb)a_{Bb} + f(BB)a_{BB} = 0$.

There is a similar relationship for variance of dominance deviations:

$$\text{Var}(D) = f(bb)d_{bb}^2 + f(Bb)d_{Bb}^2 + f(BB)d_{BB}^2.$$

where $f(bb)d_{bb} + f(Bb)d_{Bb} + f(BB)d_{BB} = 0$.

The linear regression of phenotype on genotype is shown in Figure 1.

12.3   Estimating heritability

Since only P can be observed or measured directly, heritability must be estimated from the similarities observed in subjects varying in their level of genetic or environmental similarity. The statistical analyses required to estimate the genetic and environmental components of variance depend on the sample characteristics. Briefly, better estimates are obtained using data from individuals with widely varying levels of genetic relationship - such as twins, siblings, parents and offspring, rather than from more distantly related (and therefore less similar) subjects. The standard error for heritability estimates is improved with large sample sizes.

In non-human populations it is often possible to collect information in a controlled way. For example, among farm animals it is easy to arrange for a bull to produce offspring from a large number of cows and to control environments. Such experimental control is generally not possible when gathering human data, relying on naturally occurring relationships and environments.

Studies of human heritability often utilize adoption study designs, often with identical twins who have been separated early in life and raised in different environments (see for example Fig. 2). Such individuals have identical genotypes and can be used to separate the effects of genotype and environment. A limit of this design is the common prenatal environment and the relatively low numbers of twins reared apart. A second and more common design is the twin study in which the similarity of identical and fraternal twins is used to estimate heritability. These studies can be limited by the fact that identical twins are not completely genetically identical, potentially resulting in an underestimation of heritability. Studies of twins also examine differences between twins and non-twin siblings, for instance to examine phenomena such as intrauterine competition (for example, twin-to-twin transfusion syndrome).

Heritability estimates are always relative to the genetic and environmental factors in the population, and are not absolute measurements of the contribution of genetic and environmental factors to a phenotype. Heritability estimates reflect the amount of variation in genotypic effects compared to variation in environmental effects.

Heritability can be made larger by diversifying the genetic background, e.g., by using only very out bred individuals (which increases VarG) and/or by minimizing environmental effects (decreasing VarE). The converse also holds. Due to such effects, different populations of a species might have different heritabilities for the same trait.

In observational studies, or because of evokative effects (where a genome evokes environments by its effect on them), G and E may covary: gene environment correlation. Depending on the methods used to estimate heritability, correlations between genetic factors and shared or non-shared environments may or may not be confounded with heritability.[7]

12.3.1   Common misunderstandings of heritability estimates

A common estimate of heritability is called the Heritability Index (HI), which ranges from 0 - 1. A HI index of 0 means that none of the variability between people in the study sample on the trait under investigation is due to genetic factors; an HI of 1 indicates the opposite.

Heritability estimates are often misinterpreted if it is not understood that they refer to the *proportion of variation be-*

tween individuals on a trait that is due to genetic factors. It does not indicate the degree of genetic influence on the development of a trait of an individual. For example, it is incorrect to say that since the heritability of personality traits is about .6, that means that 60% of your personality is inherited from your parents and 40% comes from the environment.

Even a highly heritable trait (such as eye color) assumes environmental inputs which are required for development: for instance temperatures and an atmosphere supporting life, etc. A more useful distinction than "nature vs. nurture" is "obligate vs. facultative"—under typical environmental ranges, what traits are more "obligate" (e.g., the nose—everyone has a nose) or more "facultative" (sensitive to environmental variations, such as specific language learned during infancy). Another useful distinction is between traits that are likely to be adaptations (such as the umbilical cord) vs. those that are byproducts of adaptations (such as the belly button), or are due to random variation (non-adaptive variation in belly button shape, e.g., convex or concave).

12.4 Estimation methods

There are essentially two schools of thought regarding estimation of heritability.

One school of thought was developed by Sewall Wright at The University of Chicago, and further popularized by C. C. Li (University of Chicago) and J. L. Lush (Iowa State University). It is based on the analysis of correlations and, by extension, regression. Path Analysis was developed by Sewall Wright as a way of estimating heritability.

The second was originally developed by R. A. Fisher and expanded at The University of Edinburgh, Iowa State University, and North Carolina State University, as well as other schools. It is based on the analysis of variance of breeding studies, using the intraclass correlation of relatives. Various methods of estimating components of variance (and, hence, heritability) from ANOVA are used in these analyses.

12.5 Regression/correlation methods of estimation

The first school of estimation uses regression and correlation to estimate heritability.

12.5.1 Selection experiments

Calculating the strength of selection, S (the difference in mean trait between the population as a whole and the se-

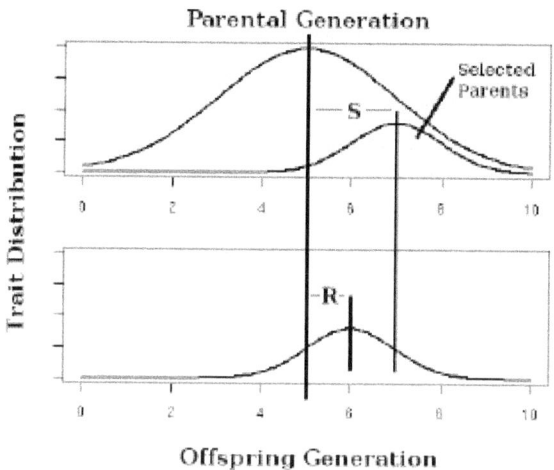

Figure 3. Strength of selection (S) and response to selection (R) in an artificial selection experiment. $h^2 = R/S$.

lected parents of the next generation, also called the *selection differential*[8]) and response to selection R (the difference in offspring and whole parental generation mean trait) in an artificial selection experiment will allow calculation of realized heritability as the response to selection relative to the strength of selection, $h^2 = R/S$ as in Fig. 3.

12.5.2 Comparison of close relatives

In the comparison of relatives, we find that in general,

$h^2 = \frac{b}{r} = \frac{t^2}{r}$ where r can be thought of as the coefficient of relatedness, b is the coefficient of regression and t the coefficient of correlation.

Parent-offspring regression

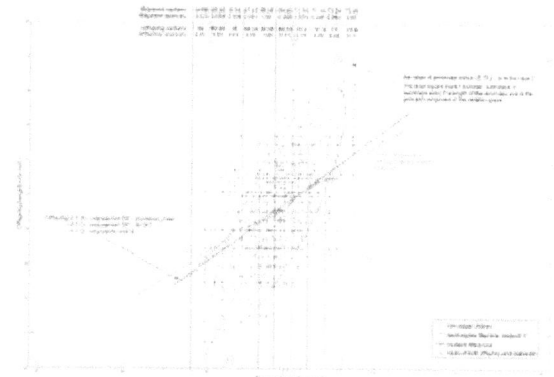

Figure 4. Sir Francis Galton's (1889) data showing the relationship between offspring height (928 individuals) as a function of mean parent height (205 sets of parents).

Heritability may be estimated by comparing parent and off-spring traits (as in Fig. 4). The slope of the line (0.57) approximates the heritability of the trait when offspring values are regressed against the average trait in the parents. If only one parent's value is used then heritability is twice the slope. (Note that this is the source of the term "regression," since the offspring values always tend to regress to the mean value for the population, *i.e.*, the slope is always less than one). This regression effect also underlies the DeFries Fulker method for analyzing twins selected for one member being affected.[9]

Sibling comparison

A basic approach to heritability can be taken using full-Sib designs: comparing similarity between siblings who share both a biological mother and a father.[10] When there is only additive gene action, this sibling phenotypic correlation is an index of *familiarity* – the sum of half the additive genetic variance plus full effect of the common environment. It thus places an upper-limit on additive heritability of twice the full-Sib phenotypic correlation. Half-Sib designs compare phenotypic traits of siblings that share one parent with other sibling groups.

Twin studies

Main article: Twin study

Heritability for traits in humans is most frequently estimated by comparing resemblances between twins (Fig. 2 & 5). "The advantage of twin studies, is that the total variance can be split up into genetic, shared or common environmental, and unique environmental components, enabling an accurate estimation of heritability".[11] Fraternal

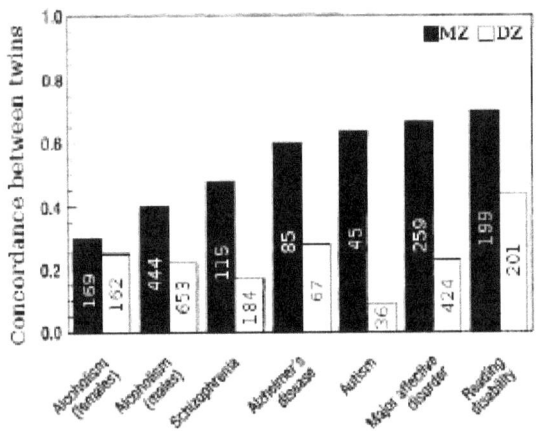

Figure 5. Twin concordances for seven psychological traits (sample size shown inside bars), with DZ being fraternal and MZ being identical twins.

or dizygotic (DZ) twins on average share half their genes (assuming there is no assortative mating for the trait), and so identical or monozygotic (MZ) twins on average are twice as genetically similar as DZ twins. A crude estimate of heritability, then, is approximately twice the difference in correlation between MZ and DZ twins, i.e. Falconer's formula $H^2 = 2(r(MZ) - r(DZ))$.

The effect of shared environment, c^2, contributes to similarity between siblings due to the commonality of the environment they are raised in. Shared environment is approximated by the DZ correlation minus half heritability, which is the degree to which DZ twins share the same genes, $c^2 = DZ - 1/2h^2$. Unique environmental variance, e^2, reflects the degree to which identical twins raised together are dissimilar, $e^2 = 1 - r(MZ)$.

The methodology of the classical twin study has been criticized, but some of these criticisms do not take into account the methodological innovations and refinements described above.

12.5.3 Extended pedigree design

While often heritability is analyzed in single generations: comparing MZ twins raised apart, or comparing the similarity of MZ and DZ twins, considerable power can be gained using more complex relationships. By studying a trait in multi-generational families, the multiple recombination of genetic and environmental effects can be decomposed using software such as ASReml and heritability estimated.[12] This design is helpful for untangling confounds such as reverse causality, maternal effects such as the prenatal environment, and confounding of genetic dominance, shared environment, and maternal gene effects[13][14] but is generally less powerful than the twin design for obtaining heritability estimates.

12.6   Analysis of variance methods of estimation

The second set of methods of estimation of heritability involves ANOVA and estimation of variance components.

12.6.1   Basic model

We use the basic discussion of Kempthorne (1957 [1969]). Considering only the most basic of genetic models, we can look at the quantitative contribution of a single locus with genotype G_i as

$$y_i = \mu + g_i + e$$

where

g_i is the effect of genotype G_i

and e is the environmental effect.

Consider an experiment with a group of sires and their progeny from random dams. Since the progeny get half of their genes from the father and half from their (random) mother, the progeny equation is

$$z_i = \mu + \tfrac{1}{2}g_i + e$$

Intraclass correlations

Consider the experiment above. We have two groups of progeny we can compare. The first is comparing the various progeny for an individual sire (called *within sire group*). The variance will include terms for genetic variance (since they did not all get the same genotype) and environmental variance. This is thought of as an *error* term.

The second group of progeny are comparisons of means of half sibs with each other (called *among sire group*). In addition to the error term as in the within sire groups, we have an addition term due to the differences among different means of half sibs. The intraclass correlation is

$$\text{corr}(z, z') = \text{corr}(\mu + \tfrac{1}{2}g + e, \mu + \tfrac{1}{2}g + e') = \tfrac{1}{4}V_g$$

since environmental effects are independent of each other.

The ANOVA

In an experiment with n sires and r progeny per sire, we can calculate the following ANOVA, using V_g as the genetic variance and V_e as the environmental variance:

The $\tfrac{1}{4}V_g$ term is the intraclass correlation among half sibs. We can easily calculate $H^2 = \frac{V_g}{V_g + V_e} = \frac{4(S-W)}{S+(r-1)W}$. The Expected Mean Square is calculated from the relationship of the individuals (progeny within a sire are all half-sibs, for example), and an understanding of intraclass correlations.

12.6.2 Model with additive and dominance terms

For a model with additive and dominance terms, but not others, the equation for a single locus is

$$y_{ij} = \mu + \alpha_i + \alpha_j + d_{ij} + e,$$

where

α_i is the additive effect of the ith allele, α_j is the additive effect of the jth allele, d_{ij} is the dominance deviation for the ij[th] genotype, and e is the environment.

Experiments can be run with a similar setup to the one given in Table 1. Using different relationship groups, we can evaluate different intraclass correlations. Using V_a as the additive genetic variance and V_d as the dominance deviation variance, intraclass correlations become linear functions of these parameters. In general,

$$= rV_a + \theta V_d.$$

where r and θ are found as

$r = P[$ alleles drawn at random from the relationship pair are identical by descent], and

$\theta = P[$ genotypes drawn at random from the relationship pair are identical by descent].

Some common relationships and their coefficients are given in Table 2.

12.6.3 Larger models

When a large, complex pedigree is available for estimating heritability, the most efficient use of the data is in a restricted maximum likelihood (REML) model. The raw data will usually have three or more data points for each individual: a code for the sire, a code for the dam and one or several trait values. Different trait values may be for different traits or for different time points of measurement.

The currently popular methodology relies on high degrees of certainty over the identities of the sire and dam; it is not common to treat the sire identity probabilistically. This is not usually a problem, since the methodology is rarely applied to wild populations (although it has been used for several wild ungulate and bird populations), and sires are invariably known with a very high degree of certainty in breeding programmes. There are also algorithms that account for uncertain paternity.

The pedigrees can be viewed using programs such as Pedigree Viewer , and analyzed with programs such as ASReml, VCE , WOMBAT or BLUPF90 family's programs

12.7 Response to selection

In selective breeding of plants and animals, the expected response to selection of a trait with known narrow-sense heritability h can be estimated using the *breeder's equation*:[15]

$$R = h^2 S$$

In this equation, the Response to Selection (R) is defined as the realized average difference between the parent generation and the next generation, and the Selection Differential (S) is defined as the average difference between the parent generation and the selected parents.

For example, imagine that a plant breeder is involved in a selective breeding project with the aim of increasing the number of kernels per ear of corn. For the sake of argument, let us assume that the average ear of corn in the parent generation has 100 kernels. Let us also assume that the selected parents produce corn with an average of 120 kernels per ear. If h^2 equals 0.5, then the next generation will produce corn with an average of $0.5(120-100) = 10$ additional kernels per ear. Therefore, the total number of kernels per ear of corn will equal, on average, 110.

Note that heritability in the above equation is equal to the ratio $Var(A)/Var(P)$ only if the genotype and the environmental noise follow Gaussian distributions .

12.8 Controversies

Heritability estimates' prominent critics, such as Steven Rose,[16] Jay Joseph,[17] and Richard Bentall, focus largely on heritability estimates in behavioral sciences and social sciences. They claim that such heritability scores are typically calculated counterintuitively to derive numerically high scores, that heritability is misinterpreted as genetically determination, and that this alleged bias distracts from other factors that researches have found more causally important, such as childhood abuse in the causation of later psychosis.[18][19]

The controversy over heritability estimates is largely via their basis in twin studies. The scarce success of molecular-genetic studies corroborating such population-genetic studies' conclusions is the *missing heritability* problem.[20] Prominent apologist Eric Turkheimer has claimed that newer molecular methods have vindicated the conventional interpretation of twin studies,[20] although it remains mostly unclear how to explain the relations between genes and behaviors.[21] Turkheimer has acknowledged that both genes and environment are heritable, that genetic contribution varies by environment, and that focus on heritability distracts from important factors.[22] Overall, however, *heritability* is a concept widely applicable.[14]

12.9 See also

- Heredity

- Heritability of IQ

12.10 References

12.10.1 Notes

[1] Wray, Naomi; Visscher, Peter (2008). "Estimating Trait Heritability". *Nature Education* **1** (1): 29. Retrieved 24 July 2015.

[2] Roff, D. A. Evolutionary quantitative genetics. Chapman and Hall, New York, NY

[3] Miller, Geoffrey (2000). *The mating mind: how sexual choice shaped the evolution of human nature*. London, Heineman. ISBN 0-434-00741-2 (also Doubleday, ISBN 0-385-49516-1) p.115

[4] Wills, C. (2007). "Principles of Population Genetics, 4th edition". *Journal of Heredity* (Book Review) **98** (4): 382–382. doi:10.1093/jhered/esm035.

- *review of*: Hartl, Daniel L.; Clark, Andrew G. (2007). Sunderland, MA: Sinauer and Associates. pp. xv + 652. ISBN 0-87893-308-5. Missing or empty |title= (help)

[5] Turkheimer, Eric (October 2000). "Three Laws of Behavior Genetics and What They Mean" (PDF). *Current Directions in Psychological Science* **9** (5): 160–164. doi:10.1111/1467-8721.00084. ISSN 0963-7214. Retrieved 29 October 2013.

[6] Kempthorne 1957

[7] Cattell RB (1960). "The multiple abstract variance analysis equations and solutions: for nature–nurture research on continuous variables". *Psychol Rev* **67** (6): 353–372. doi:10.1037/h0043487. PMID 13691636.

[8] Kempthorne 1957, p. 507; or Falconer & Mackay 1995, p. 191, for example.

[9] Defries, J. C.; Fulker, D. W. (September 1985). "Multiple regression analysis of twin data". *Behavior Genetics* **15** (5): 467–473. doi:10.1007/BF01066239. PMID 4074272.

[10] Falconer, Douglas S.; Mackay, Trudy F. C. (December 1995). *Introduction to Quantitative Genetics* (4th ed.). Longman. ISBN 978-0582243026.

[11] Gielen, M., Lindsey, P.J., Derom, C., Smeets, H.J.M., Souren, N.Y., Paulussen, A.D.C., Derom, R., & Nijhuis, J.G. (2008) "Modeling Genetic and Environmental Factors to IncreaseHeritability and Ease the Identification of Candidate Genes for Birth Weight: A Twin Study". Behavioral Genetics. 38(44-54):45. DOI 10.1007/s10519-007-9170-3

[12] Luciano, M.; Batty, G. D.; McGilchrist, M.; Linksted, P.; Fitzpatrick, B.; Jackson, C.; Pattie, A.; Dominiczak, A. F.; Morris, A. D.; Smith, B. H. (May–June 2010). "Shared genetic aetiology between cognitive ability and cardiovascular disease risk factors: Generation Scotland's Scottish family health study". *Intelligence* **38** (3): 304–313. doi:10.1016/j.intell.2010.03.002.

[13] Hill, W. G.; Goddard, M. E.; Visscher, P. M. (2008). MacKay, Trudy F. C., ed. "Data and Theory Point to Mainly Additive Genetic Variance for Complex Traits". *PLOS Genetics* **4** (2): e1000008. doi:10.1371/journal.pgen.1000008. PMC 2265475. PMID 18454194.

[14] Visscher, P. M.; Hill, W. G.; Wray, N. R. (April 2008). "Heritability in the genomics era — concepts and misconceptions" (PDF). *Nature Reviews Genetics* **9** (4): 255–266. doi:10.1038/nrg2322. PMID 18319743.

[15] Plomin, R., DeFries, J. C., & McClearn, G. E. (1990). Behavioral genetics. New York: Freeman.

[16] Rose, Steven P R (2006). "Commentary: Heritability estimates—long past their sell-by date". *International Journal of Epidemiology* **35** (3): 525–527. doi:10.1093/ije/dyl064. PMID 16645027.

[17] Jay Joseph, *The Gene Illusion* (New York: Algora, 2004), esp ch 5.

[18] Richard P Bentall. *Doctoring the Mind: Is Our Current Treatment of Mental Illness Really Any Good?* (New York: New York University Press, 2009), p 123–127.

[19] Melanie McGrath, "*Doctoring the Mind*: Review". *The Telegraph*, 2009 Jul 5.

[20] Turkheimer, Eric (2011). "Still missing". *Research in Human Development* **8** (3–4): 227–241. doi:10.1080/15427609.2011.625321.

[21] Eric Turkheimer, "Genetic prediction", *Hastings Center Report*, 2015 Sep/Oct;**45**(S1):S32–S38.

[22] Jay Joseph, *The Trouble with Twin Studies: A Reassessment of Twin Research in the Social and Behavioral Sciences* (New York & Hove: Routledge, 2015), esp p 81 [chapter summaries].

12.10.2 Books

- Falconer, D. S. and T. Mackay. 1996. Introduction to Quantitative Genetics. Ed. 4. Longman, Essex, England.

- Kempthorne, O (1957 [1969]) *An Introduction to Genetic Statistics*. John Wiley. Reprinted, 1969 by Iowa State University Press.

12.11 Further reading

- Lynch, M. & Walsh, B. 1996. *Genetics and Analysis of Quantitative Traits*. Sinauer Associates. ISBN 0-87893-481-2.

- Johnson, Wendy; Penke, Lars; Spinath, Frank M. (2011). "Understanding Heritability: What it is and What it is Not". *European Journal of Personality* **25** (4): 287–294. doi:10.1002/per.835. ISSN 0890-2070. Archived (PDF) from the original on 2011. Retrieved 15 December 2013.

12.12 External links

- Stanford Encyclopedia of Philosophy entry on Heredity and Heritability

- Quantitative Genetics Resources website, including the two volume book by Lynch and Walsh. Free access

Chapter 13

Cell fusion

Cell fusion is an important cellular process in which several uninuclear cells (cells with a single nucleus) combine to form a multinuclear cell, known as a syncytium. Cell fusion occurs during differentiation of muscle, bone and trophoblast cells, during embryogenesis, and during morphogenesis.[1] Cell fusion is a necessary event in the maturation of cells so that they maintain their specific functions throughout growth.

13.1 History

In 1847 Theodore Schwann expanded upon the theory that all living organisms are composed of cells when he added to it that discrete cells are the basis of life. Schwann observed that in certain cells the walls and cavities of the cells coalesce together. It was this observation that provided the first hint that cells fuse. It was not until 1960 that cell biologists deliberately fused cells for the first time. To fuse the cells, biologists combined isolated mouse cells, with the same kind of tissue, and induced fusion of their outer membrane using the Sendai virus (a respiratory virus in mice). Each of the fused hybrid cells contained a single nucleus with chromosomes from both fusion partners. Synkaryon became the name of this type of cell combined with a nucleus. In the late 1960s biologists successfully fused cells of different types and from different species. The hybrid products of these fusions, heterokaryon, were hybrids that maintained two or more separate nuclei. This work was headed by Henry Harris at the University of Oxford and Nils Ringertz from Sweden's Karolinska Institute. These two men are responsible for reviving the interest of cell fusion. The hybrid cells interested biologists in the area of how different kinds of cytoplasm affect different kinds of nuclei. The work conducted by Henry and Nils showed that proteins from one gene fusion affect gene expression in the other partner's nucleus, and vice versa. These hybrid cells that were created were considered forced exceptions to normal cellular integrity and it was not until 2002 that the possibility of cell fusion between cells of different types may have a real function in mammals.[2]

13.2 Two Types of Cell Fusion

There are two different types of cell fusion that can occur. These two types include homotypic and heterotypic cell fusion.

Homotypic cell fusion occurs between cells of the same type. An example of this would be osteoclasts and myofibres being fused together. Whenever the two nuclei merge a synkaryon is produced. Cell fusion normally occurs with nuclear fusion, but in the absence of nuclear fusion, the cell would be described as a binucleated heterokaryon. A heterokaryon is the melding of two or more cells into one and it may reproduce itself for several generations.[3]

Heterotypic cell fusion occurs between cells of different types, making it the exact opposite of homotypic cell fusion. The result of this fusion is also a synkaryon produced by the merging of the nuclei, and a binucleated heterokaryon in the absence of nuclear fusion. An example of this would be BMDCs being fused with parenchymatous organs.[4]

13.3 Three Methods for Fusing Cells

There are three methods that cell biologists use to fuse cells. These three ways include electrical cell fusion, polyethylene glycol cell fusion, and sendai virus induced cell fusion.

Electrical cell fusion, is an essential step in some of the most innovative methods in modern biology. This method begins when two cells are brought into contact by dielectrophoresis. Dielectrophoresis uses a highfrequency alternating current, unlike electrophoresis in which a direct current is applied. Once the cells are brought together, a pulsed voltage is applied. The pulse voltage causes the cell membrane to permeate and subsequent combining of the membranes and the cells then fuse. After this, alternative voltage is applied for a brief period of time to stabilize the process. The result of this is that the cytoplasm has mixed together and the cell membrane has completely fused. All that remains separate is the nuclei, which will fuse at a later

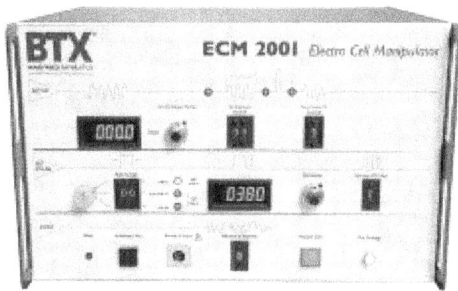

BTX ECM 2001 Electrofusion generator cell fusion applications manufactured by BTX Harvard Apparatus, Holliston MA USA

time within the cell, making the result a heterokaryon cell.[5]

Polyethylene glycol cell fusion is the simplest, but most toxic, way to fuse cells. In this type of cell fusion polyethylene glycol, PEG, acts as a dehydrating agent and fuses not only plasma membranes but also intracellular membranes. This leads to cell fusion since PEG induces cell agglutination and cell-to-cell contact. Though this type of cell fusion is the most widely used, it still has downfalls. Oftentimes PEG can cause uncontrollable fusion of multiple cells, leading to the appearance of giant polykaryons. Also, standard PEG cell fusion is poorly reproducible and different types of cells have various fusion susceptibilities. This type of cell fusion is widely used for the production of somatic cell hybrids and for nuclear transfer in mammalian cloning.[6]

Sendai virus induced cell fusion occurs in four different temperature stages. During the first stage, which lasts no longer than 10 minutes, viral adsorption takes place and the adsorbed virus can be inhibited by viral antibodies. The second stage, which is 20 minutes, is pH dependent and an addition of viral antiserum can still inhibit ultimate fusion. In the third, antibody-refractory stage, viral envelope constituents remain detectable on the surface of cells. During the fourth stage cell fusion becomes evident and HA neuraminidase and fusion factor begin to disappear. The first and second stages are the only two that are pH dependent.[7]

13.4 Cell Fusion for Human Therapy

Alternative forms of restoring organ function and replacing damaged cells are needed with donor organs and tissue for transplantation being so scarce. It is because of the scarcity that biologists have begun considering the poten-

tial for therapeutic cell fusion. Biologists have been discussing implications of the observation that cell fusion can occur with restorative effects following tissue damage or cell transplantation. Though using cell fusion for this is being talked about and worked on, there are still many challenges those who wish to implement cell fusion as a therapeutic tool face. These challenges include choosing the best cells to use for the reparative fusion, determining the best way to introduce the chosen cells into the desired tissue, discovering methods to increase the incidence in cell fusion, and ensuring that the resulting fusion products will function properly. If these challenges can be overcome then cell fusion may have therapeutic potential.[8]

13.5 Other Uses

- To study the control of cell division and gene expression.

- To Investigate malignant transformations.

- To obtain viral replication.

- For gene and chromosome mapping.

- For production of monoclonal antibodies by producing hybridoma.

- For production of Induced stem cells.

- To assess protein shuttling in what is known as a *heterokaryon fusion assay.*[9]

13.6 See also

- Interbilayer Forces in Membrane Fusion

- Fusion mechanism

- Cellular differentiation

- Lipid bilayer fusion

- Cell-cell fusogens

13.7 References

[1] "6.3. Cell fusion". Herkules.oulu.fi. Retrieved 2013-08-16.

[2] http://www.jstor.org/discover/10.2307/27858450?uid= 3739744&uid=2134&uid=369718881&uid=2&uid=70& uid=3&uid=369718871&uid=3739256&uid=60&sid= 21101968188181

[3] Cell fusion definition - Medical Dictionary definitions of popular medical terms easily defined on MedTerms

[4] Figure 1 : Inflammation as a matchmaker: revisiting cell fusion : Nature Cell Biology

[5] Principles And Applications Of Electrical Cell Fusion | Biocompare: The Buyer's Guide for Life Scientists

[6] Cancer Cell International | Full text | Cell fusion in tumor progression: the isolation of cell fusion products by physical methods

[7] Factors Affecting Cell Fusion Induced by Sendai Virus

[8] The potential of cell fusion for human therapy. [Stem Cell Rev. 2006] - PubMed - NCBI

[9] Gammal, Roseann; Baker, Krista; Heilman, Destin (2011). "Heterokaryon Technique for Analysis of Cell Type-specific Localization". *Journal of Visualized Experiments* (49). doi:10.3791/2488. ISSN 1940-087X.

13.8   Further reading

- H. Harris: *Cell fusion*, 1970, Harvard University Press, Mass.

- R. Borgens et al.: *Cell Fusion and some subcellular Properties of heterokaryons and hybrids*, Journal of Cell Biology, VOLUME 67, 1975, pages 257-280

Chapter 14

Hybrid (biology)

For other uses, see Hybrid.

In biology a hybrid, also known as cross breed, is the re-

Hercules, a "tiger", a lion/tiger hybrid.

sult of mixing, through sexual reproduction, two animals or plants of different breeds, varieties, species or genera.[1] Using genetic terminology, it may be defined as follows.[2]

1. **Hybrid** generally refers to any offspring resulting from the breeding of two genetically distinct individuals, which usually will result in a high degree of heterozygosity, though hybrid and heterozygous are not, strictly speaking, synonymous.

2. a **genetic hybrid** carries two different alleles of the same gene

3. a **structural hybrid** results from the fusion of

gametes that have differing structure in at least one chromosome, as a result of structural abnormalities

4. a **numerical hybrid** results from the fusion of gametes having different haploid numbers of chromosomes

5. a **permanent hybrid** is a situation where only the heterozygous genotype occurs, because all homozygous combinations are lethal.

From a taxonomic perspective, hybrid refers to:

1. Offspring resulting from the interbreeding between two animal species or plant species.[3] See also hybrid speciation.

2. Hybrids between different subspecies within a species (such as between the Bengal tiger and Siberian tiger) are known as **intra-specific** hybrids. Hybrids between different species within the same genus (such as between lions and tigers) are sometimes known as **interspecific** hybrids or crosses. Hybrids between different genera (such as between sheep and goats) are known as **intergeneric** hybrids. Extremely rare **interfamilial** hybrids have been known to occur (such as the guineafowl hybrids).[4] No **interordinal** (between different orders) animal hybrids are known.

3. The third type of hybrid consists of crosses between populations, breeds or cultivars within a single species. This meaning is often used in plant and animal breeding, where hybrids are commonly produced and selected, because they have desirable characteristics not found or inconsistently present in the parent individuals or populations.

14.1 Terminology

The term hybrid is derived from Latin *hybrida*, meaning the *"offspring of a tame sow and a wild boar"*, *"child of a freeman and slave"*, etc.[5] The term came into popular use in English in the 19th century, though examples of its use have been found from the early 17th century.[6]

There is a popular convention of naming hybrids by forming portmanteau words. The template for this is the naming of tiger-lion hybrids as liger and tigon in the 1920s.[7] This was playfully (but unsystematically) extended to a number of other hybrids, or hypothetical hybrids, such as beefalo (1960s), humanzee (1980s), cama (1998).

14.2   Types of hybrids

Depending on the parents, there are a number of different types of hybrids:[8]

- *Single cross hybrids* — result from the cross between two true breeding organisms and produces an F1 generation called an F1 hybrid (F1 is short for Filial 1, meaning "first offspring"). The cross between two different homozygous lines produces an F1 hybrid that is heterozygous; having two alleles, one contributed by each parent and typically one is dominant and the other recessive. Typically, the F1 generation is also phenotypically homogeneous, producing offspring that are all similar to each other.

- *Double cross hybrids* — result from the cross between two different F1 hybrids.[9]

- *Three-way cross hybrids* — result from the cross between one parent that is an F1 hybrid and the other is from an inbred line.[10]

- *Triple cross hybrids* — result from the crossing of two different three-way cross hybrids.

- *Population hybrids* — result from the crossing of plants or animals in a population with another population. These include crosses between organisms such as interspecific hybrids or crosses between different breeds.

- *Stable hybrid* – a horticultural term which typically refers to an annual plant that, if grown and bred in a small monoculture free of external pollen (e.g., an air-filtered greenhouse) will produce offspring that are "true to type" with respect to phenotype; i.e., a true breeding organism.[11]

- *Hybrid species* – results from hybrid populations evolving reproductive barriers against their parent species through hybrid speciation.[12]

14.3   Interspecific hybrids

Interspecific hybrids are bred by mating two species, normally from within the same genus. The offspring display traits and characteristics of both parents. The offspring of an interspecific cross are very often sterile; thus, hybrid sterility prevents the movement of genes from one species to the other, keeping both species distinct.[13] Sterility is often attributed to the different number of chromosomes the two species have, for example donkeys have 62 chromosomes, while horses have 64 chromosomes, and mules and hinnies have 63 chromosomes. Mules, hinnies, and other normally sterile interspecific hybrids cannot produce viable gametes, because differences in chromosome structure prevent appropriate pairing and segregation during meiosis, meiosis is disrupted, and viable sperm and eggs are not formed. However, fertility in female mules has been reported with a donkey as the father.[14]

Most often other processes occurring in plants and animals keep gametic isolation and species distinction. Species often have different mating or courtship patterns or behaviors, the breeding seasons may be distinct and even if mating does occur antigenic reactions to the sperm of other species prevent fertilization or embryo development. Hybridisation is much more common among organisms that spawn indiscriminately, like soft corals and among plants.

While it is possible to predict the genetic composition of a backcross *on average*, it is not possible to accurately predict the composition of a particular backcrossed individual, due to random segregation of chromosomes. In a species with two pairs of chromosomes, a twice backcrossed individual would be predicted to contain 12.5% of one species' genome (say, species A). However, it may, in fact, still be a 50% hybrid if the chromosomes from species A were lucky in two successive segregations, and meiotic crossovers happened near the telomeres. The chance of this is fairly high: $\left(\frac{1}{2}\right)^{(2 \times 2)} = \frac{1}{16}$ (where the "two times two" comes about from two rounds of meiosis with two chromosomes); however, this probability declines markedly with chromosome number and so the actual composition of a hybrid will be increasingly closer to the predicted composition.

14.4   Hybrid species

Main article: Hybrid speciation

While not very common, a few animal species have been recognized as being the result of hybridization. The Lonicera fly is an example of a novel animal species that resulted from natural hybridization. The American red wolf appears to be a hybrid species between gray wolf and coyote.[15] although its taxonomic status has been a subject of controversy.[16][17][18] The European edible frog appears to be a species, but is actually a semi-permanent hybrid between pool frogs and marsh frogs. The edible frog population is dependent on the presence of at least one of the parents species to be maintained.[19]

Hybrid species of plants are much more common than animals. Many of the crop species are hybrids, and hybridization appears to be an important factor in speciation in some plant groups.

14.5 Examples of hybrid animals and animal populations derived from hybrids

A mule, a domestic canary/goldfinch hybrid.

A "zonkey", a zebra/donkey hybrid.

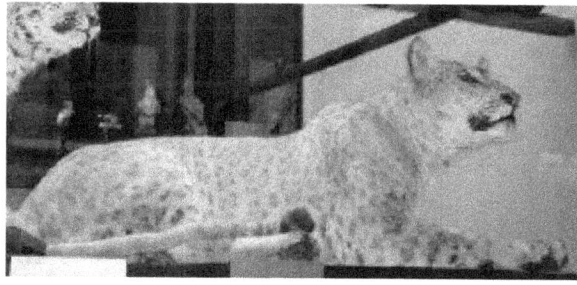

A "jaglion", a jaguar/lion hybrid.

14.5.1 Mammals

- Equid hybrids

    - Mule, a cross of female horse and a male donkey.

- Hinny, a cross between a female donkey and a male horse. Mule and hinny are examples of reciprocal hybrids.

- Zebroids

    - Zeedonk or Zonkey, a zebra/donkey cross.

    - Zorse, a zebra/horse cross

    - Zony or Zetland, a zebra/pony cross ("zony" is a generic term; "zetland" is specifically a hybrid of the Shetland pony breed with a zebra)

- hybrid ass, a cross between a donkey and an onager or Asian wild ass.

- Bovid hybrids

    - Dzo, zo or yakow; a cross between a domestic cow/bull and a yak.

    - Beefalo, a cross of an American bison and a domestic cow. This is a fertile breed; this along with genetic evidence has caused them to be recently reclassified into the same genus, *Bos*.

    - Żubroń, a hybrid between wisent (European bison) and domestic cow.

- Sheep-goat hybrid is the cross between a sheep and a goat, which belong to different genera.

- Ursid hybrids, such as the grizzly-polar bear hybrid, occur between black bears, brown bears, and polar bears.

- Felid hybrids

    - Savannah cat are a fertile **breed** developed originally from a cross between the serval [*Leptailurus serval*] and a domestic cat [*Felis catus*].

- A hybrid between a Bengal tiger and a Siberian tiger is an example of an *intra-specific* hybrid. It also includes the Indochinese tiger, Sumatran tiger too.

- Pumapards are the hybrid crosses between a puma and a leopard.

- Ligers and tigons (crosses between a lion and a tiger - the difference in name due to what species the mother and father were - ligers have a lion father and a tiger mother) and other Panthera hybrids such as the lijagulep. Various other wild cat crosses are known involving the lynx, bobcat, leopard, serval, etc.

  - Liligers are the hybrid cross between a male lion and a ligress.

- Bengals are a fertile *breed* developed originally from a cross between the Asian leopard cat [*Prionailurus bengalensis*] and the domestic cat [*Felis catus*].

- Fertile canid hybrids occur between coyotes, wolves, dingoes, jackals and domestic dogs.

- Hybrids between black and white rhinoceroses have been recognized.

- Hybrid camel, a cross between a bactrian camel and a dromedary camel[20]

- Cama, a cross between a camel and a llama, also an intergeneric hybrid.

- Wholphin, a fertile but very rare cross between a false killer whale and a bottlenose dolphin.

- At Chester Zoo in the United Kingdom, a cross between an African elephant (male) and an Asian elephant (female). The male calf was named Motty. He died of intestinal infection after twelve days.

- Bornean and Sumatran orangutan hybrids have occurred in captivity.

14.5.2 Birds

See also: Bird hybrid

- Hybrids between spotted owls and barred owls

- Cagebird breeders sometimes breed hybrids between species of finch, such as goldfinch × canary. These birds are known as mules.

- The perlin is a peregrine falcon – merlin hybrid.

- Gamebird hybrids, hybrids between gamebirds and domestic fowl, including chickens, guineafowl and peafowl, interfamilial hybrids.

- Numerous macaw hybrids and lovebird hybrids are also known in aviculture.

- Red kite × black kite: five bred unintentionally at a falconry center in England. (It is reported that the black kite (the male) refused female black kites but mated with two female red kites.)

- The mulard duck, hybrid of the domestic pekin duck and domesticated muscovy ducks.

- In Australia, New Zealand and other areas where the Pacific black duck occurs, it is hybridised by the much more aggressive introduced mallard. This is a concern to wildlife authorities throughout the affected area, as it is seen as Genetic pollution of the black duck gene pool.

- Hybridisation in gulls is a reasonably frequent occurrence in the wild.

14.5.3 Reptiles

- Hybrid iguana, a single-cross hybrid resulting from natural interbreeding between male marine iguanas and female land iguanas since the late 2000s.

- Crestoua, a cross between a Rhacodactylus Ciliatus (crested gecko) and a Rhacodactylus Chahoua.

- Colubrid snakes of the tribe Lampropeltini have been shown to produce fertile hybrid offspring.

- Hybridization between the endemic Cuban crocodile (*Crocodilus rhombifer*) and the widely distributed American crocodile (*Crocodilus acutus*) is causing conservation problems for the former species as a threat to its genetic integrity.[21]

- Saltwater crocodiles (*Crocodylus porosus*) have mated with Siamese crocodiles (*Crocodylus siamensis*) in captivity producing offspring which in many cases have grown over 20 feet (6.1 metres) in length. It is likely that wild hybridization occurred historically in parts of southeast Asia.

- Many species of boas and pythons are known to produce hybrids, such as carball (a cross between a ball python and a carpet python) or a bloodball (a cross between a blood python and a ball python) however, most of these only occur in captivity. Contrary to popular belief, boa–python hybrids are not possible due to their differing reproductive functions. Boas only produce hybrids with other species of boas, and pythons only produce hybrids with other species of pythons.

14.5.4 Amphibians

- Japanese giant salamanders and Chinese giant salamanders have created hybrids that threaten the survival of Japanese giant salamanders due to the competition for similar resources in Japan.[22]

14.5.5 Fish

- Blood parrot cichlid, which is probably created by crossing a red head cichlid and a Midas cichlid or red devil cichlid

- A group of about 50 hybrids between Australian blacktip shark and the larger common blacktip shark was found by Australia's East Coast in 2012. This is the only known case of hybridization in sharks.[23]

- Silver bream and common bream commonly produce sterile hybrids.

- Tiger muskie is a sterile hybrid between northern pike and muskellunge.

14.5.6 Insects

- Killer bees were created during an attempt to breed a strain of bees that would produce more honey and be better adapted to tropical conditions. This was done by crossing a European honey bee and an African bee.

- The *Colias eurytheme* and *C. philodice* butterflies have retained enough genetic compatibility to produce viable hybrid offspring.[24]

14.6 Hybrid plants

Many hybrids are created by humans, but natural hybrids occur as well. Plant species hybridize more readily than animal species, and the resulting hybrids are more often fertile hybrids and may reproduce, though there still exist sterile hybrids and selective hybrid elimination where the offspring are less able to survive and are thus eliminated before they can reproduce. A number of plant species are the result of hybridization and polyploidy with many plant species easily cross pollinating and producing viable seeds, the distinction between each species is often maintained by geographical isolation or differences in the flowering period. Since

A sterile hybrid between Trillium cernuum *and* T. grandiflorum

An ornamental lily hybrid known as Lilium *'Citronella'*[25]

plants hybridize frequently without much work, they are often created by humans in order to produce improved plants. These improvements can include the production of more or improved seeds, fruits or other plant parts for consumption, or to make a plant more winter or heat hardy or improve its growth and/or appearance for use in horticulture. Much work is now being done with hybrids to produce more disease resistant plants for both agricultural and horticultural crops. In many groups of plants hybridization has been used to produce larger and more showy flowers and new flower colors. Hybridization may be restricted to the desired parent species through the use of pollination bags.

Many plant genera and species have their origins in polyploidy. Autopolyploidy results from the sudden multiplication in the number of chromosomes in typical normal populations caused by unsuccessful separation of the chromosomes during meiosis. Tetraploids (plants with four sets of chromosomes rather than two) are common in a number of different groups of plants and over time these plants can differentiate into distinct species from the normal diploid line. In *Oenothera lamarchiana* the diploid species has 14 chromosomes, this species has spontaneously given rise to plants with 28 chromosomes that have been given the name *Oenothera gigas*. When hybrids are formed between the tetraploids and the diploid population, the resulting offspring tend to be sterile triploids, thus effectively stopping the intermixing of genes between the two groups of plants (unless the diploids, in rare cases, produce unreduced gametes).

Another form of polyploidy called allopolyploidy occurs when two different species mate and produce polyploid hybrids. Usually the typical chromosome number is doubled, and the four sets of chromosomes can pair up during meiosis, thus the polyploids can produce offspring. Usually, these offspring can mate and reproduce with each other but cannot back-cross with the parent species. Allopolyploids may be able to adapt to new habitats that neither of their parent species inhabited.

[26]

Sterility in a non-polyploid hybrid is often a result of chromosome number; if parents are of differing chromosome pair number, the offspring will have an odd number of chromosomes, leaving them unable to produce chromosomally balanced gametes.[27] While this is undesirable in a crop such as wheat, where growing a crop which produces no seeds would be pointless, it is an attractive attribute in some fruits. Triploid bananas and watermelons are intentionally bred because they produce no seeds (and are parthenocarpic).

14.6.1 Heterosis

Main article: heterosis

Hybrids are sometimes stronger than either parent variety, a phenomenon most common with plant hybrids, which when present is known as *hybrid vigor* (heterosis) or heterozygote advantage.[28] A transgressive phenotype is a phenotype displaying more extreme characteristics than either of the parent lines.[29] Plant breeders make use of a number of techniques to produce hybrids, including line breeding and the formation of complex hybrids. An economically important example is hybrid maize (corn), which provides a considerable seed yield advantage over open pollinated varieties. Hybrid seed dominates the commercial maize seed market in the United States, Canada and many other major maize producing countries.[30]

14.6.2 Examples of plant hybrids

The multiplication symbol × (not italicised) indicates a hybrid in the Latin binomial nomenclature. Placed before the binomial it indicates a hybrid between species from different genera (intergeneric hybrid):-

- × *Fatshedera lizei*, a hybrid between *Hedera helix* and *Fatsia japonica*

- × *Heucherella*, a hybrid genus between *Heuchera* and *Tiarella*

- × *Philageria veitchii* is a hybrid between *Lapageria rosea* and *Philesia magellanica*; it is more similar in appearance to the former

- Leyland cypress, [× *Cupressocyparis leylandii*] hybrid between Monterey cypress and Nootka cypress

- Triticale, [× *Triticosecale*] a wheat–rye hybrid

- × *Urceocharis*, a hybrid between *Eucharis* and *Urceolina*

Interspecific plant hybrids include:

- *Dianthus* × *allwoodii* (*Dianthus caryophyllus* × *Dianthus plumarius*)

- Limequat *Citrus* × *floridana*, key lime *Citrus aurantifolia* and kumquat *Citrus japonica* hybrid

- Loganberry *Rubus* × *loganobaccus*, a hybrid between raspberry *Rubus idaeus* and blackberry *Rubus ursinus*

- London plane (*Platanus orientalis* × *Platanus occidentalis*), thus forming *Platanus* × *acerifolia*

- *Magnolia* × *alba* (*Magnolia champaca* × *Magnolia montana*)

- Peppermint, a hybrid between spearmint and water mint

- *Quercus* × *warei* (*Quercus robur* × *Quercus bicolor*) 'Nadler' (marketed in the United States under the trade name Kindred Spirit hybrid oak)

- Tangelo, a hybrid of a Mandarin orange and a pomelo which may have been developed in Asia about 3,500 years ago

- Wheat; most modern and ancient wheat breeds are themselves hybrids. Bread wheat is a hexaploid hybrid of three wild grasses; durum (pasta) wheat is a tetraploid hybrid of two wild grasses

- Grapefruit, hybrid between a pomelo and the Jamaican sweet orange

Some natural hybrids:

- *Iris albicans*, a sterile hybrid which spreads by rhizome division

- Evening primrose, a flower which was the subject of famous experiments by Hugo de Vries on polyploidy and diploidy

14.7 Hybrids in nature

Hybridization between two closely related species is actually a common occurrence in nature but is also being greatly influenced by anthropogenic changes as well.[31] Hybridization is a naturally occurring genetic process where individuals from two genetically distinct populations mate.[32] As stated above, it can occur both intraspecifically, between different distinct populations within the same species, and interspecifically, between two different species. Hybrids can be either sterile/not viable or viable/fertile. This affects the kind of effect that this hybrid will have on its and other populations that it interacts with.[33] Many hybrid zones are known where the ranges of two species meet, and hybrids are continually produced in great numbers. These hybrid zones are useful as biological model systems for studying the mechanisms of speciation (Hybrid speciation). Recently DNA analysis of a bear shot by a hunter in the North West Territories confirmed the existence of naturally-occurring and fertile grizzly–polar bear hybrids.[34]

14.7.1 Anthropogenic hybridization

Changes to the environment caused by humans, such as fragmentation and Introduced species, are becoming more widespread.[35] This increases the challenges in managing certain populations that are experiencing introgression, and is a focus of conservation genetics.

Introduced species and habitat fragmentation

Humans have introduced species worldwide to environments for a long time, both intentionally such as establishing a population to be used as a biological control, and unintentionally such as accidental escapes of individuals out of agriculture. This causes drastic global effects on various populations, including through hybridization.[33][36]

When habitats become broken apart, one of two things can occur, genetically speaking. The first is that populations that were once connected can be cut off from one another, preventing their genes from interacting. Occasionally, this will result in a population of one species breeding with a population of another species as a means of surviving such as the case with the red wolves. Their population numbers being so small, they needed another means of survival. Habitat fragmentation also led to the influx of generalist species into areas where they would not have been, leading to competition and in some cases interbreeding/incorporation of a population into another. In this way, habitat fragmentation is essentially an indirect method of introducing species to an area.

14.7.2 The hybridization continuum

There is a kind of continuum with three semi-distinct categories dealing with anthropogenic hybridization: hybridization without Introgression, hybridization with widespread introgression, and essentially a Hybrid swarm.[31] Depending on where a population falls along this continuum, the management plans for that population will change. Hybridization is currently an area of great discussion within Wildlife management and habitat management. Global climate change is creating other changes such as difference in population distributions which are indirect causes for an increase in anthropogenic hybridization.

14.7.3 Consequences

Hybridization can be a less discussed way toward extinction than within detection of where a population lies along the hybrid continuum. The dispute of hybridization is how to manage the resulting hybrids. When a population experiences hybridization with substantial introgression, there still exists parent types of each set of individuals. When a complete hybrid swarm is created, all the individuals are hybrids.

14.7.4 Management of hybrids

Conservationists disagree on when is the proper time to give up on a population that is becoming a hybrid swarm or to try and save the still existing pure individuals. Once it becomes a complete mixture, we should look to conserve those hybrids to avoid their loss.[31] Most leave it as a case-by-case basis, depending on detecting of hybrids within the group. It is nearly impossible to regulate hybridization via policy because hybridization can occur beneficially when it occurs "naturally" and there is the matter of protecting those previously mentioned hybrid swarms because if they are the only remaining evidence of prior species, they need to be conserved as well.[31]

14.7.5 Expression of parental traits in hybrids

When two distinct types of organisms breed with each other, the resulting hybrids typically have intermediate traits (e.g., one parent has red flowers, the other has white, and the hybrid, pink flowers).[37] Commonly, hybrids also combine traits seen only separately in one parent or the other (e.g., a bird hybrid might combine the yellow head of one parent with the orange belly of the other).[37]

In a hybrid, any trait that falls outside the range of parental variation is termed heterotic. Heterotic hybrids do have new traits, that is, they are not intermediate. *Positive heterosis* produces more robust hybrids, they might be stronger or bigger; while the term *negative heterosis* refers to weaker or smaller hybrids.[38] Heterosis is common in both animal and plant hybrids. For example, hybrids between a lion and a tigress ("ligers") are much larger than either of the two progenitors, while a tigon (lioness × tiger) is smaller. Also the hybrids between the common pheasant (*Phasianus colchicus*) and domestic fowl (*Gallus gallus*) are larger than either of their parents, as are those produced between the common pheasant and hen golden pheasant (*Chrysolophus pictus*).[39] Spurs are absent in hybrids of the former type, although present in both parents.[40]

14.7.6 Genetic mixing and extinction

Main article: Genetic pollution

Regionally developed ecotypes can be threatened with extinction when new alleles or genes are introduced that alter that ecotype. This is sometimes called genetic mixing.[41] Hybridization and introgression of new genetic material can lead to the replacement of local genotypes if the hybrids are more fit and have breeding advantages over the indigenous ecotype or species. These hybridization events can result from the introduction of non native genotypes by humans or through habitat modification, bringing previously isolated species into contact. Genetic mixing can be especially detrimental for rare species in isolated habitats, ultimately affecting the population to such a degree that none of the originally genetically distinct population remains.[42][43]

14.7.7 Effect on biodiversity and food security

Main articles: biodiversity and food security

In agriculture and animal husbandry, the Green Revolution's use of conventional hybridization increased yields by breeding "high-yielding varieties". The replacement of locally indigenous breeds, compounded with unintentional cross-pollination and crossbreeding (genetic mixing), has reduced the gene pools of various wild and indigenous breeds resulting in the loss of genetic diversity.[44] Since the indigenous breeds are often well-adapted to local extremes in climate and have immunity to local pathogens this can be a significant genetic erosion of the gene pool for future breeding. Therefore, commercial plant geneticists strive to breed "widely adapted" cultivars to counteract this tendency.[45]

14.8 Limiting factors

A number of conditions exist that limit the success of hybridization, the most obvious is great genetic diversity between most species. But in animals and plants that are more closely related hybridization barriers can include morphological differences, differing times of fertility, mating behaviors and cues, physiological rejection of sperm cells or the developing embryo.

In plants, barriers to hybridization include blooming period differences, different pollinator vectors, inhibition of pollen tube growth, somatoplastic sterility, cytoplasmic-genic male sterility and structural differences of the chromosomes.[46]

14.9 Mythical, legendary and religious hybrids

Main article: Mythological hybrid

Ancient folktales often contain mythological creatures, sometimes these are described as hybrids (e.g., hippogriff as the offspring of a griffin and a horse, and the Minotaur which is the offspring of Pasiphaë and a white bull). More often they are kind of chimera, i.e., a composite of the physical attributes of two or more kinds of animals, mythical beasts, and often humans, with no suggestion that they are the result of interbreeding, e.g., harpies, mermaids, and centaurs.

In the Bible, the Old Testament contains several passages

which talk about a first generation of hybrid giants who were known as the Nephilim.[47] The Book of Genesis (6:4) states that "the sons of God went to the daughters of humans and had children by them". As a result, the offspring was born as hybrid giants who became mighty heroes of old and legendary famous figures of ancient times.[48] In addition, the Book of Numbers (13:33) says that the descendants of Anak came from the Nephilim, whose bodies looked exactly like men, but with an enormous height. According to the apocryphal Book of Enoch the Nephilim were wicked sons of fallen angels who had lusted with attractive women.[49]

14.10 See also

- Artificial selection
- Bird hybrids
- Cabbit
- Canid hybrid
- Chimera (genetics)
- Chloroplast capture (botany)
- Felid hybrids
- F1 hybrids
- Genetic admixture
- Genetic erosion
- Grex (horticulture)
- Heterosis (hybrid vigor)
- Hybrid lovebird
- Hybrid (mythology)
- Hybrid name (botany)
- Hybrid speciation
- Hybrid swarm
- Hybrid zone
- Hybrot
- Inbreeding
- Interspecific pregnancy
- Intraspecific breeding
- Macropod hybrids

- Purebred
- Selective breeding
- Sheep-goat hybrids
- Species barrier
- Synergy

14.11 References

[1] "Hybrid". Merriam Webster. Retrieved 2014-06-12.

[2] Rieger, R.; Michaelis A.; Green, M. M. (1991). *Glossary of Genetics* (5th ed.). Springer-Verlag. ISBN 0-387-52054-6 page 256

[3] Keeton, William T. 1980. Biological science. New York: Norton. ISBN 0-393-95021-2 page A9.

[4] Ghigi A. 1936. "Galline di faraone e tacchini" Milano (Ulrico Hoepli)

[5] askoxford.com

[6] Oxford English Dictionary Online. Oxford University Press 2007.

[7] "When the sire is a lion the result is termed a Liger, whilst the converse is a Tigon." Edward George Boulenger, *World natural history*. B. T. Batsford ltd., 1937, p. 40.

[8] Wricke, Gunter, and Eberhard Weber. 1986. *Quantitative genetics and selection in plant breeding*. Berlin: W. de Gruyter. Page 257.

[9] J. O. Rawlings, C. Clark Cockerham *Analysis of Double Cross Hybrid Populations*. J. O. Rawlings, C. Clark Cockerham Biometrics. Vol. 18, No. 2 (Jun., 1962), pp. 229-244 doi:10.2307/2527461

[10] Roy, Darbeshwar. 2000. *Plant breeding analysis and exploitation of variation*. Pangbourne, UK: Alpha Science International. Page 446.

[11] Toogood, A.(ed.) (1999). *Plant Propagation* (1st American ed.). American Horticultural Society. ISBN 0-7894-5520-X page 21

[12] Arnold, M.L. (1996). *Natural Hybridization and Evolution*. New York: Oxford University Press. p. 232. ISBN 978-0-19-509975-1.

[13] Keeton, William T. 1980. *Biological science*. New York: Norton. ISBN 0-393-95021-2 Page 800

[14] Rong, R; Chandley, A. C.; Song, J; McBeath, S; Tan, P. P.; Bai, Q; Speed, R. M. (1988). "A fertile mule and hinny in China". *Cytogenetics and cell genetics* **47** (3): 134–9. doi:10.1159/000132531. PMID 3378453.

[15] Esch, Mary. "Study: Eastern wolves are hybrids with coyotes". Associated Press.

[16] Rutledge, Linda Y.; Wilson, Paul J.; Klütsch, Cornelya F.C.; Patterson, Brent R.; White, Bradley N. (2012). "Conservation genomics in perspective: A holistic approach to understanding Canis evolution in North America" (PDF). Biological Conservation 155: 186–192. doi:10.1016/j.biocon.2012.05.017. Retrieved 2013-07-01.

[17] Chambers, Steven M.; Fain, Steven R.; Fazio, Bud; Amaral, Michael (2012). "An account of the taxonomy of North American wolves from morphological and genetic analyses". North American Fauna 77: 1–67. doi:10.3996/nafa.77.0001. Retrieved 2013-07-02.

[18] Dumbacher, J., Review of Proposed Rule Regarding Status of the Wolf Under the Endangered Species Act, NCEAS (January 2014)

[19] Frost, Grant, Faivovich, Bain, Haas, Haddad, de Sá, Channing, Wilkinson, Donnellan, Raxworthy, Campbell, Blotto, Moler, Drewes, Nussbaum, Lynch, Green, and Wheeler 2006. The amphibian tree of life. Bulletin of the American Museum of Natural History. Number 297. New York. Issued March 15, 2006.

[20] R.W. Bulliet The Camel and the Wheel (Cambridge Mass. '75) 164-75

[21] http://www.savingwildplaces.com/swp-home/swp-crocodile/8287793?preview=&psid=&ph=class%2525253dawc-148772

[22] Amphibians.org May 2014 Godzilla vs. Godzilla—How the Chinese Giant Salamander is taking a toll on its Japanese Comic Counterpart

[23] Voloder, Dubravka (3 January 1012). "Print Email Facebook Twitter More World-first hybrid sharks found off Australia". ABC News. Retrieved 5 January 2012.

[24] Grula, John W.; Taylor, Orley R. (1980). "The Effect of X-Chromosome Inheritance on Mate-Selection Behavior in the Sulfur Butterflies, Colias eurytheme and C. Philodice". Evolution 34 (4): 688–95. doi:10.2307/2408022.

[25] "Lilium Hybrids". Pacific Bulb Society. Retrieved 2015-03-22.

[26] Warschefsky, E.; Penmetsa, R. V.; Cook, D. R.; von Wettberg, E. J. B. (8 October 2014). "Back to the wilds: Tapping evolutionary adaptations for resilient crops through systematic hybridization with crop wild relatives". American Journal of Botany 101 (10): 1791–1800. doi:10.3732/ajb.1400116. PMID 25326621.

[27] University of Colorado Principles of Genetics (MCDB 2150) Lecture 33: Chromosomal changes: Monosomy, Trisomy, Polyploidy, Structural Changes

[28] "Evaluating the utility of Arabidopsis thaliana as a model for understanding heterosis in hybrid crops", Euphytica (Springer Netherlands) 156 (1-2), July 2007: 157–171, doi:10.1007/s10681-007-9362-1, ISSN 0014-2336

[29] Rieseberg, Loren H.; Margaret A. Archer; Robert K. Wayne (July 1999). "Transgressive segregation, adaptation and speciation". Heredity 83 (4): 363–372. doi:10.1038/sj.hdy.6886170. PMID 10583537.

[30] Smith C. Wayne. Corn: Origin, History, Technology, and Production. Wiley Series in Crop Science, 2004, p. 332.

[31] Allendorf, Fred W.; R.F. Leary; P. Spruell; J.K. Wenburg (November 2001). "The problems with hybrids: setting conservation guidelines". TRENDS in Ecology & Evolution 16 (11): 613–622. doi:10.1016/S0169-5347(01)02290-X.

[32] Allendorf, Fred (2007). Conservation and the Genetics of Populations. Malden, MA: Blackwell Publishing. p. 534.

[33] Allendorf, Fred (2007). Conservation and the Genetics of Populations. Malden, MA: Blackwell Publishing. pp. 421–448.

[34] "Hybrid bear shot dead in Canada". BBC News. 13 May 2006.

[35] Ehrlich, Paul; John Holdren (26 March 1971). "Impact of population Growth". Science 171 (3977): 1212–1216. doi:10.1126/science.171.3977.1212.

[36] Vitousek, Peter; Carla M. D'Antonio; Lloyd L. Loope; Marcel Rejmánek; Randy Westbrooks (1997). "Introduced Species: A Significant Component of Human-cause Global Change". New Zealand Journal of Ecology 21 (1): 1–16.

[37] McCarthy, Eugene M. 2006. Handbook of Avian Hybrids of the World. Oxford: Oxford University Press. Pp. 16-17.

[38] McCarthy, Eugene M. 2006. Handbook of Avian Hybrids of the World. Oxford: Oxford University Press. P. 17.

[39] Darwin, C. 1868. Variation of Animals and Plants under Domestication, vol. II, p. 125

[40] Spicer, J. W. G. 1854. Note on hybrid gallinaceous birds. The Zoologist, 12: 4294-4296 (see p. 4295).

[41] Mooney, H. A.; Cleland, E. E. (2001). "The evolutionary impact of invasive species". Proc Natl Acad Sci U S A. 98 (10): 5446–5451. doi:10.1073/pnas.091093398. PMC 33232. PMID 11344292.

[42] Rhymer, JM; Simberloff, D (1996). "Extinction by Hybridization and Introgression". Annual Review of Ecology and Systematics 27: 83–109. doi:10.1146/annurev.ecolsys.27.1.83.

[43] Brad M. Potts, Robert C. Barbour, Andrew B. Hingston (2001) Genetic Pollution from Farm Forestry using eucalypt species and hybrids; A report for the RIRDC/L&WA/FWPRDC; Joint Venture Agroforestry

Program: RIRDC Publication No 01/114; RIRDC Project No CPF - 3A; ISBN 0-642-58336-6; ISSN 1440-6845; Australian Government, Rural Industrial Research and Development Corporation

[44] Devinder Sharma "Genetic Pollution: The Great Genetic Scandal"; Bulletin 28. hosted by www.farmedia.org

[45] Troyer, A. Forrest. *Breeding Widely Adapted Cultivars: Examples from Maize.* Encyclopedia of Plant and Crop Science, 27 February 2004.

[46] "Barriers to hybridization of Solanum bulbocastanumDun. and S. VerrucosumSchlechtd. and structural hybridity in their F1 plants". *Euphytica* (Springer Netherlands) **25** (1). January 1976: 1–10. doi:10.1007/BF00041523. ISSN 0014-2336

[47] James L. Kugel (2009), "Traditions of the Bible: A Guide to the Bible As It Was at the Start of the Common Era". Harvard University Press. p. 198

[48] James L. Kugel (1997), The Bible as it was, Harvard University Press. p. 110

[49] Gregory A. Boyd, God at War: The Bible & Spiritual Conflict, p. 177

14.12   External links

- Artificial Hybridisation – Artificial Hybridisation in orchids

- Domestic Fowl Hybrids

- Hybrid Mammals

- Hybridisation in animals Evolution Revolution: Two Species Become One. Study Says (nationalgeographic.com)

- Hybrids of wildcats with domestic cats

- Scientists Create Butterfly Hybrid – Creation of new species through hybridization was thought to be common only in plants, and rare in animals.

- What is a human admixed embryo?

- Video of Mirror carp & Gold fish spawning at a fishing venue in France

Chapter 15

Recombinant DNA

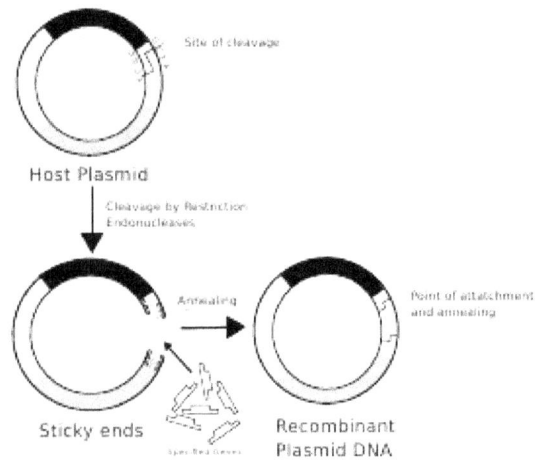

Construction of recombinant DNA, in which a foreign DNA fragment is inserted into a plasmid vector. In this example, the gene indicated by the white color is inactivated upon insertion of the foreign DNA fragment.

Recombinant DNA (rDNA) molecules are DNA molecules formed by laboratory methods of genetic recombination (such as molecular cloning) to bring together genetic material from multiple sources, creating sequences that would not otherwise be found in the genome. Recombinant DNA is possible because DNA molecules from all organisms share the same chemical structure. They differ only in the nucleotide sequence within that identical overall structure.

15.1 Introduction

Recombinant DNA is the general name for a piece of DNA that has been created by the combination of at least two strands. Recombinant DNA molecules are sometimes called **chimeric DNA**, because they can be made of material from two different species, like the mythical chimera. R-DNA technology uses palindromic sequences and leads to the production of sticky and blunt ends.

The DNA sequences used in the construction of recombinant DNA molecules can originate from any species. For example, plant DNA may be joined to bacterial DNA, or human DNA may be joined with fungal DNA. In addition, DNA sequences that do not occur anywhere in nature may be created by the chemical synthesis of DNA, and incorporated into recombinant molecules. Using recombinant DNA technology and synthetic DNA, literally any DNA sequence may be created and introduced into any of a very wide range of living organisms.

Proteins that can result from the expression of recombinant DNA within living cells are termed recombinant proteins. When recombinant DNA encoding a protein is introduced into a host organism, the recombinant protein is not necessarily produced. Expression of foreign proteins requires the use of specialized expression vectors and often necessitates significant restructuring by foreign coding sequences.

Recombinant DNA differs from genetic recombination in that the former results from artificial methods in the test tube, while the latter is a normal biological process that results in the remixing of existing DNA sequences in essentially all organisms.

15.2 Creating recombinant DNA

Main article: Molecular cloning

Molecular cloning is the laboratory process used to create recombinant DNA.[1][2][3][4] It is one of two widely used methods, along with polymerase chain reaction (PCR) used to direct the replication of any specific DNA sequence chosen by the experimentalist. The fundamental difference between the two methods is that molecular cloning involves replication of the DNA within a living cell, while PCR replicates DNA in the test tube, free of living cells.

Formation of recombinant DNA requires a cloning vector, a DNA molecule that replicates within a living cell. Vectors are generally derived from plasmids or viruses, and repre-

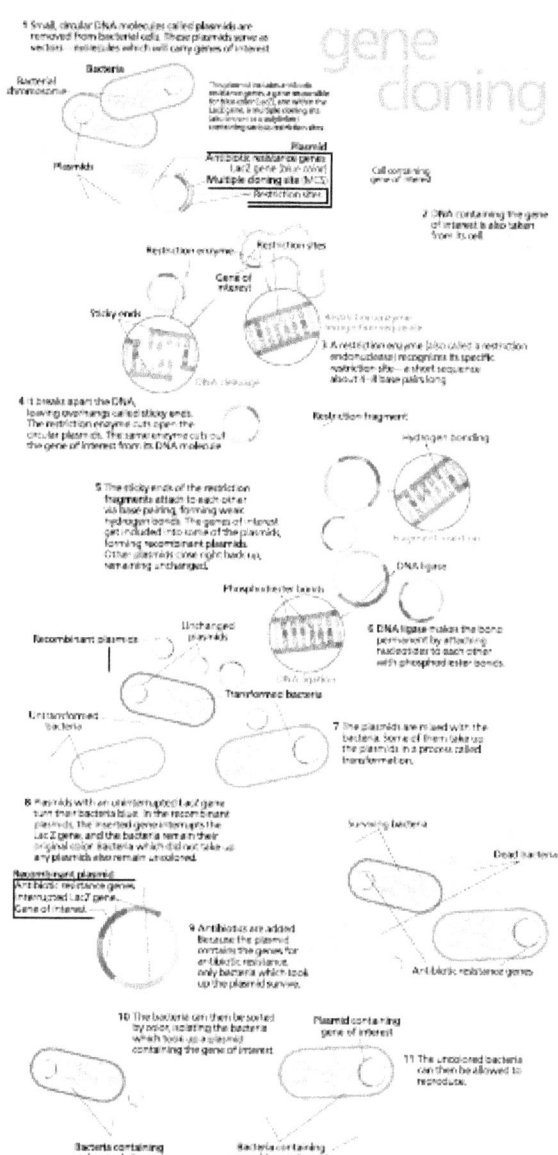

tor DNA. (3) Preparation of DNA to be cloned. (4) Creation of recombinant DNA. (5) Introduction of recombinant DNA into the host organism, (6) Selection of organisms containing recombinant DNA, and (7) Screening for clones with desired DNA inserts and biological properties.[4] *These steps are described in some detail in a related article (molecular cloning).*

15.3 Expression of recombinant DNA

Main article: Protein production

Following transplantation into the host organism, the foreign DNA contained within the recombinant DNA construct may or may not be expressed. That is, the DNA may simply be replicated without expression, or it may be transcribed and translated at a recombinant protein is produced. Generally speaking, expression of a foreign gene requires restructuring the gene to include sequences that are required for producing an mRNA molecule that can be used by the host's translational apparatus (e.g. promoter, translational initiation signal, and transcriptional terminator).[6] Specific changes to the host organism may be made to improve expression of the ectopic gene. In addition, changes may be needed to the coding sequences as well, to optimize translation, make the protein soluble, direct the recombinant protein to the proper cellular or extracellular location, and stabilize the protein from degradation.[7]

15.4 Properties of organisms containing recombinant DNA

In most cases, organisms containing recombinant DNA have apparently normal phenotypes. That is, their appearance, behavior and metabolism are usually unchanged, and the only way to demonstrate the presence of recombinant sequences is to examine the DNA itself, typically using a polymerase chain reaction (PCR) test.[8] Significant exceptions exist, and are discussed below.

If the rDNA sequences encode a gene that is expressed, then the presence of RNA and/or protein products of the recombinant gene can be detected, typically using RT-PCR or western hybridization methods.[8] Gross phenotypic changes are not the norm, unless the recombinant gene has been chosen and modified so as to generate biological activity in the host organism.[9] Additional phenotypes that are encountered include toxicity to the host organism induced by the recombinant gene product, especially if it is over-

sent relatively small segments of DNA that contain necessary genetic signals for replication, as well as additional elements for convenience in inserting foreign DNA, identifying cells that contain recombinant DNA, and, where appropriate, expressing the foreign DNA. The choice of vector for molecular cloning depends on the choice of host organism, the size of the DNA to be cloned, and whether and how the foreign DNA is to be expressed.[5] The DNA segments can be combined by using a variety of methods, such as restriction enzyme/ligase cloning or Gibson assembly.

In standard cloning protocols, the cloning of any DNA fragment essentially involves seven steps: (1) Choice of host organism and cloning vector, (2) Preparation of vec-

expressed or expressed within inappropriate cells or tissues. In some cases, recombinant DNA can have deleterious effects even if it is not expressed. One mechanism by which this happens is insertional inactivation, in which the rDNA becomes inserted into a host cell's gene. In some cases, researchers use this phenomenon to "knock out" genes to determine their biological function and importance.[10] Another mechanism by which rDNA insertion into chromosomal DNA can affect gene expression is by inappropriate activation of previously unexpressed host cell genes. This can happen, for example, when a recombinant DNA fragment containing an active promoter becomes located next to a previously silent host cell gene, or when a host cell gene that functions to restrain gene expression undergoes insertional inactivation by recombinant DNA.

15.5   Applications of recombinant DNA technology

A group of GloFish fluorescent fish

Recombinant DNA is widely used in biotechnology, medicine and research. Today, recombinant proteins and other products that result from the use of rDNA technology are found in essentially every western pharmacy, doctor's or veterinarian's office, medical testing laboratory, and biolog-

ical research laboratory. In addition, organisms that have been manipulated using recombinant DNA technology, as well as products derived from those organisms, have found their way into many farms, supermarkets, home medicine cabinets, and even pet shops, such as those that sell GloFish and other genetically modified animals.

The most common application of recombinant DNA is in basic research, in which the technology is important to most current work in the biological and biomedical sciences.[8] Recombinant DNA is used to identify, map and sequence genes, and to determine their function. rDNA probes are employed in analyzing gene expression within individual cells, and throughout the tissues of whole organisms. Recombinant proteins are widely used as reagents in laboratory experiments and to generate antibody probes for examining protein synthesis within cells and organisms.[2]

Many additional practical applications of recombinant DNA are found in industry, food production, human and veterinary medicine, agriculture, and bioengineering.[2] Some specific examples are identified below.

Recombinant chymosin

Found in rennet, chymosin is an enzyme required to manufacture cheese. It was the first genetically engineered food additive used commercially. Traditionally, processors obtained chymosin from rennet, a preparation derived from the fourth stomach of milk-fed calves. Scientists engineered a non-pathogenic strain (K-12) of *E. coli* bacteria for large-scale laboratory production of the enzyme. This microbiologically produced recombinant enzyme, identical structurally to the calf derived enzyme, costs less and is produced in abundant quantities. Today about 60% of U.S. hard cheese is made with genetically engineered chymosin. In 1990, FDA granted chymosin "generally-recognized-as-safe" (GRAS) status based on data showing that the enzyme was safe.[11]

Recombinant human insulin Almost completely replaced insulin obtained from animal sources (e.g. pigs and cattle) for the treatment of insulin-dependent diabetes. A variety of different recombinant insulin preparations are in widespread use.[12] Recombinant insulin is synthesized by inserting the human insulin gene into *E. coli*, or yeast (saccharomyces cerevisiae)[13] which then produces insulin for human use.[14]

Recombinant human growth hormone (HGH, somatotropin) Administered to patients whose pituitary glands generate insufficient quantities to support normal growth and development. Before recombinant HGH became available, HGH for therapeutic use was obtained from pituitary glands of cadavers. This unsafe practice led to some patients developing Creutzfeldt–Jakob disease. Recombinant HGH eliminated this problem, and is now used therapeutically.[15] It has also been misused as a performance enhancing drug by athletes and others.[16] DrugBank entry

Recombinant blood clotting factor VIII A blood-clotting protein that is administered to patients with forms of the bleeding disorder hemophilia, who are unable to produce factor VIII in quantities sufficient to support normal blood coagulation.[17] Before the development of recombinant factor VIII, the protein was obtained by processing large quantities of human blood from multiple donors, which carried a very high risk of transmission of blood borne infectious diseases, for example HIV and hepatitis B. DrugBank entry

Recombinant hepatitis B vaccine Hepatitis B infection is controlled through the use of a recombinant hepatitis B vaccine, which contains a form of the hepatitis B virus surface antigen that is produced in yeast cells. The development of the recombinant subunit vaccine was an important and necessary development because hepatitis B virus, unlike other common viruses such as polio virus, cannot be grown in vitro. Vaccine information from Hepatitis B Foundation

Diagnosis of infection with HIV Each of the three widely used methods for diagnosing HIV infection has been developed using recombinant DNA. The antibody test (ELISA or western blot) uses a recombinant HIV protein to test for the presence of antibodies that the body has produced in response to an HIV infection. The DNA test looks for the presence of HIV genetic material using reverse transcription polymerase chain reaction (RT-PCR). Development of the RT-PCR test was made possible by the molecular cloning and sequence analysis of HIV genomes. HIV testing page from US Centers for Disease Control (CDC)

Golden rice A recombinant variety of rice that has been engineered to express the enzymes responsible for β-carotene biosynthesis.[9] This variety of rice holds substantial promise for reducing the incidence of vitamin A deficiency in the world's population.[18] Golden rice is not currently in use, pending the resolution of regulatory and intellectual property[19] issues.

Herbicide-resistant crops Commercial varieties of important agricultural crops (including soy, maize/corn, sorghum, canola, alfalfa and cotton) have been developed that incorporate a recombinant gene that results in resistance to the herbicide glyphosate (trade name

Roundup), and simplifies weed control by glyphosate application.[20] These crops are in common commercial use in several countries.

Insect-resistant crops *Bacillus thuringeiensis* is a bacterium that naturally produces a protein (Bt toxin) with insecticidal properties.[18] The bacterium has been applied to crops as an insect-control strategy for many years, and this practice has been widely adopted in agriculture and gardening. Recently, plants have been developed that express a recombinant form of the bacterial protein, which may effectively control some insect predators. Environmental issues associated with the use of these transgenic crops have not been fully resolved.[21]

15.6 History of recombinant DNA

The idea of recombinant DNA was first proposed by Peter Lobban, a graduate student of Prof. Dale Kaiser in the Biochemistry Department at Stanford University Medical School.[22] The first publications describing the successful production and intracellular replication of recombinant DNA appeared in 1972 and 1973.[23][24][25][26] Stanford University applied for a US patent on recombinant DNA in 1974, listing the inventors as Stanley N. Cohen and Herbert W. Boyer; this patent was awarded in 1980.[27] The first licensed drug generated using recombinant DNA technology was human insulin, developed by Genentech and Licensed by Eli Lilly and Company.[28]

15.7 Controversy

Scientists associated with the initial development of recombinant DNA methods recognized that the potential existed for organisms containing recombinant DNA to have undesirable or dangerous properties. At the 1975 Asilomar Conference on Recombinant DNA, these concerns were discussed and a voluntary moratorium on recombinant DNA research was initiated for experiments that were considered particularly risky. This moratorium was widely observed until the National Institutes of Health (USA) developed and issued formal guidelines for rDNA work. Today, recombinant DNA molecules and recombinant proteins are usually not regarded as dangerous. However, concerns remain about some organisms that express recombinant DNA, particularly when they leave the laboratory and are introduced into the environment or food chain. These concerns are discussed in the articles on genetically modified organisms and genetically modified food controversies.

15.8   See also

- Asilomar conference on recombinant DNA

- Genetic engineering

- Genetically modified organism

- Recombinant virus

- Vector DNA

- Biomolecular engineering

- Recombinant DNA Technology

15.9   References

[1] Campbell, Neil A. & Reece, Jane B.. (2002). *Biology (6th ed.)*. San Francisco: Addison Wesley. pp. 375–401. ISBN 0-201-75054-6.

[2] Peter Walter; Alberts, Bruce; Johnson, Alexander S.; Lewis, Julian; Raff, Martin C.; Roberts, Keith (2008). *Molecular Biology of the Cell (5th edition, Extended version)*. New York: Garland Science. ISBN 0-8153-4111-3.. Fourth edition is available online through the NCBI Bookshelf: link

[3] Berg, Jeremy Mark; Tymoczko, John L.; Stryer, Lubert (2010). *Biochemistry, 7th ed. (Biochemistry (Berg))*. W.H. Freeman & Company. ISBN 1-4292-2936-5. Fifth edition available online through the NCBI Bookshelf: link

[4] Watson, James D. (2007). *Recombinant DNA: Genes and Genomes: A Short Course*. San Francisco: W.H. Freeman. ISBN 0-7167-2866-4.

[5] Russell, David W.; Sambrook, Joseph (2001). *Molecular cloning: a laboratory manual*. Cold Spring Harbor, N.Y: Cold Spring Harbor Laboratory. ISBN 0-87969-576-5.

[6] Hannig, G.; Makrides, S. (1998). "Strategies for optimizing heterologous protein expression in Escherichia coli". *Trends in Biotechnology* 16 (2): 54–60. doi:10.1016/S0167-7799(97)01155-4. PMID 9487731.

[7] Brondyk, W. H. (2009). "Chapter 11 Selecting an Appropriate Method for Expressing a Recombinant Protein". *Methods in enzymology*. Methods in Enzymology 463: 131–147. doi:10.1016/S0076-6879(09)63011-1. ISBN 9780123745361. PMID 19892171.

[8] Brown, Terry (2006). *Gene Cloning and DNA Analysis: an Introduction*. Cambridge, MA: Blackwell Pub. ISBN 1-4051-1121-6.

[9] Ye, X.; Al-Babili, S.; Klöti, A.; Zhang, J.; Lucca, P.; Beyer, P.; Potrykus, I. (2000). "Engineering the provitamin A (beta-carotene) biosynthetic pathway into (carotenoid-free) rice endosperm". *Science* 287 (5451): 303–305. doi:10.1126/science.287.5451.303. PMID 10634784.

[10] Koller, B. H.; Smithies, O. (1992). "Altering Genes in Animals by Gene Targeting". *Annual Review of Immunology* 10: 705–730. doi:10.1146/annurev.iy.10.040192.003421. PMID 1591000.

[11] Donna U. Vogt and Mickey Parish. (1999) Food Biotechnology in the United States: Science, Regulation, and Issues

[12] Gualandi-Signorini, A.; Giorgi, G. (2001). "Insulin formulations--a review". *European review for medical and pharmacological sciences* 5 (3): 73–83. PMID 12004916.

[13] #Insulin aspart

[14] DrugBank: Insulin Regular (DB00030)

[15] Von Fange, T.; McDiarmid, T.; MacKler, L.; Zolotor, A. (2008). "Clinical inquiries: Can recombinant growth hormone effectively treat idiopathic short stature?". *The Journal of family practice* 57 (9): 611–612. PMID 18786336.

[16] Fernandez, M.; Hosey, R. (2009). "Performance-enhancing drugs snare nonathletes, too". *The Journal of family practice* 58 (1): 16–23. PMID 19141266.

[17] Manco-Johnson, M. J. (2010). "Advances in the Care and Treatment of Children with Hemophilia". *Advances in Pediatrics* 57 (1): 287–294. doi:10.1016/j.yapd.2010.08.007. PMID 21056743.

[18] Paine, J. A.; Shipton, C. A.; Chaggar, S.; Howells, R. M.; Kennedy, M. J.; Vernon, G.; Wright, S. Y.; Hinchliffe, E.; Adams, J. L.; Silverstone, A. L.; Drake, R. (2005). "Improving the nutritional value of Golden Rice through increased pro-vitamin a content". *Nature Biotechnology* 23 (4): 482–487. doi:10.1038/nbt1082. PMID 15793573.

[19] Deccan Herald. " Foreign group roots for 'golden rice' in India". March 18, 2015 http://www.deccanherald.com/content/466247/foreign-group-roots-golden-rice.html

[20] Funke, T.; Han, H.; Healy-Fried, M.; Fischer, M.; Schönbrunn, E. (2006). "Molecular basis for the herbicide resistance of Roundup Ready crops". *Proceedings of the National Academy of Sciences* 103 (35): 13010–13015. doi:10.1073/pnas.0603638103. PMC 1559744. PMID 16916934.

[21] Mendelsohn, M.; Kough, J.; Vaituzis, Z.; Matthews, K. (2003). "Are Bt crops safe?". *Nature Biotechnology* 21 (9): 1003–1009. doi:10.1038/nbt0903-1003. PMID 12949561.

[22] Lear, J. (1978). *Recombinant DNA: The Untold Story*. New York: Crown Publishers. p. 43.

[23] Jackson, D.; Symons, R.; Berg, P. (1972). "Biochemical method for inserting new genetic information into DNA of Simian Virus 40: Circular SV40 DNA molecules containing lambda phage genes and the galactose operon of Escherichia coli". *Proceedings of the National Academy of Sciences of the United States of America* 69 (10): 2904–2909. doi:10.1073/pnas.69.10.2904. PMC 389671. PMID 4342968.

[24] Mertz, J. E.; Davis, R. W. (1972). "Cleavage of DNA by R 1 restriction endonuclease generates cohesive ends". *Proceedings of the National Academy of Sciences of the United States of America* **69** (11): 3370–4. doi:10.1073/pnas.69.11.3370. PMC 389773. PMID 4343968.

[25] Lobban, P.; Kaiser, A. (1973). "Enzymatic end-to end joining of DNA molecules". *Journal of Molecular Biology* **78** (3): 453–471. doi:10.1016/0022-2836(73)90468-3. PMID 4754844.

[26] Cohen, S.; Chang, A.; Boyer, H.; Helling, R. (1973). "Construction of biologically functional bacterial plasmids in vitro". *Proceedings of the National Academy of Sciences of the United States of America* **70** (11): 3240–3244. doi:10.1073/pnas.70.11.3240. PMC 427208. PMID 4594039.

[27] Hughes, S. (2001). "Making dollars out of DNA. The first major patent in biotechnology and the commercialization of molecular biology, 1974-1980". *Isis; an international review devoted to the history of science and its cultural influences* **92** (3): 541–575. doi:10.1086/385281. PMID 11810894.

[28] Johnson, I. S. (1983). "Human insulin from recombinant DNA technology". *Science* **219** (4585): 632–637. doi:10.1126/science.6337396. PMID 6337396.

15.9.1 Further reading

- Judson, Horace F. 1979. *The Eighth Day of Creation: Makers of the Revolution in Biology*. Touchstone Books, ISBN 0-671-22540-5. 2nd edition: Cold Spring Harbor Laboratory Press. 1996 paperback: ISBN 0-87969-478-5.

- Micklas, David. 2003. *DNA Science: A First Course*. Cold Spring Harbor Press: ISBN 978-0-87969-636-8.

- Rasmussen, Nicolas, *Gene Jockeys: Life Science and the rise of Biotech Enterprise*, Johns Hopkins University Press, (Baltimore), 2014. ISBN 978-1-42141-340-2.

- Rosenfeld, Israel. 2010. *DNA: A Graphic Guide to the Molecule that Shook the World*. Columbia University Press: ISBN 978-0-231-14271-7.

- Schultz, Mark and Zander Cannon. 2009. *The Stuff of Life: A Graphic Guide to Genetics and DNA*. Hill and Wang: ISBN 0-8090-8947-5.

- Watson, James. 2004. *DNA: The Secret of Life*. Random House: ISBN 978-0-09-945184-6.

15.10 External links

- Recombinant DNA fact sheet (from University of New Hampshire)

- Plasmids in Yeasts (Fact sheet from San Diego State University)

- Animation illustrating construction of recombinant DNA and foreign protein production by recombinant bacteria

- Recombinant DNA research at UCSF and commercial application at Genentech Edited transcript of 1994 interview with Herbert W. Boyer. Living history project. Oral history.

- Recombinant Protein Purification Principles and Methods Handbook

Chapter 16

Vector (molecular biology)

In molecular cloning, a **vector** is a DNA molecule used as a vehicle to artificially carry foreign genetic material into another cell, where it can be replicated and/or expressed. A vector containing foreign DNA is termed recombinant DNA. The four major types of vectors are plasmids, viral vectors, cosmids, and artificial chromosomes. Of these, the most commonly used vectors are plasmids. Common to all engineered vectors are an origin of replication, a multicloning site, and a selectable marker.

The vector itself is generally a DNA sequence that consists of an insert (transgene) and a larger sequence that serves as the "backbone" of the vector. The purpose of a vector which transfers genetic information to another cell is typically to isolate, multiply, or express the insert in the target cell. Vectors called expression vectors (expression constructs) specifically are for the expression of the transgene in the target cell, and generally have a promoter sequence that drives expression of the transgene. Simpler vectors called transcription vectors are only capable of being transcribed but not translated: they can be replicated in a target cell but not expressed, unlike expression vectors. Transcription vectors are used to amplify their insert.

Insertion of a vector into the target cell is usually called transformation for bacterial cells, transfection for eukaryotic cells, although insertion of a viral vector is often called transduction.

16.1 Characteristics

16.1.1 Plasmids

Main article: Plasmid vector

Plasmids are double-stranded and generally circular DNA sequences that are capable of automatically replicating in a host cell. Plasmid vectors minimalistically consist of an origin of replication that allows for semi-independent replication of the plasmid in the host. Plasmids are found widely in many bacteria, for example in *Escherichia coli*, but may also be found in a few eukaryotes, for example in yeast such as *Saccharomyces cerevisiae*.[1] Bacterial plasmids may be conjugative/transmissible and non-conjugative:

- conjugative: mediate DNA transfer through conjugation and therefore spread rapidly among the bacterial cells of a population: e.g., F plasmid, many R and some col plasmids.

- nonconjugative- do not mediate DNA through conjugation. e.g., many R and col plasmids.

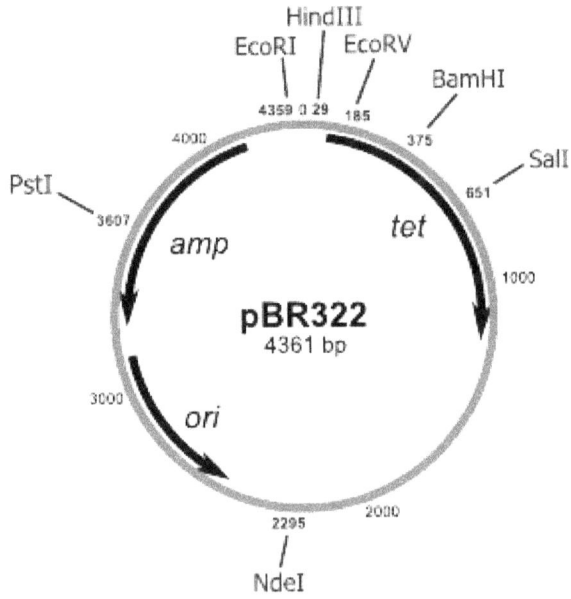

The pBR322 plasmid is one of the first plasmids widely used as a cloning vector.

Plasmids with specially-constructed features are commonly used in laboratory for cloning purposes. These plasmid are generally non-conjugative but may have many more features, notably a "multiple cloning site" where multiple restriction enzyme cleavage sites allow for the insertion of a transgene insert. The bacteria containing the plasmids can generate millions of copies of the vector within the bacteria in hours, and the amplified vectors can be extracted from the bacteria for further manipulation. Plasmids may be used specifically as transcription vectors and such plasmids may lack crucial sequences for protein expression. Plasmids used for protein expression, called expression vectors, would include elements for translation of protein, such as a ribosome binding site, start and stop codons.

16.1.2 Viral vectors

Main article: Viral vector

Viral vectors are generally genetically engineered viruses carrying modified viral DNA or RNA that has been rendered noninfectious, but still contain viral promoters and also the transgene, thus allowing for translation of the transgene through a viral promoter. However, because viral vectors frequently are lacking infectious sequences, they require helper viruses or packaging lines for large-scale transfection. Viral vectors are often designed for permanent incorporation of the insert into the host genome, and thus leave distinct genetic markers in the host genome after incorporating the transgene. For example, retroviruses leave a characteristic retroviral integration pattern after insertion that is detectable and indicates that the viral vector has incorporated into the host genome.

16.2 Transcription

Transcription is a necessary component in all vectors: the premise of a vector is to multiply the insert (although expression vectors later also drive the translation of the multiplied insert). Thus, even stable expression is determined by stable transcription, which generally depends on promoters in the vector. However, expression vectors have a variety of expression patterns: constitutive (consistent expression) or inducible (expression only under certain conditions or chemicals). This expression is based on different promoter activities, not post-transcriptional activities. Thus, these two different types of expression vectors depend on different types of promoters.

Viral promoters are often used for constitutive expression in plasmids and in viral vectors because they normally force constant transcription in many cell lines and types reliably.

Inducible expression depends on promoters that respond to the induction conditions: for example, the murine mammary tumor virus promoter only initiates transcription after dexamethasone application and the *Drosophilia* heat shock promoter only initiates after high temperatures.

16.3 Expression

Main article: Expression vector

Expression vectors produce proteins through the transcription of the vector's insert followed by translation of the mRNA produced, they therefore require more components than the simpler transcription-only vectors. Expression in different host organism would require different elements, although they share similar requirements, for example a promoter for initiation of transcription, a ribosomal binding site for translation initiation, and termination signals.

16.3.1 Prokaryotes expression vector

- Promoter - commonly used inducible promoters are promoters derived from *lac* operon and the T7 promoter. Other strong promoters used include Trp promoter and Tac-Promoter, which a hybrid of both the Trp and Lac Operon promoters.

- Ribosome binding site (RBS) Follows the promoter, and promotes efficient translation of the protein of interest.

- Translation initiation site - Shine-Dalgarno sequence enclosed in the RBS, 8 base-pairs upstream of the AUG start codon.

16.3.2 Eukaryotes expression vector

Eukaryote expression vectors require sequences that encode for:

- Polyadenylation tail: Creates a polyadenylation tail at the end of the transcribed pre-mRNA that protects the mRNA from exonucleases and ensures transcriptional and translational termination: stabilizes mRNA production.

- Minimal UTR length: UTRs contain specific characteristics that may impede transcription or translation, and thus the shortest UTRs or none at all are encoded for in optimal expression vectors.

- Kozak sequence: Vectors should encode for a Kozak sequence in the mRNA, which assembles the ribosome for translation of the mRNA.

16.4   Features

Modern artificially-constructed vectors contain essential components as well as other additional features:

- Origin of replication: Necessary for the replication and maintenance of the vector in the host cell.

- Promoter: Promoters are used to drive the transcription of the vector's transgene as well as the other genes in the vector such as the antibiotic resistance gene. Some cloning vectors need not have a promoter for the cloned insert but it is an essential component of expression vectors so that the cloned product may be expressed.

- Cloning site: This may be a multiple cloning site or other features that allow for the insertion of foreign DNA into the vector through ligation.

- Genetic markers: Genetic markers for viral vectors allow for confirmation that the vector has integrated with the host genomic DNA.

- Antibiotic resistance: Vectors with antibiotic-resistance open reading frames allow for survival of cells that have taken up the vector in growth media containing antibiotics through antibiotic selection.

- Epitope: Vector contains a sequence for a specific epitope that is incorporated into the expressed protein. Allows for antibody identification of cells expressing the target protein.

- Reporter genes: Some vectors may contain a reporter gene that allow for identification of plasmid that contains inserted DNA sequence. An example is *lacZ-α* which codes for the N-terminus fragment of β-galactosidase, an enzyme that digests galactose. A multiple cloning site is located within *lacZ-α*, and an insert successfully ligated into the vector will disrupt the gene sequence, resulting in an inactive β-galactosidase. Cells containing vector with an insert may be identified using blue/white selection by growing cells in media containing an analogue of galactose (X-gal). Cells expressing β-galactosidase (therefore doesn't contain an insert) appear as blue colonies. White colonies would be selected as those that may contain an insert. Other commonly used reporters include green fluorescent protein and luciferase.

- Targeting sequence: Expression vectors may include encoding for a targeting sequence in the finished protein that directs the expressed protein to a specific organelle in the cell or specific location such as the periplasmic space of bacteria.

- Protein purification tags: Some expression vectors include proteins or peptide sequences that allows for easier purification of the expressed protein. Examples include polyhistidine-tag, glutathione-S-transferase, and maltose binding protein. Some of these tags may also allow for increased solubility of the target protein. The target protein is fused to the protein tag, but a protease cleavage site positioned in the polypeptide linker region between the protein and the tag allows the tag to be removed later.

16.5   See also

- Plasmid
- Viral vector
- Cloning vector
- Expression vector
- Minicircle
- Recombinant DNA
- Naked DNA
- Vector (epidemiology), an organism that transmits disease

16.6   References

[1] T. A. Brown (2010). "Chapter 2 - Vectors for Gene Cloning: Plasmids and Bacteriophages". *Gene Cloning and DNA Analysis: An Introduction* (6th ed.). Wiley-Blackwell. ISBN 978-1405181730.

- Freshney, Ian R. *Culture of Animal Cells: A manual of basic technique*. John Wiley & Sons, Inc., Hoboken, New Jersey. ISBN 978-0-471-45329-1

16.7   External links

- Waksman Scholars introduction to vectors
- A comparison of vectors in use for clinical gene transfer
- Gene Transport Unit

Chapter 17

Transcription (genetics)

This article is about a genetic process. For other uses, see Transcription.

Transcription is the first step of gene expression, in which

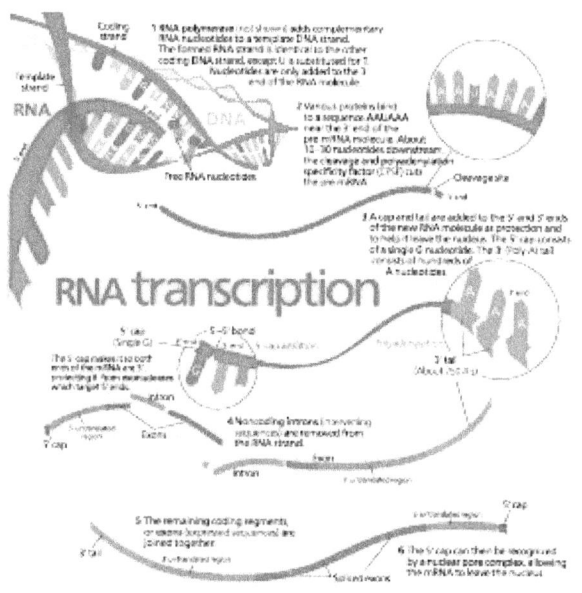

Simplified diagram of mRNA synthesis and processing. Enzymes not shown.

a particular segment of DNA is copied into RNA (mRNA) by the enzyme RNA polymerase.

Both RNA and DNA are nucleic acids, which use base pairs of nucleotides as a complementary language. The two can be converted back and forth from DNA to RNA by the action of the correct enzymes. During transcription, a DNA sequence is read by an RNA polymerase, which produces a complementary, antiparallel RNA strand called a primary transcript.

Transcription proceeds in the following general steps:

1. One or more sigma factor protein binds to the RNA polymerase holoenzyme, allowing it to bind to promoter DNA.

2. RNA polymerase creates a transcription bubble, which separates the two strands of the DNA helix. This is done by breaking the hydrogen bonds between complementary DNA nucleotides.

3. RNA polymerase adds matching RNA nucleotides to the complementary nucleotides of one DNA strand.

4. RNA sugar-phosphate backbone forms with assistance from RNA polymerase to form an RNA strand.

5. Hydrogen bonds of the untwisted RNA-DNA helix break, freeing the newly synthesized RNA strand.

6. If the cell has a nucleus, the RNA may be further processed. This may include polyadenylation, capping, and splicing.

7. The RNA may remain in the nucleus or exit to the cytoplasm through the nuclear pore complex.

The stretch of DNA transcribed into an RNA molecule is called a *transcription unit* and encodes at least one gene. If the gene transcribed encodes a protein, messenger RNA (mRNA) will be transcribed; the mRNA will in turn serve as a template for the protein's synthesis through translation. Alternatively, the transcribed gene may encode for either non-coding RNA (such as microRNA), ribosomal RNA (rRNA), transfer RNA (tRNA), or other enzymatic RNA molecules called ribozymes.[1] Overall, RNA helps synthesize, regulate, and process proteins; it therefore plays a fundamental role in performing functions within a cell.

In virology, the term may also be used when referring to mRNA synthesis from an RNA molecule (i.e., RNA replication). For instance, the genome of a negative-sense single-stranded RNA (ssRNA -) virus may be template for a positive-sense single-stranded RNA (ssRNA +). This is because the positive-sense strand contains the information needed to translate the viral proteins for viral replication afterwards. This process is catalysed by a viral RNA replicase.[2]

17.1   Background

A DNA transcription unit encoding for a protein may contain both a *coding sequence*, which will be translated into the protein, and *regulatory sequences*, which direct and regulate the synthesis of that protein. The regulatory sequence before ("upstream" from) the coding sequence is called the five prime untranslated region (5'UTR); the sequence after ("downstream" from) the coding sequence is called the three prime untranslated region (3'UTR).[1]

As opposed to DNA replication, transcription results in an RNA complement that includes the nucleotide uracil (U) in all instances where thymine (T) would have occurred in a DNA complement.

Only one of the two DNA strands serve as a template for transcription. The antisense strand of DNA is read by RNA polymerase from the 3' end to the 5' end during transcription (3' → 5'). The complementary RNA is created in the opposite direction, in the 5' → 3' direction, matching the sequence of the sense strand with the exception of switching uracil for thymine. This directionality is because RNA polymerase can only add nucleotides to the 3' end of the growing mRNA chain. This use of only the 3' → 5' DNA strand eliminates the need for the Okazaki fragments that are seen in DNA replication.[1] This removes the need for an RNA primer to initiate RNA synthesis, as is the case in DNA replication.

The *non*-template sense strand of DNA is called the coding strand, because its sequence is the same as the newly created RNA transcript (except for the substitution of uracil for thymine). This is the strand that is used by convention when presenting a DNA sequence.

Transcription has some proofreading mechanisms, but they are fewer and less effective than the controls for copying DNA; therefore, transcription has a lower copying fidelity than DNA replication.[3]

17.2   Major steps

Transcription is divided into *pre-initiation, initiation, promoter clearance, elongation* and *termination*.[1]

17.2.1   Pre-initiation

In **eukaryotes**, a core promoter sequence in the DNA must be present for RNA polymerase to initiate transcription. Promoters are regions of DNA that promote transcription - in eukaryotes, they are found at −30, −75, and −90 base pairs upstream from the transcription start site (abbreviated to TSS). Transcription factors are proteins that bind

to these promoter sequences and facilitate the binding of RNA Polymerase.[4]

The most characterized type of core promoter in eukaryotes is a short DNA sequence known as a TATA box, found 25-30 base pairs upstream from the TSS.[4] The TATA box, as a core promoter, is the binding site for a transcription factor known as TATA-binding protein (TBP) (which is itself a subunit of another transcription factor, called Transcription Factor II D (TFIID)). After TFIID binds to the TATA box via the TBP, five more transcription factors and RNA polymerase combine around the TATA box in a series of stages to form a preinitiation complex. One transcription factor, Transcription factor II H, has two components with helicase activity and so is involved in the separating of opposing strands of double-stranded DNA to form the initial transcription bubble. However, the preinitiation complex produces only a low transcription rate on its own. Other proteins known as activators and repressors, along with any associated coactivators or corepressors, are responsible for modulating transcription rate.[4]

Thus, preinitiation complex contains:.

1. Core Promoter Sequence

2. Transcription Factors

3. RNA Polymerase

4. Activators and Repressors.

The transcription preinitiation in **archaea** is, in essence, homologous to that of eukaryotes, but is much less complex.[5] The archaeal preinitiation complex assembles at a TATA-box binding site; however, in archaea, this complex is composed of only RNA polymerase II, TBP, and TFB (the archaeal homologue of eukaryotic transcription factor II B (TFIIB)).[6][7]

17.2.2   Initiation

Simple diagram of transcription initiation. RNAP = RNA polymerase

In **bacteria**, transcription begins with the binding of RNA polymerase to the promoter in DNA. RNA polymerase is a core enzyme consisting of five subunits: 2 α subunits, 1 β subunit, 1 β' subunit, and 1 ω subunit. At the start of initiation, the core enzyme is associated with a sigma factor that aids in finding the appropriate −35 and −10 base pairs upstream of promoter sequences.[8] When the sigma factor and RNA polymerase combine, they form a holoenzyme.

Transcription initiation is more complex in **eukaryotes**. Eukaryotic RNA polymerase does not directly recognize the core promoter sequences. Instead, a collection of proteins called transcription factors mediate the binding of RNA polymerase and the initiation of transcription. Only after certain transcription factors are attached to the promoter does the RNA polymerase bind to it. The completed assembly of transcription factors and RNA polymerase bind to the promoter, forming a transcription initiation complex. Transcription in the archaea domain is similar to transcription in eukaryotes.[9]

17.2.3 Promoter clearance

After the first bond is synthesized, the RNA polymerase must clear the promoter. During this time there is a tendency to release the RNA transcript and produce truncated transcripts. This is called *abortive initiation* and is common for both eukaryotes and prokaryotes.[10]

In **prokaryotes**, abortive initiation continues to occur until an RNA product of a threshold length of approximately 10 nucleotides is synthesized, at which point promoter escape occurs and a transcription elongation complex is formed. The σ factor is released according to a stochastic model.[11] Mechanistically, promoter escape occurs through a scrunching mechanism, where the energy built up by DNA scrunching provides the energy needed to break interactions between RNA polymerase holoenzyme and the promoter.[12]

In **eukaryotes**, after several rounds of 10nt abortive initiation, promoter clearance coincides with the TFIIH's phosphorylation of serine 5 on the carboxy terminal domain of RNAP II, leading to the recruitment of capping enzyme (CE).[13][14] The exact mechanism of how CE induces promoter clearance in eukaryotes is not yet known.

17.2.4 Elongation

Simple diagram of transcription elongation

One strand of the DNA, the *template strand* (or noncoding strand), is used as a template for RNA synthesis. As transcription proceeds, RNA polymerase traverses the template strand and uses base pairing complementarity with the DNA template to create an RNA copy. Although RNA polymerase traverses the template strand from 3' → 5', the coding (non-template) strand and newly formed RNA can also be used as reference points, so transcription can be described as occurring 5' → 3'. This produces an RNA molecule from 5' → 3', an exact copy of the coding strand (except that thymines are replaced with uracils, and the nucleotides are composed of a ribose (5-carbon) sugar where DNA has deoxyribose (one fewer oxygen atom) in its sugar-phosphate backbone).

mRNA transcription can involve multiple RNA polymerases on a single DNA template and multiple rounds of transcription (amplification of particular mRNA), so many mRNA molecules can be rapidly produced from a single copy of a gene.

Elongation also involves a proofreading mechanism that can replace incorrectly incorporated bases. In eukaryotes, this may correspond with short pauses during transcription that allow appropriate RNA editing factors to bind. These pauses may be intrinsic to the RNA polymerase or due to chromatin structure.

17.2.5 Termination

Main article: Terminator (genetics)

Bacteria use two different strategies for transcription termination - *Rho-independent termination* and *Rho-dependent termination*. In Rho-independent transcription termination, also called intrinsic termination, RNA transcription stops when the newly synthesized RNA molecule forms a G-C-rich hairpin loop followed by a run of Us. When the hairpin forms, the mechanical stress breaks the weak rU-dA bonds, now filling the DNA-RNA hybrid. This pulls the poly-U transcript out of the active site of the RNA polymerase, in effect, terminating transcription. In the "Rho-dependent" type of termination, a protein factor called "Rho" destabilizes the interaction between the template and the mRNA, thus releasing the newly synthesized mRNA from the elongation complex.[15]

Transcription termination in **eukaryotes** is less understood but involves cleavage of the new transcript followed by template-independent addition of adenines at its new 3' end, in a process called polyadenylation.[16]

17.3   Inhibitors

Transcription inhibitors can be used as antibiotics against, for example, pathogenic bacteria (antibacterials) and fungi (antifungals).   An example of such an antibacterial is rifampicin, which inhibits prokaryotic DNA transcription into mRNA by inhibiting DNA-dependent RNA polymerase by binding its beta-subunit.   8-Hydroxyquinoline is an antifungal transcription inhibitor.[17] The effects of histone methylation may also work to inhibit the action of transcription.

17.4   Transcription factories

Main article: Transcription factories

Active transcription units are clustered in the nucleus, in discrete sites called transcription factories or euchromatin. Such sites can be visualized by allowing engaged polymerases to extend their transcripts in tagged precursors (Br-UTP or Br-U) and immuno-labeling the tagged nascent RNA. Transcription factories can also be localized using fluorescence in situ hybridization or marked by antibodies directed against polymerases. There are ~10,000 factories in the nucleoplasm of a HeLa cell, among which are ~8,000 polymerase II factories and ~2,000 polymerase III factories. Each polymerase II factory contains ~8 polymerases. As most active transcription units are associated with only one polymerase, each factory usually contains ~8 different transcription units. These units might be associated through promoters and/or enhancers, with loops forming a 'cloud' around the factor.[18]

17.5   History

A molecule that allows the genetic material to be realized as a protein was first hypothesized by François Jacob and Jacques Monod. Severo Ochoa won a Nobel Prize in Physiology or Medicine in 1959 for developing a process for synthesizing RNA in vitro with polynucleotide phosphorylase, which was useful for cracking the genetic code. RNA synthesis by RNA polymerase was established in vitro by several laboratories by 1965; however, the RNA synthesized by these enzymes had properties that suggested the existence of an additional factor needed to terminate transcription correctly.

In 1972, Walter Fiers became the first person to actually prove the existence of the terminating enzyme.

Roger D. Kornberg won the 2006 Nobel Prize in Chem-istry "for his studies of the molecular basis of eukaryotic transcription".[19]

17.6   Measuring and detecting transcription

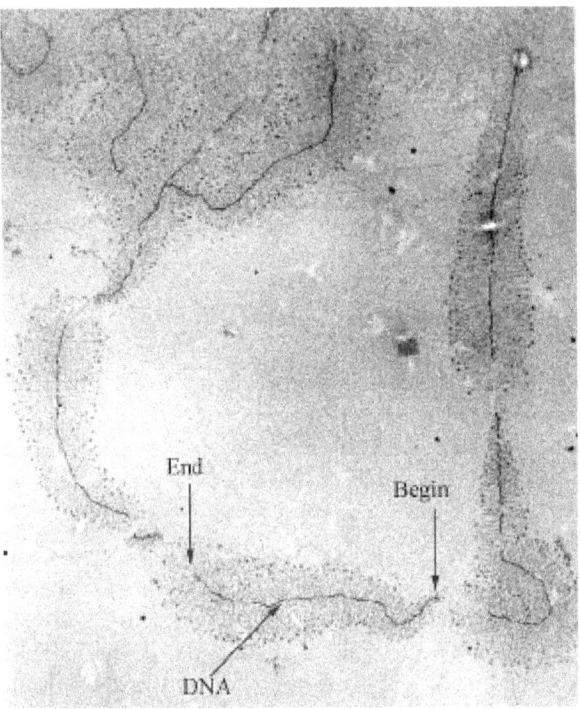

Electron micrograph of transcription of ribosomal RNA. The forming ribosomal RNA strands are visible as branches from the main DNA strand.

Transcription can be measured and detected in a variety of ways:

- Nuclear Run-on assay: measures the relative abundance of newly formed transcripts

- RNase protection assay and ChIP-Chip of RNAP: detect active transcription sites

- RT-PCR: measures the absolute abundance of total or nuclear RNA levels, which may however differ from transcription rates

- DNA microarrays: measures the relative abundance of the global total or nuclear RNA levels; however, these may differ from transcription rates

- In situ hybridization: detects the presence of a transcript

- MS2 tagging: by incorporating RNA stem loops, such as MS2, into a gene, these become incorporated into newly synthesized RNA. The stem loops can then be detected using a fusion of GFP and the MS2 coat protein, which has a high affinity, sequence-specific interaction with the MS2 stem loops. The recruitment of GFP to the site of transcription is visualised as a single fluorescent spot. This new approach has revealed that transcription occurs in discontinuous bursts, or pulses (see Transcriptional bursting). With the notable exception of in situ techniques, most other methods provide cell population averages, and are not capable of detecting this fundamental property of genes.[20]

- Northern blot: the traditional method, and until the advent of RNA-Seq, the most quantitative

- RNA-Seq: applies next-generation sequencing techniques to sequence whole transcriptomes, which allows the measurement of relative abundance of RNA, as well as the detection of additional variations such as fusion genes, post-transcriptional edits and novel splice sites

17.7 Reverse transcription

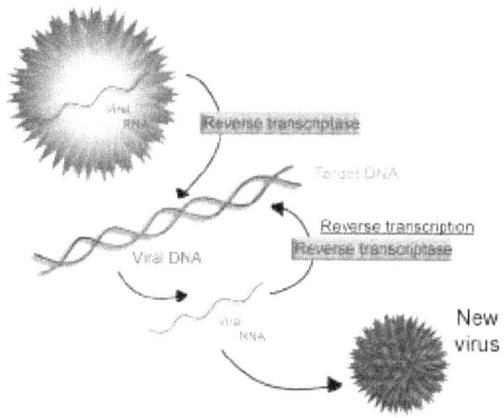

Scheme of reverse transcription

Some viruses (such as HIV, the cause of AIDS), have the ability to transcribe RNA into DNA. HIV has an RNA genome that is *reverse transcribed* into DNA. The resulting DNA can be merged with the DNA genome of the host cell. The main enzyme responsible for synthesis of DNA from an RNA template is called reverse transcriptase.

In the case of HIV, reverse transcriptase is responsible for synthesizing a complementary DNA strand (cDNA) to the viral RNA genome. The enzyme ribonuclease H then digests the RNA strand, and reverse transcriptase synthesises a complementary strand of DNA to form a double helix DNA structure ("cDNA"). The cDNA is integrated into the host cell's genome by the enzyme integrase, which causes the host cell to generate viral proteins that reassemble into new viral particles. In HIV, subsequent to this, the host cell undergoes programmed cell death, or apoptosis of T cells.[21] However, in other retroviruses, the host cell remains intact as the virus buds out of the cell.

Some eukaryotic cells contain an enzyme with reverse transcription activity called telomerase. Telomerase is a reverse transcriptase that lengthens the ends of linear chromosomes. Telomerase carries an RNA template from which it synthesizes a repeating sequence of DNA, or "junk" DNA. This repeated sequence of DNA is called a telomere and can be thought of as a "cap" for a chromosome. It is important because every time a linear chromosome is duplicated, it is shortened. With this "junk" DNA or "cap" at the ends of chromosomes, the shortening eliminates some of the non-essential, repeated sequence rather than the protein-encoding DNA sequence, that is farther away from the chromosome end.

Telomerase is often activated in cancer cells to enable cancer cells to duplicate their genomes indefinitely without losing important protein-coding DNA sequence. Activation of telomerase could be part of the process that allows cancer cells to become *immortal*. The immortalizing factor of cancer via telomere lengthening due to telomerase has been proven to occur in 90% of all carcinogenic tumors *in vivo* with the remaining 10% using an alternative telomere maintenance route called ALT or Alternative Lengthening of Telomeres.[22]

17.8 See also

- Crick's central dogma - DNA is transcribed to RNA, which is translated to polypeptides (polypeptides cannot "reverse translate" into RNA or DNA)

- Eukaryotic transcription

- Gene regulation

- Bacterial transcription

- RNA Polymerase

- Reverse transcription - process viruses use to make DNA from RNA

- Splicing - process of removing introns from precursor messenger RNA (pre-mRNA) to make messenger RNA (mRNA)

- Translation - process of decoding RNA to form polypeptides

- Transcription factor

17.9   References

[1] Eldra P. Solomon, Linda R. Berg, Diana W. Martin. *Biology, 8th Edition, International Student Edition*. Thomson Brooks/Cole. ISBN 978-0495317142

[2] "Tentative identification of RNA-dependent RNA polymerases of dsRNA viruses and their relationship to positive strand RNA viral polymerases". *FEBS Letters* **252**: 42–46. July 1989. doi:10.1016/0014-5793(89)80886-5. PMID 2759231.

[3] Berg J, Tymoczko JL, Stryer L (2006). *Biochemistry* (6th ed.). San Francisco: W. H. Freeman. ISBN 0-7167-8724-5.

[4] Basic Medical Biochemistry, 4th edition, Marks. Chapter 14

[5] Littlefield, O., Korkhin, Y., and Sigler, P.B. (1999). "The structural basis for the oriented assembly of a TBP/TFB/promoter complex". *PNAS* **96** (24): 13668–13673. doi:10.1073/pnas.96.24.13668. PMC 24122. PMID 10570130.

[6] Hausner, W., Michael Thomm, M. (2001). "Events during Initiation of Archaeal Transcription: Open Complex Formation and DNA-Protein Interactions". *Journal of Bacteriology* **183** (10): 3025–3031. doi:10.1128/JB.183.10.3025-3031.2001. PMC 95201. PMID 11325929.

[7] Qureshi, SA; Bell, SD; Jackson, SP (1997). "Factor requirements for transcription in the archaeon Sulfolobus shibatae". *EMBO Journal* **16** (10): 2927–2936. doi:10.1093/emboj/16.10.2927. PMC 1169900. PMID 9184236.

[8] Raven, Peter H. (2011). *Biology* (9th ed.). New York: McGraw-Hill. pp. 278–301. ISBN 978-0-07-353222-6.

[9] Mohamed Ouhammouch, Robert E. Dewhurst, Winfried Hausner, Michael Thomm, and E. Peter Geiduschek (2003). "Activation of archaeal transcription by recruitment of the TATA-binding protein". *Proceedings of the National Academy of Sciences of the United States of America* **100** (9): 5097–5102. doi:10.1073/pnas.0837150100. PMC 154304. PMID 12692306.

[10] Goldman, S.; Ebright, R.; Nickels, B. (May 2009). "Direct detection of abortive RNA transcripts in vivo". *Science* **324** (5929): 927–928. doi:10.1126/science.1169237. PMC 2718712. PMID 19443781.

[11] Raffaelle, M.; Kanin, E. I.; Vogt, J.; Burgess, R. R.; Ansari, A. Z. (2005). "Holoenzyme Switching and Stochastic Release of Sigma Factors from RNA Polymerase in Vivo". *Molecular Cell* **20** (3): 357–366. doi:10.1016/j.molcel.2005.10.011. PMID 16285918.

[12] Revyakin, A.; Liu, C.; Ebright, R.; Strick, T. (2006). "Abortive initiation and productive initiation by RNA polymerase involve DNA scrunching". *Science* **314** (5802): 1139–1143. doi:10.1126/science.1131398. PMC 2754787. PMID 17110577.

[13] Mandal, S. S.; Chu, C.; Wada, T.; Handa, H.; Shatkin, A. J.; Reinberg, D. (2004). "Functional interactions of RNA-capping enzyme with factors that positively and negatively regulate promoter escape by RNA polymerase II". *Proceedings of the National Academy of Sciences* **101** (20): 7572–7577. doi:10.1073/pnas.0401493101. PMC 419647. PMID 15136722.

[14] Goodrich, J. A.; Tjian, R. (1994). "Transcription factors IIE and IIH and ATP hydrolysis direct promoter clearance by RNA polymerase II". *Cell* **77** (1): 145–156. doi:10.1016/0092-8674(94)90242-9. PMID 8156590.

[15] Richardson, J (2002). "Rho-dependent termination and ATPases in transcript termination". *Biochimica et Biophysica Acta* **1577** (2): 251–260. doi:10.1016/S0167-4781(02)00456-6.

[16] Lykke-Andersen, S; Jensen, TH (2007). "Overlapping pathways dictate termination of RNA polymerase II transcription". *Biochimie* **89** (10): 1177–82. doi:10.1016/j.biochi.2007.05.007.

[17] 8-Hydroxyquinoline info from SIGMA-ALDRICH. Retrieved Feb 2012

[18] Papantonis, A (2012-10-26). "TNFα signals through specialized factories where responsive coding and miRNA genes are transcribed". *Nature EMBO J.*

[19] "Chemistry 2006". *Nobel Foundation*. Retrieved March 29, 2007.

[20] Raj, A.; van Oudenaarden, A. (2008). "Nature, nurture, or chance: stochastic gene expression and its consequences". *Cell* **135**: 216–26. doi:10.1016/j.cell.2008.09.050. PMC 3118044. PMID 18957198.

[21] Kolesnikova I. N. (2000). "Some patterns of apoptosis mechanism during HIV-infection". *Dissertation* (in Russian). Retrieved February 20, 2011.

[22] ALT and Telomerase from Nature. Retrieved May 2010

17.10   External links

- Interactive Java simulation of transcription initiation. From Center for Models of Life at the Niels Bohr Institute.

- Interactive Java simulation of transcription interference--a game of promoter dominance in bacterial virus. From Center for Models of Life at the Niels Bohr Institute.

- Biology animations about this topic under Chapter 15 and Chapter 18

- Virtual Cell Animation Collection. Introducing Transcription

- Easy to use DNA transcription site

Chapter 18

Transformation (genetics)

Not to be confused with an unrelated process called malignant transformation which occurs in the progression of cancer.

In molecular biology, **transformation** is the genetic alter-

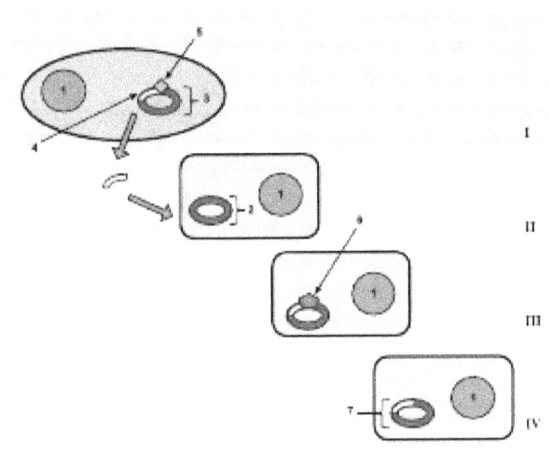

In this image, a gene from bacterial cell 1 is moved from bacterial cell 1 to bacterial cell 2. This process of bacterial cell 2 taking up new genetic material is called transformation.

ation of a cell resulting from the direct uptake and incorporation of exogenous genetic material (exogenous DNA) from its surroundings through the cell membrane(s). Transformation occurs naturally in some species of bacteria, but it can also be effected by artificial means in other cells. For transformation to happen, bacteria must be in a state of competence, which might occur as a time-limited response to environmental conditions such as starvation and cell density.

Transformation is one of three processes by which exogenous genetic material may be introduced into a bacterial cell, the other two being conjugation (transfer of genetic material between two bacterial cells in direct contact) and transduction (injection of foreign DNA by a bacteriophage virus into the host bacterium).

"Transformation" may also be used to describe the inser-

tion of new genetic material into nonbacterial cells, including animal and plant cells; however, because "transformation" has a special meaning in relation to animal cells, indicating progression to a cancerous state, the term should be avoided for animal cells when describing introduction of exogenous genetic material. Introduction of foreign DNA into eukaryotic cells is often called "transfection".[1]

18.1 History

Transformation was first demonstrated in 1928 by British bacteriologist Frederick Griffith. Griffith discovered that a strain of *Streptococcus pneumoniae* could be made virulent after being exposed to heat-killed virulent strains. Griffith hypothesized that some "transforming principle" from the heat-killed strain was responsible for making the harmless strain virulent. In 1944 this "transforming principle" was identified as being genetic by Oswald Avery, Colin MacLeod, and Maclyn McCarty. They isolated DNA from a virulent strain of *S. pneumoniae* and using just this DNA were able to make a harmless strain virulent. They called this uptake and incorporation of DNA by bacteria "transformation" (See Avery-MacLeod-McCarty experiment). The results of Avery et al.'s experiments were at first skeptically received by the scientific community and it was not until the development of genetic markers and the discovery of other methods of genetic transfer (conjugation in 1947 and transduction in 1953) by Joshua Lederberg that Avery's experiments were accepted.[2]

It was originally thought that *Escherichia coli*, a commonly used laboratory organism, was refractory to transformation. However, in 1970, Morton Mandel and Akiko Higa showed that *E. coli* may be induced to take up DNA from bacteriophage λ without the use of helper phage after treatment with calcium chloride solution.[3] Two years later in 1972, Stanley Cohen, Annie Chang and Leslie Hsu showed that CaCl

2 treatment is also effective for transformation of plasmid DNA.[4] The method of transformation by Mandel and Higa

was later improved upon by Douglas Hanahan.[5] The discovery of artificially induced competence in *E. coli* created an efficient and convenient procedure for transforming bacteria which allows for simpler molecular cloning methods in biotechnology and research, and it is now a routinely used laboratory procedure.

Transformation using electroporation was developed in the late 1980s, increasing the efficiency of in-vitro transformation and increasing the number of bacterial strains that could be transformed.[6] Transformation of animal and plant cells was also investigated with the first transgenic mouse being created by injecting a gene for a rat growth hormone into a mouse embryo in 1982.[7] In 1907 a bacterium that caused plant tumors, *Agrobacterium tumefaciens*, was discovered and in the early 1970s the tumor-inducing agent was found to be a DNA plasmid called the Ti plasmid.[8] By removing the genes in the plasmid that caused the tumor and adding in novel genes, researchers were able to infect plants with *A. tumefaciens* and let the bacteria insert their chosen DNA into the genomes of the plants.[9] Not all plant cells are susceptible to infection by *A. tumefaciens*, so other methods were developed, including electroporation and micro-injection.[10] Particle bombardment was made possible with the invention of the Biolistic Particle Delivery System (gene gun) by John Sanford in the 1980s.[11][12][13]

18.2 Methods and mechanisms

18.2.1 Definitions

Bacterial transformation may be referred to as a stable genetic change brought about by the uptake of naked DNA (DNA without associated cells or proteins) to increase DNA quantity; competence refers to the state of being able to take up exogenous DNA from the environment. There are two forms of transformation and competence: natural and artificial.

Natural transformation

Natural transformation is a bacterial adaptation for DNA transfer that depends on the expression of numerous bacterial genes whose products appear to be responsible for this process.[14][15] In general, transformation is a complex, energy-requiring developmental process. In order for a bacterium to bind, take up and recombine exogenous DNA into its chromosome, it must become competent, that is, enter a special physiological state. Competence development in *Bacillus subtilis* requires expression of about 40 genes.[16] The DNA integrated into the host chromosome is usually

(but with rare exceptions) derived from another bacterium of the same species, and is thus homologous to the resident chromosome.

In *B. subtilis* the length of the transferred DNA is greater than 1271 kb (more than 1 million bases).[17] The length transferred is likely double stranded DNA and is often more than a third of the total chromosome length of 4215 kb.[18] It appears that about 7-9% of the recipient cells take up an entire chromosome.[19]

The capacity for natural transformation appears to occur in a number of prokaryotes, and thus far 67 prokaryotic species (in seven different phyla) are known to undergo this process.[15]

Competence for transformation is typically induced by high cell density and/or nutritional limitation, conditions associated with the stationary phase of bacterial growth. Transformation in *Haemophilus influenzae* occurs most efficiently at the end of exponential growth as bacterial growth approaches stationary phase.[20] Transformation in *Streptococcus mutans*, as well as in many other streptococci, occurs at high cell density and is associated with biofilm formation.[21] Competence in *B. subtilis* is induced toward the end of logarithmic growth, especially under conditions of amino acid limitation.[22]

Transformation, as an adaptation for DNA repair

Competence is specifically induced by DNA damaging conditions. For instance, transformation is induced in *Streptococcus pneumoniae* by the DNA damaging agents mitomycin C (a DNA crosslinking agent) and fluoroquinolone (a topoisomerase inhibitor that causes double-strand breaks).[23] In *B. subtilis*, transformation is increased by UV light, a DNA damaging agent.[24] In *Helicobacter pylori*, ciprofloxacin, which interacts with DNA gyrase and introduces double-strand breaks, induces expression of competence genes, thus enhancing the frequency of transformation[25] Using *Legionella pneumophila*, Charpentier et al.[26] tested 64 toxic molecules to determine which of these induce competence. Of these only six, all DNA damaging agents, caused strong induction. These DNA damaging agents were mitomycin C (which causes DNA inter-strand crosslinks), norfloxacin, ofloxacin and nalidixic acid (inhibitors of DNA gyrase that cause double-strand breaks [27]), bicyclomycin (causes single- and double-strand breaks[28]), and hydroxyurea (induces DNA base oxidation[29]). UV light also induced competence in L. pneumophila. Charpentier et al.[26] suggested that competence for transformation probably evolved as a DNA damage response.

Logarithmically growing bacteria differ from stationary phase bacteria with respect to the number of genome copies present in the cell, and this has implications for the capability to carry out an important DNA repair process. During logarithmic growth, two or more copies of any particular region of the chromosome may be present in a bacterial cell, as cell division is not precisely matched with chromosome replication. The process of homologous recombinational repair (HRR) is a key DNA repair process that is especially effective for repairing double-strand damages, such as double-strand breaks. This process depends on a second homologous chromosome in addition to the damaged chromosome. During logarithmic growth, a DNA damage in one chromosome may be repaired by HRR using sequence information from the other homologous chromosome. Once cells approach stationary phase, however, they typically have just one copy of the chromosome, and HRR requires input of homologous template from outside the cell by transformation.[30]

To test whether the adaptive function of transformation is repair of DNA damages, a series of experiments were carried out using *B. subtilis* irradiated by UV light as the damaging agent (reviewed by Michod et al.[31] and Bernstein et al.[30]) The results of these experiments indicated that transforming DNA acts to repair potentially lethal DNA damages introduced by UV light in the recipient DNA. The particular process responsible for repair was likely HRR. Transformation in bacteria can be viewed as a primitive sexual process, since it involves interaction of homologous DNA from two individuals to form recombinant DNA that is passed on to succeeding generations. Bacterial transformation in prokaryotes may have been the ancestral process that gave rise to meiotic sexual reproduction in eukaryotes (see Wikipedia articles Evolution of sexual reproduction; Meiosis.)

Natural competence

Main article: Natural competence

About 1% of bacterial species are capable of naturally taking up DNA under laboratory conditions; more may be able to take it up in their natural environments. DNA material can be transferred between different strains of bacteria, in a process that is called horizontal gene transfer. Some species upon cell death release their DNA to be taken up by other cells, however transformation works best with DNA from closely related species. These naturally competent bacteria carry sets of genes that provide the protein machinery to bring DNA across the cell membrane(s). The transport of the exogeneous DNA into the cells may require proteins that are involved in the assembly of type IV pili and type II secretion system, as well as DNA translocase complex at the cytoplasmic membrane.[14]

Due to the differences in structure of the cell envelope between Gram-positive and Gram-negative bacteria, there are some differences in the mechanisms of DNA uptake in these cells, however most of them share common features that involve related proteins. The DNA first binds to the surface of the competent cells on a DNA receptor, and passes through the cytoplasmic membrane via DNA translocase.[32] Only single-stranded DNA may pass through, the other strand being degraded by nucleases in the process. The translocated single-stranded DNA may then be integrated into the bacterial chromosomes by a RecA-dependent process. In Gram-negative cells, due to the presence of an extra membrane, the DNA requires the presence of a channel formed by secretins on the outer membrane. Pilin may be required for competence, but its role is uncertain.[33] The uptake of DNA is generally non-sequence specific, although in some species the presence of specific DNA uptake sequences may facilitate efficient DNA uptake.[34]

Artificial competence

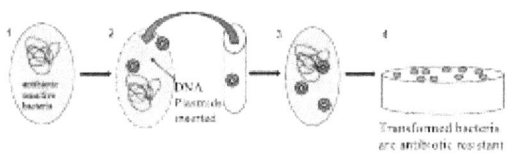

Schematic of bacterial transformation — for which artificial competence must first be induced.

Artificial competence can be induced in laboratory procedures that involve making the cell passively permeable to DNA by exposing it to conditions that do not normally occur in nature.[35] Typically the cells are incubated in a solution containing divalent cations (often calcium chloride) under cold conditions, before being exposed to a heat pulse (heat shock).

It has been found[36] that growth of Gram negative bacteria in 20 mM Mg reduces the number of protein-to-lipopolysaccharide bonds by increasing the ratio of ionic to covalent bonds, which increases membrane fluidity, facilitating transformation. The role of lipopolysaccharides here are verified from the observation that shorter O-side chains are more effectively transformed — perhaps because of improved DNA accessibility.

The surface of bacteria such as *E. coli* is negatively charged due to phospholipids and lipopolysaccharides on its cell surface, and the DNA is also negatively charged. One function of the divalent cation therefore would be to shield the charges by coordinating the phosphate groups and other negative charges, thereby allowing a DNA molecule to adhere to the cell surface.

DNA entry into *E. coli* cells is through channels known as zones of adhesion or Bayer's junction, with a typical cell carrying as many as 400 such zones. Their role was established when cobalamine (which also uses these channels) was found to competitively inhibit DNA uptake. Another type of channel implicated in DNA uptake consists of poly (HB):poly P:Ca. In this poly (HB) is envisioned to wrap around DNA (itself a polyphosphate), and is carried in a shield formed by Ca ions.[36]

It is suggested that exposing the cells to divalent cations in cold condition may also change or weaken the cell surface structure, making it more permeable to DNA. The heat-pulse is thought to create a thermal imbalance across the cell membrane, which forces the DNA to enter the cells through either cell pores or the damaged cell wall.

Electroporation is another method of promoting competence. In this method the cells are briefly shocked with an electric field of 10-20 kV/cm, which is thought to create holes in the cell membrane through which the plasmid DNA may enter. After the electric shock, the holes are rapidly closed by the cell's membrane-repair mechanisms.

18.2.2 Yeast

Most species of yeast, including *Saccharomyces cerevisiae*, may be transformed by exogenous DNA in the environment. Several methods have been developed to facilitate this transformation at high frequency in the lab.[37]

- Yeast cells may be treated with enzymes to degrade their cell walls, yielding spheroplasts. These cells are very fragile but take up foreign DNA at a high rate.[38]

- Exposing intact yeast cells to alkali cations such as those of cesium or lithium allows the cells to take up plasmid DNA.[39] Later protocols adapted this transformation method, using lithium acetate, polyethylene glycol, and single-stranded DNA.[40] In these protocols, the single-stranded DNA preferentially binds to the yeast cell wall, preventing plasmid DNA from doing so and leaving it available for transformation.[41]

- Electroporation: Formation of transient holes in the cell membranes using electric shock; this allows DNA to enter as described above for bacteria.[42]

- Enzymatic digestion[43] or agitation with glass beads[44] may also be used to transform yeast cells.

Efficiency. Different yeast genera and species take up foreign DNA with different efficiencies.[45] Also, most transformation protocols have been developed for baker's yeast,

S. cerevisiae, and thus may not be optimal for other species. Even within one species, different strains have different transformation efficiencies, sometimes different by 3 orders of magnitude. For instance, when S. cerevisiae strains were transformed with 10 ug of plasmid YEp13, the strain DKD-5D-H yielded between 550 and 3115 colonies while strain OS1 yielded less than 5 colonies.[46]

18.2.3 Plants

A number of methods are available to transfer DNA into plant cells. Some vector-mediated methods are:

- *Agrobacterium*-mediated transformation is the easiest and most simple plant transformation. Plant tissue (often leaves) are cut into small pieces, e.g. 10x10mm, and soaked for 10 minutes in a fluid containing suspended *Agrobacterium*. The bacteria will attach to many of the plant cells exposed by the cut. The plant cells secrete wound-related phenolic compounds which in turn act to upregulate the virulence operon of the Agrobacterium. The virulence operon includes many genes that encode for proteins that are part of a Type IV secretion system that exports from the bacterium proteins and DNA (delineated by specific recognition motifs called border sequences and excised as a single strand from the virulence plasmid) into the plant cell through a structure called a pilus. The transferred DNA (called T-DNA) is piloted to the plant cell nucleus by nuclear localization signals present in the Agrobacterium protein VirD2, which is covalently attached to the end of the T-DNA at the Right border (RB). Exactly how the T-DNA is integrated into the host plant genomic DNA is an active area of plant biology research. Assuming that a selection marker (such as an antibiotic resistance gene) was included in the T-DNA, the transformed plant tissue can be cultured on selective media to produce shoots. The shoots are then transferred to a different medium to promote root formation. Once roots begin to grow from the transgenic shoot, the plants can be transferred to soil to complete a normal life cycle (make seeds). The seeds from this first plant (called the T1, for first transgenic generation) can be planted on a selective (containing an antibiotic), or if an herbicide resistance gene was used, could alternatively be planted in soil, then later treated with herbicide to kill wildtype segregants. Some plants species, such as *Arabidopsis thaliana* can be transformed by dipping the flowers or whole plant, into a suspension of *Agrobacterium tumefaciens*, typically strain C58 (C=Cherry, 58=1958, the year in which this particular strain of *A. tumefaciens* was isolated from a cherry tree in an orchard at Cornell

University in Ithaca, New York). Though many plants remain recalcitrant to transformation by this method, research is ongoing that continues to add to the list the species that have been successfully modified in this manner.

- Viral transformation (transduction): Package the desired genetic material into a suitable plant virus and allow this modified virus to infect the plant. If the genetic material is DNA, it can recombine with the chromosomes to produce transformant cells. However, genomes of most plant viruses consist of single stranded RNA which replicates in the cytoplasm of infected cell. For such genomes this method is a form of transfection and not a real transformation, since the inserted genes never reach the nucleus of the cell and do not integrate into the host genome. The progeny of the infected plants is virus-free and also free of the inserted gene.

Some vector-less methods include:

- Gene gun: Also referred to as particle bombardment, microprojectile bombardment, or biolistics. Particles of gold or tungsten are coated with DNA and then shot into young plant cells or plant embryos. Some genetic material will stay in the cells and transform them. This method also allows transformation of plant plastids. The transformation efficiency is lower than in *Agrobacterium*-mediated transformation, but most plants can be transformed with this method.

- Electroporation: Formation of transient holes in cell membranes using electric pulses of high field strength; this allows DNA to enter as described above for bacteria.[47]

18.2.4 Animals

Introduction of DNA into animal cells is usually called transfection, and is discussed in the corresponding article.

18.3 Practical aspects of transformation in molecular biology

Further information: Transformation efficiency

The discovery of artificially induced competence in bacteria allow bacteria such as *Escherichia coli* to be used as a convenient host for the manipulation of DNA as well as expressing proteins. Typically plasmids are used for transformation in *E. coli*. In order to be stably maintained in the

cell, a plasmid DNA molecule must contain an origin of replication, which allows it to be replicated in the cell independently of the replication of the cell's own chromosome.

The efficiency with which a competent culture can take up exogenous DNA and express its genes is known as transformation efficiency and is measured in colony forming unit (cfu) per µg DNA used. A transformation efficiency of 1×10^8 cfu/µg for a small plasmid like pUC19 is roughly equivalent to 1 in 2000 molecules of the plasmid used being transformed.

In calcium chloride transformation, the cells are prepared by chilling cells in the presence of Ca2+ (in CaCl$_2$ solution), making the cell become permeable to plasmid DNA. The cells are incubated on ice with the DNA, and then briefly heat-shocked (e.g., at 42 °C for 30–120 seconds). This method works very well for circular plasmid DNA. Non-commercial preparations should normally give 10^6 to 10^7 transformants per microgram of plasmid; a poor preparation will be about 10^4/µg or less, but a good preparation of competent cells can give up to ~10^8 colonies per microgram of plasmid.[48] Protocols, however, exist for making supercompetent cells that may yield a transformation efficiency of over 10^9.[49] The chemical method, however, usually does not work well for linear DNA, such as fragments of chromosomal DNA, probably because the cell's native exonuclease enzymes rapidly degrade linear DNA. In contrast, cells that are naturally competent are usually transformed more efficiently with linear DNA than with plasmid DNA.

The transformation efficiency using the CaCl$_2$ method decreases with plasmid size, and electroporation therefore may be a more effective method for the uptake of large plasmid DNA.[50] Cells used in electroporation should be prepared first by washing in cold double-distilled water to remove charged particles that may create sparks during the electroporation process.

18.3.1 Selection and screening in plasmid transformation

Because transformation usually produces a mixture of relatively few transformed cells and an abundance of non-transformed cells, a method is necessary to select for the cells that have acquired the plasmid. The plasmid therefore requires a selectable marker such that those cells without the plasmid may be killed or have their growth arrested. Antibiotic resistance is the most commonly used marker for prokaryotes. The transforming plasmid contains a gene that confers resistance to an antibiotic that the bacteria are otherwise sensitive to. The mixture of treated cells is cultured on media that contain the antibiotic so that only transformed cells are able to grow. Another method of selection is the use of certain auxotrophic markers that can compensate for an inability to metabolise certain amino acids, nucleotides, or sugars. This method requires the use of suitably mutated strains that are deficient in the synthesis or utility of a particular biomolecule, and the transformed cells are cultured in a medium that allows only cells containing the plasmid to grow.

In a cloning experiment, a gene may be inserted into a plasmid used for transformation. However, in such experiment, not all the plasmids may contain a successfully inserted gene. Additional techniques may therefore be employed further to screen for transformed cells that contain plasmid with the insert. Reporter genes can be used as markers, such as the *lacZ* gene which codes for β-galactosidase used in blue-white screening. This method of screening relies on the principle of α-complementation, where a fragment of the *lacZ* gene (*lacZα*) in the plasmid can complement another mutant *lacZ* gene (*lacZΔM15*) in the cell. Both genes by themselves produce non-functional peptides, however, when expressed together, as when a plasmid containing *lacZ-α* is transformed into a *lacZΔM15* cells, they form a functional β-galactosidase. The presence of an active β-galactosidase may be detected when cells are grown in plates containing X-gal, forming characteristic blue colonies. However, the multiple cloning site, where a gene of interest may be ligated into the plasmid vector, is located within the *lacZα* gene. Successful ligation therefore disrupts the *lacZα* gene, and no functional β-galactosidase can form, resulting in white colonies. Cells containing successfully ligated insert can then be easily identified by its white coloration from the unsuccessful blue ones.

Other commonly used reporter genes are green fluorescent protein (GFP), which produces cells that glow green under blue light, and the enzyme luciferase, which catalyzes a reaction with luciferin to emit light. The recombinant DNA may also be detected using other methods such as nucleic acid hybridization with radioactive RNA probe, while cells that expressed the desired protein from the plasmid may also be detected using immunological methods.

18.4 References

[1] Alberts, Bruce; et al. (2002). *Molecular Biology of the Cell*. New York: Garland Science. p. G:35. ISBN 978-0-8153-4072-0.

[2] Lederberg, Joshua (1994). *The Transformation of Genetics by DNA: An Anniversary Celebration of AVERY, MACLEOD and MCCARTY(1944) in* Anecdotal, Historical and Critical Commentaries on Genetics. The Rockfeller University, New York, New York 10021-6399. PMID 8150273.

[3] Mandel, Morton; Higa, Akiko (1970). "Calcium-dependent bacteriophage DNA infection". *Journal of Molecular Biology* **53** (1): 159–162. doi:10.1016/0022-2836(70)90051-3. PMID 4922220.

[4] Cohen, Stanley; Chang, Annie; Hsu, Leslie (1972). "Nonchromosomal Antibiotic Resistance in Bacteria: Genetic Transformation of Escherichia coli by R-Factor DNA". *Proceedings of the National Academy of Sciences* **69** (8): 2110–4. doi:10.1073/pnas.69.8.2110. PMC 426879. PMID 4559594.

[5] Hanahan, D. (1983). "Studies on transformation of Escherichia coli with plasmids". *Journal of Molecular Biology* **166** (4): 557–580. doi:10.1016/S0022-2836(83)80284-8. PMID 6345791.

[6] Wirth R, Friesenegger A, Fiedler S (March 1989). "Transformation of various species of gram-negative bacteria belonging to 11 different genera by electroporation". *Mol. Gen. Genet.* **216** (1): 175–7. doi:10.1007/BF00332248. PMID 2659971.

[7] Palmiter, Richard; Ralph L. Brinster; Robert E. Hammer; Myrna E. Trumbauer; Michael G. Rosenfeld; Neal C. Birnberg; Ronald M. Evans (1982). "Dramatic growth of mice that develop from eggs microinjected with metallothionein–growth hormone fusion genes". *Nature* **300** (5893): 611–5. doi:10.1038/300611a0. PMID 6958982.

[8] Nester, Eugene. "Agrobacterium: The Natural Genetic Engineer (100 Years Later)". Retrieved 14 January 2011.

[9] Zambryski, P.; Joos, H.; Genetello, C.; Leemans, J.; Montagu, M. V.; Schell, J. (1983). "Ti plasmid vector for the introduction of DNA into plant cells without alteration of their normal regeneration capacity". *The EMBO Journal* **2** (12): 2143–2150. PMC 555426. PMID 16453482.

[10] Peters, Pamela. "Transforming Plants - Basic Genetic Engineering Techniques". Retrieved 28 January 2010.

[11] "Biologists invent gun for shooting cells with DNA" (PDF). *Cornell Chronicle*. 14 May 1987. p. 3.

[12] Sanford JC, et al. (1987). "Delivery of substances into cells and tissues using a particle bombardment process". *Journal of Particulate Science and Technology* **5**: 27–37. doi:10.1080/02726358708904533.

[13] Klein TM, Wolf ED, Wu R, Sanford JC (7 May 1987). "High-velocity microprojectiles for delivering nucleic acids into living cells". *Nature* **327** (6117): 70–73. doi:10.1038/327070a0.

[14] Chen I, Dubnau D (March 2004). "DNA uptake during bacterial transformation" (PDF). *Nat. Rev. Microbiol.* **2** (3): 241–9. doi:10.1038/nrmicro844. PMID 15083159.

[15] Johnsborg O, Eldholm V, Håvarstein LS (December 2007). "Natural genetic transformation: prevalence, mechanisms and function". *Res. Microbiol.* **158** (10): 767–78. doi:10.1016/j.resmic.2007.09.004. PMID 17997281.

[16] Solomon JM, Grossman AD (April 1996). "Who's competent and when: regulation of natural genetic competence in bacteria". *Trends Genet.* **12** (4): 150–5. doi:10.1016/0168-9525(96)10014-7. PMID 8901420.

[17] Saito Y, Taguchi H, Akamatsu T (March 2006). "Fate of transforming bacterial genome following incorporation into competent cells of Bacillus subtilis: a continuous length of incorporated DNA". *J. Biosci. Bioeng.* **101** (3): 257–62. doi:10.1263/jbb.101.257. PMID 16716928.

[18] Saito Y, Taguchi H, Akamatsu T (April 2006). "DNA taken into Bacillus subtilis competent cells by lysed-protoplast transformation is not ssDNA but dsDNA". *J. Biosci. Bioeng.* **101** (4): 334–9. doi:10.1263/jbb.101.334. PMID 16716942.

[19] Akamatsu T, Taguchi H (April 2001). "Incorporation of the whole chromosomal DNA in protoplast lysates into competent cells of *Bacillus subtilis*". *Biosci. Biotechnol. Biochem.* **65** (4): 823–9. doi:10.1271/bbb.65.823. PMID 11388459.

[20] Goodgal SH, Herriott RM (July 1961). "Studies on transformations of *Hemophilus influenzae*. I. Competence". *J. Gen. Physiol.* **44** (6): 1201–27. doi:10.1085/jgp.44.6.1201. PMC 2195138. PMID 13707010.

[21] Aspiras MB, Ellen RP, Cvitkovitch DG (September 2004). "ComX activity of *Streptococcus mutans* growing in biofilms". *FEMS Microbiol. Lett.* **238** (1): 167–74. doi:10.1016/j.femsle.2004.07.032. PMID 15336418.

[22] Anagnostopoulos C, Spizizen J (May 1961). "Requirements for transformation in *Bacillus subtilis*". *J. Bacteriol.* **81** (5): 741–6. PMC 279084. PMID 16561900.

[23] Claverys JP, Prudhomme M, Martin B (2006). "Induction of competence regulons as a general response to stress in gram-positive bacteria". *Annu. Rev. Microbiol.* **60**: 451–75. doi:10.1146/annurev.micro.60.080805.142139. PMID 16771651.

[24] Michod RE, Wojciechowski MF, Hoelzer MA (January 1988). "DNA repair and the evolution of transformation in the bacterium *Bacillus subtilis*". *Genetics* **118** (1): 31–9. PMC 1203263. PMID 8608929.

[25] Dorer MS, Fero J, Salama NR (2010). Blanke, Steven R. ed. "DNA damage triggers genetic exchange in Helicobacter pylori". *PLoS Pathog.* **6** (7): e1001026. doi:10.1371/journal.ppat.1001026. PMC 2912397. PMID 20686662.

[26] Charpentier X, Kay E, Schneider D, Shuman HA (March 2011). "Antibiotics and UV radiation induce competence for natural transformation in *Legionella pneumophila*". *J. Bacteriol.* **193** (5): 1114–21. doi:10.1128/JB.01146-10. PMC 3067580. PMID 21169481.

[27] Albertini S, Chételat AA, Miller B, et al. (July 1995). "Genotoxicity of 17 gyrase- and four mammalian topoisomerase II-poisons in prokaryotic and eukaryotic test systems". *Mutagenesis* **10** (4): 343–51. doi:10.1093/mutage/10.4.343. PMID 7476271.

[28] Washburn RS, Gottesman ME (January 2011). "Transcription termination maintains chromosome integrity". *Proc. Natl. Acad. Sci. U.S.A.* **108** (2): 792–7. doi:10.1073/pnas.1009564108. PMC 3021805. PMID 21183718.

[29] Sakano K, Oikawa S, Hasegawa K, Kawanishi S (November 2001). "Hydroxyurea induces site-specific DNA damage via formation of hydrogen peroxide and nitric oxide". *Jpn. J. Cancer Res.* **92** (11): 1166–74. doi:10.1111/j.1349-7006.2001.tb02136.x. PMID 11714440.

[30] Bernstein H, Bernstein C, Michod RE (2012). DNA repair as the primary adaptive function of sex in bacteria and eukaryotes. In: *DNA Repair: New Research*. Editors: Sakura Kimura and Sora Shimizu. Nova Sci. Publ., Hauppauge, N.Y. Chapter 1: 1–49. ISBN 978-1-62100-808-8 https://www.novapublishers.com/catalog/product_info.php?products_id=31918

[31] Michod RE, Bernstein H, Nedelcu AM (May 2008). "Adaptive value of sex in microbial pathogens". *Infect. Genet. Evol.* **8** (3): 267–85. doi:10.1016/j.meegid.2008.01.002. PMID 18295550. http://www.hummingbirds.arizona.edu/Faculty/Michod/Downloads/IGE%20review%20sex.pdf

[32] Lacks, S.; Greenberg, B.; Neuberger, M. (1974). "Role of a Deoxyribonuclease in the Genetic Transformation of Diplococcus pneumoniae". *Proceedings of the National Academy of Sciences of the United States of America* **71** (6): 2305–2309. doi:10.1073/pnas.71.6.2305. PMC 388441. PMID 4152205.

[33] Long, C. D.; Tobiason, D. M.; Lazio, M. P.; Kline, K. A.; Seifert, H. S. (2003). "Low-Level Pilin Expression Allows for Substantial DNA Transformation Competence in Neisseria gonorrhoeae". *Infection and immunity* **71** (11): 6279–6291. doi:10.1128/iai.71.11.6279-6291.2003. PMC 219589. PMID 14573647.

[34] Sisco, K. L.; Smith, H. O. (1979). "Sequence-specific DNA uptake in Haemophilus transformation". *Proceedings of the National Academy of Sciences of the United States of America* **76** (2): 972–976. doi:10.1073/pnas.76.2.972. PMC 383110. PMID 311478.

[35] Donahue RA, Bloom FR (July 1998). "Large-volume transformation with high-throughput efficiency chemically competent cells" (PDF). *Focus* **20** (2): 54–56. OCLC 12352630.

[36] Srivastava, Sheela (2013). *Genetics of Bacteria* (PDF). India: Springer-Verlag. doi:10.1007/978-81-322-1090-0. ISBN 978-81-322-1089-4.

[37] Kawai, Shigeyuki; Hashimoto, Wataru; Murata, Kousaku (1 November 2010). "Transformation of Saccharomyces cerevisiae and other fungi: methods and possible underlying mechanism.". *Bioengineered Bugs* **1** (6): 395–403. doi:10.4161/bbug.1.6.13257.

[38] Hinnen, A; Hicks, JB; Fink, GR (April 1978). "Transformation of yeast.". *Proceedings of the National Academy of Sciences of the United States of America* **75** (4): 1929–33. doi:10.1073/pnas.75.4.1929. PMID 347451.

[39] Ito, H; Fukuda, Y; Murata, K; Kimura, A (January 1983). "Transformation of intact yeast cells treated with alkali cations". *Journal of Bacteriology* **153** (1): 163–8. PMC 217353. PMID 6336730.

[40] Gietz, RD; Woods, RA (2002). "Transformation of yeast by lithium acetate/single-stranded carrier DNA/polyethylene glycol method". *Methods in enzymology*. Methods in Enzymology **350**: 87–96. doi:10.1016/S0076-6879(02)50957-5. ISBN 9780121822538. PMID 12073338.

[41] Gietz, RD; Schiestl, RH; Willems, AR; Woods, RA (Apr 15, 1995). "Studies on the transformation of intact yeast cells by the LiAc/SS-DNA/PEG procedure". *Yeast* **11** (4): 355–60. doi:10.1002/yea.320110408. PMID 7785336.

[42] Schiestl, Robert H.; Manivasakam, P.; Woods, Robin A.; Gietzt, R.Daniel (1 August 1993). "Introducing DNA into Yeast by Transformation". *Methods* **5** (2): 79–85. doi:10.1006/meth.1993.1011.

[43] Spencer, F.; Ketner, G.; Connelly, C.; Hieter, P. (1 August 1993). "Targeted Recombination-Based Cloning and Manipulation of Large DNA Segments in Yeast". *Methods* **5** (2): 161–175. doi:10.1006/meth.1993.1021.

[44] Costanzo, MC; Fox, TD (November 1988). "Transformation of yeast by agitation with glass beads". *Genetics* **120** (3): 667–70. PMC 1203545. PMID 3066683.

[45] Dohmen, R. J.; Strasser, A. W.; Höner, C. B.; Hollenberg, C. P. (1991). "An efficient transformation procedure enabling long-term storage of competent cells of various yeast genera". *Yeast* **7** (7): 691–2. doi:10.1002/yea.320070704. PMID 1776359.

[46] Hayama, Y; Fukuda, Y; Kawai, S; Hashimoto, W; Murata, K (2002). "Extremely simple, rapid and highly efficient transformation method for the yeast Saccharomyces cerevisiae using glutathione and early log phase cells". *Journal of bioscience and bioengineering* **94** (2): 166–71. PMID 16233287.

[47] V.Singh and D.K.Jain (2014). *ISC BIOLOGY*. Nageen Prakashan. p. 840.

[48] Bacterial Transformation

[49] Inoue, H.; Nojima, H.; Okayama, H. (1990). "High efficiency transformation of Escherichia coli with plasmids". *Gene* **96** (1): 23–28. doi:10.1016/0378-1119(90)90336-P. PMID 2265755.

[50] Donahue RA, Bloom FR (September 1998). "Transformation efficiency of *E. coli* electroporated with large plasmid DNA" (PDF). *Focus* **20** (3): 77–78. Archived from the original on September 3, 2011.

18.5 External links

- Bacterial Transformation (a Flash Animation)
- "Ready, aim, fire!" At the Max Planck Institute for Molecular Plant Physiology in Potsdam-Golm plant cells are 'bombarded' using a particle gun

Chapter 19

Gene gun

PDS-1000/He Particle Delivery System

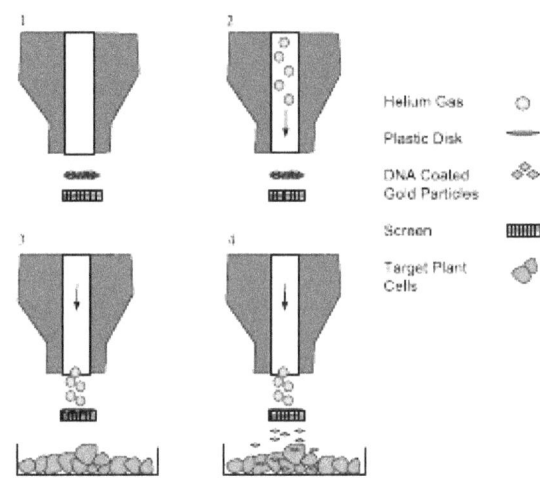

A gene gun is used for injecting cells with genetic information, it is also known as biolistic particle delivery system. Gene guns can be used effectively on most cells but are mainly used on plant cells. Step 1 The gene gun apparatus is ready to fire. Step 2 When the gun is turned on and the helium flows through. Step 3 The helium moving the disk with DNA coated particles toward the screen. Step 4 The helium having pushed the particles moving through the screen and moving to the target cells to transform the cells.

A **gene gun** or a **biolistic particle delivery system**, originally designed for plant transformation, is a device for injecting cells with genetic information: the inserted genetic material are termed transgenes. The payload is an elemental particle of a heavy metal coated with plasmid DNA. This technique is often simply referred to as **bioballistics** or **biolistics**.

This device is able to transform almost any type of cell, including plants, and is not limited to genetic material of the nucleus: it can also transform organelles, including plastids.

19.1 Gene gun design

The gene gun was originally a Crosman air pistol modified to fire dense tungsten particles. It was invented by John C Sanford, Ed Wolf and Nelson Allen at Cornell University,[1][2][3] and Ted Klein of DuPont, between 1983 and 1986. The original target was onions (chosen for their large cell size) and it was used to deliver particles coated with a marker gene.[4] Genetic transformation was then proven when the onion tissue expressed the gene.

The earliest custom manufactured gene guns (fabricated by Nelson Allen) used a 22 caliber nail gun cartridge to propel an extruded polyethylene cylinder (bullet) down a 22 cal. Douglas barrel. A droplet of the tungsten powder and genetic material was placed on the bullet and shot down the barrel at a lexan "stopping" disk with a petri dish below. The bullet welded to the disk and the genetic information blasted into the sample in the dish with a doughnut effect (devastation in the middle, a ring of good transformation and little around the edge). The gun was connected to a vacuum pump and was under vacuum while firing. The early design was put into limited production by a Rumsey-Loomis (a local machine shop then at Mecklenburg Rd in Ithaca, NY, USA). Later the design was refined by removing the "surge tank" and changing to nonexplosive propellants. DuPont added a plastic extrusion to the exterior to visually improve the machine for mass production to the scientific community. Biorad contracted with Dupont to manufacture and distribute the device. Improvements include the use of helium propellant and a multi-disk-collision delivery mechanism. Other heavy metals such as gold and silver are also used. Gold may be favored because it has better uniformity than tungsten and tungsten can be toxic to cells, but its use may be limited due to availability and cost.

19.2 Biolistic construct design

A construct is a piece of DNA inserted into the target's genome, including parts that are intended to be removed later.[5] All biolistic transformations require a construct to proceed and while there is great variation among biolistic constructs, they can be broadly sorted into two categories: those which are designed to transform eukaryotic nuclei, and those designed to transform prokaryotic-type genomes such as mitochondria, plasmids or plastids.[5]

Those meant to transform prokaryotic genomes generally have the gene or genes of interest, at least one promoter and terminator sequence, and a reporter gene; which is a gene used to ease detection or removal of those cells which didn't integrate the construct into their DNA.[5] These genes may each have their own promoter and terminator, or be grouped to produce multiple gene products from one transcript, in which case binding sites for translational machinery should be placed between each to ensure maximum translational efficiency. In any case the entire construct is flanked by regions called border sequences which are similar in sequence to locations within the genome, this allows the construct to target itself to a specific point in the existing genome.[5]

Constructs meant for integration into a eukaryotic nucleus follow a similar pattern except that: the construct contains no border sequences because the sequence rearrangement that prokaryotic constructs rely on rarely occurs in eukaryotes; and each gene contained within the construct must be expressed by its own copy of a promoter and terminator sequence.[5]

Though the above designs are generally followed, there are exceptions. For example, the construct might include a Cre-Lox system to selectively remove inserted genes; or a prokaryotic construct may insert itself downstream of a promoter, allowing the inserted genes to be governed by a promoter already in place and eliminating the need for one to be included in the construct.[5]

19.3 Application

Gene guns are so far mostly applied for plant cells. However, there is much potential use in humans and other animals as well.

19.3.1 Plants

The target of a gene gun is often a callus of undifferentiated plant cells growing on gel medium in a Petri dish. After the gold particles have impacted the dish, the gel and callus are largely disrupted. However, some cells were not obliterated in the impact, and have successfully enveloped a DNA coated gold particle, whose DNA eventually migrates to and integrates into a plant chromosome.

Cells from the entire Petri dish can be re-collected and selected for successful integration and expression of new DNA using modern biochemical techniques, such as a using a tandem selectable gene and northern blots.

Selected single cells from the callus can be treated with a series of plant hormones, such as auxins and gibberellins, and each may divide and differentiate into the organized, specialized, tissue cells of an entire plant. This capability of total re-generation is called totipotency. The new plant that originated from a successfully shot cell may have new genetic (heritable) traits.

The use of the gene gun may be contrasted with the use of *Agrobacterium tumefaciens* and its Ti plasmid to insert genetic information into plant cells. See transformation for different methods of transformation in different species.

19.3.2 Humans and other animals

Gene guns have also been used to deliver DNA vaccines.

The delivery of plasmids into rat neurons through the use of a gene gun, specifically DRG neurons, is also used as a pharmacological precursor in studying the effects of neurodegenerative diseases such as Alzheimer's disease.

The gene gun has become a common tool for labeling subsets of cells in cultured tissue. In addition to being able to transfect cells with DNA plasmids coding for fluorescent proteins, the gene gun can be adapted to deliver a wide variety of vital dyes to cells.[6]

Gene gun bombardment has also been used to transform *Caenorhabditis elegans*, as an alternative to microinjection.[7]

19.4 Advantages

Biolistics has proven to be a versatile method of genetic modification and it is generally preferred to engineer transformation-resistant crops, such as cereals. Notably, *Bt* maize is a product of biolistics.[5] Plastid transformation has also seen great success with particle bombardment when compared to other current techniques, such as *Agrobacterium* mediated transformation, which have difficulty targeting the vector to and stably expressing in the chloroplast.[5][8] In addition, there are no reports of a chloroplast silencing a transgene inserted with a gene gun.[9] Additionally, with only one firing of a gene gun, a skilled technician can generate two transformed organisms.[8] This technology has even allowed for modification of specific tissues *in situ*, although this is likely to damage large numbers of cells and transform only some, rather than all, cells of the tissue.[10]

19.5 Limitations

However, biolistics introduces the construct randomly into the target cells. Thus the altered DNA sequences may be transformed into whatever genomes are present in the cell, be they nuclear, mitochondrial, plasmid or any others, in any combination, though proper construct design may mitigate this. Another issue is that the gene inserted may be overexpressed when the construct is inserted multiple times in either the same or different locations of the genome.[5] This is due to the ability of the constructs to give and take genetic information from other constructs, causing some to carry no transgene and others to carry multiple copies; the number of copies inserted depends on both how many copies of the transgene an inserted construct has, and how many were inserted.[5] Also, because eukaryotic constructs rely on illegitimate recombination, a process by which the transgene is integrated into the genome without similar genetic sequences, and not homologous recombination, which inserts at similar sequences, they cannot be targeted to specific locations within the genome.[5]

19.6 References

[1] Segelken, Roger (14 May 1987). "Biologist invent gun for shooting cells with DNA" (PDF). *Cornell Chronicle*. p. 3. Retrieved 5 June 2014.

[2] Sanford, J.C.; Klein, T.M.; Wolf, E.D.; Allen, N. (1987). "Delivery of substances into cells and tissues using a particle bombardment process". *Particulate Science and Technology* **5** (1): 27–37. doi:10.1080/02726358708904533.

[3] Klein, T.M.; Wolf, E.D.; Wu, R.; Sanford, J.C. (May 1987). "High-velocity microprojectiles for delivering nucleic acids into living cells". *Nature* **327**: 70–73. doi:10.1038/327070a0.

[4] Segelken, Roger. "The Gene Shotgun". Cornell University College of Agriculture and Life Sciences. Archived from the original on 26 April 2010. Retrieved 5 June 2014.

[5] Slater, Adrian; Scott, Nigel; Fowler, Mark (2008). *Plant Biotechnology: the genetic manipulation of plants* (2 ed.). Oxford, New York, USA: Oxford University Press Inc. ISBN 978-0-19-928261-6.

[6] Gan, Wen-Biao; Grutzendler, Jaime; Wong, Wai Thong; Wong, Rachel O.L.; Lichtman, Jeff W (2000). "Multicolor "DiOlistic" Labeling of the Nervous System Using Lipophilic Dye Combinations". *Neuron* **27** (2): 219–25. doi:10.1016/S0896-6273(00)00031-3. PMID 10985343.

[7] Praitis, Vida (2006). "Creation of Transgenic Lines Using Microparticle Bombardment Methods" **351**: 93–108. doi:10.1385/1-59745-151-7:93. ISBN 1-59745-151-7.

[8] Sanford, John (April 28, 2006). "Biolistic plant transformation". *Physiologia Plantarum* **79** (1): 206–209. doi:10.1111/j.1399-3054.1990.tb05888.x. Retrieved 16 October 2015.

[9] Kikkert, Julie; Vidal, Jose; Reisch, Bruce (2005). "Stable transformation of plant cells by particle bombardment/biolistics". *Methods of Molecular Biology* **286**: 61–78. doi:10.1385/1-59259-827-7:061. PMID 15310913. Retrieved 16 October 2015.

[10] Hayward, M.D.; Bosemark, N.O.; Romagosa, T. (2012). *Plant Breeding: Principles and Prospects*. Springer Science & Business Media. p. 131. ISBN 9789401115247.

19.7 Further reading

- O'Brien, J; Holt, M; Whiteside, G; Lummis, SC; Hastings, MH (2001). "Modifications to the hand-held Gene Gun: improvements for in vitro Biolistic transfection of organotypic neuronal tissue". *Journal of Neuroscience Methods* **112** (1): 57–64. doi:10.1016/S0165-0270(01)00457-5. PMID 11640958.

19.8 External links

- John O'Brien presents...Gene Gun Barrels for more information about biolistics

Chapter 20

Agrobacterium

Agrobacterium is a genus of Gram-negative bacteria established by H. J. Conn that uses horizontal gene transfer to cause tumors in plants. *Agrobacterium tumefaciens* is the most commonly studied species in this genus. *Agrobacterium* is well known for its ability to transfer DNA between itself and plants, and for this reason it has become an important tool for genetic engineering.

The *Agrobacterium* genus is quite heterogeneous. Recent taxonomic studies have reclassified all of the *Agrobacterium* species into new genera, such as *Ahrensia*, *Pseudorhodobacter*, *Ruegeria*, and *Stappia*,[1][2] but most species have been controversially reclassified as *Rhizobium* species.[3][4][5]

20.1 Plant pathogen

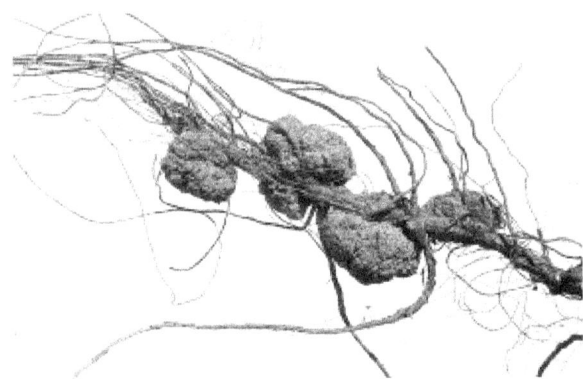

The large growths on these roots are galls induced by Agrobacterium *sp.*

A. tumefaciens causes crown-gall disease in plants. The disease is characterised by a tumour-like growth or gall on the infected plant, often at the junction between the root and the shoot. Tumors are incited by the conjugative transfer of a DNA segment (T-DNA) from the bacterial tumour-inducing (Ti) plasmid. The closely related species, *A. rhizogenes*, induces root tumors, and carries the dis-tinct Ri (root-inducing) plasmid. Although the taxonomy of *Agrobacterium* is currently under revision it can be generalised that 3 biovars exist within the genus, *A. tumefaciens*, *A. rhizogenes*, and *A. vitis*. Strains within *A. tumefaciens* and *A. rhizogenes* are known to be able to harbour either a Ti or Ri-plasmid, whilst strains of *A. vitis*, generally restricted to grapevines, can harbour a Ti-plasmid. Non-*Agrobacterium* strains have been isolated from environmental samples which harbour a Ri-plasmid whilst laboratory studies have shown that non-*Agrobacterium* strains can also harbour a Ti-plasmid. Many environmental strains of *Agrobacterium* possess neither a Ti nor Ri-plasmid. These strains are of course avirulent.

The plasmid T-DNA is integrated semi-randomly into the genome of the host cell,[6] and the tumor morphology genes on the T-DNA are expressed, causing the formation of a gall. The T-DNA carries genes for the biosynthetic enzymes for the production of unusual amino acids, typically octopine or nopaline. It also carries genes for the biosynthesis of the plant hormones, auxin and cytokinins, and for the biosynthesis of opines, providing a carbon and nitrogen source for the bacteria that most other micro-organisms can't use, giving *Agrobacterium* a selective advantage.[7] By altering the hormone balance in the plant cell, the division of those cells cannot be controlled by the plant, and tumors form. The ratio of auxin to cytokinin produced by the tumor genes determines the morphology of the tumor (root-like, disorganized or shoot-like).

20.2 *Agrobacterium* in humans

Although generally seen as an infection in plants, *Agrobacterium* can be responsible for opportunistic infections in humans with weakened immune systems,[8][9] but has not been shown to be a primary pathogen in otherwise healthy individuals. One of the earliest associations of human disease caused by *Agrobacterium radiobacter* was reported by Dr. J. R. Cain in Scotland (1988).[10] A later study suggested that *Agrobacterium* attaches to and genetically transforms several types of human cells by integrating its T-DNA into the human cell genome. The study was conducted using cultured human tissue and did not draw any conclusions regarding related biological activity in nature.[11]

20.3   Uses in biotechnology

See also: horizontal gene transfer

The ability of *Agrobacterium* to transfer genes to plants and fungi is used in biotechnology, in particular, genetic engineering for plant improvement. A modified Ti or Ri plasmid can be used. The plasmid is 'disarmed' by deletion of the tumor inducing genes; the only essential parts of the T-DNA are its two small (25 base pair) border repeats, at least one of which is needed for plant transformation. Marc Van Montagu and Jozef Schell at the University of Ghent (Belgium) discovered the gene transfer mechanism between *Agrobacterium* and plants, which resulted in the development of methods to alter *Agrobacterium* into an efficient delivery system for gene engineering in plants.[12][13] A team of researchers led by Dr Mary-Dell Chilton were the first to demonstrate that the virulence genes could be removed without adversely affecting the ability of *Agrobacterium* to insert its own DNA into the plant genome (1983).

The genes to be introduced into the plant are cloned into a plant transformation vector that contains the T-DNA region of the disarmed plasmid, together with a selectable marker (such as antibiotic resistance) to enable selection for plants that have been successfully transformed. Plants are grown on media containing antibiotic following transformation, and those that do not have the T-DNA integrated into their genome will die. An alternative method is agroinfiltration.

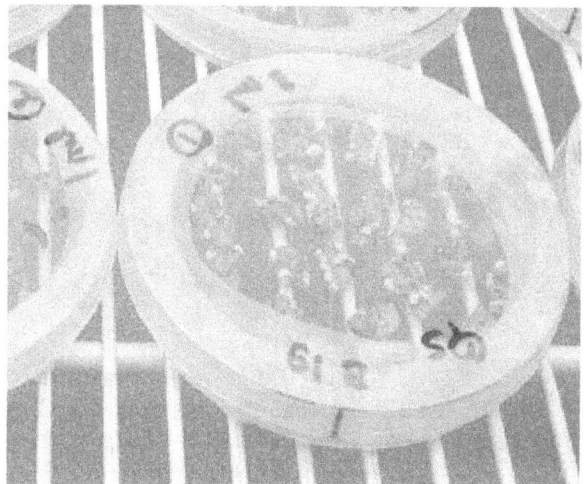

Plant (S. chacoense*) transformed using* Agrobacterium. *Transformed cells start forming calluses on the side of the leaf pieces*

Transformation with *Agrobacterium* can be achieved in two ways. Protoplasts or alternatively leaf-discs can be incubated with the *Agrobacterium* and whole plants regenerated using plant tissue culture. A common transformation protocol for *Arabidopsis* is the floral-dip method: the flowers are dipped in an *Agrobacterium* culture, and the bacterium transforms the germline cells that make the female gametes. The seeds can then be screened for antibiotic resistance (or another marker of interest), and plants that have not integrated the plasmid DNA will die when exposed to the correct condition of antibiotic.

Agrobacterium does not infect all plant species, but there are several other effective techniques for plant transformation including the gene gun.

Agrobacterium is listed as being the vector of genetic material that was transferred to these USA GMOs:[14]

- Soybean

- Cotton

- Corn

- Sugar Beet

- Alfalfa

- Wheat

- Rapeseed Oil (Canola)

- Creeping bentgrass (for animal feed)

- Rice (Golden Rice)

20.4   Genomics

The sequencing of the genomes of several species of *Agrobacterium* has permitted the study of the evolutionary history of these organisms and has provided information on the genes and systems involved in pathogenesis, biological control and symbiosis. One important finding is the possibility that chromosomes are evolving from plasmids in many of these bacteria. Another discovery is that the diverse chromosomal structures in this group appear to be capable of supporting both symbiotic and pathogenic lifestyles. The availability of the genome sequences of *Agrobacterium* species will continue to increase, resulting in substantial insights into the function and evolutionary history of this group of plant-associated microbes.[15]

20.5   See also

- Agroinfiltration

- Marc Van Montagu

- *Rhizobium rhizogenes* (formerly *Agrobacterium rhizogenes*)

20.6   References

[1] Uchino, Yoshihito; Yokota, Akira; Sugiyama, Junta (1997). "Phylogenetic position of the marine subdivision of Agrobacterium species based on 16S rRNA sequence analysis". *The Journal of General and Applied Microbiology* **43** (4): 243–247. doi:10.2323/jgam.43.243. PMID 12501326.

[2] Uchino, Yoshihito; Hirata, Aiko; Yokota, Akira; Sugiyama, Junta (1998). "Reclassification of marine Agrobacterium species: Proposals of *Stappia stellulata* gen. nov., comb. Nov., *Stappia aggregata* sp. nov., nom. Rev., *Ruegeria atlantica* gen. nov., comb. Nov., Ruegeria gelatinovora comb. Nov., Ruegeria algicola comb. Nov., and *Ahrensia kieliense* gen. nov., sp. nov., nom. Rev". *The Journal of General and Applied Microbiology* **44** (3): 201–210. doi:10.2323/jgam.44.201. PMID 12501429.

[3] Young, J. M.; Kuykendall, L. D.; Martinez-Romero, E; Kerr, A; Sawada, H (2001). "A revision of *Rhizobium* Frank 1889, with an emended description of the genus, and the inclusion of all species of *Agrobacterium* Conn 1942 and *Allorhizobium undicola* de Lajudie *et al.* 1998 as new combinations: *Rhizobium radiobacter, R. rhizogenes, R. rubi, R. undicola,* and *R. vitis*". *Int J Syst Evol Microbiol* **51** (Pt 1): 89–103. doi:10.1099/00207713-51-1-89. PMID 11211278.

[4] Farrand, S. K.; Van Berkum, P. B.; Oger, P (2003). "Agrobacterium is a definable genus of the family Rhizobiaceae". *International Journal of Systematic and Evolutionary Microbiology* **53** (5): 1681–7. doi:10.1099/ijs.0.02445-0. PMID 13130068.

[5] Young, J. M.; Kuykendall, L. D.; Martinez-Romero, E; Kerr, A; Sawada, H (2003). "Classification and nomenclature of *Agrobacterium* and *Rhizobium*—a reply to Farrand *et al.* (2003)". *International Journal of Systematic and Evolutionary Microbiology* **53** (5): 1689–95. doi:10.1099/ijs.0.02762-0. PMID 13130069.

[6] Francis, Kirk E.; Spiker, Steven (2004). "Identification of Arabidopsis thaliana transformants without selection reveals a high occurrence of silenced T-DNA integrations". *The Plant Journal* **41** (3): 464–77. doi:10.1111/j.1365-313X.2004.02312.x. PMID 15659104.

[7] Pitzschke, Andrea; Hirt, Heribert (2010). "New insights into an old story: Agrobacterium-induced tumour formation in plants by plant transformation". *The EMBO Journal* **29** (6): 1021–32. doi:10.1038/emboj.2010.8. PMC 2845280. PMID 20150897.

[8] Hulse, M.; Johnson, S.; Ferrieri, P. (1993). "Agrobacterium Infections in Humans: Experience at One Hospital and Review". *Clinical Infectious Diseases* **16** (1): 112–7. doi:10.1093/clinids/16.1.112. PMID 8448285.

[9] Dunne Jr, W. M.; Tillman, J; Murray, J. C. (1993). "Recovery of a strain of Agrobacterium radiobacter with a mucoid phenotype from an immunocompromised child with bacteremia". *Journal of clinical microbiology* **31** (9): 2541–3. PMC 265809. PMID 8408587.

[10] Cain, John Raymond (1988). "A case of septicaemia caused by Agrobacterium radiobacter". *Journal of Infection* **16** (2): 205–6. doi:10.1016/s0163-4453(88)94272-7. PMID 3351321.

[11] Kunik, T.; Tzfira, T; Kapulnik, Y; Gafni, Y; Dingwall, C; Citovsky, V (2001). "Genetic transformation of HeLa cells by *Agrobacterium*". *Proceedings of the National Academy of Sciences* **98** (4): 1871–6. Bibcode:2001PNAS...98.1871K. doi:10.1073/pnas.041327598. JSTOR 3054968. PMC 29349. PMID 11172043.

[12] Schell, J.; Van Montagu, M. (1977). "The Ti-Plasmid of Agrobacterium Tumefaciens, A Natural Vector for the Introduction of NIF Genes in Plants?". In Hollaender, Alexander; Burris, R. H.; Day, P. R.; Hardy, R. W. F.; Helinski, D. R.; Lamborg, M. R.; Owens, L.; Valentine, R. C. *Genetic Engineering for Nitrogen Fixation*. Basic Life Sciences **9**. pp. 159–79. doi:10.1007/978-1-4684-0880-5_12. ISBN 978-1-4684-0882-9. PMID 336023.

[13] Joos, H; Timmerman, B; Montagu, M. V.; Schell, J (1983). "Genetic analysis of transfer and stabilization of Agrobacterium DNA in plant cells". *The EMBO Journal* **2** (12): 2151–60. PMC 555427. PMID 16453483.

[14] The FDA List of Completed Consultations on Bioengineered Foods Archived May 13, 2008, at the Wayback Machine.

[15] Setubal, Joao C.; WOod, Derek; Burr, Thomas; Farrand, Stephen K.; Goldman, Barry S.; Goodner, Brad; Otten, Leon; Slater, Steven (2009). "The Genomics of *Agrobacterium*: Insights into its Pathogenicity, Biocontrol, and Evolution". In Jackson, Robert W. *Plant Pathogenic Bacteria: Genomics and Molecular Biology*. Caister Academic Press. pp. 91–112. ISBN 978-1-904455-37-0.

20.7   External links

- Kyndt, Tina; Quispe, Dora; Zhai, Hong; Jarret, Robert; Ghislain, Marc; Liu, Qingchang; Gheysen, Godelieve; Kreuze, Jan F. (2015). "The genome of cultivated sweet potato contains *Agrobacterium* T-DNAs with expressed genes: An example of a naturally transgenic food crop". *Proceedings of the National Academy of Sciences* **112**:

201419685. doi:10.1073/pnas.1419685112. Lay summary – *Phys.org* (April 21, 2015).

- Current taxonomy of *Agrobacterium* species, and new *Rhizobium* names

- Agrobacteria is used as gene ferry - Plant transformation with *Agrobacterium*]

Chapter 21

Transfection

Transfection is the process of deliberately introducing nucleic acids into cells. The term is often used for non-viral methods in eukaryotic cells.[1] It may also refer to other methods and cell types, although other terms are preferred: "transformation" is more often used to describe non-viral DNA transfer in bacteria and non-animal eukaryotic cells, including plant cells. In animal cells, transfection is the preferred term as transformation is also used to refer to progression to a cancerous state (carcinogenesis) in these cells. Transduction is often used to describe virus-mediated DNA transfer.

The word *transfection* is a blend of *trans-* and *infection*. Genetic material (such as supercoiled plasmid DNA or siRNA constructs), or even proteins such as antibodies, may be transfected.

Transfection of animal cells typically involves opening transient pores or "holes" in the cell membrane to allow the uptake of material. Transfection can be carried out using calcium phosphate, by electroporation, by cell squeezing or by mixing a cationic lipid with the material to produce liposomes, which fuse with the cell membrane and deposit their cargo inside.

Transfection can result in unexpected morphologies and abnormalities in target cells.

21.1   Terminology

The meaning of the term has evolved.[2] The original meaning of transfection was "infection by transformation." i.e., introduction of DNA (or RNA) from a prokaryote-infecting virus or bacteriophage into cells, resulting in an infection. Because the term transformation had another sense in animal cell biology (a genetic change allowing long-term propagation in culture, or acquisition of properties typical of cancer cells), the term transfection acquired, for animal cells, its present meaning of a change in cell properties caused by introduction of DNA.

21.2   Methods

There are various methods of introducing foreign DNA into a eukaryotic cell: some rely on physical treatment (electroporation, cell squeezing, nanoparticles, magnetofection), other on chemical materials or biological particles (viruses) that are used as carriers.

21.2.1   Chemical-based transfection

Chemical-based transfection can be divided into several kinds: cyclodextrin,[3] polymers,[4] liposomes, or nanoparticles [5] (with or without chemical or viral functionalization. See below).

- One of the cheapest methods uses **calcium phosphate**, originally discovered by F. L. Graham and A. J. van der Eb in 1973[6] (see also [7]). HEPES-buffered saline solution (HeBS) containing phosphate ions is combined with a calcium chloride solution containing the DNA to be transfected. When the two are combined, a fine precipitate of the positively charged calcium and the negatively charged phosphate will form, binding the DNA to be transfected on its surface. The suspension of the precipitate is then added to the cells to be transfected (usually a cell culture grown in a monolayer). By a process not entirely understood, the cells take up some of the precipitate, and with it, the DNA. This process has been a preferred method of identifying many oncogenes.[8]

- Other methods use **highly branched organic compounds**, so-called dendrimers, to bind the DNA and get it into the cell.

- A very efficient method is the inclusion of the DNA to be transfected in **liposomes**, i.e. small, membrane-bounded bodies that are in some ways similar to the structure of a cell and can actually fuse with the cell membrane, releasing the DNA into the cell. For eukaryotic cells, transfection is better achieved using

cationic liposomes (or mixtures), because the cells are more sensitive. See lipofection for more details.

- Another method is the use of **cationic polymers** such as DEAE-dextran or polyethylenimine. The negatively charged DNA binds to the polycation and the complex is taken up by the cell via endocytosis.

21.2.2 Non-chemical methods

- Electroporation (gene electrotransfer) is a popular method, where transient increase in the permeability of cell membrane is achieved when the cells are exposed to short pulses of an intense electric field.

- Cell squeezing is a method invented in 2012 by Armon Sharei, Robert Langer and Klavs Jensen at MIT. It enables delivery of molecules into cells by a gentle squeezing of the cell membrane. It is a high throughput vector-free microfluidic platform for intracellular delivery. It eliminates the possibility of toxicity or off-target effects as it does not rely on exogenous materials or electrical fields.[9]

- Sonoporation uses high-intensity ultrasound to induce pore formation in cell membranes. This pore formation is attributed mainly to the cavitation of gas bubbles interacting with nearby cell membranes since it is enhanced by the addition of ultrasound contrast agent, a source of cavitation nuclei.

- Optical transfection is a method where a tiny (~1 μm diameter) hole is transiently generated in the plasma membrane of a cell using a highly focused laser. This technique was first described in 1984 by Tsukakoshi et al., who used a frequency tripled Nd:YAG to generate stable and transient transfection of normal rat kidney cells.[10] In this technique, one cell at a time is treated, making it particularly useful for single cell analysis.

- Protoplast fusion is a technique in which transformed bacterial cells are treated with lysozyme in order to remove the cell wall. Following this, fusogenic agents (e.g., Sendai virus, PEG, electroporation) are used in order to fuse the protoplast carrying the gene of interest with the target recipient cell. A major disadvantage of this method is that bacterial components are non-specifically introduced into the target cell as well.

- Impalefection is a method of introducing DNA bound to a surface of a nanofiber that is inserted into a cell. This approach can also be implemented with arrays of nanofibers that are introduced into large numbers of cells and intact tissue.

- Hydrodynamic delivery is a method used in mice and rats, but to a lesser extent in larger animals, in which DNA most often in plasmids (including transposons) can be delivered to the liver using hydrodynamic injection that involves infusion of a relatively large volume in the blood in less than 10 seconds; nearly all of the DNA is expressed in the liver by this procedure.[11][12][13]

21.2.3 Particle-based methods

- A direct approach to transfection is the gene gun, where the DNA is coupled to a nanoparticle of an inert solid (commonly gold), which is then "shot" directly into the target cell's nucleus.

- Magnetofection, or magnet-assisted transfection, is a transfection method that uses magnetic force to deliver DNA into target cells. Nucleic acids are first associated with magnetic nanoparticles. Then, application of magnetic force drives the nucleic acid particle complexes towards and into the target cells, where the cargo is released.[14]

- Impalefection is carried out by impaling cells by elongated nanostructures and arrays of such nanostructures such as carbon nanofibers or silicon nanowires which have been functionalized with plasmid DNA.

- Another particle-based method of transfection is known as particle bombardment. The nucleic acid is delivered through membrane penetration at a high velocity, usually connected to microprojectiles.[1]

21.2.4 Viral methods

DNA can also be introduced into cells using viruses as a carrier. In such cases, the technique is called viral transduction, and the cells are said to be transduced. Adenoviral vectors can be useful for viral transfection methods because they can transfer genes into a wide variety of human cells and have high transfer rates.[1] Lentiviral vectors are also helpful due to their ability to transduce cells not currently undergoing mitosis.

21.2.5 Other (and hybrid) methods

Other methods of transfection include nucleofection, which has proved very efficient in transfection of the THP-1 cell line, creating a viable cell line that was able to be differentiated into mature macrophages,[15] heat shock.

21.3   Stable and transient transfection

For some applications of transfection, it is sufficient if the transfected genetic material is only transiently expressed. Since the DNA introduced in the transfection process is usually not integrated into the nuclear genome, the foreign DNA will be diluted through mitosis or degraded. Cell lines expressing the Epstein–Barr virus (EBV) nuclear antigen 1 (EBNA1) or the SV40 large-T antigen, allow episomal amplification of plasmids containing the viral EBV (293E) or SV40 (293T) origins of replication, greatly reducing the rate of dilution.[16]

If it is desired that the transfected gene actually remain in the genome of the cell and its daughter cells, a stable transfection must occur. To accomplish this, a marker gene is co-transfected, which gives the cell some selectable advantage, such as resistance towards a certain toxin. Some (very few) of the transfected cells will, by chance, have integrated the foreign genetic material into their genome. If the toxin is then added to the cell culture, only those few cells with the marker gene integrated into their genomes will be able to proliferate, while other cells will die. After applying this selective stress (selection pressure) for some time, only the cells with a stable transfection remain and can be cultivated further.

A common agent for selecting stable transfection is Geneticin, also known as G418, which is a toxin that can be neutralized by the product of the neomycin resistance gene.

21.4   RNA transfection

Main article: RNA transfection

RNA can also be transfected into cells to transiently express its coded protein, or to study RNA decay kinetics. The latter application is referred as siRNA transfection or **RNA silencing**, and has become a major application in research (to replace the "knock-down" experiments, to study the expression of proteins, i.e. of Endothelin-1[17]) with potential applications in gene-therapy.

A limitation of the silencing approach rely on the toxicity of the transfection for cells, and its suspected effect on the expression of other genes/proteins.

21.5   See also

- Protofection

- Transformation
- Transduction
- Cationic liposome
- Nucleofection
- Magnet assisted transfection
- Impalefection

21.6   References

[1] "Transfection". *Protocols and Applications Guide*. Promega.

[2] "Transfection" at *Dorland's Medical Dictionary*

[3] Menuel S, Fontanay S, Clarot I, Duval R.E, Diez L. Marsura A (2008). "Synthesis and Complexation Ability of a Novel Bis- (guanidinium)-tetrakis-(β-cyclodextrin) Dendrimeric Tetrapod as a Potential Gene Delivery (DNA and siRNA) System. Study of Cellular siRNA Transfection". *Bioconjugate Chem*. **19** (12): 2357–62. doi:10.1021/bc800193p. PMID 19053312.

[4] Fischer D, von Harpe A, Kunath K, Petersen H, Li YX, Kissel T (2002). "Copolymers of ethylene imine and N-(2-hydroxyethyl)-ethylene imine as tools to study effects of polymer structure on physicochemical and biological properties of DNA complexes". *Bioconjugate Chem*. **13** (5): 1124–33. doi:10.1021/bc025550w.

[5] "Nanoparticle Based Transfection Reagents". *Biology Transfection Research Resource*. Transfection.ws.

[6] Graham FL, van der Eb AJ (1973). "A new technique for the assay of infectivity of human adenovirus 5 DNA". *Virology* **52** (2): 456–67. doi:10.1016/0042-6822(73)90341-3. PMID 4705382.

[7] Bacchetti S, Graham F (1977). "Transfer of the gene for thymidine kinase to thymidine kinase-deficient human cells by purified herpes simplex viral DNA". *Proc Natl Acad Sci USA* **74** (4): 1590–4. doi:10.1073/pnas.74.4.1590. PMC 430836. PMID 193108.

[8] Kriegler, Michael (1991). *Transfer and Expression: A Laboratory Manual*. W. H. Freeman. pp. 96–97. ISBN 0716770040.

[9] Sharei A, Zoldan J, Adamo A, Sim WY, Cho N, Jackson E, Mao S, Schneider S, Han MJ, Lytton-Jean A, Basto PA, Jhunjhunwala S, Lee J, Heller DA, Kang JW, Hartoularos GC, Kim KS, Anderson DG, Langer R, Jensen KF (February 2013). "A vector-free microfluidic platform for intracellular delivery". *Proc. Natl. Acad. Sci. U.S.A.* **110** (6): 2082–7. doi:10.1073/pnas.1218705110. PMC 3568376. PMID 23341631.

[10] Tsukakoshi M, Kurata S, Nomiya Y, et al. (1984). "A Novel Method of DNA Transfection by Laser Microbeam Cell Surgery". *Applied Physics B: Photophysics and Laser Chemistry* **35** (3): 135–140. Bibcode:1984ApPhB..35..135T. doi:10.1007/BF00697702.

[11] Zhang G, Budker V, Wolff JA (July 1999). "High levels of foreign gene expression in hepatocytes after tail vein injections of naked plasmid DNA". *Hum. Gene Ther.* **10** (10): 1735–7. doi:10.1089/10430349950017734. PMID 10428218.

[12] Zhang G, Vargo D, Budker V, Armstrong N, Knechtle S, Wolff JA (October 1997). "Expression of naked plasmid DNA injected into the afferent and efferent vessels of rodent and dog livers". *Hum. Gene Ther.* **8** (15): 1763–72. doi:10.1089/hum.1997.8.15-1763. PMID 9358026.

[13] Bell JB, Podetz-Pedersen KM, Aronovich EL, Belur LR, McIvor RS, Hackett PB (2007). "Preferential delivery of the Sleeping Beauty transposon system to livers of mice by hydrodynamic injection". *Nat Protoc* **2** (12): 3153–65. doi:10.1038/nprot.2007.471. PMC 2548418. PMID 18079715.

[14] "Magnetofection — Magnetic assisted transfection & transduction". OzBiosciences—The art of delivery systems.

[15] Schnoor M, Buers I, Sietmann A, et al. (May 2009). "Efficient non-viral transfection of THP-1 cells". *J. Immunol. Methods* **344** (2): 109–15. doi:10.1016/j.jim.2009.03.014. PMID 19345690.

[16] Durocher Y, Perret S, Kamen A (January 2002). "High-level and high-throughput recombinant protein production by transient transfection of suspension-growing human 293-EBNA1 cells". *Nucleic Acids Res.* **30** (2): E9. doi:10.1093/nar/30.2.e9. PMC 99848. PMID 11788735.

[17] Mawji IA, Marsden PA (June 2006). "RNA transfection is a versatile tool to investigate endothelin-1 posttranscriptional regulation". *Exp. Biol. Med. (Maywood)* **231** (6): 704–8. PMID 16740984.

21.7 External links

- Transfection at the US National Library of Medicine Medical Subject Headings (MeSH)

- Biology Research Resource — Articles and Forums about Transfection

- Bonetta L (2005). "The inside scoop—evaluating gene delivery methods". *Nature Methods* **2** (11): 875–883. doi:10.1038/nmeth1105-875.

- Research in optical transfection at the University of St Andrews

Chapter 22

Electroporation

Electroporation, or **electropermeabilization**, is a microbiology technique in which an electrical field is applied to cells in order to increase the permeability of the cell membrane, allowing chemicals, drugs, or DNA to be introduced into the cell.[1] In microbiology, the process of electroporation is often used to transform bacteria, yeast, or plant protoplasts by introducing new coding DNA. If bacteria and plasmids are mixed together, the plasmids can be transferred into the bacteria after electroporation. Several hundred volts across a distance of several millimeters are typically used in this process. Afterwards, the cells have to be handled carefully until they have had a chance to divide, producing new cells that contain reproduced plasmids. This process is approximately ten times more effective than chemical transformation.[1][2]

Electroporation is also highly efficient for the introduction of foreign genes into tissue culture cells, especially mammalian cells. For example, it is used in the process of producing knockout mice, as well as in tumor treatment, gene therapy, and cell-based therapy. The process of introducing foreign DNA into eukaryotic cells is known as transfection. Electroporation is highly effective for transfecting cells in suspension using electroporation cuvettes. Electroporation has proven efficient for use on tissues in vivo, for in utero applications as well as in ovo transfection. Adherent cells can also be transfected using electroporation, providing researchers with an alternative to trypsinizing their cells prior to transfection.

22.1 Laboratory practice

Electroporation is performed with **electroporators**, purpose-built appliances which create an electrostatic field in a cell solution. The cell suspension is pipetted into a glass or plastic cuvette which has two aluminum electrodes on its sides. For bacterial electroporation, typically a suspension of around 50 microliters is used. Prior to electroporation, this suspension of bacteria is mixed with the plasmid to be transformed. The mixture is pipetted into the cuvette, the

Cuvettes for electroporation. These are plastic with aluminium electrodes and a blue lid. They hold a maximum of 400 μl.

voltage and capacitance are set, and the cuvette is inserted into the electroporator. The process requires direct contact between the electrodes and the suspension. Immediately after electroporation, one milliliter of liquid medium is added to the bacteria (in the cuvette or in an Eppendorf tube), and the tube is incubated at the bacteria's optimal temperature for an hour or more to allow recovery of the cells and expression of the plasmid, followed by bacterial culture on agar plates.

The success of the electroporation depends greatly on the purity of the plasmid solution, especially on its salt content. Solutions with high salt concentrations might cause an electrical discharge (known as arcing), which often reduces the viability of the bacteria. For a further detailed investigation of the process, more attention should be paid to the output impedance of the porator device and the input impedance of the cells suspension (e.g. salt content).

Since the cell membrane is not able to pass current (except in ion channels), it acts as an electrical capacitor. Subjecting membranes to a high-voltage electric field results in their temporary breakdown, resulting in pores that are large enough to allow macromolecules (such as DNA) to enter or leave the cell. [3]

22.2 Medical applications

Main article: Irreversible electroporation

The first research looking at how electroporation might be used on human cells was conducted by researchers at Eastern Virginia Medical School and Old Dominion University, and published in 2003.[4]

The first successful treatment of malignant cutaneous tumors implanted in mice was completed in 2007 by a group of scientists who achieved complete tumor ablation in 12 out of 13 mice. They accomplished this by sending 80 pulses of 100 microseconds at 0.3 Hz with an electrical field magnitude of 2500 V/cm to treat the cutaneous tumors.[5]

A higher voltage of electroporation was found in pigs to irreversibly destroy target cells within a narrow range while leaving neighboring cells unaffected, and thus represents a promising new treatment for cancer, heart disease and other disease states that require removal of tissue.[6] Irreversible electroporation (IRE) has since proven effective in treating human cancer, with surgeons at Johns Hopkins and other institutions now using the technology to treat pancreatic cancer previously thought to be unresectable.[7]

A recent technique called non-thermal irreversible electroporation (N-TIRE) has proven successful in treating many different types of tumors and other unwanted tissue. This procedure is done using small electrodes (about 1mm in diameter), placed either inside or surrounding the target tissue to apply short, repetitive bursts of electricity at a predetermined voltage and frequency. These bursts of electricity increase the resting transmembrane potential (TMP), so that nanopores form in the plasma membrane. When the electricity applied to the tissue is above the electric field threshold of the target tissue, the cells become permanently permeable from the formation of nanopores. As a result, the cells are unable to repair the damage and die due to a loss of homeostasis.[8] N-TIRE is unique to other tumor ablation techniques in that it does not create thermal damage to the tissue around it.

Contrastingly, reversible electroporation occurs when the electricity applied with the electrodes is below the electric field threshold of the target tissue. Because the electricity applied is below the cells' threshold, it allows the cells to repair their phospholipid bilayer and continue on with their normal cell functions. Reversible electroporation is typically done with treatments that involve getting a drug or gene (or other molecule that is not normally permeable to the cell membrane) into the cell. Not all tissue has the same electric field threshold; therefore careful calculations need to be made prior to a treatment to ensure safety and efficacy.[9]

One major advantage of using N-TIRE is that, when done correctly according to careful calculations, it only affects the target tissue. Proteins, the extracellular matrix, and critical structures such as blood vessels and nerves are all unaffected and left healthy by this treatment. This allows for a quicker recovery, and facilitates a more rapid replacement of dead tumor cells with healthy cells.[10]

Before doing the procedure, scientists must carefully calculate exactly what needs to be done, and treat each patient on an individual case-by-case basis. To do this, imaging technology such as CT scans and MRI's are commonly used to create a 3D image of the tumor. From this information, they can approximate the volume of the tumor and decide on the best course of action including the insertion site of electrodes, the angle they are inserted in, the voltage needed, and more, using software technology. Often, a CT machine will be used to help with the placement of electrodes during the procedure, particularly when the electrodes are being used to treat tumors in the brain.[11]

The entire procedure is very quick, typically taking about five minutes. The success rate of these procedures is high and is very promising for future treatment in humans. One disadvantage to using N-TIRE is that the electricity delivered from the electrodes can stimulate muscle cells to contract, which could have lethal consequences depending on the situation. Therefore, a paralytic agent must be used when performing the procedure. The paralytic agents that have been used in such research are successful; however, there is always some risk, albeit slight, when using anesthetics.

A more recent technique has been developed called high-frequency irreversible electroporation (H-FIRE). This technique uses electrodes to apply bipolar bursts of electricity at a high frequency, as opposed to unipolar bursts of electricity at a low frequency. This type of procedure has the same tumor ablation success as N-TIRE. However, it has one distinct advantage, H-FIRE does not cause muscle contraction in the patient and therefore there is no need for a paralytic agent.[12]

22.2.1 Drug and gene delivery

Electroporation can also be used to help deliver drugs or genes into the cell by applying short and intense electric pulses that transiently permeabilize cell membrane, thus allowing transport of molecules otherwise not transported through a cellular membrane. This procedure is referred to as electrochemotherapy when the molecules to be transported are chemotherapeutic agents or gene electrotransfer when the molecule to be transported is DNA. Scientists from Karolinska Institutet and the University of Oxford use electroporation of exosomes to deliver siRNAs, antisense oligonucleotides, chemotherapeutic agents and proteins specifically to neurons after inject them systemically (in blood). Because these exosomes are able to cross the blood brain barrier this protocol could solve the issue of poor delivery of medications to the central nervous system and cure Alzheimer's, Parkinson's Disease and brain cancer among other diseases.[13]

22.3   Physical mechanism

Further information: Lipid bilayer mechanics

Electroporation allows cellular introduction of large highly charged molecules such as DNA which would never passively diffuse across the hydrophobic bilayer core.[1] This phenomenon indicates that the mechanism is the creation of nm-scale water-filled holes in the membrane. Although electroporation and dielectric breakdown both result from application of an electric field, the mechanisms involved are fundamentally different. In dielectric breakdown the barrier material is ionized, creating a conductive pathway. The material alteration is thus chemical in nature. In contrast, during electroporation the lipid molecules are not chemically altered but simply shift position, opening up a pore which acts as the conductive pathway through the bilayer as it is filled with water.

Electroporation is a dynamic phenomenon that depends on the local transmembrane voltage at each point on the cell membrane. It is generally accepted that for a given pulse duration and shape, a specific transmembrane voltage threshold exists for the manifestation of the electroporation phenomenon (from 0.5 V to 1 V). This leads to the definition of an electric field magnitude threshold for electroporation (E_{th}). That is, only the cells within areas where $E \geqq E_{th}$ are electroporated. If a second threshold (E_{ir}) is reached or surpassed, electroporation will compromise the viability of the cells, i.e., irreversible electroporation (IRE).[14]

Electroporation is a multi-step process with several distinct phases.[15] First, a short electrical pulse must be applied. Typical parameters would be 300–400 mV for < 1 ms across the membrane (note- the voltages used in cell experiments are typically much larger because they are being applied across large distances to the bulk solution so the resulting field across the actual membrane is only a small fraction of the applied bias). Upon application of this potential the membrane charges like a capacitor through the migration of ions from the surrounding solution. Once the

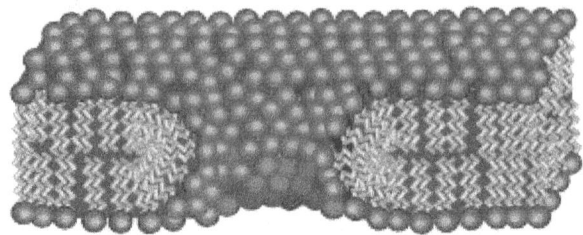

Schematic showing the theoretical arrangement of lipids in a hydrophobic pore (top) and a hydrophilic pore (bottom).

critical field is achieved there is a rapid localized rearrangement in lipid morphology. The resulting structure is believed to be a "pre-pore" since it is not electrically conductive but leads rapidly to the creation of a conductive pore.[16] Evidence for the existence of such pre-pores comes mostly from the "flickering" of pores, which suggests a transition between conductive and insulating states.[17] It has been suggested that these pre-pores are small (~3 Å) hydrophobic defects. If this theory is correct, then the transition to a conductive state could be explained by a rearrangement at the pore edge, in which the lipid heads fold over to create a hydrophilic interface. Finally, these conductive pores can either heal, resealing the bilayer or expand, eventually rupturing it. The resultant fate depends on whether the critical defect size was exceeded[18] which in turn depends on the applied field, local mechanical stress and bilayer edge energy.

22.4   References

[1] Neumann, E; Schaefer-Ridder, M; Wang, Y; Hofschneider, PH (1982). "Gene transfer into mouse lyoma cells by electroporation in high electric fields". *The EMBO Journal* **1** (7): 841–5. PMC 553119. PMID 6329708.

[2] Sugar, I.P.; Neumann, E. (1984). "Stochastic model for electric field-induced membrane pores electroporation". *Biophysical Chemistry* **19** (3): 211–25. doi:10.1016/0301-4622(84)87003-9. PMID 6722274.

[3] Potter, Huntington; Heller, Richard (2003-05-01). "Transfection by Electroporation". *Current*

protocols in molecular biology / edited by Frederick M. Ausubel ... [et al.] CHAPTER: Unit–9.3. doi:10.1002/0471142727.mb0903s62. ISSN 1934-3639. PMC 2975437. PMID 18265334.

[4] Beebe SJ, Fox PM, Rec LJ, Willis EL, Schoenbach KH (August 2003). "Nanosecond, high-intensity pulsed electric fields induce apoptosis in human cells". FASEB J. 17 (11): 1493–5. doi:10.1096/fj.02-0859fje. PMID 12824299.

[5] Al-Sakere, Bassim; André, Franck; Bernat, Claire; Connault, Elisabeth; Opolon, Paule; Davalos, Rafael V.; Rubinsky, Boris; Mir, Lluis M. (2007). Isalan, Mark, ed. "Tumor Ablation with Irreversible Electroporation". PLoS ONE 2 (11): e1135. Bibcode:2007PLoSO...2.1135A. doi:10.1371/journal.pone.0001135. PMC 2065844. PMID 17989772.

[6] Sarah Yang (2007-02-12). "New medical technique punches holes in cells, could treat tumors". Retrieved 2007-12-13.

[7] "A Potential Boon for Pancreatic Cancer Patients". Johns Hopkins Surgery: News From the Johns Hopkins Department of Surgery. 2014-06-23.

[8] Garcia, Paulo A.; Rossmeisl, John H.; Davalos, Rafael V. (2011). "Electrical conductivity changes during irreversible electroporation treatment of brain cancer". 2011 Annual International Conference of the IEEE Engineering in Medicine and Biology Society: 739–42. doi:10.1109/IEMBS.2011.6090168. ISBN 978-1-4577-1589-1. PMID 22254416.

[9] Garcia, P A; Neal, Robert E; Rossmeisl, John H; Davalos, R V (2010). "Non-thermal irreversible electroporation for deep intracranial disorders". 2010 Annual International Conference of the IEEE Engineering in Medicine and Biology: 2743–6. doi:10.1109/IEMBS.2010.5626371. ISBN 978-1-4244-4123-5. PMID 21095962.

[10] Garcia, Paulo A.; Rossmeisl, John H.; Neal, Robert E.; Ellis, Thomas L.; Olson, John D.; Henao-Guerrero, Natalia; Robertson, John; Davalos, Rafael V. (2010). "Intracranial Nonthermal Irreversible Electroporation: In Vivo Analysis". The Journal of Membrane Biology 236 (1): 127–36. doi:10.1007/s00232-010-9284-z. PMID 20668843.

[11] Neal, R E; Garcia, P A; Rossmeisl, J H; Davalos, R V (2010). "A study using irreversible electroporation to treat large, irregular tumors in a canine patient". 2010 Annual International Conference of the IEEE Engineering in Medicine and Biology: 2747–50. doi:10.1109/IEMBS.2010.5626372. ISBN 978-1-4244-4123-5. PMID 21095963.

[12] Arena, Christopher B; Sano, Michael B; Rossmeisl, John H; Caldwell, John L; Garcia, Paulo A; Rylander, Marissa; Davalos, Rafael V (2011). "High-frequency irreversible electroporation (H-FIRE) for non-thermal ablation without muscle contraction". BioMedical Engineering OnLine 10: 102. doi:10.1186/1475-925X-10-102. PMC 3258292. PMID 22104372.

[13] El-Andaloussi S, Lee Y, Lakhal-Littleton S, Li J, Seow Y, Gardiner C, Alvarez-Erviti L, Sargent IL, Wood MJ (December 2012). "Exosome-mediated delivery of siRNA in vitro and in vivo". Nat Protoc 7 (12): 2112–26. doi:10.1038/nprot.2012.131. PMID 23154783.

[14] Ivorra, Antoni; Rubinsky, Boris. "Gels with predetermined conductivity used in electroporation of tissue USPTO Application #: 20080214986 — Class: 604 21 (USPTO)".

[15] Weaver, James C.; Chizmadzhev, Yu.A. (1996). "Theory of electroporation: A review". Bioelectrochemistry and Bioenergetics 41 (2): 135–60. doi:10.1016/S0302-4598(96)05062-3.

[16] Becker, S. M.; Kuznetsov, A. V. (2007). "Local Temperature Rises Influence in Vivo Electroporation Pore Development: A Numerical Stratum Corneum Lipid Phase Transition Model". Journal of Biomechanical Engineering 129 (5): 712–21. doi:10.1115/1.2768380. PMID 17887897.

[17] Melikov, Kamran C.; Frolov, Vadim A.; Shcherbakov, Arseniy; Samsonov, Andrey V.; Chizmadzhev, Yury A.; Chernomordik, Leonid V. (2001). "Voltage-Induced Nonconductive Pre-Pores and Metastable Single Pores in Unmodified Planar Lipid Bilayer". Biophysical Journal 80 (4): 1829–36. Bibcode:2001BpJ....80.1829M. doi:10.1016/S0006-3495(01)76153-X. PMC 1301372. PMID 11259296.

[18] Joshi, R.; Schoenbach, K. (2000). "Electroporation dynamics in biological cells subjected to ultrafast electrical pulses: A numerical simulation study". Physical Review E 62 (1 Pt B): 1025–33. Bibcode:2000PhRvE..62.1025J. doi:10.1103/PhysRevE.62.1025. PMID 11088559.

Chapter 23

Microinjection

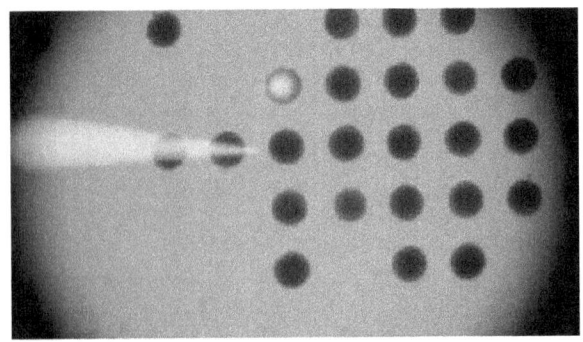

Microinjection of a fluorescent dye into Ciona intestinalis *eggs positioned in a microwell array.*

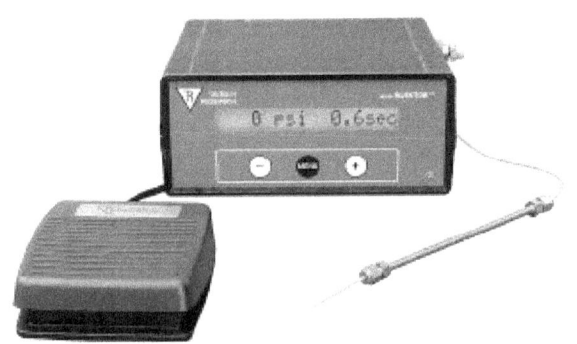

A microinjection controller designed by Tritech Research for controlling the pressure applied to a hollow glass needle with a microscopic tip, placed in a brass needle holder, to regulate the delivery of substances like DNA into cells, stem cells into embryos, and sperm into eggs.

Microinjection is the use of a glass micropipette to inject a liquid substance at a microscopic or borderline macroscopic level. The target is often a living cell but may also include intercellular space. Microinjection is a simple mechanical process usually involving an inverted microscope with a magnification power of around 200x (though sometimes it is performed using a dissecting stereo microscope at 40–50x or a traditional compound upright microscope at similar power to an inverted model).

For processes such as cellular or pronuclear injection the target cell is positioned under the microscope and two micromanipulators—one holding the pipette and one holding a microcapillary needle usually between 0.5 and 5 μm in diameter (larger if injecting stem cells into an embryo)—are used to penetrate the cell membrane and/or the nuclear envelope.[1] In this way the process can be used to introduce a vector into a single cell. Microinjection can also be used in the cloning of organisms, in the study of cell biology and viruses, and for treating male subfertility through intracytoplasmic sperm injection (ICSI, "*IK-see*").

23.1 History

The use of microinjection as a biological procedure began in the early twentieth century, though even through the 1970s it was not commonly used. By the 1990s, however, its use had escalated significantly and it is now considered a common laboratory technique, along with vesicle fusion, electroporation, chemical transfection, and viral transduction, for introducing a small amount of a substance into a small target.[2]

23.2 Basic types

There are two basic types of microinjection systems. The first is called a *constant flow system* and the second is called a *pulsed flow system*. In a constant flow system, which is relatively simple and inexpensive though clumsy and outdated, a constant flow of a sample is delivered from a micropipette and the amount of the sample which is injected is determined by how long the needle remains in the cell. This system typically requires a regulated pressure source, a capillary holder, and either a coarse or a fine micromanipulator. A pulsed flow system, however, allows for greater control and consistency over the amount of sample injected: the most common arrangement for ICSI injection includes an Eppendorf "Femtojet" injector coupled with an Eppendorf "InjectMan", though procedures involving other targets usually take advantage of much less expensive equipment of similar capability. Because of its increased control over needle placement and movement and in addition to the increased precision over the volume of substance delivered, the pulsed flow technique usually results in less damage to the receiving cell than the constant flow technique. However, the Eppendorf line, at least, has a complex user interface and its particular system components are usually much more expensive than those necessary to create a constant flow system or than other pulsed flow injection systems.[3]

23.3 Pronuclear injection

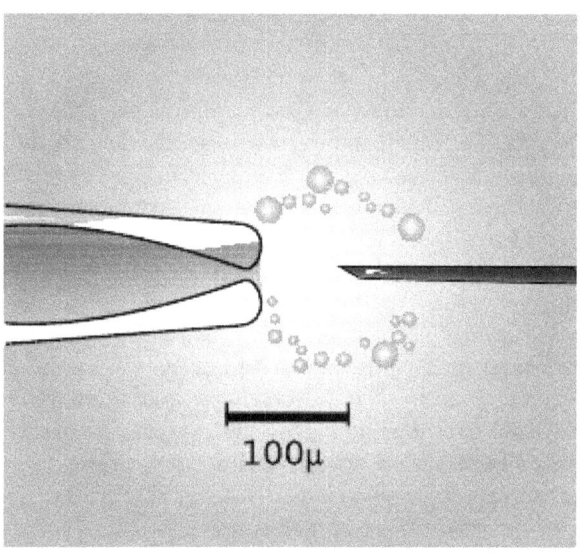

Diagram of the intracytoplasmic sperm injection of a human egg. Micromanipulator on the left holds egg in position while microinjector on the right delivers a single sperm cell.

Pronuclear injection is a technique used to create transgenic organisms by injecting genetic material into the nucleus of a fertilized oocyte. This technique is commonly used to study the role of genes using mouse animal models.

23.3.1 Pronuclear injection in mice

The pronuclear injection of mouse sperm is one of the two most common methods for producing transgenic animals (along with the genetic engineering of embryonic stem cells).[4] In order for pronuclear injection to be successful, the genetic material (typically linear DNA) must be injected while the genetic material from the oocyte and sperm are separate (i.e., the pronuclear phase).[5] In order to obtain these oocytes, mice are commonly superovulated using gonadotrophins.[6] Once plugging has occurred, oocytes

are harvested from the mouse and injected with the genetic material. The oocyte is then implanted in the oviduct of a pseudopregnant animal.[5] While efficiency varies, 10-40% of mice born from these implanted oocytes may contain the injected construct.[6] Transgenic mice can then be bred to create transgenic lines.

23.4 See also

- Transgenesis
- Genetically modified mouse
- *Biology: The Unity and Diversity of Life*

23.5 References

[1] David B. Burr; Matthew R. Allen (11 June 2013). *Basic and Applied Bone Biology*. Academic. p. 157. ISBN 978-0-12-391459-0. Retrieved 15 July 2013.

[2] Juan Carlos Lacal; Rosario Perona; James Feramisco (11 June 1999). *Microinjection*. Springer. p. 9. ISBN 978-3-7643-6019-1. Retrieved 13 July 2013.

[3] Robert D. Goldman; David L. Spector (1 January 2005). *Live Cell Imaging: A Laboratory Manual*. CSHL. p. 54. ISBN 978-0-87969-683-2. Retrieved 15 July 2013.

[4] Heinz Peter Nasheuer (2010). *Genome Stability and Human Diseases*. Springer. p. 328. ISBN 978-90-481-3471-7. Retrieved 15 July 2013.

[5] Mullin, Ann. "Pronuclear Injection". Tulane University.

[6] "Pronuclear Injection". UC San Diego. Retrieved 21 February 2013.

Chapter 24

Viral transformation

Viral transformation is the change in growth, phenotype, or indefinite reproduction of cells caused by the introduction of inheritable material. Through this process, a virus causes harmful transformations of an in vivo cell or cell culture. The term can also be understood as DNA transfection using a viral vector.

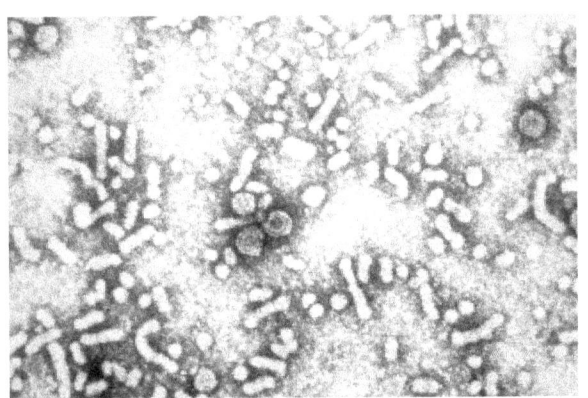

Figure 1: Hepatitis-B virions

Viral transformation can occur both naturally and medically. Natural transformations can include viral cancers, such as human papillomavirus (HPV) and T-cell Leukemia virus type I. Hepatitis B and C are also the result of natural viral transformation of the host cells. Viral transformation can also be induced for use in medical treatments.

Cells that have been virally transformed can be differentiated from untransformed cells through a variety of growth, surface, and intracellular observations. The growth of transformed cells can be impacted by a loss of growth limitation caused by cell contact, less oriented growth, and high saturation density. Transformed cells can lose their tight junctions, increase their rate of nutrient transfer, and increase their protease secretion. Transformation can also affect the cytoskeleton and change in the quantity of signal molecules.

24.1 Type

See also: Malignant transformation

There are three types of viral infections that can be considered under the topic of viral transformation. These are cytocidal, persistent, and transforming infections. Cytocidal infections can cause fusion of adjacent cells, disruption of transport pathways including ions and other cell signals, disruption of DNA, RNA and protein synthesis, and nearly always leads to cell death. Persistent infections involve viral material that lays dormant within a cell until activated by some stimulus. This type of infection usually causes few obvious changes within the cell but can lead to long chronic diseases. Transforming infections are also referred to as malignant transformation. This infection causes a host cell to become malignant and can be either cytocidal (usually in the case of RNA viruses) or persistent (usually in the case of DNA viruses). Cells with transforming infections undergo immortalization and inherit the genetic material to produce tumors. Since the term cytocidal, or cytolytic, refers to cell death, these three infections are not mutually exclusive. Many transforming infections by DNA tumor viruses are also cytocidal.[1]

Table 1: Cellular effects of viral infections[1]

24.1.1 Cytocidal infections

Cytocidal infections are often associated with changes in cell morphology, physiology and are thus important for the complete viral replication and transformation. *Cytopathic Effects*, often include a change in cell's morphology such as fusion with adjacent cells to form polykaryocytes as well as the synthesis of nuclear and cytoplasmic inclusion bodies. *Physiological changes* include the insufficient movement of ions, formation of secondary messengers, and activation of cellular cascades to continue cellular activity. *Biochemically*, many viruses inhibit the synthesis of host DNA, RNA, proteins directly or even interfere with protein-protein, DNA-protein, RNA-protein interactions at the subcellular level. *Genotoxicity* involves breaking, fragmenting, or rearranging chromosomes of the host. Lastly, *biologic effects* include the viruses' ability to affect the activity of antigens and immunologloglobulins in the host cell.[1]

There are two types of cytocidal infections, productive and abortive. In productive infections, additional infectious viruses are produced. Abortive infections do not produce infectious viruses. One example of a productive cytocidal infection is the herpes virus.[2]

24.1.2 Persistent infections

There are three types of persistent infections, latent, chronic and slow, in which the virus stays inside the host cell for prolonged periods of time. During *latent infections* there is minimal to no expression of infected viral genome. The genome remains within the host cell until the virus is ready for replication. *Chronic infections* have similar cellular effects as acute cytocidal infections but there is a limited number of progeny and viruses involved in transformation. Lastly, *slow infections* have a longer incubation period in which no physiological, morphological or subcellular changes may be involved.[1]

24.1.3 Transforming infections

Transformation infections is limited to abortive or restrictive infections.[1] This constitutes the broadest category of infections as it can include both cytocidal and persistent infection. Viral transformation is most commonly understood as transforming infections, so the remainder of the article focuses on detailing transforming infections.

24.2 Process

In order for a cell to be transformed by a virus, the viral DNA must be entered into the host cell. The simplest consideration is viral transformation of a bacterial cell. This process is called lysogeny. As shown in Figure 2, a bacteriophage lands on a cell and pins itself to the cell. The phage can then penetrate the cell membrane and inject the viral DNA into the host cell. The viral DNA can then either lay dormant until stimulated by a source such as UV light or it can be immediately taken up by the host's genome. In either case the viral DNA will replicate along with the original host DNA during cell replication causing two cells to now be infected with the virus. The process will continue to propagate more and more infected cells.[3] This process is in contrast to the lytic cycle where a virus only uses the

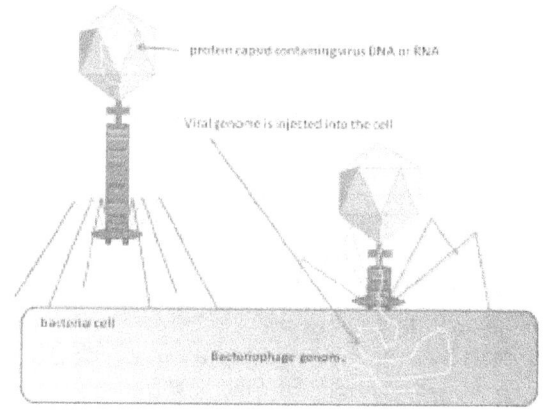

Figure 2: Phage injecting its genome into bacterial cell

host cell's replication machinery to replicate itself before destroying the host cell.[4]

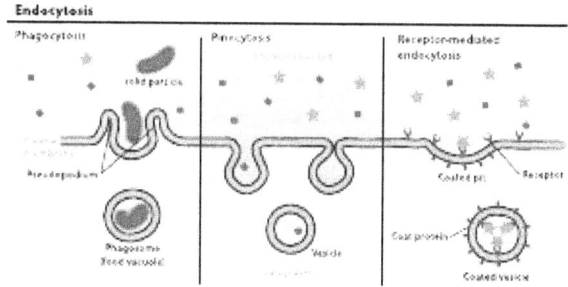

Figure 3: Examples of endocytosis

The process is similar in animal cells. In most cases, rather than viral DNA being injected into an animal cell, a section of the membrane encases the virus and the cell then absorbs both the virus and the encasing section of the membrane into the cell. This process, called endocytosis, is shown in Figure 3.[5]

24.3 Transformation of the host cell

Viral transformation disrupts the normal expression of the host cell's genes in favor of expressing a limited number of viral genes. The virus also can disrupt communication between cells and cause cells to divide at an increased rate.[6]

24.3.1 Physiological

Viral transformation can impose characteristically determinable features upon a cell. Typical phenotypic changes include high saturation density, anchorage-independent growth, loss of contact inhibition, loss of orientated growth, immortalization, disruption of the cell's cytoskeleton.

24.3.2   Biochemical

Viral genes are expressed through the use of the host cell's replication machinery; therefore, many viral genes have promoters that support binding of many transcription factors found naturally in the host cells. These transcription factors along with the virus' own proteins can repress or activate genes from both the virus and the host cell's genome. Many viruses can also increase the production of the cell's regulatory proteins.[1]

24.3.3   Genetic

Depending on the virus, a variety of genetic changes can occur in the host cell. In the case of a lytic cycle virus, the cell will only survive long enough to the replication machinery to be used to create additional viral units. In other cases, the viral DNA will be persist within the host cell and replicate as the cell replicates. This viral DNA can either be incorporated into the host cell's genetic material or persist as a separate genetic vector. Either case can lead to damage of the host cell's chromosomes. It is possible that the damage can be repaired; however, the most common result is an instability in the original genetic material or suppression or alteration of the gene expression.[1]

24.4   Assays

See also: Virus quantification

An assay is an analytic tool often used in a laboratory setting in order to assess or measure some quality of a target entity.[7] In virology, assays can be used to differentiate between transformed and non-transformed cells. Varying the assay used, changes the selective pressure on the cells and therefore can change what properties are selected in the transformed cells.[6]

Three common assays used are the focus forming assay, the Anchorage independent growth assay, and the reduced serum assay.

The focus forming assay (FFA) is used to grow cells containing a transforming oncogene on a monolayer of non-transformed cells. The transformed cells will form raised, dense spots on the sample as they grow without contact inhibition.[8] This assay is highly sensitive compared to other assays used for viral analysis, such as the yield reduction assay.[9]

An example of the Anchorage independent growth assay is the soft agar assay. The assay is assessing the cells' ability to grow in a gel or viscous fluid. Transformed cells can grow in this environment and are considered anchorage independent. Cells that can only grow when attached to a solid surface are anchorage dependent untransformed cells. This assay is considered one of the most stringent for detection of malignant transformation.[10]

In a reduced serum assay, cells are assayed by exploiting the changes in cell serum requirements. Non-transformed cells require at least a 5% serum medium in order to grow; however, transformed cells can grow in an environment with significantly less serum.[6]

24.5   Examples of Natural Transformation

Natural transformation is the viral transformation of cells without the interference of medical science. This is the most commonly considered form of viral transformation and includes many cancers and diseases, such as HIV, Hepatitis B, and T-cell Leukemia virus type I.

24.5.1   Viral oncogenesis

See also: Oncovirus

As many as 20% of human tumors are caused by viruses.[11] Some such viruses that are commonly recognized include HPV, T-cell Leukemia virus type I, and hepatitis B.

Viral oncogenesis are most common with DNA and RNA tumor viruses, most frequently the retroviruses.[12] There are two types of oncogenic retroviruses: acute transforming viruses and non-acute transforming viruses. Acute transforming viruses induce a rapid tumor growth since they carry viral oncogenes in their DNA/RNA to induce such growth. An example of an acute transforming virus is the Rous Sarcoma Virus (RSV) that carry the v-src oncogene. v-Src is part of the c-src, which is a cellular proto-oncogene that stimulates rapid cell growth and expansion. A non-acute transforming virus on the other hand induces a slow tumor growth, since it does not carry any viral oncogenes. It induces tumor growth by transcriptionally activating the proto-oncogenes particularly the long terminal repeat (LTR) in the proto-oncogenes.[12]

Viral Oncogonesis through transformation can occur via 2 mechanisms:[1]

1. The tumor virus can introduce and express a "transforming" gene either through the integration of DNA

or RNA into the host genome.

2. The tumor virus can alter expression on preexisting genes of the host.

One or both of these mechanisms can occur in the same host cell.

Hepatitis B

The Hepatitis B viral protein X is believed to cause hepatocellular carcinoma through transformation, typically of liver cells. The viral DNA is incorporated into the host cell's genome causing rapid cell replication and tumor growth.[13]

Papillomaviruses

Papillomaviruses typically target epithelial cells and cause everything from warts to cervical cancer. When human papillomavirus (HPV) transforms a cell, it interferes with the function of cellular proteins while degrading other cellular proteins.[14]

Herpesviruses

The herpesviruses, Kaposi's sarcoma-associated herpesvirus and Epstein-Barr virus, are believed to cause cancer in humans, such as Kaposi's sarcoma, Burkitt's lymphoma, and nasopharyngeal carcinoma. Although genes have been identified in these viruses that cause transformation, the manner in which the virus transforms and replicates the host cell is not understood.[14]

Retroviruses

The retroviruses include T-cell Leukemia virus type I, HIV, and Rous Sarcoma Virus (RSV). The viral gene tax is expressed when the T-cell Leukemia virus transforms a cell altering the expression of cellular growth control genes and causing the transformed cells to become cancerous. HIV works differently by not directly causing cells to become cancerous but by instead making those infected more susceptible to lymphoma and Kaposi's sarcoma. Many other retroviruses contain the three genes, gag, pol, and env, which do not directly cause transformation or tumor formation.[14]

24.5.2 HIV

Human immunodeficiency virus is a viral infection that targets the lymph nodes. HIV binds to the immune CD4 cell

and reverse transcriptase alters the host cell genome to allow integration of the viral DNA via integrase. The virus replicates using the host cell's machinery and then leaves the cell to infect additional cells via budding.[15]

24.6 Medical Applications

There are many applications in which viral transformation can be artificially induced in a cell culture in order to treat an illness or other condition. A cell culture is infected with a virus causing the transformation; transformed cells can then be used to either produce treatments or be directly introduced into the body.

24.6.1 Personalized type I interferons

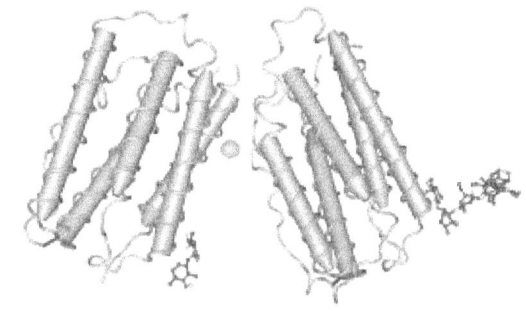

Figure 4: Type I Human Interferon

Type I interferons (IFNs) are used to treat a wide variety of medical conditions including hepatitis C, cancers, viral and inflammatory diseases. IFNs can either be extracted from a natural source, such as cultured human cells or blood leukocytes, or they can be manufactured with recombinant DNA technologies. Most of these IFN treatments have a low response rate.[16]

The use of viral transformation of the Epstein-Barr virus (EBV) has been recommended to create personalized IFNs. In this process, primary B lymphocytes are transformed with EBV. These cells can then be used to produce IFNs specific for the patient from which the B lymphocytes were extracted. This personalization decreases the likelihood of an antibody response and therefore increases the effectiveness of the treatment.[16]

24.6.2 Cancer treatments

When a virus transforms a cell it often causes cancer by either altering the cells' existing genome or introducing additional genetic material which causes cells to uncontrollably replicate.[11] It is rarely considered that what causes so much harm also has the capability of reversing the process and slowing the cancer growth or even leading to remission. Viruses transform host cells in order to survive and replicate; however, the immune responses of the host cell are typically compromised during transformation making transformed cells more susceptible to other viruses.[17]

The idea of using viruses to treat cancers was first introduced in 1951 when a 4-year-old boy suddenly went into a temporary remission from leukemia while he had chickenpox. This led to research in the 1990s where scientists worked to create a strain of the herpes simplex virus strong enough to infect and transform tumor cells but weak enough to leave healthy cells unharmed. Treating patients with viral transformation has the possibility of treating patients more safely and more effectively than using traditional methods, such as chemotherapy. Viruses used in the treatment of cancer gain strength and increase their effectiveness as the multiply in the body while causing only minor side effects, such as nausea, fatigue, and aches.[17]

24.7 References

[1] Baron, ed. by Samuel (1996). *Medical microbiology* (4. ed.). Galveston, Texas: Univ. of Texas Medical Branch. ISBN 0963117211.

[2] Huang, CR; Lin, SS; Chou, MY; Ho, CC; Wang, L; Lee, YL; Chen, CS; Yang, CC (2005). "Demonstration of different modes of cell death upon herpes simplex virus 1 infection in different types of oral cells.". *Acta virologica* **49** (1): 7–15. PMID 15929393.

[3] "Lysogeny". *Encyclopaedia Britannica*. Retrieved 8 April 2014.

[4] "Two Life Cycles of a Virus". Retrieved 8 April 2014.

[5] "The cycle of infection". Encyclopaedia Britannica. Retrieved 8 April 2014.

[6] Heaphy, Shaun. "Viral Transformation of Cells". University of Cape Town. Retrieved 25 March 2014.

[7] Assay. *Wikipedia, The Free Encyclopedia*. Accessed 2014-03-25.

[8] Andrews, David. "Transformation assays: focus forming assay" (PDF). Andrews Lab. Retrieved 25 March 2014.

[9] Winship, Timothy R (11 Dec 1979). "A Sensitive Method for Quantification of Vesicular Stomatitis Virus Defective Interfering Particles: Focus Forming Assay" (PDF). *Journal of General Virology* **48** (1). doi:10.1099/0022-1317-48-1-237. Retrieved 25 March 2014.

[10] Provost, Joseph. "Soft Agar Assay for Cology Formation" (PDF). Wallert and Provost Lab. Retrieved 25 March 2014.

[11] Dayaram, T; Marriott, SJ (Aug 2008). "Effect of transforming viruses on molecular mechanisms associated with cancer.". *Journal of cellular physiology* **216** (2): 309–14. doi:10.1002/jcp.21439. PMID 18366075.

[12] Fan, Hung (June 15, 2011). "Cell Transformation by RNA Viruses: An Overview". *Viruses* **3** (12): 858–860. doi:10.3390/v3060858.

[13] Schaefer, S; Gerlich, WH (1995). "In vitro transformation by hepatitis B virus DNA.". *Intervirology* **38** (3-4): 143–54. PMID 8682609.

[14] Cooper, Geoffrey M. (2000). *The cell : a molecular approach* (2nd ed.). Washington: ASM Press. ISBN 0878931023.

[15] "HIV Life Cycle". U.S. Department of Health and Human Services. Retrieved 7 May 2014.

[16] Xu, Dongsheng; Zhang, Luwen (June 2010). "Viral transformation for production of personalized type I interferons". *Biotechnology Journal* **5** (6): 578–581. doi:10.1002/biot.201000038.

[17] Nuwer, Rachel (19 Mar 2012). "Viruses Recruited as Killers of Tumors". *New York Times*. Retrieved 6 May 2014.

Chapter 25

Lipofection

Lipofection (or liposome transfection) is a technique used to inject genetic material into a cell by means of liposomes, which are vesicles that can easily merge with the cell membrane since they are both made of a phospholipid bilayer. Lipofection generally uses a positively charged (cationic) lipid to form an aggregate with the negatively charged (anionic) genetic material.[1] A net positive charge on this aggregate has been assumed to increase the effectiveness of transfection through the negatively charged phospholipid bilayer.[1] This transfection technology performs the same tasks as other biochemical procedures utilizing polymers, DEAE dextran, calcium phosphate, and electroporation. The main advantages of lipofection are its high efficiency, its ability to transfect all types of nucleic acids in a wide range of cell types, its ease of use, reproducibility, and low toxicity. In addition, this method is suitable for all transfection applications (transient, stable, co-transfection, reverse, sequential or multiple transfections). High throughput screening assay has also shown good efficiency in some in vivo models.

25.1 See also

- Lipofectamine

25.2 References

- Felgner PL, Gadek TR, Holm M, et al. (November 1987). "Lipofection: a highly efficient, lipid-mediated DNA-transfection procedure". *Proc. Natl. Acad. Sci. U.S.A.* **84** (21): 7413–7. doi:10.1073/pnas.84.21.7413. PMC 299306. PMID 2823261.

- Felgner JH, Kumar R, Sridhar CN, et al. (January 1994). "Enhanced gene delivery and mechanism studies with a novel series of cationic lipid formulations". *J. Biol. Chem.* **269** (4): 2550–61. PMID 8300583.

25.3 Notes

[1] http://www.freepatentsonline.com/EP1129064.pdf

Chapter 26

Transgenesis

Transgenesis is the process of introducing an exogenous gene—called a transgene—into a living organism so that the organism will exhibit a new property and transmit that property to its offspring. Transgenesis can be facilitated by liposomes, enzymes, plasmid vectors, viral vectors, pronuclear injection, protoplast fusion, and ballistic DNA injection.

Transgenic organisms are able to express foreign genes because the genetic code is similar for all organisms. This means that a specific DNA sequence will code for the same protein in all organisms. Due to this similarity in protein sequence, scientists can cut DNA at these common protein points and add other genes. An example of this is the "super mice" of the 1980s. These mice were able to produce the human protein tPA to treat blood clots.

26.1 Using plasmids from bacteria

The most common type of transgenesis research is done with bacteria and viruses which are able to replicate foreign DNA.[1] The plasmid DNA is cut using restriction enzymes, while the DNA to be copied is also cut with the same restriction enzyme, producing complementary sticky-ends. This allows the foreign DNA to hybridise with the plasmid DNA and be sealed by DNA ligase enzyme, creating a genetic code not normally found in nature. Altered DNA is inserted into plasmids for replication.[2]

26.2 Gene transfer technology

26.2.1 DNA microinjection

The Desired gene construct is injected in the pronucleus of a reproductive cell using a glass needle around 0.5 to 5 micrometers in diameter. The manipulated cell is cultured in vitro to develop to a specific embryonic phase, is then transferred to a recipient female. DNA microinjection does not

have a high success rate (roughly 2% of all injected subjects), even if the new DNA is incorporated in the genome, if it is not accepted by the germ-line the new traits will not appear in their offspring. If DNA is injected in multiple sites the chances of over-expression increase.[3]

26.2.2 Retrovirus-mediated gene transfer

A retrovirus is a virus that carries its genetic material in the form of RNA rather than DNA. Retroviruses are used as vectors to transfer genetic material into the host cell. The result is a chimera, an organism consisting of tissues or parts of diverse genetic constitution. Chimeras are inbred for as many as 20 generations until homozygous genetic offspring are born.[3]

26.2.3 Restriction enzyme mediated integration

Restriction enzyme mediated integration (REMI) is a technique for integrating DNA (linearised plasmid) into the genome sites that have been generated by the same restriction enzyme used for the DNA linearisation. The plasmid integration occurs at the corresponding sites in the genome, often by regenerating the recognition sites by same the restriction enzyme used for plasmid linearisation.

26.2.4 Stem cell transgenesis

Multipotent stem cell transgenesis

Multipotent stem cells can only differentiate into a limited number of therapeutically useful cell types, nevertheless their safety and relative lack of complexity to us have resulted in the vast majority of current personalized cellular therapeutics involving multipotent stem cells (typically mesenchymal stem cells from adipose tissue).[4]

Pluripotent stem cell transgenesis

Transgenic vectors can be delivered randomly, or targeted to a specific genomic location, such as a safe harbor . Scientists have performed research and technology development to provide the tools necessary to permit safe and effective pluripotent stem cell (PSC) transgenesis.[5][6][7][8][9][10][11][12]

Totipotent stem cell transgenesis

The manipulated gene construct is inserted into totipotent stem cells, cells which can develop into any specialized cell. Cells containing the desired DNA are incorporated into the host's embryo, resulting in a chimeric animal. Unlike the other two methods of injection which require live transgenic offspring for testing, embryonic cell transfer can be tested at the cell stage.

26.3 Applications

26.3.1 Pharming

Main article: Pharming (genetics)

Pharming, a portmanteau of "farming" and "pharmaceutical", refers to the use of genetic engineering to insert genes that code for useful pharmaceuticals into host animals or plants that would otherwise not express those genes. Pharming has gained application in biotechnology since the development of transgenic "super mice" in 1982. "Super mice" were genetically altered to produce the human drug, tPA (tissue plasminogen activator to treat blood clots), in 1987.[2] Since then, "super mice" pharming has come a long way. Using RNA interference, scientists have produced a cow whose milk contains increased amounts of casein, a protein used to make cheese and other foods, and almost no beta-lactoglobulin, a component in milk whey protein that causes allergies.[13]

Pharming examples:[14]

- Haemoglobin as a blood substitute

- Human protein C anticoagulant

- Alpha-1 antitrypsin (AAT) for treatment of AAT deficiency

- Insulin for diabetes treatment

- Vaccines (antigens)

- Growth hormones for treatment of deficiencies

- Factor VIII blood clotting factor

- Factor IX blood clotting factor

- Fibrinogen blood clotting factor

- Lactoferrin as an infant formula additive

26.3.2 Medical

Transgenesis can be used to neutralize genes that would normally prevent xenotransplantation. For example, a protein found in pigs can cause humans to reject their transplanted organs. This protein can be replaced by a similar human genome to prevent the rejection.[15]

26.4 Ethical concerns

Transgenesis has created certain ethical concerns. Examples include rights for animals that have been improved intellectually, legal ramifications, and possible health risks.[16]

26.5 Diagram

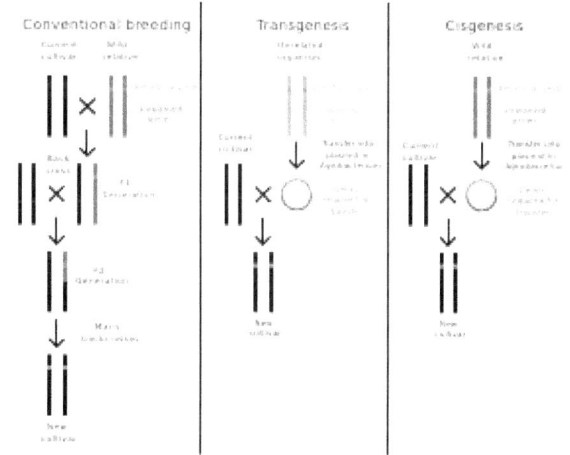

A diagram comparing the genetic changes achieved through conventional plant breeding, transgenesis and cisgenesis

.

Note: New genotypes created with transgenic technologies also require multiple backcrossings. Furthermore, backcrossing does not account for the majority of time required to create, field test and release/commercialize a new variety.

26.6 References

[1] "Mousepox Case Study — Module 4.0". Federation of American Scientists. History of Transgenics. Archived from the original on July 15, 2007.

[2] Redway, Keith. "Transgenic organisms". *Gene Manipulation & Recombinant DNA*. University of Westminster. Retrieved June 28, 2014.

[3] Margawati, Endang Tri (January 2003). "Transgenic Animals: Their Benefits To Human Welfare". *Actionbioscience*. Retrieved June 29, 2014.

[4] clinicaltrials.org

[5] Capecchi MR (June 2005). "Gene targeting in mice: functional analysis of the mammalian genome for the twenty-first century". *Nat. Rev. Genet.* **6** (6): 507–12. doi:10.1038/nrg1619. PMID 15931173.

[6] Cong L, Ran FA, Cox D, et al. (February 2013). "Multiplex genome engineering using CRISPR/Cas systems". *Science* **339** (6121): 819–23. doi:10.1126/science.1231143. PMC 3795411. PMID 23287718.

[7] DiCarlo JE, Norville JE, Mali P, Rios X, Aach J, Church GM (April 2013). "Genome engineering in *Saccharomyces cerevisiae* using CRISPR-Cas systems". *Nucleic Acids Res.* **41** (7): 4336–43. doi:10.1093/nar/gkt135. PMC 3627607. PMID 23460208.

[8] Friedland AE, Tzur YB, Esvelt KM, Colaiácovo MP, Church GM, Calarco JA (August 2013). "Heritable genome editing in *C. elegans* via a CRISPR-Cas9 system". *Nat. Methods* **10** (8): 741–3. doi:10.1038/nmeth.2532. PMC 3822328. PMID 23817069.

[9] Hwang WY, Fu Y, Reyon D, et al. (March 2013). "Efficient genome editing in zebrafish using a CRISPR-Cas system". *Nat. Biotechnol.* **31** (3): 227–9. doi:10.1038/nbt.2501. PMC 3686313. PMID 23360964.

[10] Nguyen HN, Reijo Pera RA (2008). "Metaphase spreads and spectral karyotyping of human embryonic stem cells". *CSH Protoc*: pdb.prot5047. PMID 21356916.

[11] Mali P, Yang L, Esvelt KM, et al. (February 2013). "RNA-guided human genome engineering via Cas9". *Science* **339** (6121): 823–6. doi:10.1126/science.1232033. PMC 3712628. PMID 23287722.

[12] Xue H, Wu J, Li S, Rao MS, Liu Y (March 2014). "Genetic Modification in Human Pluripotent Stem Cells by Homologous Recombination and CRISPR/Cas9 System". *Methods Mol. Biol.* doi:10.1007/7651_2014_73. PMID 24615461.

[13] Lopatto, Elizabeth (October 1, 2012). "Gene-Modified Cow Makes Milk Rich in Protein, Study Finds". *Bloomberg Businessweek* (New York City).

[14] Buy M (1997). "Transgenic Animals". *CCAC Resource Supplement*. Canadian Council on Animal Care (CCAC).

[15] "Actionbioscience | Transgenic Animals: Their Benefits To Human Welfare". actionbioscience.org. Retrieved November 29, 2014.

[16] "Actionbioscience | Ethical Issues in Genetic Engineering and Transgenics". actionbioscience.org. Retrieved November 29, 2014.

Chapter 27

Cisgenesis

Potatoes after treatment with Phytophthora infestans. *The normal potatoes have blight but the cisgenic potatoes are healthy*

Cisgenesis is a product designation for a category of genetically engineered plants. A variety of classification schemes have been proposed[1] that order genetically modified organisms based on the nature of introduced genotypical changes, rather than the process of genetic engineering.

Cisgenesis (from "same" and "beginning") is one term for organisms that have been engineered using a process in which genes are artificially transferred between organisms that could otherwise be conventionally bred.[2][3] Unlike in transgenesis, genes are only transferred between closely related organisms.[4] However, while future technologies may allow genomes to be directly edited within an individual organism, currently nucleic acid sequences must be isolated and introduced using the same technologies that are used to produce transgenic organisms. The term was first introduced in 2000 by Henk J. Schouten and Henk Jochemsen,[5] and in 2004 a PhD thesis by Jan Schaart of Wageningen University in 2004, discussing making strawberries less susceptible to *Botrytis cinerea*.

In Europe, currently, this process is governed by the same laws as transgenesis but researchers at Wageningen University in the Netherlands feel that this should be changed and regulated in the same way as conventionally bred plants. However, other scientists, writing in Nature Biotechnology, have disagreed. [3] In 2012 the European Food Safety Authority (EFSA) issued a report with their risk assessment of cisgenic and intragenic plants. They compared the hazards associated with plants produced by cisgenesis and intragenesis with those obtained either by conventional plant breeding techniques or transgenesis. The EFSA concluded that "similar hazards can be associated with cisgenic and conventionally bred plants, while novel hazards can be associated with intragenic and transgenic plants."[6]

Cisgenesis has been applied to transfer of natural resistance genes to the devastating disease *Phytophthora infestans* in potato[7] and scab *(Venturia inaequalis)* in apple. [8][9]

Cisgenesis and transgenesis use artificial gene transfer, which results in less extensive change to an organism's genome than mutagenesis, which was widely used before genetic engineering was developed.[10]

Some people believe that cisgenesis should not face as much regulatory oversight as genetic modification created through transgenesis as it is possible, if not practical, to transfer alleles among closely related species even by traditional crossing. The primary biological advantage of cisgenesis is that it does not disrupt favorable heterozygous states, particularly in asexually propagated crops such as potato, which do not breed true to seed. One application of cisgenesis is to create blight resistant potato plants by transferring known resistance loci wild genotypes into modern, high yielding varieties.[11]

The Dutch government has proposed to exclude cisgenic plants from the European GMO Regulation, in view of the safety of cisgenic plants compared to classically bred plants, and their contribution to durable food production.[12]

27.1   Related classification scheme

A related classification scheme proposed by Kaare Nielsen is:[1]

27.2   Diagram

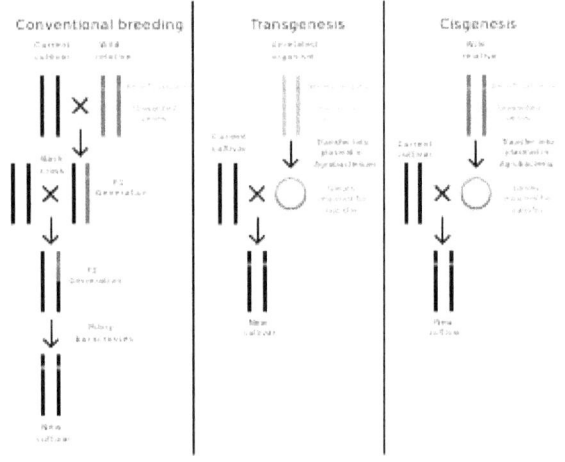

A diagram comparing the genetic changes achieved through conventional plant breeding, transgenesis and cisgenesis

.

27.3   References

[1] Nielsen, K. M. (2003). "Transgenic organisms—time for conceptual diversification". *Nature Biotechnology* **21** (3): 227–228. doi:10.1038/nbt0303-227. PMID 12610561.

[2] Cisgenesis definitions cisgenesis.com

[3] Schubert, D.; Williams, D. (2006). "'Cisgenic' as a product designation". *Nature Biotechnology* **24** (11): 1327–9. doi:10.1038/nbt1106-1327. PMID 17093469.

[4] MacKenzie D (2 August 2008). "How the humble potato could feed the world". *New Scientist* (2667): 30–33.

[5] Jochemsen H, ed. (2000). *Toetsen en begrenzen: een ethische en politieke beoordeling van de moderne biotechnologie.* ISBN 9072016327.

[6] Staff (2012) Scientific opinion addressing the safety assessment of plants developed through cisgenesis and intragenesis EFSA Panel on Genetically Modified Organisms. Parma, Italy. Retrieved 1 October 2012

[7] Park T-H, Vleeshouwers VGAA, Jacobsen E; et al. (2009). "Molecular breeding for resistance to *Phytophthora infestans* (Mont.) de Bary in potato (*Solanum tuberosum L.*): a perspective of cisgenesis". *Plant Breeding* **128**: 109–117. doi:10.1111/j.1439-0523.2008.01619.x.

[8] Vanblaere T, Flachowsky H, Gessler C, Broggini GA (January 2014). "Molecular characterization of cisgenic lines of apple 'Gala' carrying the Rvi6 scab resistance gene". *Plant Biotechnol. J.* **12** (1): 2–9. doi:10.1111/pbi.12110. PMID 23998808.

[9] Joshi SG, Schaart JG, Groenwold R, Jacobsen E, Schouten HJ, Krens FA (April 2011). "Functional analysis and expression profiling of HcrVf1 and HcrVf2 for development of scab resistant cisgenic and intragenic apples". *Plant Mol. Biol.* **75** (6): 579–91. doi:10.1007/s11103-011-9749-1. PMC 3057008. PMID 21293908.

[10] Schouten, H.; Krens, F.; Jacobsen, E. (2006). "Do cisgenic plants warrant less stringent oversight?". *Nature Biotechnology* **24** (7): 753. doi:10.1038/nbt0706-753. PMID 16841052.

[11] Jacobsen, E.; Schouten, H. J. (2008). "Cisgenesis, a New Tool for Traditional Plant Breeding, Should be Exempted from the Regulation on Genetically Modified Organisms in a Step by Step Approach". *Potato Research* **51**: 75–88. doi:10.1007/s11540-008-9097-y. Free version

[12] "Brief aan Eurocommissaris d.d. 18 december 2013 over Nieuwe veredelingstechnieken in de biotechnologie" [Letter to Commissioner dated December 18, 2013 on new breeding techniques in biotechnology] (in Dutch). 2014-01-06.

Chapter 28

Gene knockout

A **gene knockout** (abbreviation: **KO**) is a genetic technique in which one of an organism's genes is made inoperative ("knocked out" of the organism). Also known as **knockout organisms** or simply **knockouts**, they are used in learning about a gene that has been sequenced, but which has an unknown or incompletely known function. Researchers draw inferences from the difference between the knockout organism and normal individuals.

The term also refers to the process of creating such an organism, as in "knocking out" a gene. The technique is essentially the opposite of a gene knockin. Knocking out two genes simultaneously in an organism is known as a **double knockout (DKO)**. Similarly the terms **triple knockout (TKO)** and **quadruple knockouts (QKO)** are used to describe three or four knocked out genes, respectively.

28.1   Method

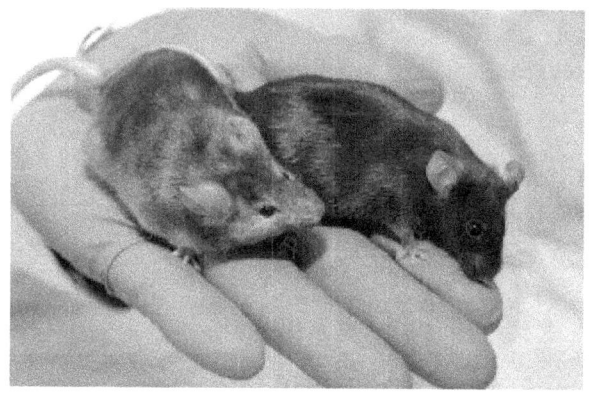

A laboratory mouse in which a gene affecting hair growth has been knocked out (left), is shown next to a normal lab mouse.

Knockout is accomplished through a combination of techniques, beginning in the test tube with a plasmid, a bacterial artificial chromosome or other DNA construct, and proceeding to cell culture. Individual cells are genetically transfected with the DNA construct. Often the goal is to create a transgenic animal that has the altered gene. If so, embryonic stem cells are genetically transformed and inserted into early embryos. Resulting animals with the genetic change in their germline cells can then often pass the gene knockout to future generations.

To create knockout moss, transfection of protoplasts is the preferred method. Such transformed *Physcomitrella*-protoplasts directly regenerate into fertile moss plants. Eight weeks after transfection, the plants can be screened for gene targeting via PCR.[1]

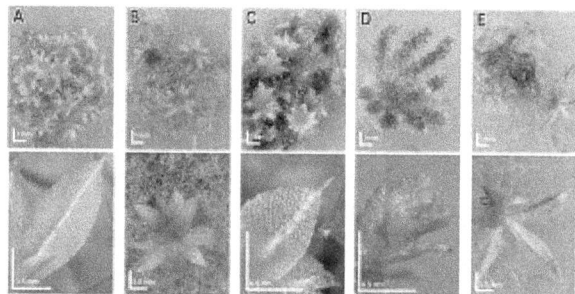

Wild-type Physcomitrella and knockout mosses: *Deviating phenotypes induced in gene-disruption library transformants. Physcomitrella wild-type and transformed plants were grown on minimal Knop medium to induce differentiation and development of gametophores. For each plant, an overview (upper row; scale bar corresponds to 1 mm) and a close-up (bottom row; scale bar equals 0.5 mm) are shown. A: Haploid wild-type moss plant completely covered with leafy gametophores and close-up of wild-type leaf. B–D: Different mutants.*[2]

The construct is engineered to recombine with the target gene, which is accomplished by incorporating sequences from the gene itself into the construct. Recombination then occurs in the region of that sequence within the gene, resulting in the insertion of a foreign sequence to disrupt the gene. With its sequence interrupted, the altered gene in most cases will be translated into a nonfunctional protein, if it is translated at all.

A conditional knockout allows gene deletion in a tissue or time specific manner. This is done by introducing short se-

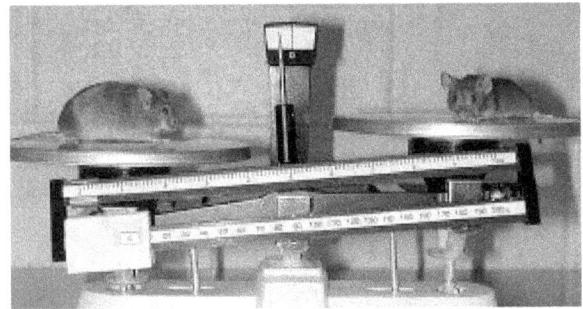

A knockout mouse (left) that is a model of obesity, compared with a normal mouse.

quences called loxP sites around the gene. These sequences will be introduced into the germ-line via the same mechanism as a knock-out. This germ-line can then be crossed to another germline containing Cre-recombinase which is a viral enzyme that can recognize these sequences, recombines them and deletes the gene flanked by these sites.

Because the desired type of DNA recombination is a rare event in the case of most cells and most constructs, the foreign sequence chosen for insertion usually includes a reporter. This enables easy selection of cells or individuals in which knockout was successful. Sometimes the DNA construct inserts into a chromosome without the desired homologous recombination with the target gene. To eliminate such cells, the DNA construct often contains a second region of DNA that allows such cells to be identified and discarded.

In diploid organisms, which contain two alleles for most genes, and may as well contain several related genes that collaborate in the same role, additional rounds of transformation and selection are performed until every targeted gene is knocked out. Selective breeding may be required to produce homozygous knockout animals.

Gene knockin is similar to gene knockout, but it replaces a gene with another instead of deleting it.

28.2   Use

Knockouts are primarily used to understand the role of a specific gene or DNA region by comparing the knockout organism to a wildtype with a similar genetic background.

Knockouts organisms are also used as screening tools in the development of drugs, to target specific biological processes or deficiencies by using a specific knockout, or to understand the mechanism of action of a drug by using a library of knockout organisms spanning the entire genome, such as in *Saccharomyces cerevisiae*.[3]

28.3   See also

- Gene knockdown
- Conditional gene knockout
- Gene knockin
- Knockout mouse
- Knockout moss
- Germline
- Gene silencing
- Planned extinction
- Recombineering

28.4   References

[1] Ralf Reski (1998): Physcomitrella and Arabidopsis: the David and Goliath of reverse genetics. Trends Plant in Science 3, 209-210

[2] Egener et al.   BMC Plant Biology 2002 2:6 doi:10.1186/1471-2229-2-6

[3] http://www-sequence.stanford.edu/group/yeast_deletion_project/deletions3.html

28.5   External links

- Diagram of targeted gene replacement
- Frontiers in Bioscience Gene Knockout Database (available on archive only)
- International Knockout Mouse Consortium
- KOMP Repository

Chapter 29

Gene knockdown

Gene knockdown refers to an experimental technique by which the expression of one or more of an organism's genes are reduced. The reduction can occur either through genetic modification or by treatment with a reagent such as a short DNA or RNA oligonucleotide that has a sequence complementary to either gene or an mRNA transcript.[1]

29.1   Gene knockdowns vs. transient knockdowns

If genetic modification of DNA is done, the result is called "knockdown organism." If the change in gene expression is caused by an oligonucleotide binding to an mRNA or temporarily binding to a gene, this leads to a temporary change in gene expression that does not modify the chromosomal DNA, and the result is referred to as a "transient knockdown".[1]

In a transient knockdown, the binding of this oligonucleotide to the active gene or its transcripts causes decreased expression through a variety of processes. Binding can occur either through the blocking of transcription (in the case of gene-binding), the degradation of the mRNA transcript (e.g. by small interfering RNA (siRNA)) or RNase-H dependent antisense), or through the blocking of either mRNA translation, pre-mRNA splicing sites, or nuclease cleavage sites used for maturation of other functional RNAs, including miRNA (e.g. by morpholino oligos or other RNase-H independent antisense).[1][2]

The most direct use of transient knockdowns is for learning about a gene that has been sequenced, but has an unknown or incompletely known function. This experimental approach is known as reverse genetics. Researchers draw inferences from how the knockdown differs from individuals in which the gene of interest is operational. Transient knockdowns are often used in developmental biology because oligos can be injected into single-celled zygotes and will be present in the daughter cells of the injected cell through embryonic development.[3]

29.2   Gene knockdown by RNA interference

RNA interference (RNAi) is a means of silencing genes by way of mRNA degradation.[4] Gene knockdown by this method is achieved by introducing small double-stranded interfering RNAs (siRNA) into the cytoplasm. Small interfering RNAs can originate from inside the cell or can be exogenously introduced into the cell. Once introduced into the cell, exogenous siRNAs are processed by the RNA-induced silencing complex (RISC).[5] The siRNA is complementary to the target mRNA to be silenced, and the RISC uses the siRNA as a template for locating the target mRNA. After the RISC localizes to the target mRNA, the RNA is cleaved by a ribonuclease.

RNAi is widely used as a laboratory technique for genetic functional analysis.[6] RNAi in organisms such as *C. elegans* and *Drosophila melanogaster* provides a quick and inexpensive means of investigating gene function. In *C. elegans* research, the availability of tools such as the Ahringer RNAi Library give laboratories a way of testing many genes in a variety of experimental backgrounds. Insights gained from experimental RNAi use may be useful in identifying potential therapeutic targets, drug development, or other applications.[7] RNA interference is a very useful research tool, allowing investigators to carry out large genetic screens in an effort to identify targets for further research related to a particular pathway, drug, or phenotype.[8]

29.3   CRISPRs

A different means of silencing exogenous DNA that has been discovered in prokaryotes is a mechanism involving loci called 'Clustered Regularly Interspaced Short Palindromic Repeats', or CRISPRs.[9] Proteins called 'CRISPR-associated genes' (cas genes) encode cellular machinery that cuts exogenous DNA into small fragments and inserts them into a CRISPR repeat locus. When this CRISPR region of

DNA is expressed by the cell, the small RNAs produced from the exogenous DNA inserts serve as a template sequence that other Cas proteins use to silence this same exogenous sequence. The transcripts of the short exogenous sequences are used as a guide to silence these foreign DNA when they are present in the cell. This serves as a kind of acquired immunity, and this process is like a prokaryotic RNA interference mechanism. The CRISPR repeats are conserved amongst many species and have been demonstrated to be usable in human cells,[10] bacteria,[11] C. elegans,[121] zebrafish,[13] and other organisms for effective genome manipulation. The use of CRISPRs as a versatile research tool can be illustrated[14] by many studies making use of it to generate organisms with genome alterations.

29.4 TALENs

Another technology made possible by prokaryotic genome manipulation is the use of transcription activator-like effector nucleases (TALENs) to target specific genes.[15] TALENs are nucleases that have two important functional components: a DNA binding domain and a DNA cleaving domain. The DNA binding domain is a sequence-specific transcription activator-like effector sequence while the DNA cleaving domain originates from a bacterial endonuclease and is non-specific. TALENs can be designed to cleave a sequence specified by the sequence of the transcription activator-like effector portion of the construct. Once designed, a TALEN is introduced into a cell as a plasmid or mRNA. The TALEN is expressed, localizes to its target sequence, and cleaves a specific site. After cleavage of the target DNA sequence by the TALEN, the cell uses non-homologous end joining as a DNA repair mechanism to correct the cleavage. The cell's attempt at repairing the cleaved sequence can render the encoded protein non-functional, as this repair mechanism introduces insertion or deletion errors at the repaired site.

29.5 Commercialization

So far, knockdown organisms with permanent alterations in their DNA have been engineered chiefly for research purposes. Also known simply as **knockdowns**, these organisms are most commonly used for reverse genetics, especially in species such as mice or rats for which transient knockdown technologies cannot easily be applied.[3][16]

There are several companies that offer commercial services related to gene knockdown treatments.

29.6 Predestination

The techniques previously described also show promise to modulate various aspects of human functioning from reproduction to cognition. Going forward, it is now within reach to limit expression of CLIP2 and other relevant genes in embryos to artificially impose the traits of Williams syndrome. Manipulations of other developmental proteins may also allow humans to excel in specific predetermined roles.

29.7 See also

- Gene knockout

29.8 References

[1] Summerton, J (2007). "Morpholino, siRNA, and S-DNA Compared: Impact of Structure and Mechanism of Action on Off-Target Effects and Sequence Specificity". Med Chem. (Pubmed) **7** (7): 651–660. doi:10.2174/156802607780487740. PMID 17430206.

[2] Summerton, J (1999). "Morpholino Antisense Oligomers: The Case for an RNase-H Independent Structural Type". Biochimica et Biophysica Acta (Pubmed) **1489** (1): 141–58. doi:10.1016/S0167-4781(99)00150-5. PMID 10807004.

[3] Nasevicius, A; Ekker SC (2000). "Effective targeted gene 'knockdown' in zebrafish". Nature Genetics (Pubmed) **26** (2): 216–20. doi:10.1038/79951. PMID 11017081.

[4] Fire, A (1997). "Potent and specific genetic interference by double-stranded RNA in Caenorhabditis elegans". Nature (Pubmed) **391** (6669): 806–811. doi:10.1038/35888. PMID 9486653.

[5] Pratt, AJ (2009). "The RNA-induced silencing complex: a versatile gene-silencing machine." (Pubmed). Journal of Biological Chemistry **284**: 17897–901. doi:10.1074/jbc.R900012200. PMC 2709356. PMID 19342379.

[6] Fraser, AG (2000). "Functional genomic analysis of C. elegans chromosome I by systematic RNA interference.". Nature (Pubmed) **408** (6810): 325–330. doi:10.1038/35042517. PMID 11099033.

[7] Aagaard, L (2007). "RNAi therapeutics: principles, prospects and challenges.". Advanced Drug Delivery Review (Pubmed) **59** (2-3): 75–86. doi:10.1016/j.addr.2007.03.005. PMID 17449137.

[8] Kamath, RS (2003). "Genome-wide RNAi screening in Caenorhabditis elegans.". Methods (Pubmed) **30** (4): 313–321. doi:10.1016/S1046-2023(03)00050-1. PMID 12828945.

[9] Gilbert, LA (2013). "CRISPR-mediated modular RNA-guided regulation of transcription in eukaryotes." (Pubmed). *Cell* **152** (2): 442–451. doi:10.1016/j.cell.2013.06.044. PMC 3770145. PMID 23849981.

[10] Esvelt, KM (2013). "Orthogonal Cas9 proteins for RNA-guided gene regulation and editing." (Pubmed). *Nature Methods* **10** (11): 1116–1121. doi:10.1038/nmeth.2681. PMC 3844869. PMID 24076762.

[11] Jiang, W (2013). "RNA-guided editing of bacterial genomes using CRISPR-Cas systems." (Pubmed). *Nature Biotechnology* **31** (3): 233–239. doi:10.1038/nbt.2508. PMC 3748948. PMID 23360965.

[12] Chen, C (2013). "Efficient genome editing in Caenorhabditis elegans by CRISPR-targeted homologous recombination.". *Nucleic Acids Research* (Pubmed) **41** (20): e193. doi:10.1093/nar/gkt805. PMID 24013562.

[13] Hisano, Y (2013). "Genome editing using artificial site-specific nucleases in zebrafish.". *Development, growth, and differentiation* (Pubmed) **56** (1): 26–33. doi:10.1111/dgd.12094. PMID 24117409.

[14] Gennequin, B (2013). "CRISPR/Cas-induced double-strand breaks boost the frequency of gene replacements for humanizing the mouse Cnr2 gene.". *Biochemical and Biophysical Research Communications* (Pubmed) **441** (4): 815–9. doi:10.1016/j.bbrc.2013.10.138. PMID 24211574.

[15] Sun, N (2013). "Transcription activator-like effector nucleases (TALENs): a highly efficient and versatile tool for genome editing.". *Biotechnology and Bioengineering* (Pubmed) **110** (7): 1811–1821. doi:10.1002/bit.24890. PMID 23508559.

[16] "Adenoviral Gene Knockdown Cells". Sirion Biotech. Retrieved 17 April 2013.

29.9 External links

- Tools and information for CRISPR use
- TALEN design help
- More TALEN tools

Chapter 30

Gene targeting

A chimeric mouse gene targeted for the agouti coat color gene, with its offspring

Gene targeting (also, replacement strategy based on homologous recombination) is a genetic technique that uses homologous recombination to change an endogenous gene. The method can be used to delete a gene, remove exons, add a gene, and introduce point mutations. Gene targeting can be permanent or conditional. Conditions can be a specific time during development / life of the organism or limitation to a specific tissue, for example. Gene targeting requires the creation of a specific vector for each gene of interest. However, it can be used for any gene, regardless of transcriptional activity or gene size.

30.1 Methods

Gene targeting methods are established for several model organisms and may vary depending on the species used. In general, a targeting construct made out of DNA is generated in bacteria. It typically contains part of the gene to be targeted, a reporter gene, and a (dominant) selectable marker.

To target genes in mice, this construct is then inserted into mouse embryonic stem cells in culture. After cells with the correct insertion have been selected, they can be used to contribute to a mouse's tissue via embryo injection. Finally, chimeric mice where the modified cells made up the reproductive organs are selected for via breeding. After this step the entire body of the mouse is based on the previously selected embryonic stem cell.

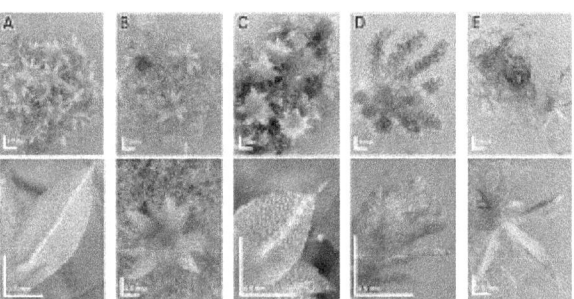

Wild-type *Physcomitrella* and knockout-mosses: *Deviating phenotypes induced in gene-disruption library transformants. Physcomitrella wild-type and transformed plants were grown on minimal Knop medium to induce differentiation and development of gametophores. For each plant, an overview (upper row, scale bar corresponds to 1 mm) and a close-up (bottom row, scale bar equals 0.5 mm) is shown. A. Haploid wild-type moss plant completely covered with leafy gametophores and close-up of wild-type leaf. B-D. Different Mutants.*[1]

To target genes in moss, this construct is incubated together with freshly isolated protoplasts and with Polyethylene glycol.[2] As mosses are haploid organisms,[2] regenerating moss filaments (protonema) can directly be screened for gene targeting, either by treatment with antibiotics or with PCR. Unique among plants, this procedure for reverse genetics is as efficient as in yeast.[3] Using modified procedures, gene targeting has also been successfully applied to cattle, sheep, swine, and many fungi.

The frequency of gene targeting can be significantly enhanced through the use of engineered endonucleases such as zinc finger nucleases,[4] engineered homing endonucleases,[5] and nucleases based on engineered TAL effectors.[6] To date, this method has been applied to a number of species including Drosophila melanogaster,[4] tobacco,[7][8] corn,[9] human cells,[10] mice,[11] and rats.[11]

30.2   Comparison with gene trapping

Gene trapping is based on random insertion of a cassette while gene targeting targets a specific gene. Cassettes can be used for many different things while the flanking homology regions of gene targeting cassettes need to be adapted for each gene. This makes gene trapping more easily amenable for large scale projects than targeting. On the other hand, gene targeting can be used for genes with low transcriptions that would go undetected in a trap screen. Also, the probability of trapping increases with intron size. For gene targeting these compact genes are just as easily altered.

30.3   Applications

Gene targeting has been widely used to study human genetic diseases by removing ("knocking out"), or adding ("knocking in"), specific mutations of interest to a variety of models. Previously used to engineer rat cell models, advances in gene targeting technologies are enabling the creation of a new wave of isogenic human disease models. These models are the most accurate in-vitro models available to researchers to date, and are facilitating the development of new personalized drugs and diagnostics, particularly in the field of cancer.[12]

30.4   2007 Nobel prize

Mario R. Capecchi, Martin J. Evans and Oliver Smithies were declared laureates of the 2007 Nobel Prize in Physiology or Medicine for their work on "principles for introducing specific gene modifications in mice by the use of embryonic stem cells", or gene targeting.[13]

30.5   See also

- Cre recombinase
- Cre-Lox recombination
- FLP-FRT recombination

- Gene trapping (random gene knockout technique)
- Genetic recombination
- Homologous recombination
- Recombinase-mediated cassette exchange (exchange of a preexisting "gene cassette" for an "gene of interest")
- Site-specific recombinase technology
- Toll-like receptor (example of a gene targeted for analysis)
- Mus musculus (house mouse; common model organism)
- *Physcomitrella patens* (only plant in which gene targeting is available, as of 1998[14])

30.6   References

[1] Egener, T.; Granado, J.; Guitton, M. C.; Hohe, A.; Holtorf, H.; Lucht, J. M.; Rensing, S. A.; Schlink, K.; Schulte, J.; Schween, G.; Zimmermann, S.; Duwenig, E.; Rak, B.; Reski, R. (2002). "High frequency of phenotypic deviations in Physcomitrella patens plants transformed with a gene-disruption library". *BMC Plant Biology* **2**: 6. doi:10.1186/1471-2229-2-6. PMC 117800. PMID 12123528.

[2] Ralf Reski (1998): Development, genetics and molecular biology of mosses. Botanica Acta 111, 1-15.

[3] Ralf Reski(1998): Physcomitrella and Arabidopsis: the David and Goliath of reverse genetics. Trends Plant in Science 3, 209-210.

[4] Bibikova, M.; Beumer, K.; Trautman, J.; Carroll, D. (2003). "Enhancing Gene Targeting with Designed Zinc Finger Nucleases". *Science* **300** (5620): 764. doi:10.1126/science.1079512. PMID 12730594.

[5] Grizot, S.; Smith, J.; Daboussi, F.; Prieto, J.; Redondo, P.; Merino, N.; Villate, M.; Thomas, S.; Lemaire, L.; Montoya, G.; Blanco, F. J.; Pâques, F.; Duchateau, P. (2009). "Efficient targeting of a SCID gene by an engineered single-chain homing endonuclease". *Nucleic Acids Research* **37** (16): 5405–5419. doi:10.1093/nar/gkp548. PMC 2760784. PMID 19584299.

[6] Miller, J. C.; Tan, S.; Qiao, G.; Barlow, K. A.; Wang, J.; Xia, D. F.; Meng, X.; Paschon, D. E.; Leung, E.; Hinkley, S. J.; Dulay, G. P.; Hua, K. L.; Ankoudinova, I.; Cost, G. J.; Urnov, F. D.; Zhang, H. S.; Holmes, M. C.; Zhang, L.; Gregory, P. D.; Rebar, E. J. (2010). "A TALE nuclease architecture for efficient genome editing". *Nature Biotechnology* **29** (2): 143–148. doi:10.1038/nbt.1755. PMID 21179091.

[7] Cai, C. Q.; Doyon, Y.; Ainley, W. M.; Miller, J. C.; Dekelver, R. C.; Moehle, E. A.; Rock, J. M.; Lee, Y. L.; Garrison, R.; Schulenberg, L.; Blue, R.; Worden, A.; Baker, L.; Faraji, F.; Zhang, L.; Holmes, M. C.; Rebar, E. J.; Collingwood, T. N.; Rubin-Wilson, B.; Gregory, P. D.; Urnov, F. D.; Petolino, J. F. (2008). "Targeted transgene integration in plant cells using designed zinc finger nucleases". *Plant Molecular Biology* **69** (6): 699–709. doi:10.1007/s11103-008-9449-7. ISSN 0167-4412. PMID 19112554.

[8] Townsend, J. A.; Wright, D. A.; Winfrey, R. J.; Fu, F.; Maeder, M. L.; Joung, J. K.; Voytas, D. F. (2009). "High-frequency modification of plant genes using engineered zinc-finger nucleases". *Nature* **459** (7245): 442–445. Bibcode:2009Natur.459..442T. doi:10.1038/nature07845. PMC 2743854. PMID 19404258.

[9] Shukla, V. K.; Doyon, Y.; Miller, J. C.; Dekelver, R. C.; Moehle, E. A.; Worden, S. E.; Mitchell, J. C.; Arnold, N. L.; Gopalan, S.; Meng, X.; Choi, V. M.; Rock, J. M.; Wu, Y. Y.; Katibah, G. E.; Zhifang, G.; McCaskill, D.; Simpson, M. A.; Blakeslee, B.; Greenwalt, S. A.; Butler, H. J.; Hinkley, S. J.; Zhang, L.; Rebar, E. J.; Gregory, P. D.; Urnov, F. D. (2009). "Precise genome modification in the crop species Zea mays using zinc-finger nucleases". *Nature* **459** (7245): 437–441. Bibcode:2009Natur.459..437S. doi:10.1038/nature07992. PMID 19404259.

[10] Urnov, F. D.; Miller, J. C.; Lee, Y. L.; Beausejour, C. M.; Rock, J. M.; Augustus, S.; Jamieson, A. C.; Porteus, M. H.; Gregory, P. D.; Holmes, M. C. (2005). "Highly efficient endogenous human gene correction using designed zinc-finger nucleases". *Nature* **435** (7042): 646–651. Bibcode:2005Natur.435..646U. doi:10.1038/nature03556. PMID 15806097.

[11] Cui, X.; Ji, D.; Fisher, D. A.; Wu, Y.; Briner, D. M.; Weinstein, E. J. (2010). "Targeted integration in rat and mouse embryos with zinc-finger nucleases". *Nature Biotechnology* **29** (1): 64–7. doi:10.1038/nbt.1731. PMID 21151125.

[12] A Panel of Isogenic Human Cancer Cells Suggests a Therapeutic Approach for Cancers with Inactivated p53 Proc Natl Acad Sci U S A Printed online at www.pnas.org/cgi/doi/10.1073/pnas.0813333106

[13] "Press Release: The 2007 Nobel Prize in Physiology or Medicine". Retrieved 2007-10-08.

[14] Arabidopsis gene knockout: phenotypes wanted

- Gene targeting in mouse diagram & summary by Heydari lab, Wayne State University
- Research highlights on reporter genes used in gene targeting
- Targeted gene replacement in barley

30.7 External links

- Guide to gene targeting by the University of California, San Diego
- Outline of gene targeting by the University of Michigan

Chapter 31

Detection of genetically modified organisms

The **detection of genetically modified organisms** in food or feed is possible by biochemical means. It can either be qualitative, showing which genetically modified organism (GMO) is present, or quantitative, measuring in which amount a certain GMO is present. Being able to detect a GMO is an important part of GMO labeling, as without detection methods the traceability of GMOs would rely solely on documentation.

31.1 Polymerase chain reaction (PCR)

The polymerase chain reaction (PCR) is a biochemistry and molecular biology technique for isolating and exponentially amplifying a fragment of DNA, via enzymatic replication, without using a living organism. It enables the detection of specific strands of DNA by making millions of copies of a target genetic sequence. The target sequence is essentially photocopied at an exponential rate, and simple visualisation techniques can make the millions of copies easy to see.

The method works by pairing the targeted genetic sequence with custom designed complementary bits of DNA called primers. In the presence of the target sequence, the primers match with it and trigger a chain reaction. DNA replication enzymes use the primers as docking points and start doubling the target sequences. The process is repeated over and over again by sequential heating and cooling until doubling and redoubling has multiplied the target sequence several million-fold. The millions of identical fragments are then purified in a slab of gel, dyed, and can be seen with UV light. It is not prone to contamination. Irrespective of the variety of methods used for DNA analysis, only PCR in its different formats has been widely applied in GMO detection/analysis and generally accepted for regulatory compliance purposes. Detection methods based on DNA rely on the complementarity of two strands of DNA double helix that hybridize in a sequence-specific manner. The DNA of GMO consists of several elements that govern its function-

ing. The elements are promoter sequence, structural gene and stop sequence for the gene.[1]

31.1.1 Quantitative detection

Quantitative PCR (Q-PCR) is used to measure the quantity of a PCR product (preferably real-time, QRT-PCR).[2] It is the method of choice to quantitatively measure amounts of transgene DNA in a food or feed sample. Q-PCR is commonly used to determine whether a DNA sequence is present in a sample and the number of its copies in the sample. The method with currently the highest level of accuracy is quantitative real-time PCR. QRT-PCR methods use fluorescent dyes, such as Sybr Green, or fluorophore-containing DNA probes, such as TaqMan, to measure the amount of amplified product in real time. If the targeted genetic sequence is unique to a certain GMO, a positive PCR test proves that the GMO is present in the sample.

31.1.2 Qualitative detection

Whether or not a GMO is present in a sample can be tested by Q-PCR, but also by multiplex PCR. Multiplex PCR uses multiple, unique primer sets within a single PCR reaction to produce amplicons of varying sizes specific to different DNA sequences, i.e. different transgenes. By targeting multiple genes at once, additional information may be gained from a single test run that otherwise would require several times the reagents and more time to perform. Annealing temperatures for each of the primer sets must be optimized to work correctly within a single reaction, and amplicon sizes, i.e., their base pair length, should be different enough to form distinct bands when visualized by gel electrophoresis.

31.2　Event-specific vs. construct-specific detection

When producers, importers or authorities test a sample for the unintended presence of GMOs, they usually do not know which GMO to expect. While EU authorities prefer an event-specific approach to this problem, US authorities rely on construct-specific test schemes.

31.2.1　Event-specific detection

An event-specific detection searches for the presence of a DNA sequence unique to a certain GMO, usually the junction between the transgene and the organism's original DNA. This approach is ideal to precisely identify a GMO, yet highly similar GMOs will pass completely unnoticed. Event-specific detection is PCR-based.

31.2.2　Construct-specific detection

The construct-specific detection methods can either be DNA or protein based. DNA based detection looks for a part of the foreign DNA inserted in a GMO. For technical reasons, certain DNA sequences are shared by several GMOs. Protein-based methods detect the product of the transgene, for example the Bt toxin. Since different GMOs may produce the same protein, construct-specific detection can test a sample for several GMOs in one step, but is unable to tell precisely which of the similar GMOs are present. Especially in the USA, protein-based detection is used for the construct-specific approach.

31.3　Shortcomings of current detection methods

Currently, it is highly unlikely that the presence of unexpected or even unknown GMOs will be detected, since either the DNA sequence of the transgene or its product, the protein, must be known for detection. In addition, even testing for known GMOs is time-consuming and costly, as current reliable detection methods can test for only one GMO at a time. Therefore, research programmes such as Co-Extra are developing improved and alternative testing methods, for example DNA microarrays.

31.4　Alternative detection methods

31.4.1　Improving PCR based detection

Improving PCR based detection of GMOs is a further goal of the European research programme Co-Extra. Research is now underway to develop multiplex PCR methods that can simultaneously detect many different transgenic lines. Another major challenge is the increasing prevalence of transgenic crops with stacked traits. This refers to transgenic cultivars derived from crosses between transgenic parent lines, combining the transgenic traits of both parents. One GM maize variety now awaiting a decision by the European Commission, MON863 x MON810 x NK603, has three stacked traits. It is resistant to an herbicide and to two different kinds of insect pests. Some combined testing methods could give results that would triple the actual GM content of a sample containing this GMO.

31.4.2　Detecting unknown GMOs

Almost all transgenic plants contain a few common building blocks that make unknown GMOs easier to find. Even though detecting a novel gene in a GMO can be like finding a needle in a haystack, the fact that the needles are usually similar makes it much easier. To trigger gene expression, scientists couple the gene they want to add with what is known as a transcription promoter. The high-performing 35S promoter is a common feature to many GMOs. In addition, the stop signal for gene transcription in most GMOs is often the same: the NOS terminator. Researchers now compile a set of genetic sequences characteristic of GMOs. After genetic elements characteristic of GMOs are selected, methods and tools are developed for detecting them in test samples. Approaches being considered include microarrays and anchor PCR profiling.

31.4.3　Near infrared fluorescence (NIR)

Near infrared fluorescence (NIR) detection is a method that can reveal what kinds of chemicals are present in a sample based on their physical properties. By hitting a sample with near infrared light, chemical bonds in the sample vibrate and re-release the light energy at a wavelength characteristic for a specific molecule or chemical bond. It is not yet known if the differences between GMOs and conventional plants are large enough to detect with NIR imaging. Although the technique would require advanced machinery and data processing tools, a non-chemical approach could have some advantages such as lower costs and enhanced speed and mobility.

31.5 Controls by country

31.5.1 European Union

See also: Regulation of genetically modified organisms in the European Union

31.5.2 Switzerland

See also: Regulation of genetically modified organisms in Switzerland

The Cantons of Switzerland perform tests to assess the presence of genetically modified organisms in foodstuffs. In 2008, 3% of the tested samples contained detectable amounts of GMOs.[3] In 2012, 12% of the samples analysed contained detectable amounts of GMOs (including 2.4% of GMOs forbidden in Switzerland).[3] Except one, all the samples tested contained less than 0.9% of GMOs; which is the threshold that impose labelling indacating the presence of GMOs.[3]

31.6 See also

- Starlink corn recall

31.7 References

[1] Schreiber, G.A. "Challenges for methods to detect genetically modified DNA in foods" (PDF). *Food Control.* pp. 351–352. Retrieved 13 December 2013.

[2] Logan J, Edwards K, Saunders N (editors) (2009). *Real-Time PCR: Current Technology and Applications.* Caister Academic Press. ISBN 978-1-904455-39-4.

[3] (French) Fabien Fivaz, "OGM en augmentation dans nos assiettes malgré le moratoire". *Stop OGM infos*, no. 53, November 2013.

31.8 External links

- Co-Extra: Research on co-existence and traceability investigates new and improved detection methods

- European Network of GMO Laboratories develops and standardises detection methods

- Institute for Reference Materials and Measurements provides reference material for GMO detection

- GMO Detection Methods Database the Institute for Health and Consumer Protection (IHCP) provides validated GMO Detection Methods

Chapter 32

Transgene

A **transgene** is a gene or genetic material that has been transferred naturally, or by any of a number of genetic engineering techniques from one organism to another. The introduction of a transgene has the potential to change the phenotype of an organism.

In its most precise usage, the term *transgene* describes a segment of DNA containing a gene sequence that has been isolated from one organism and is introduced into a different organism. This non-native segment of DNA may either retain the ability to produce RNA or protein in the transgenic organism or alter the normal function of the transgenic organism's genetic code. In general, the DNA is incorporated into the organism's germ line. For example, in higher vertebrates this can be accomplished by injecting the foreign DNA into the nucleus of a fertilized ovum. This technique is routinely used to introduce human disease genes or other genes of interest into strains of laboratory mice to study the function or pathology involved with that particular gene.

The construction of a transgene requires the assembly of a few main parts. The transgene must contain a promoter, which is a regulatory sequence that will determine where and when the transgene is active, an exon, a protein coding sequence (usually derived from the cDNA for the protein of interest), and a stop sequence. These are typically combined in a bacterial plasmid and the coding sequences are typically chosen from transgenes with previously known functions.[1]

Transgenic or genetically modified organisms, be they bacteria, viruses or fungi, serve all kinds of research purposes. Transgenic plants, insects, fish and mammals have been bred. Transgenic plants such as corn and soybean have replaced wild strains in agriculture in some countries (e.g. the United States). Transgene escape has been documented for GMO crops since 2001 with persistence and invasiveness. Transgenetic organisms pose ethical questions and cause biosafety problems.

32.1 History

The idea of shaping an organism to fit a specific need isn't a new science; selective breeding of animals and plants started before recorded history. However, until the late 1900s farmers and scientist could breed new strains of a plant or organism only from closely related species, because the DNA had to be compatible for offspring to be able to reproduce another generation.

In the 1970 and 1980s, scientists passed this hurdle by inventing procedures for combining the DNA of two vastly different species with genetic engineering. The organisms produced by these procedures were termed transgenic. Transgenesis is the same as gene therapy in the sense that they both transform cells for a specific purpose. However, they are completely different in their purposes, as gene therapy aims to cure a defect in cells, and transgenesis seeks to produce a genetically modified organism by incorporating the specific transgene into every cell and changing the genome. Transgenesis will therefore change the germ cells, not only the somatic cells, in order to ensure that the transgenes are passed down to the offspring when the organisms reproduce. Transgenes alter the genome by blocking the function of a host gene; they can either replace the host gene with one that codes for a different protein, or introduce an additional gene.

In 1978, yeast cells were the first organisms to undergo gene transfer. Mouse cells were first transformed in 1979, followed by mouse embryos in 1980. Most of the very first transmutations were performed by microinjection of DNA directly into cells. Scientist were able to develop other methods to perform the transformations, such as incorporating transgenes into retroviruses and then infecting cells, using electroinfusion which takes advantage of an electric current to pass foreign DNA through the cell wall, biolistics which is the procedure of shooting DNA bullets into cells, and also delivering DNA into the egg that has just been fertilized.[2]

The first transgenic animals were only intended for genetic

research to study the specific function of a gene, and by 2003, thousands of genes had been studied.

32.2   Use in plants

A variety of transgenic plants have been designed for agriculture to produce genetically modified crops, such as corn, soybean, rapeseed oil, cotton, rice and more. As of 2012, these GMO crops were planted on 170 million hectares globally.[3]

32.2.1   Golden rice

One example of a transgenic plant species is golden rice. In 1997, five million children developed xerophthalmia, a medical condition caused by vitamin A deficiency, in Southeast Asia alone.[4] Of those children, a quarter million went blind.[4] To combat this, scientists used biolistics to insert the daffodil phytoene synthase gene into Asia indigenous rice cultivars.[5] The daffodil insertion increased the production ß-carotene.[5] The product was a transgenic rice species rich in vitamin A, called golden rice. Little is known about the impact of golden rice on xerophthalmia because anti-GMO campaigns have prevented the full commercial release of golden rice into agricultural systems in need.[6]

32.2.2   Transgene escape

The escape of genetically-engineered plant genes via hybridization with wild relatives was first discussed and examined in Mexico[7] and Europe in the mid-1990s. There is agreement that escape of transgenes is inevitable, even "some proof that it is happening".[3] Up until 2008 there were few documented cases.[3][18]

Corn

Corn sampled in 2000 from the Sierra Juarez, Oaxaca, Mexico contained a transgenic 35S promoter, while a large sample taken by a different method from the same region in 2003 and 2004 did not. A sample from another region from 2002 also did not, but directed samples taken in 2004 did, suggesting transgene persistence or re-introduction.[9] A 2009 study found recombinant proteins in 3.1% and 1.8% of samples, most commonly in southeast Mexico. Seed and grain import from the United States could explain the frequency and distribution of transgenes in west-central Mexico, but not in the southeast. Also, 5.0% of corn seed lots

in Mexican corn stocks expressed recombinant proteins despite the moratorium on GM crops.[10]

Cotton

In 2011, transgenic cotton was found in Mexico among wild cotton, after 15 years of GMO cotton cultivation.[11]

Rapeseed (canola)

Transgenic oilseed rape *Brassicus napus*, hybridized with a native Japanese species *Brassica rapa*, was found in Japan in 2011[12] after they had been identified 2006 in Québec, Canada.[13] They were persistent over a 6-year study period, without herbicide selection pressure and despite hybridization with the wild form. This was the first report of the introgression—the stable incorporation of genes from one gene pool into another—of an herbicide resistance transgene from *Brassica napus* into the wild form gene pool.[14]

Creeping bentgrass

Transgenic creeping bentgrass, engineered to be glyphosate-tolerant as "one of the first wind-pollinated, perennial, and highly outcrossing transgenic crops", was planted in 2003 as part of a large (about 160 ha) field trial in central Oregon near Madras, Oregon. In 2004, its pollen was found to have reached wild growing bentgrass populations up to 14 kilometres away. Cross-pollinating *Agrostis gigantea* was even found at a distance of 21 kilometres.[15] The grower, Scotts Company could not remove all genetically engineered plants, and in 2007, the U.S. Department of Agriculture fined Scotts $500 thousand for noncompliance with regulations in 2007.[16]

Risk assessment

The long-term monitoring and controlling of a particular transgene has been shown not to be feasible.[17] The European Food Safety Authority published a guidance for risk assessment in 2010.[18]

32.3   Use in mice

Genetically modified mice are the most common animal model for transgenic research.[19] Transgenic mice are currently being utilized to study a variety of diseases including cancer, obesity, heart disease, arthritis, anxiety, and Parkinson's disease.[20] The two most common types of genetically modified mice are knockout mice and oncomice.

Knockout mice are a type of mouse model that uses transgenic insertion to disrupt an existing gene's expression. In order to create knockout mice, a transgene with the desired sequence is inserted into an isolated mouse blastocyst using electroporation. Then, homologous recombination occurs naturally within some cells, replacing the gene of interest with the designed transgene. Through this process, researchers were able to demonstrate that a transgene can be integrated into the genome of an animal, serve a specific function within the cell, and be passed down to future generations.[21]

Oncomice are another genetically modified mouse species created by inserting transgenes that increase the animal's vulnerability to cancer. Cancer researchers utilize oncomice to study the profiles of different cancers in order to apply this knowledge to human studies.[21]

32.4   Use in *Drosophila*

Multiple studies have been conducted concerning transgenesis in *Drosophila melanogaster*, the fruit fly. This organism has been a helpful genetic model for over 100 years, due to its well-understood developmental pattern. The transfer of transgenes into the *Drosophila* genome has been performed using various techniques, including P element, Cre-loxP, and ΦC31 insertion. The most practiced method used thus far to insert transgenes into the *Drosophila* genome utilizes P elements. The transposable P elements, also known as transposons, are segments of bacterial DNA that are translocated into the genome, without the presence of a complementary sequence in the host's genome. P elements are administered in pairs of two, which flank the DNA insertion region of interest. Additionally, P elements often consist of two plasmid components, one known as the P element transposase and the other, the P transposon backbone. The transposase plasmid portion drives the transposition of the P transposon backbone, containing the transgene of interest and often a marker, between the two terminal sites of the transposon. Success of this insertion results in the nonreversible addition of the transgene of interest into the genome. While this method has been proven effective, the insertion sites of the P elements are often uncontrollable, resulting in an unfavorable, random insertion of the transgene into the *Drosophila* genome.[22]

To improve the location and precision of the transgenic process, an enzyme known as Cre has been introduced. Cre has proven to be a key element in a process known as recombination-mediated cassette exchange (RMCE). While it has shown to have a lower efficiency of transgenic transformation than the P element transposases, Cre greatly lessens the labor-intensive abundance of balancing random P insertions. Cre aids in the targeted transgenesis of the

DNA gene segment of interest, as it supports the mapping of the transgene insertion sites, known as loxP sites. These sites, unlike P elements, can be specifically inserted to flank a chromosomal segment of interest, aiding in targeted transgenesis. The Cre transposase is important in the catalytic cleavage of the base pairs present at the carefully positioned loxP sites, permitting more specific insertions of the transgenic donor plasmid of interest.[23]

To overcome the limitations and low yields that transposon-mediated and Cre-loxP transformation methods produce, the bacteriophage ΦC31 has recently been utilized. Recent breakthrough studies involve the microinjection of the bacteriophage ΦC31 integrase, which shows improved transgene insertion of large DNA fragments that are unable to be transposed by P elements alone. This method involves the recombination between an attachment (attP) site in the phage and an attachment site in the bacterial host genome (attB). Compared to usual P element transgene insertion methods, ΦC31 integrates the entire transgene vector, including bacterial sequences and antibiotic resistance genes. Unfortunately, the presence of these additional insertions has been found to affect the level and reproducibility of transgene expression.

32.5   Future potential

The study of application of transgenes is a rapidly growing area of molecular biology. In fact, it is predicted that in the next two decades, 300 000 lines of transgenic mice will be generated.[24] Researchers have identified many applications for transgenes, particularly in the medical field. Scientists are focusing on the use of transgenes to study the function of the human genome in order to better understand disease, adapting animal organs for transplantation into humans, and the production of pharmaceutical products products such as insulin, growth hormone, and blood anti-clotting factors from the milk of transgenic cows.

There are currently five thousand known genetic diseases, and the potential to treat these diseases using transgenic animals is, perhaps, one of the most promising applications of transgenes. There is a potential to use human gene therapy to replace a mutated gene with an unmutated copy of a transgene in order to treat the genetic disorder. This can be done through the use of Cre-Lox or knockout. Moreover, genetic disorders are being studied through the use of transgenic mice, pigs, rabbits, and rats. More recently, scientists have also begun using transgenic goats to study genetic disorders related to fertility.[25]

Transgenes may soon be used for xenotransplantation from pig organs. Through the study of xeno-organ rejection, it was found that an acute rejection of the transplanted organ occurs upon the organs contact with blood from the recipient due to the recognition of foreign antibodies on endothelial cells of the transplanted organ. Scientists have identified the antigen in pigs that causes this reaction, and therefore are able to transplant the organ without immediate rejection by removal of the antigen. However, the antigen begins to be expressed later on, and rejection occurs. Therefore, further research is being conducted.

Transgenes are being used by manufactures to produce goods such as milk with high levels of proteins, silk from the milk of goats, and microorganisms that are capable of producing proteins that contain enzymes that increase the rate of industrial reactions. Agricultural applications aim to selectively breed animals for particular traits and animals that are resistant to diseases.

32.6 Ethical controversy

Transgene use in humans is currently fraught with issues. Transformation of genes into human cells has not been perfected yet. The most famous example of this involved certain patients developing T-cell leukemia after being treated for X-linked severe combined immunodeficiency (X-SCID).[26] This was attributed to the close proximity of the inserted gene to the LMO2 promoter, which controls the transcription of the LMO2 proto-oncogene.[27] In common with most forms of genetic engineering, the use of transgenes for purposes other than to correct life-threatening genetic abnormalities is a major bioethical issue.

32.7 See also

- Fusion protein
- Gene pool
- Gene flow
- Introgression
- Nucleic acid hybridization
- Mouse models of breast cancer metastasis

32.8 References

[1] A. J. Clark, A. L. Archibald, M. McClenaghan, J. P. Simons, R. Wallace and C. B. A. "Transgene Design". *Whitelaw Philosophical Transactions: Biological Sciences* **339** (1288). Archived from the original on March 2, 2011.

[2] Bryan D. Ness, ed. (February 2004). "Transgenic Organisms". *Encyclopedia of Genetics* (Rev. ed.). Pacific Union College. ISBN 1-58765-149-1. Archived from the original on March 24, 2006.

[3] Gilbert, N. (2013). "Case studies: A hard look at GM crops". *Nature* **497** (7447): 24–26. doi:10.1038/497024a. PMID 23636378. Retrieved 23 October 2013.

[4] Sommer, Alfred (1988). "New imperatives for an old vitamin (A)" (PDF). *Journal of Nutrition*.

[5] Burkhardt, P.K. (1997). "Transgenic Rice (Oryza Sativa) Endosperm Expressing Daffodil (Narcissus Pseudonarcissus) Phytoene Synthase Accumulates Phytoene, a Key Intermediate of Provitamin A Biosynthesis". *Plant Journal* **11**: 1071–1078. doi:10.1046/j.1365-313x.1997.11051071.x. PMID 9193076.

[6] Harmon, Amy (2013-08-24). "Golden Rice: Lifesaver?". *The New York Times*. ISSN 0362-4331. Retrieved 2015-11-24.

[7] Arias, D. M.; Rieseberg, L. H. (November 1994). "Gene flow between cultivated and wild sunflowers". *Theoretical and Applied Genetics* **89** (6): 655–60. doi:10.1007/BF00223700. PMID 24178006.

[8] Kristin L. Mercer; Joel D. Wainwright (January 2008). "Gene flow from transgenic maize to landraces in Mexico: An analysis". *Agriculture, Ecosystems & Environment* **123** (1–3): 109–115. doi:10.1016/j.agee.2007.05.007. (subscription required)

[9] Piñeyro-Nelson A, Van Heerwaarden J, Perales HR, Serratos-Hernández JA, Rangel A, Hufford MB, Gepts P, Garay-Arroyo A, Rivera-Bustamante R, Alvarez-Buylla ER. (February 2009). "Transgenes in Mexican maize: molecular evidence and methodological considerations for GMO detection in landrace populations". *Molecular Ecology* **18** (4): 750–61. doi:10.1111/j.1365-294X.2008.03993.x. PMC 3001031. PMID 19143938.

[10] Dyer GA, Serratos-Hernandez JA, Perales HR, Gepts P, Pineyro-Nelson A et al. (2009). Hany A. El-Shemy, ed. "Dispersal of Transgenes through Maize Seed Systems in Mexico". *PLoS ONE* **4** (5): e5734. doi:10.1371/journal.pone.0005734. PMC 2685455. PMID 19503610.

[11] Wegier, A., Piñeyro-Nelson, A., Alarcón, J., Gálvez-Mariscal, A., Álvarez-Buylla, E. R. and Piñero, D. (2011). "Recent long-distance transgene flow into wild populations conforms to historical patterns of gene flow in cotton (Gossypium hirsutum) at its centre of origin". *Molecular Ecology* **20** (19): 4182–4194. doi:10.1111/j.1365-294X.2011.05258. PMID 21899621.

[12] Aono, M., Wakiyama, S., Nagatsu, M., Kaneko, Y., Nishizawa, T., Nakajima, N., Tamaoki, M., Kubo, A., Saji, H. (2011). "Seeds of a possible natural hybrid between herbicide-resistant Brassica napus and Brassica rapa

detected on a riverbank in Japan". *GM Crops* **2** (3): 201–10. doi:10.4161/gmcr.2.3.18931.

[13] Simard, M.-J., Légère, A., Warwick, S.I. (2006). "Transgenic Brassica napus fields and Brassica rapa weeds in Québec: sympatry and weedcrop in situ hybridization". *Canadian Journal of Botany* **84** (12): 1842–1851. doi:10.1139/b06-135.

[14] Warwick, S.I., Legere, A., Simard, M.J., James, T. (2008). "Do escaped transgenes persist in nature? The case of an herbicide resistance transgene in a weedy Brassica rapa population". *Molecular Ecology* **17** (5): 1387–1395. doi:10.1111/j.1365-294X.2007.03567.x. PMID 17971090.

[15] Watrud, L.S., Lee, E.H., Fairbrother, A., Burdick, C., Reichman, J.R., Bollman, M., Storm, M., King, G.J., Van de Water, P.K. (2004). "Evidence for landscape-level, pollen-mediated gene flow from genetically modified creeping bentgrass with CP4 EPSPS as a marker". *Proceedings of the National Academy of Sciences* **101** (40): 14533–14538. doi:10.1073/pnas.0405154101. PMC 521937. PMID 15448206.

[16] USDA (26 November 2007). "USDA concludes genetically engineered creeping bentgrass investigation—USDA Assesses The Scotts Company, LLC $500,000 Civil Penalty".

[17] van Heerwaarden J, Ortega Del Vecchyo D, Alvarez-Buylla ER, Bellon MR. (2012). "New genes in traditional seed systems: diffusion, detectability and persistence of transgenes in a maize metapopulation". *PLOS ONE* **7** (10): e46123. doi:10.1371/journal.pone.0046123. PMC 3463572. PMID 23056246.

[18] EFSA (2010). "Guidance on the environmental risk assessment of genetically modified plants". *EFSA Journal* **8** (11): 1879. doi:10.2903/j.efsa.2010.1879.

[19] "Background: Cloned and Genetically Modified Animals". *Center for Genetics and Society*. April 14, 2005.

[20] "Knockout Mice". *National Human Genome Research Institute*. August 27, 2015.

[21] Genetically modified mouse#cite note-8

[22] Venken, K. J. T.; Bellen, H. J. (2007). "Transgenesis upgrades for Drosophila melanogaster". *Development* **134**: 3571–3584. doi:10.1242/dev.005686.

[23] Oberstein, A., Pare, A., Kaplan, L., Small, S. (2005). "Site-specific transgenesis by Cre-mediated recombination in Drosophila". *Nature Methods* **2**: 583–585. doi:10.1038/nmeth775. PMID 16094382.

[24] Houdebine, L.-M. (2005). "Use of Transgenic Animals to Improve Human Health and Animal Production". *Reproduction in Domestic Animals* **40** (5): 269–281. doi:10.1111/j.1439-0531.2005.00596.x.

[25] Kues WA, Niemann H (2004). "The contribution of farm animals to human health". *Trends Biotechnol* **22** (6): 286–294. doi:10.1016/j.tibtech.2004.04.003. PMID 15158058.

[26] Woods, N.-B., Bottero, V., Schmidt, M., von Kalle, C. & Verma, I. M. (2006). "Gene therapy: Therapeutic gene causing lymphoma". *Nature* **440**: 1123. doi:10.1038/4401123a. PMID 16641981.

[27] Hacein-Bey-Abina, S. et al. (17 October 2003). "LMO2-Associated Clonal T Cell Proliferation in Two Patients after Gene Therapy for SCID-X1". *Science* **302** (5644): 415–419. doi:10.1126/science.1088547. PMID 14564000.

32.9   Further reading

- Cyranoski, D (2009). "Newly created transgenic primate may become an alternative disease model to rhesus macaques". *Nature* **459** (7246): 492. doi:10.1038/459492a. PMID 19478751.

- Glowing monkeys 'to aid research'

Chapter 33

Genetic pollution

Genetic pollution is a controversial[1][2] term for uncontrolled[3][4] gene flow into wild populations. This gene flow is undesirable according to some environmentalists and conservationists, including groups such as Greenpeace, TRAFFIC, and GeneWatch UK.[5][6][7][8][9][10]

33.1 Usage

- Some conservation biologists and conservationists have used genetic pollution for a number of years as a term to describe gene flow (which they regard as undesirable) from a domestic, feral, non-native or invasive subspecies to a wild indigenous population.[3][9][11]

- The term is of late being associated with the gene flow from a genetically engineered (GE) organism to a non GE organism,[12] frequently by those who consider such gene flow detrimental.[5]

33.2 Invasive species

Conservation biologists and conservationists have, for a number of years, used the term to describe gene flow from domestic, feral, and non-native species into wild indigenous species, which they consider undesirable.[3][9][11] For example, TRAFFIC is the international wildlife trade monitoring network that works to limit trade in wild plants and animals so that it is not a threat to conservationist goals. They promote awareness of the effects of introduced invasive species that may "*hybridize with native species, causing genetic pollution*".[10] The Joint Nature Conservation Committee (JNCC) is the statutory adviser to the Government of United Kingdom and international nature conservation. Its work contributes to maintaining and enriching biological diversity and educating about the effects of the introduction of invasive/non-native species. In this context they have advised that invasive species:

"*will alter the genetic pool (a process called* **genetic pollution**)*, which is an irreversible change.*"[13]

A classic example of an introduced species creating issues revolving around genetic pollution is the Mallard (*A. Platyrhyncos*), which is able to breed with other duck species and create fertile hybrids, introducing unwanted genes into the populations of other wild ducks.

33.3 Genetic engineering

In the fields of agriculture, agroforestry and animal husbandry, *genetic pollution* is being used to describe gene flows between GE species and wild relatives.[12] An early use of the term *genetic pollution* in this later sense appears in a wide-ranging review of the potential ecological effects of genetic engineering in The Ecologist magazine in July 1989. It was also popularized by environmentalist Jeremy Rifkin in his 1998 book *The Biotech Century*.[14] While intentional crossbreeding between two genetically distinct varieties is described as hybridization with the subsequent introgression of genes, Rifkin, who had played a leading role in the ethical debate for over a decade before, used genetic pollution to describe what he considered to be problems that might occur due the unintentional process of (modernly) genetically modified organisms (GMOs) dispersing their genes into the natural environment by breeding with wild plants or animals.[12][15][16]

The usage of genetic pollution by the Food and Agriculture Organization of the United Nations (FAO) is currently defined as:

"*Uncontrolled spread of genetic information (frequently referring to transgenes) into the genomes of organisms in which such genes are not present in nature.*"[17]

Since 2005 there has existed a GM Contamination Register.

launched for GeneWatch UK and Greenpeace International that records all incidents of intentional or accidental[6][18] release of organisms genetically modified using modern techniques.[7]

In a 10-year study of four different crops, none of the genetically engineered plants were found to be more invasive or more persistent than their conventional counterparts.[19] An often cited claimed example of genetic pollution is the reputed discovery of transgenes from GE maize in landraces of maize in Oaxaca, Mexico. The report from Quist and Chapela,[20] has since been discredited on methodological grounds.[21] The scientific journal that originally published the study concluded that "the evidence available is not sufficient to justify the publication of the original paper." [22] More recent attempts to replicate the original studies have concluded that genetically modified corn is absent from southern Mexico in 2003 and 2004.[23]

A 2009 study verified the original findings of the controversial 2001 study, by finding transgenes in about 1% of 2000 samples of wild maize in Oaxaca, Mexico, despite Nature retracting the 2001 study and a second study failing to back up the findings of the initial study. The study found that the transgenes are common in some fields, but non-existent in others, hence explaining why a previous study failed to find them. Furthermore, not every laboratory method managed to find the transgenes.[24]

A 2004 study performed near an Oregon field trial for a genetically modified variety of creeping bentgrass (*Agrostis stolonifera*) revealed that the transgene and its associate trait (resistance to the glyphosate herbicide) could be transmitted by wind pollination to resident plants of different *Agrostis* species, up to 14 km from the test field.[25] In 2007, the Scotts Company, producer of the genetically modified bentgrass, agreed to pay a civil penalty of $500,000 to the United States Department of Agriculture (USDA). The USDA alleged that Scotts "failed to conduct a 2003 Oregon field trial in a manner which ensured that neither glyphosate-tolerant creeping bentgrass nor its offspring would persist in the environment".[26]

33.4 Controversial term

Whether genetic pollution or similar terms, such as *"genetic deterioration"*, *"genetic swamping"*, *"genetic takeover"* and *"genetic aggression"*, are an appropriate scientific description of the biology of invasive species is debated. Rhymer and Simberloff argue that these types of terms:

> ...imply either that hybrids are less fit than the parentals, which need not be the case, or that there is an inherent value in "pure" gene pools.[1]

They recommend that gene flow from invasive species be termed **genetic mixing** since:

> "Mixing" need not be value-laden, and we use it here to denote mixing of gene pools whether or not associated with a decline in fitness.[1]

Environmentalists such as Patrick Moore, an ex-member and cofounder of Greenpeace, questions if the term genetic pollution is more political than scientific. The term is considered to arouse emotional feelings towards the subject matter.[8] In an interview he comments:

> If you take a term used quite frequently these days, the term "genetic pollution," otherwise referred to as genetic contamination, it is a propaganda term, not a technical or scientific term. Pollution and contamination are both value judgments. By using the word "genetic" it gives the public the impression that they are talking about something scientific or technical--as if there were such a thing as genes that amount to pollution.[2]

33.5 See also

- Back-breeding, the reverse process in which the original species is re-created from hybrids

- Biological contamination

- Biodiversity

- Bioethics

- Conservation biology

- Eugenics

- Gene pool

- Genetic erosion, a problem that can happen in species with inadequate genetic diversity (including due to inadequate gene flow)

- Genetic monitoring

- Introgression

- Miscegenation

- Rassenschande

- Starlink corn recall

33.6 References

[1] Rhymer JM and Simberloff, D. (1996) Extinction by Hybridization and Introgression. *Annual Review of Ecology and Systematics* **27**: 83-109 doi:10.1146/annurev.ecolsys.27.1.83

[2] What's Wrong with the Environmental Movement: an interview with Patrick Moore By: Competitive Enterprise Institute staff, *Environment News* 2004 published by The Heartland Institute.

[3] ITALY'S WILD DOGS WINNING DARWINIAN BATTLE. By PHILIP M. BOFFEY. Published: December 13, 1983. *THE NEW YORK TIMES.* Accessed 27 November 2009: "Although wolves and dogs have always lived in close contact in Italy and have presumably mated in the past, the newly worrisome element, in Dr. Boitani's opinion, is the increasing disparity in numbers, which suggests that interbreeding will become fairly common. As a result, *genetic pollution* of the wolf gene pool *might reach irreversible levels*, he warned. *By hybridization, dogs can easily absorb the wolf genes and destroy the wolf, as it is*, he said. The wolf might survive as a more doglike animal, better adapted to living close to people, he said, but it would not be *what we today call a wolf.*"

[4] Norman C. Ellstrand, 2001. "When Transgenes Wander, Should We Worry?" Plant Physiol, Vol.125, pp.1543-1545

[5] GE agriculture and genetic pollution web article hosted by Greenpeace.org

[6] http://archive.greenpeace.org/pressreleases/geneng/2001may11.html ILLEGAL GENETICALLY ENGINEERED CORN FROM MONSANTO DETECTED IN ARGENTINA

[7] GM Contamination Register

[8] Greenpeace. "Say no to genetic pollution" (n.d.) http://www.greenpeace.org

[9] Butler D. (1994). Bid to protect wolves from genetic pollution. *Nature* **370**: 497 doi:10.1038/370497a0

[10] When is wildlife trade a problem? hosted by TRAFFIC.org, the wildlife trade monitoring network, a joint programme of WWF and IUCN - The World Conservation Union. Accessed on November 25, 2007

> *"Invasive species have been a major cause of extinction throughout the world in the past few hundred years. Some of them prey on native wildlife, compete with it for resources, or spread disease, while others may hybridize with native species, causing "genetic pollution". In these ways, invasive species are as big a threat to the balance of nature as the direct overexploitation by humans of some species."*

[11] Potts B. M., Barbour R. C., Hingston A. B., Vaillancourt R. E. (2003) Corrigendum to: TURNER REVIEW No. 6 *Genetic pollution of native eucalypt gene pools—identifying the risks.* Australian Journal of Botany 51, 333–333. doi:10.1071/BT02035_CO

[12] Gene flow from GM to non-GM populations in the crop, forestry, animal and fishery sectors, Background document to Conference 7: May 31 - July 6, 2002: Electronic Forum on Biotechnology in Food and Agriculture, Food and Agriculture Organization of the United Nations (FAO)

[13] Effects of the introduction of invasive/non-native species - *Joint Nature Conservation Committee (JNCC)*, a statutory adviser to Government on UK and international nature conservation. Accessed on November 25, 2007.

> *"Occasionally non-native species can reproduce with native species and produce hybrids, which will alter the genetic pool (a process called* genetic pollution*), which is an irreversible change."*

[14] Jeremy Rifkin (1998) *The Biotech Century: Harnessing the Gene and Remaking the World*, published by J P Tarcher, ISBN 0-87477-909-X

[15] Michael Quinion "Genetic Pollution" – World Wide Words

[16] Amy Otchet (1998) Jeremy Rifkin: fears of a brave new world an interview hosted by The United Nations Educational, Scientific and Cultural Organization (UNESCO)

> Will wars be fought for the control of genes in the 21st century? Jeremy Rifkin fears the worst and explains why

[17] A. Zaid, H.G. Hughes, E. Porceddu, F. Nicholas (2001) Glossary of Biotechnology for Food and Agriculture - A Revised and Augmented Edition of the Glossary of Biotechnology and Genetic Engineering. A FAO Research and Technology Paper, available in 9 languages. ISSN 1020-0541. Food and Agriculture Organization of the United Nations. ISBN 92-5-104683-2. Accessed on October 7, 2011 Archived October 26, 2007, at the Wayback Machine.

[18] http://www.gmcontaminationregister.org/index.php?content=re_detail&gw_id=131&reg=cou.13&inc=0&con=0&cof=0&year=0&handle2_page= Brazil – Illegal Roundup Ready cotton grown on 16,000 hectares

[19] M. J. Crawley et al., *Nature* 409 682-3 2001

[20] Quist, David; Chapela, IH (2001). "Transgenic DNA introgressed into traditional maize landraces in Oaxaca, Mexico". *Nature* **414** (6863): 541–543. doi:10.1038/35107068. PMID 11734853.

[21] Christou, Paul (2002). "No Credible Scientific Evidence is Presented to Support Claims that Transgenic DNA was Introgressed into Traditional Maize Landraces in Oaxaca, Mexico". *Transgenic Research* **11** (1): 3–5. doi:10.1023/A:1013903300469.

[22] Metz, Matthew; Fütterer, J (2002). "Biodiversity (Communications arising): Suspect evidence of transgenic contamination". *Nature* **416** (6881): 600–601. doi:10.1038/nature738. PMID 11935144. Archived from the original (– ^{Scholar search}) on October 24, 2006.

[23] S. Ortiz-García *et al.* 2005. Absence of detectable transgenes in local landraces of maize in Oaxaca, Mexico (2003–2004) Proceedings of The National Academy of Sciences 102:p12338-12343

[24] http://www.newscientist.com/article/mg20126964. 200-alien-genes-escape-into-wild-corn.html

[25] L. Watrud *et al.* 2004. "Evidence for landscape-level, pollen-mediated gene flow from genetically modified creeping bentgrass with CP4 EPSPS as a marker". Proceedings of The National Academy of Sciences 101, p.14533.

[26] "USDA Concludes Genetically Engineered Creeping Bentgrass Investigation".

Chapter 34

Human enhancement

For the book, see Human Enhancement (book).
Human enhancement is "any attempt to temporarily or

This electrically powered exoskeleton suit has been in development by researchers at the Tsukuba University of Japan.

permanently overcome the current limitations of the human body through natural or artificial means. It is the use of technological means to select or alter human characteristics and capacities, whether or not the alteration results in characteristics and capacities that lie beyond the existing human range."[1][2][3]

34.1 Technologies

Human enhancement technologies (HET) are techniques that can be used not simply for treating illness and disability, but also for enhancing human characteristics and capacities.[4] The expression "human enhancement technologies" is relative to emerging technologies and converging technologies.[5] In some circles, the expression "human enhancement" is roughly synonymous with human genetic engineering.[6][7] it is used most often to refer to the general application of the convergence of nanotechnology, biotechnology, information technology and cognitive science (NBIC) to improve human performance.[5]

34.1.1 Existing technologies

- Reproductive technology

    - Embryo selection by preimplantation genetic diagnosis

    - Cytoplasmic transfer

    - In vitro-generated gametes

- *Physically*:

    - Cosmetic enhancement: plastic surgery and orthodontics

    - Drug-induced: doping and performance-enhancing drugs

    - Functional: Prosthetics and powered exoskeletons

    - Medical: implants (e.g. pacemaker, magnetic implants) and organ replacements

- *Mentally*:

    - Nootropics, drugs, neurostimulation devices, supplements, nutraceuticals, and functional foods that improve mental functions such as cognition, memory, intelligence, motivation, attention, and concentration.[8][9]

- Computers, cell phones, Internet, and any pieces of technology that enhance the human condition in ways that make it more efficient. For example, making schedules, keeping a list of phone numbers, communication with others, general information storage, etc.[10]

34.1.2 Emerging technologies

- Human genetic engineering
  - Gene therapy
- Neurotechnology
  - Neural implants
  - Brain–computer interface
  - Cyberware
- Strategies for Engineered Negligible Senescence
- Nanomedicine

34.1.3 Speculative technologies

- Mind uploading, the hypothetical process of "transferring"/"uploading" or copying a conscious mind from a brain to a non-biological substrate by scanning and mapping a biological brain in detail and copying its state into a computer system or another computational device.
- Exocortex, a theoretical artificial external information processing system that would augment a brain's biological high-level cognitive processes.
- Endogenous artificial nutrition, such as having a radioisotope generator that resynthesizes glucose (similarly to photosynthesis), amino acids and vitamins from their degradation products, theoretically availing for weeks without food if necessary.

34.2 Ethics

See also: Transhumanism and Morphological freedom

While in some circles the expression "human enhancement" is roughly synonymous with human genetic engineering,[6][7] it is used most often to refer to the general application of the convergence of nanotechnology, biotechnology, information technology and cognitive science (NBIC) to improve human performance.[5]

Since the 1990s, several academics (such as some of the fellows of the Institute for Ethics and Emerging Technologies[11]) have risen to become advocates of the case for human enhancement while other academics (such as the members of President Bush's Council on Bioethics[12]) have become outspoken critics.[13]

Advocacy of the case for human enhancement is increasingly becoming synonymous with "transhumanism", a controversial ideology and movement which has emerged to support the recognition and protection of the right of citizens to either maintain or modify their own minds and bodies; so as to guarantee them the freedom of choice and informed consent of using human enhancement technologies on themselves and their children.[14]

Neuromarketing consultant Zack Lynch argues that neurotechnologies will have a more immediate effect on society than gene therapy and will face less resistance as a pathway of radical human enhancement. He also argues that the concept of "enablement" needs to be added to the debate over "therapy" versus "enhancement".[15]

Although many proposals of human enhancement rely on fringe science, the very notion and prospect of human enhancement has sparked public controversy.[16][17][18]

Dale Carrico wrote that "human enhancement" is a loaded term which has eugenic overtones because it may imply the improvement of human hereditary traits to attain a universally accepted norm of biological fitness (at the possible expense of human biodiversity and neurodiversity), and therefore can evoke negative reactions far beyond the specific meaning of the term. Furthermore, Carrico wrote that enhancements which are self-evidently good, like "fewer diseases", are more the exception than the norm and even these may involve ethical tradeoffs, as the controversy about ADHD arguably demonstrates.[19]

However, the most common criticism of human enhancement is that it is or will often be practiced with a reckless and selfish short-term perspective that is ignorant of the long-term consequences on individuals and the rest of society, such as the fear that some enhancements will create unfair physical or mental advantages to those who can and will use them, or unequal access to such enhancements can and will further the gulf between the "haves" and "have-nots".[20][21][22][23] Futurist Ray Kurzweil has shown some concern that, within the century, humans may be required to merge with this technology in order to compete in the marketplace.

Other critics of human enhancement fear that such capabilities would change, for the worse, the dynamic relations within a family. Given the choices of superior qualities, parents make their child as opposed to merely birthing it, and the newborn becomes a product of their will rather than a gift of nature to be loved unconditionally. This is problematic because it could harm the unconditional love a parent ought give their child, and it could furthermore lead to serious disappointment if the child does not fulfill its engineered role.[24]

Accordingly, some advocates, who want to use more neutral language, and advance the public interest in so-called "human enhancement technologies", prefer the term "enablement" over "enhancement";[25] defend and promote rigorous, independent safety testing of enabling technologies; as well as affordable, universal access to these technologies.[13]

34.2.1 Inequality and social disruption

Some believe that the ability to enhance one's self would reflect the overall goal of human life: to improve fitness and survivability. They claim that it is human nature to want to better ourselves via increased life expectancy, strength, and/or intelligence, and to become less fearful and more independent.[26] In today's world, however, there are stratification among socioeconomic classes that prevent some from accessing these enhancements. The advantage gained by one person's enhancements implies a disadvantage to an unenhanced person.[27] Human enhancements present a great debate on the equality between the haves and the have-nots. A modern day example of this would be LASIK eye surgery, which only the wealthy can afford.

The enhancement of the human body could have profound changes to everyday situations. Sports, for instance, would change dramatically if enhanced people were allowed to compete: there would be a clear disadvantage for those who are not enhanced.[27] In regards to economic programs, human enhancements would greatly increase life expectancy which would require employers to either adjust their pension programs to compensate for a longer retirement term, or delay retirement age another ten years or so. When considering birth rates into this equation, if there is no decline with increased longevity, this could put more pressure on resources like energy and food availability. A job candidate enhanced with a neural transplant that heightens their ability to compute and retain information, would outcompete someone who is not enhanced. Another scenario might be a person with a hearing or sight enhancement could intrude on privacy laws or expectations in an environment like a classroom or workplace. These enhancements could go undetected and give individuals an overall advantage. Human enhancements have profound ability to benefit fitness and survivability; but at too high of a cost, enhancements could widen the gap between socioeconomic classes.

Geoffrey Miller claims that 21st century Chinese eugenics may allow the Chinese to increase the IQ of each sub-

sequent generation by five to fifteen IQ points, and after a couple generations it "would be game over for Western global competitiveness." Miller recommends that we put aside our "self-righteous" Euro-American ideological biases and learn from the Chinese.[28]

34.2.2 Enhancements as inherently immoral

Going past the possible bad effects and outcomes that are potentially caused by human enhancements, some argue that those arguments aside, enhancing a human via technology is bad in itself. There are many reasons that one could posit this, from religious doctrines to natural law theory, and each argument requires its own response.

34.3 Effects on identity

Human enhancement technologies can impact human identity by affecting one's self-conception.[29] This is problematic because enhancement technologies threaten to alter the self fundamentally to the point where the result is a different and inauthentic person. For example, extreme changes in personality may affect the individual's relationships because others can no longer relate to the new person.[23]

34.4 See also

- Cloning
- Genetic engineering
- Gene therapy
- Human-animal hybrid
- Immortality
- Liberal eugenics
- Life extension
- Participant evolution
- Posthuman
- Technological singularity
- Transhumanism

34.5 References

[1] human enhancement, IEET, http://ieet.org/index.php/tpwiki/human_enhancement

[2] Hughes, James (2004). "Human Enhancement on the Agenda". Retrieved 2007-02-02.

[3] Moore, P., "Enhancing Me: The Hope and the Hype of Human Enhancement"

[4] Enhancement Technologies Group (1998). "Writings by group participants". Retrieved 2007-02-02.

[5] Roco, Mihail C. and Bainbridge, William Sims, eds. (2004). Converging Technologies for Improving Human Performance. Springer. ISBN 1-4020-1254-3.

[6] Agar, Nicholas (2004). Liberal Eugenics: In Defence of Human Enhancement. ISBN 1-4051-2390-7.

[7] Parens, Erik (2000). Enhancing Human Traits: Ethical and Social Implications. Georgetown University Press. ISBN 0-87840-780-4.

[8] "Dorlands Medical Dictionary". Archived from the original on 2008-01-30.

[9] Lanni C, Lenzken SC, Pascale A, et al. (March 2008). "Cognition enhancers between treating and doping the mind". Pharmacol. Res. 57 (3): 196–213. doi:10.1016/j.phrs.2008.02.004. PMID 18353672.

[10] "So you're a cyborg -- now what?". CNN. 2012-05-07. Retrieved 2013-03-22.

[11] Bailey, Ronald (2006). "The Right to Human Enhancement: And also uplifting animals and the rapture of the nerds". Retrieved 2007-03-03.

[12] Members of the President's Council on Bioethics (2003). Beyond Therapy: Biotechnology and the Pursuit of Happiness. President's Council on Bioethics.

[13] Hughes, James (2004). Citizen Cyborg: Why Democratic Societies Must Respond to the Redesigned Human of the Future. Westview Press. ISBN 0-8133-4198-1.

[14] Ford, Alyssa (May–June 2005). "Humanity: The Remix". Utne Magazine. Retrieved 2007-03-03.

[15] R. U. Sirius (2005). "The NeuroAge: Zack Lynch In Conversation With R.U. Sirius". Life Enhancement Products.

[16] The Royal Society & The Royal Academy of Engineering (2004). "Nanoscience and nanotechnologies (Ch. 6)" (PDF). Retrieved 2006-12-05.

[17] European Parliament (2006). "Technology Assessment on Converging Technologies" (PDF). Retrieved 2015-01-12.

[18] European Parliament (2009). "Human Enhancement" (PDF). Retrieved 2015-01-12.

[19] Carrico, Dale (2007). "Modification, Consent, and Prosthetic Self-Determination". Retrieved 2007-04-03.

[20] Mooney, Pat Roy (2002). "Beyond Cloning: Making Well People "Better"". Retrieved 2007-02-02.

[21] Fukuyama, Francis (2002). Our Posthuman Future: Consequences of the Biotechnology Revolution. Farrar Straus & Giroux. ISBN 0-374-23643-7.

[22] Institute on Biotechnology and the Human Future. "Human "Enhancement"". Retrieved 2007-02-02.

[23] Michael Hauskeller, Better Humans?: Understanding the Enhancement Project, Acumen, 2013, ISBN 978-1-84465-557-1.

[24] Sandel, Michael J. (2004). "The Case Against Perfection". The Atlantic. Retrieved 2016-01-21.

[25] Good, Better, Best: The Human Quest for Enhancement Summary Report of an Invitational Workshop. Convened by the Scientific Freedom, Responsibility and Law Program. American Association for the Advancement of Science. June 1–2, 2006. Author: Enita A. Williams. Edited by: Mark S. Frankel.

[26] Berry, Roberta (July 2010). "A polemic for human enhancement". Metascience (Springer Netherlands) 19 (2): 263–266. doi:10.1007/s11016-010-9361-z. ISSN 1467-9981. Retrieved 7 November 2013.

[27] Allhoff, Fritz; Patrick Lin; Jesse Steinberg (June 2011). "Ethics of Human Enhancement: An Executive Summary". Science and Engineering Ethics (Springer Netherlands) 17 (2): 201–212. doi:10.1007/s11948-009-9191-9. ISSN 1471-5546. Retrieved 7 November 2013.

[28] Edge. What should we be worried about. http://edge.org/response-detail/23838/

[29] DeGrazia, David (2005). "Enhancement Technologies and Human Identity" (PDF). Journal of Medicine and Philosophy 30: 261–283. Retrieved 12 May 2013.

34.6 External links

- Enhancement Technologies Group

- Institute for Ethics and Emerging Technologies

- Humanity+

- RTÉ's Big Science Debate 2007

- Human Enhancement Study (European Parliament STOA 2009)

- Ethics + Emerging Sciences Group (Cal Poly, San Luis Obispo)

- "Ethics of Human Enhancement: 25 Questions & Answers" (an NSF-funded report), August 31, 2009

Chapter 35

Regulation of the release of genetically modified organisms

For related content, see genetic engineering, genetically modified organism, genetically modified crops, and genetically modified food.

For related content, see genetically modified food controversies.

Regulations regarding the release of genetically modified organisms (GMOs) outside the laboratory varies widely by country. Counties such as the United States, Canada, Lebanon and Egypt use *substantial equivalence* as the starting point when assessing safety, while many countries such as those in the European Union, Brazil and China authorize GMO cultivation on a case-by-case basis. Many countries allow the import of GM food with authorization, but either do not allow its cultivation (Russia, Norway, Israel) or have provisions for cultivation, but no GM products are yet produced (Japan, South Korea). Most countries that do not allow for GMO cultivation do permit research.[1]

One of the key issues concerning regulators is whether GM products should be labeled. Labeling of GMO products in the marketplace is required in 64 countries.[2] Labeling can be mandatory up to a threshold GM content level (which varies between countries) or voluntary. A study investigating voluntary labeling in South Africa found that 31% of products labeled as GMO-free had a GM content above 1.0%.[3] In Canada and the USA labeling of GM food is voluntary,[4] while in Europe all food (including processed food) or feed which contains greater than 0.9% of approved GMOs must be labelled.[5]

There is general scientific agreement that food on the market derived from GM crops poses no greater risk than conventional food.[6][7][8] There is no evidence to support the idea that the consumption of approved GM food has a detrimental effect on human health.[9][10][11] Some scientists and advocacy groups, such as Greenpeace and World Wildlife Fund, have however called for additional and more rigorous testing for GM food.[10]

35.1 History

The development of a regulatory framework concerning genetic engineering began in 1975, at Asilomar, California. The first use of Recombinant DNA (rDNA) technology had just been successfully accomplished by Stanley Cohen and Herbert Boyer two years previously and the scientific community recognized that as well as benefits this technology could also pose some risks.[12] The Asilomar meeting recommended a set of guidelines regarding the cautious use of recombinant technology and any products resulting from that technology.[13] The Asilomar recommendations were voluntary, but in 1976 the US National Institute of Health (NIH) formed a rDNA advisory committee.[14] This was followed by other regulatory offices (the United States Department of Agriculture (USDA), Environmental Protection Agency (EPA) and Food and Drug Administration (FDA)), effectively making all rDNA research tightly regulated in the USA.[15] In 1982 the Organization for Economic Co-operation and Development (OECD) released a report into the potential hazards of releasing genetically modified organisms into the environment as the first transgenic plants were being developed.[16] As the technology improved and genetically modified organisms moved from model organisms to potential commercial products the USA established a committee at the Office of Science and Technology (OSTP) to develop mechanisms to regulate the developing technology.[15] In 1986 the OSTP assigned regulatory approval of genetically modified plants in the US to the USDA, FDA and EPA.[17]

The basic concepts for the safety assessment of foods derived from GMOs have been developed in close collaboration under the auspices of the Organisation for Economic Co-operation and Development (OECD) and the United Nations' World Health Organisation (WHO) and Food and Agricultural Organisation (FAO). A first joint FAO/WHO consultation in 1990 resulted in the publication of the report 'Strategies for Assessing the Safety of Foods Produced by Biotechnology' in 1991.[18] Building on that, an international consensus was reached by the OECD's Group of National Experts on Safety in Biotechnology, for assessing biotechnology in general, including field testing GM crops.[19] That Group met again in Bergen, Norway in 1992 and reached consensus on principles for evaluating the safety of GM food; its report, 'The safety evaluation of foods derived by modern technology – concepts and principles' was published in 1993.[20] That report recommends conducting the safety assessment of a GM food on a case-by-case basis through comparison to an existing food with a long history of safe use. This basic concept has been refined in subsequent workshops and consultations organized by the OECD, WHO, and FAO, and the OECD in particular has taken the lead in acquiring data and developing standards for conventional foods to be used in assessing substantial equivalence.[21][22] In 2003 the Codex Alimentarius Commission of the FAO/WHO adopted a set of "Principles and Guidelines on foods derived from biotechnology" to help countries coordinate and standardize regulation of GM food to help ensure public safety and facilitate international trade,[23] and updated its guidelines for import and export of food in 2004.[24]

35.2   Substantial equivalence

Main article: Substantial equivalence

"Substantial equivalence" is a starting point for the safety assessment for GM foods that is widely used by national and international agencies - including the Canadian Food Inspection Agency, Japan's Ministry of Health and Welfare and the U.S. Food and Drug Administration, the United Nation's Food and Agriculture Organization, the World Health Organization and the OECD.[25]

A quote from FAO, one of the agencies that developed the concept, is useful for defining it: "Substantial equivalence embodies the concept that if a new food or food component is found to be substantially equivalent to an existing food or food component, it can be treated in the same manner with respect to safety (i.e., the food or food component can be concluded to be as safe as the conventional food or food component)".[26] The concept of substantial equivalence also recognises the fact that existing foods often contain toxic components (usually called antinutrients) and are still able to be consumed safely - in practice there is some tolerable chemical risk taken with all foods, so a comparative method for assessing safety needs to be adopted. For instance, potatoes and tomatoes can contain toxic levels of respectively, solanine and alpha-tomatine alkaloids.[27][28]

To decide if a modified product is substantially equivalent, the product is tested by the manufacturer for unexpected changes in a limited set of components such as toxins, nutrients, or allergens that are present in the unmodified food. The manufacturer's data is then assessed by a regulatory agency, such as the U.S. Food and Drug Administration. That data, along with data on the genetic modification itself and resulting proteins (or lack of protein), is submitted to regulators. If regulators determine that the submitted data show no significant difference between the modified and unmodified products, then the regulators will generally not require further food safety testing. However, if the product has no natural equivalent, or shows significant differences from the unmodified food, or for other reasons that regulators may have (for instance, if a gene produces a protein that had not been a food component before), the regulators may require that further safety testing be carried out.[20]

A 2003 review in *Trends in Biotechnology* identified seven main parts of a standard safety test:[29]

1. Study of the introduced DNA and the new proteins or metabolites that it produces;

2. Analysis of the chemical composition of the relevant plant parts, measuring nutrients, anti-nutrients as well as any natural toxins or known allergens;

3. Assess the risk of gene transfer from the food to microorganisms in the human gut;

4. Study the possibility that any new components in the food might be allergens;

5. Estimate how much of a normal diet the food will make up;

6. Estimate any toxicological or nutritional problems revealed by this data in light of data on equivalent foods;

7. Additional animal toxicity tests if there is the possibility that the food might pose a risk.

There has been discussion about applying new biochemical concepts and methods in evaluating substantial equivalence, such as **metabolic profiling** and **protein profiling**. These concepts refer, respectively, to the complete measured biochemical spectrum (total fingerprint) of compounds (metabolites) or of proteins present in a food or crop. The goal would be to compare overall the **biochemical profile** of a new food to an existing food to see if the new food's profile falls within the **range of natural variation** already exhibited by the profile of existing foods or crops. However, these techniques are not considered sufficiently evaluated, and standards have not yet been developed, to apply them.[30]

There are controversies over the definition and application of substantial equivalence. See section in genetically modified food controversies.

35.3 By continent

35.3.1 Africa

In 2010, after nine years of talks, the Common Market for Eastern and Southern Africa (COMESA) produced a draft policy on GM technology, which was sent to all 19 national governments for consultation in September 2010. Under the proposed policy, new GM crops would be scientifically assessed by COMESA. If the GM crop was deemed safe for the environmental and human health, permission would be granted for the crop to be grown in all 19 member countries, although the final decision would be left to each individual country.[31]

In 2012, South Africa was the major commercial grower of genetically modified crops in Africa, with smaller amounts grown in Burkina Faso (maize), Egypt (cotton) and Sudan (cotton).[32][33]Kenya passed laws in 2011,[34] and Ghana[35] and Nigeria[36] passed laws in 2012 which allowed the production and importation of GM crops. By 2013 Cameroon, Malawi and Uganda had approved trials of genetically altered crops.[32] A study investigating voluntary labeling in South Africa found that 31% of products labeled GMO-free had a GM content above 1.0%.[3] 2011 studies for Uganda showed that transgenic bananas had a high potential to reduce rural poverty but that urban consumers with a relatively higher income might reject the introduction.[37][38]

In 2002, Zambia cut off the flow of genetically modified food (mostly maize) from UN's World Food Programme on the basis of the Cartagena Protocol.[39] This left the population without food aid during a famine.[40] In December 2005 the Zambian government changed its mind in the face of further famine and allowed the importation of GM maize.[41] However, the Zambian Minister for Agriculture Mundia Sikatana insisted in 2006, that the ban on genetically modified maize remained, saying "We do not want GM (genetically modified) foods and our hope is that all of us can continue to produce non-GM foods."[42][43]

35.3.2 Asia

India and China are the two largest producers of genetically modified products in Asia.[44] India currently only grows GM cotton, while China produces GM varieties of cotton, poplar, petunia, tomato, papaya and sweet pepper. Cost of enforcement of regulations in India are generally higher,

possibly due to the greater influence farmers and small seed firms have on policy makers, while the enforcement of regulations was more effective in China.[45] Other Asian countries that grew GM crops in 2011 were Pakistan, the Philippines and Myanmar.[44][46] GM crops were approved for commercialisation in Bangladesh in 2013 and in Vietnam and Indonesia in 2014.[47]

China

GM crops in China go through three phases of field trials (pilot field testing, environmental release testing, and pre-production testing) before they are submitted to the Office of Agricultural Genetic Engineering Biosafety Administration (OAGEBA) for assessment.[48] Producers must apply to OAGEBA at each stage of the field tests. The Chinese Ministry of Science and Technology developed the first biosafety regulations for GM products in 1993 and they were updated in 2001.[49] The 75 member National Biosafety Committee evaluates all applications, although OAGEBA has the final decision. Most of the National Biosafety Committee are involved in biotechnology leading to criticisms that they do not represent a wide enough range of public concerns.[48]

India

The release of transgenic crops in India is governed by the Indian Environment Protection Act, which was enacted in 1986. The Institutional Biosafety Committee (IBSC), Review Committee on Genetic Manipulation (RCGM) and Genetic Engineering Approval Committee (GEAC) all review any genetically modified organism to be released, with transgenic crops also needing permission from the Ministry of Agriculture.[50] India regulators cleared the Bt brinjal, a genetically modified eggplant, for commercialisation in October 2009. Following opposition from some scientists, farmers and environmental groups a moratorium was imposed on its release in February 2010.[51][52]

Official Reports on GMO There have been four official reports on GMO in India till August 2013 :

1. The 'Jairam Ramesh Report' - February 2010, imposing an indefinite moratorium on Bt Brinjal [53]

2. The Sopory Committee Report - August 2012 [54]

3. The Parliamentary Standing Committee (PSC) Report on GM crops - August 2012[55]

4. Final Report of The Technical Expert Committee established by Supreme Court - July 2013[56]

Japan

Two laws regulate food safety and food quality in Japan, the Food Sanitation Law passed in 1947 and the Law Concerning Standardization and Proper Labeling of Agricultural and Forestry Products passed in 1950. The Food Sanitation Law has been amended and updated many times; an amendment dealing with pre-market approval and labeling of GMOs was passed in 2000 and came into effect in 2001. Japan passed laws to implement the Cartagena Protocol on Biosafety in September 2003 which came into effect in February 2004 - the Law Concerning the Conservation and Sustainable Use of Biological Diversity through Regulations on the Use of Living Modified Organisms (Law No. 97 of 2003).[57][58]:6

Authority for approvals for various uses of genetically modified organisms is divided in Japan. The Ministry of the Environment has final approval for all uses of GMOs, but crops for commercial use and live vaccines for animals first go through the Ministry of Agriculture, Forestry and Fisheries; viruses for gene therapy and other medical applications first go through the Ministry of Health, Labor and Welfare; field trials of GM crops and recombinant DNA used in biotechnology research first goes through the Ministry of Education, Culture, Sports, Science and Technology; and uses in the process of production of industrial enzymes, etc. goes through the Ministry of Economy, Trade and Industry.[58]:8

Japan has not approved any commodity GM crops to be grown in Japan, but does allow import of agricultural products made from GM crops and food made of imported GM ingredients.[57] Japan does however allow cultivation of GM flowers (e.g. Blue roses).[58]:3

GM foods must undergo a safety assessment prior to being awarded certification for distribution to the domestic market. The Food Safety Commission (FSC) performs food and feed safety risk assessments.[59]

Certain GM food must be labeled, but this is limited to designated genetically modified agricultural products, which are soybean, corn, potato, rapeseed, cottonseed, alfalfa and beet, and is limited to 32 processed foods which contain soybean, corn and potato, alfalfa and beet, in which recombinant DNA or the resulting protein still exists even after processing. However, processed food in which recombinant DNA or protein is dissolved in or removed during processing, such as soy sauce, soybean oil, corn flakes, millet jelly, corn oil, rapeseed oil, cottonseed oil, and others, do not have to be labeled.[57]

Japan does not require traceability, and allows negative labeling ("GMO-free" and the like).[57]

Philippines

The Philippines bans all GMOs recently overturning existing Department of Agriculture regulations. A petition filed on May 17, 2013 by environmental group Greenpeace Southeast Asia and farmer-scientist coalition Masipag (Magsasaka at Siyentipiko sa Pagpapaunlad ng Agrikultura) asked the appellate court to stop the planting of Bt eggplant in test fields, saying the impacts of such an undertaking to the environment, native crops and human health are still unknown. The Court of Appeals granted the petition, citing the precautionary principle stating "when human activities may lead to threats of serious and irreversible damage to the environment that is scientifically plausible but uncertain, actions shall be taken to avoid or diminish the threat." [60] Respondents filed a motion for reconsideration in June 2013 and on September 20, 2013 the Court of Appeals chose to uphold their May decision saying the bt talong field trials violate the people's constitutional right to a "balanced and healthful ecology." [61][62] The Supreme Court on Tuesday, December 8, 2015 permanently stopped the field testing for Bt (Bacillus thuringiensis) talong (eggplant), upholding the decision of the Court of Appeals which stopped the field trials for the genetically modified eggplant.[63] The Philippines Supreme Court also took the unprecedented step and invalidated the Department of Agriculture administrative order allowing the field testing, propagation and commercialization, and importation of GMOs.[64]

35.3.3   Europe

See also: Regulation of genetically modified organisms in the European Union and Regulation of genetically modified organisms in Switzerland

Until the 1990s, Europe's regulation was less strict than in the United States, one turning point being cited as the export of the United States' first GM-containing soy harvest in 1996. The GM soy made up about 2% of the total harvest at the time, and EuroCommerce and European food retailers required that it be separated.[65] In 1998, the use of MON810, a Bt expressing maize conferring resistance to the European corn borer, was approved for commercial cultivation in Europe. Shortly thereafter, the EU enacted a *de facto* moratorium on new approvals of GMOs pending new regulatory laws passed in 2003.

Those new laws provided the European Union (EU) with possibly the most stringent GMO regulations in the world.[5] All GMOs, along with irradiated food, are considered "new food" and subject to extensive, case-by-case, science based food evaluation by the European Food Safety Authority

(EFSA). The criteria for authorization fall in four broad categories: "safety," "freedom of choice," "labelling," and "traceability."[66] The EFSA reports to the European Commission who then draft a proposal for granting or refusing the authorisation. This proposal is submitted to the Section on GM Food and Feed of the Standing Committee on the Food Chain and Animal Health and if accepted it will be adopted by the EC or passed on to the Council of Agricultural Ministers. Once in the Council it has three months to reach a qualified majority for or against the proposal, if no majority is reached the proposal is passed back to the EC who will then adopt the proposal.[5][67] However, even after authorization, individual EU member states can ban individual varieties under a 'safeguard clause' if there are "justifiable reasons" that the variety may cause harm to humans or the environment. The member state must then supply sufficient evidence that this is the case.[68] The Commission is obliged to investigate these cases and either overturn the original registrations or request the country to withdraw its temporary restriction. The laws of the EU also stipulated that member nations establish coexistence regulations.[69] In many cases national coexistence regulations include minimum distances between fields of GM crops and non-GM crops. The distances for GM maize from non-GM maize for the six largest biotechnology countries are: France: 50 meters, Britain: 110 meters for grain maize and 80 for silage maize, Netherlands: 25 meters in general and 250 for organic or GM-free fields, Sweden: 15–50 meters, Finland: data not available, and Germany: 150 meters and 300 from organic fields.[70] Larger minimum distance requirements discriminate against adoption of GM crops by smaller farms.[71][72][73]

In 2006, the World Trade Organization concluded that the EU moratorium, which had been in effect from 1998 to 2004,[74] had violated international trade rules.[75][76] The moratorium had not affected previously approved crops. The only crop authorised for cultivation before the moratorium was Monsanto's MON 810. The next approval for cultivation was the Amflora potato for industrial applications in 2010[77][78] which was grown in Germany, Sweden and the Czech Republic that year.[79]

The slow pace of approval has been criticized as endangering European food safety[80][81] although as of 2012, the EU has authorized the use of 48 genetically modified organisms. Most of these were for use in animal feed (it was reported in 2012 that the EU imports about 30 million tons a year of GM crops for animal consumption.[82]), food or food additives. 26 of these were varieties of maize.[83] In July 2012 the EU gave approval for an Irish trial cultivation of potatoes resistant to the blight that caused the Great Irish Famine.[84]

The safeguard clause mentioned above has been applied by many member states in various circumstances, and in April 2011 there were 22 active bans in place across six member states: Austria, France, Germany, Luxembourg, Greece, and Hungary.[85] However, on review many of these have been considered scientifically unjustified.[68][86]

- In January 2005, the Hungarian government announced a ban on importing and planting of genetic modified maize seeds, which was subsequently authorized by the EU.

- In February 2008 the French government used the safeguard clause to ban the cultivation of MON810 after Senator Jean-François Le Grand, chairman of a committee set up to evaluate biotechnology, said there were "serious doubts" about the safety of the product[87] (although this ban was declared illegal in 2011 by the European Court of Justice and the French Conseil d'État[88]). The French farm ministry reinstated the ban in 2012, but this was rejected by the EFSA.[89]

- In 2009 German Federal Minister Ilse Aigner announced an immediate halt to cultivation and marketing of MON810 maize under the safeguard clause.[90]

- In March 2010, Bulgaria imposed a complete ban on genetically modified crop growing either commercially or for trials.[91] The cabinet of Boyko Borisov initially imposed a 5-year moratorium, but later extended it to a permanent ban after widespread public protests against the introduction of genetically modified crops in the country. And in recent years, France and several other European countries banned cultivation of Monsanto's MON-810 corn and similar genetically modified food crops.

- Since January 2013 Poland's government placed a ban on Monsanto's GM corn, MON 810 and has launched a communication campaign with farmers', announcing they will now be strictly monitoring farms for GM corn crops. Poland is the eighth EU member to ban the production of GMOs although they have been approved by European Food Safety Authority. Europe is not against the use of GM crops when it comes to laboratory research, they are working to regulate the field.[92]

In 2012, the European Food Safety Authority (EFSA) Panel on Genetically Modified Organisms (GMO) released a "Scientific opinion addressing the safety assessment of plants developed through cisgenesis and intragenesis" in a response to a request from the European Commission.[93] The opinion was, that while "the frequency of unintended changes may differ between breeding techniques and their occurrence cannot be predicted and needs to be assessed case by case," "similar hazards can be associated with cisgenic and conventionally bred plants, while novel hazards can be associated with intragenic and transgenic plants." In other words, cisgenic genetic engineering approaches should be considered similar in risk to conventional breeding approaches, each of which are less risky than transgenic approaches.

In 2014 a panel of experts set up by the UK Biotechnology and Biological Sciences Research Council argued that "A regulatory system based on the characteristics of a novel crop, by whatever method it has been produced, would provide a more effective and robust regulation than current EU processes , which consider new crop varieties differently depending on the method used to produce them." They said that new forms of "genome editing" allow targeting specific sites and making precise changes in the DNA of crops. In the future it would become increasingly difficult if not impossible to tell which method has been used (conventional breeding or genetic engineering) to produce a novel crop. They proposed that existing EU regulatory system should be replaced with a more logical system like that used for new medicines.[94]

In 2015 Germany, Poland, France, Scotland and several other member states opted out of cultivating GMO crops in their territory.[95]

Labeling and traceability

The regulations concerning the import and sale of GMOs for human and animal consumption grown outside the EU involve providing freedom of choice to the farmers and consumers.[96] All food (including processed food) or feed which contains greater than 0.9% of approved GMOs must be labelled. Twice GMOs unapproved by the EC have arrived in the EU and been forced to return to their port of origin.[5] The first was in 2006 when a shipment of rice from America containing an experimental GMO variety (LLRice601) not meant for commercialisation arrived at Rotterdam. The second in 2009 when trace amounts of a GMO maize approved in the US were found in a "non-GM" soy flour cargo.[5]

The coexistence has raised significant concern in many European countries and so EU law also requires that all GM food be traceable to its origin, and that all food with GM content greater than 0.9% be labelled.[97] Due to high demand from European consumers for freedom of choice between GM and non-GM foods, EU regulations require measures to avoid mixing of foods and feed produced from GM crops and conventional or organic crops, which can be done via isolation distances or biological containment strategies.[98][99] (Unlike the US, European countries require labeling of GM food.) European research programs

such as Co-Extra, Transcontainer, and SIGMEA are investigating appropriate tools and rules for traceability. The OECD has introduced a "unique identifier" which is given to any GMO when it is approved, which must be forwarded at every stage of processing.[100] Such measures are generally not used in North America because they are very costly and the industry admits of no safety-related reasons to employ them.[101] The EC has issued guidelines to allow the coexistence of GM and non-GM crops through buffer zones (where no GM crops are grown).[98] These are regulated by individual countries and vary from 15 meters in Sweden to 800 meters in Luxembourg.[5] All food (including processed food) or feed which contains greater than 0.9% of approved GMOs must be labelled.

A 5-digit price look-up code beginning with the digit 8 indicates genetically modified food.[102] However the absence of the "8" does not necessarily indicate the food is not genetically modified since no retailer to date has elected to use the digit in voluntarily labeling genetically modified foods.[103]

35.3.4   North America

See also: Genetic engineering in the United States

As of 2002 the United States, Canada, and Mexico did not require labeling of genetically modified foods.[104]

Canada

Mainland Canada is one of the world's largest producers of GM canola[105] and also grows GM maize, soybean and sugarbeet.[33] Health Canada, under the Food and Drugs Act, and the Canadian Food Inspection Agency[106] are responsible for evaluating the safety and nutritional value of genetically modified foods. Environmental assessments of biotechnology-derived plants are carried out by the CFIA's Plant Biosafety Office (PBO).[107] The Canadian regulatory system is based on whether a product has novel features regardless of method of origin. In other words, a product is regulated as GM if it carries some trait not previously found in the species whether it was generated using traditional breeding methods (e.g. selective breeding, cell fusion, mutation breeding) or genetic engineering.[108][109][110] Canadian law requires that manufacturers and importers submit detailed scientific data to Health Canada for safety assessments for approval. This data includes: information on how the GM plant was developed; nucleic acid data that characterizes the genetic change; composition and nutritional data of the novel food compared to the original non-modified food' potential for new toxins; and potential for being an allergen. A decision is then made whether to approve the product for release along with any restrictions or requirements. Labeling of foods as products of Genetic Engineering or not products of Genetic Engineering is voluntary.[4][111] The Canadian regulations were reviewed by the Canadian Biotechnology Advisory Committee between 1999 and 2003, with the conclusion that the current level of regulation was satisfactory. The committee was accused by environmental and citizen groups of not representing the full spectrum of public interests by only having one member of the board of 20 representing non-governmental organisations and for being too closely aligned to industry groups.[112]

Mexico

In February 2005, after consulting the Mexican Academy of Sciences, Mexico's senate passed a law allowing to plant and sell genetically modified cotton and soybean.[113] The law requires all genetically modified products to be labelled according to guidelines issued by the Mexican Ministry of Health. In 2009, the government enacted statutory provisions for the regulation of genetically modified maize.[114] Mexico is the center of diversity for maize and concerns had been raised about the impact genetically modified maize could have on local strains.[115][116] In 2013, a federal judge ordered Mexico's SAGARPA (Secretaría de Agricultura, Ganadería, Desarrollo Rural, Pesca, y Alimentación), which is Mexico's Secretary of Agriculture, and SEMARNAT (Secretaría de Medio Ambiente y Recursos Naturales), equivalent of the EPA, to temporarily halt any new GMO corn permits, accepting a lawsuit brought by opponents of the crop.[117]

United States

Federal regulation The USA is the largest commercial grower of genetically modified crops in the world.[118]

United States regulatory policy is governed by the Coordinated Framework for Regulation of Biotechnology[119] This regulatory policy framework that was developed under the Presidency of Ronald Reagan to ensure safety of the public and to ensure the continuing development of the fledgling biotechnology industry without overly burdensome regulation.[120] The policy as it developed had three tenets: "(1) U.S. policy would focus on the product of genetic modification (GM) techniques, not the process itself, (2) only regulation grounded in verifiable scientific risks would be tolerated, and (3) GM products are on a continuum with existing products and, therefore, existing statutes are sufficient to review the products."[120] In 2015 the Obama administration announced that it would update the way the government regulated genetically modified crops.[121]

For a genetically modified organism to be approved for release, it must be assessed under the Plant Protection Act by the Animal and Plant Health Inspection Service (APHIS) agency within the US Department of Agriculture (USDA) and may also be assessed by the Food and Drug Administration (FDA) and the Environmental protection agency (EPA), depending on the intended use of the organism. The USDA evaluates the plant's potential to become a weed. The FDA has a voluntary consultation process with the developers of genetically engineered plants. The Federal Food, Drug, and Cosmetic Act, which outlines FDA's responsibilities, does not require pre-market clearance of food, including genetically modified food plants.[122][123] The EPA regulates genetically modified plants with pesticide properties, as well as agrochemical residues.[124] Most genetically modified plants are reviewed by at least two of the agencies, with many subject to all three.[15][125] Within the organization are departments that regulate different areas of GM food including, the Center for Food Safety and Applied Nutrition (CFSAN,) and the Center for Biologics Evaluation and Research (CBER).[124] As of 2008, all developers of genetically modified crops in the US had made use of the voluntary process.[126] Final approval can still be denied by individual counties within each state. In 2004, Mendocino County, California became the first county to impose a ban on the "Propagation, Cultivation, Raising, and Growing of Genetically Modified Organisms", the measure passing with a 57% majority.[127] In May, 2014 Jackson and Josephine Counties in Southern Oregon passed initiatives similar to that passed by Mendocino County; both passing by 2 to 1 margins.[128]

Several laws govern the US regulatory agencies. These laws are statutes the agencies review when determining the safety of a particular GM food. These laws include:[124]

- The Federal Insecticide, Fungicide, and Rodenticide Act (FIFRA) (EPA);

- The Toxic Substances Control Act (TSCA) (EPA);

- The Federal Food, Drug, and Cosmetic Act (FFDCA) (FDA and EPA);

- The Plant Protection Act (PPA) (USDA);

- The Virus-Serum-Toxin Act (VSTA) (USDA);

- The Public Health Service Act (PHSA)(FDA);

- The Dietary Supplement Health and Education Act (DSHEA) (FDA)

- The Meat Inspection Act (MIA)(USDA);

- The Poultry Products Inspection Act (PPIA) (USDA);

- The Egg Products Inspection Act (EPIA) (USDA); and

- The National Environmental Protection Act (NEPA).

State regulation   Several states have passed regulations concerning labelling of GM food; Connecticut passed a GMO labeling bill in May 2013, but the bill will only be triggered after four other states enact similar legislation.[129] On January 9, 2014, Maine's governor signed a bill requiring labeling for foods made with GMO's, with a similar triggering mechanism as Connecticut's bill.[130] In May 2014 Vermont passed a law requiring labeling of food containing ingredients derived from genetically modified organisms.[131][132] A federal judge ruled Maui's GMO ban invalid.[133]

35.3.5 South America

Brazil and Argentina are the 2nd and 3rd largest producers of GM food behind the USA.[133]

The Argentine government was one of the first to accept GM food. Assessment of GM products for release is provided by the National Agricultural Biotechnology Advisory Committee (environmental impact), the National Service of Health and Agrifood Quality (food safety) and the National Agribusiness Direction (effect on trade), with the final decision made by the Secretariat of Agriculture, Livestock, Fishery and Food.[134] The government is looking to tighten the current law which allows farmers to keep seed without paying royalties in a bid to encourage more private investment.[135]

In Brazil the National Biosafety Technical Commission is responsible for assessing environmental and food safety and prepares guidelines for transport, importation and field experiments involving GM products. The Council of Ministers evaluates the commercial and economical issues with release.[134] The National Biosafety Technical Commission has 27 members and includes 12 scientists, 9 ministerial representatives and 6 other specialists.

Honduras,[136] Costa Rica,[137] Colombia,[138] Bolivia,[139] Paraguay,[140] Chile,[141] and Uruguay[142] also allow GM crops to be grown.

Venezuela banned genetically modified seeds in 2004.[143] in 2008. Ecuador prohibited genetically engineered crops and seeds in its 2008 Constitution, approved by 64% of the population in a referendum[144] (although Ecuadorian President Rafael Correa said in 2012 that this was "a mistake".[145] Peru has banned transgenic crops.[145]

35.3.6 Oceania

Malaysia, New Zealand, and Australia require labeling so consumers can exercise choice between foods that have genetically modified, conventional or organic origins.[146]

Australia

Genetic engineering in Australia was originally (since 1987) overseen by the Genetic Manipulation Advisory Committee, before the Office of the Gene Technology Regulator (OGTR) and Food Standards Australia New Zealand took over in 2001.[147][148] The OTGR is a Commonwealth Government Authority within the Department of Health and Ageing and reports directly to Parliament through a Ministerial Council on Gene Technology and has legislative powers.[147][149] It was established as part of the Gene Technology Act 2003 and operates according to the Gene Technology Regulations 2001. The OGTR reports directly to Parliament through a Ministerial Council on Gene Technology and has legislative powers.[147][149] The OGTR decides on license applications for the release of all genetically modified organisms, while regulation is provided by the Therapeutic Goods Administration for GM medicines or Food Standards Australia New Zealand for GM food. The individual state governments are then able to assess the impact of release on markets and trade and apply further legislation to control approved genetically modified products.[148]

Genetically modified cotton, canola, and carnations are grown in Australia.[150][151] Genetically modified cotton has been grown commercially in New South Wales and Queensland since 1996.[152] GM canola was approved in 2003[153] and was first grown in 2008[154] and was first approved in Western Australia in 2010.[155]

In 2011 genetically modified plants were grown in all states except South Australia and Tasmania, who have extended their moratoriums until 2019 and 2014.[156] The Queensland and Northern Territory Governments have not implemented any further legislation beyond the national level, but several other states placed bans on planting certain GM crops.[148] In 2007 the New South Wales government extended a blanket moratorium on GM food crops until 2011, but allowed groups to apply for exemptions. New South Wales approved GM Canola for commercial cultivation in 2008, while the Victorian government let the moratorium on GM Canola expire in 2007.[152] Western Australia passed the Genetically Modified Crops Free Areas Act in 2003 and was declared a GM free area in 2004. In 2008 an exception was made for the commercial cultivation of GM cotton in the Ord River Irrigation Areas.[156] Trials of GM canola were carried out in 2003 and in 2010 the Western Australian government allowed the commercialisation of GM canola.[153]

New Zealand

As of 2004 no genetically modified food was grown in New Zealand, and no medicines containing live genetically modified organisms have been approved for use.[157] However, medicines manufactured using genetically modified organisms that do not contain live organisms have been approved for sale, and imported foods with genetically modified components are sold. In 2000 the Government appointed a Royal Commission to report on issues relating to genetically modified organisms (GMOs). The Report of the Royal Commission on Genetic Modification, released in July 2001, concluded that New Zealand should keep its options open with regard to genetic engineering and to proceed carefully in order to minimise and manage any risks. Field trials have been carried out with GM pine trees and brassicas.[158][159] Food Standards Australia New Zealand (FSANZ) must approve any food produced from GM crops, or made using genetically engineered enzymes, before it can be marketed in Australia or New Zealand. FSANZ makes a list of such approvals available on its website.[160]

35.4   See also

- Genetic engineering

- Genetically modified crops

- Genetically modified food

- Genetically modified food controversies

- Genetically modified organisms

35.5   References

[1] http://www.loc.gov/law/help/restrictions-on-gmos/

[2] Hallenbeck, Terri (2014-04-27). "How GMO labeling came to pass in Vermont". *Burlington Free Press*. Retrieved 2014-05-28.

[3] Gerda M. Botha and Christopher D. Viljoen (2009) "South Africa: A case study for voluntary GM labelling" Food Chemistry 112(4):1060–1064

[4] The Regulation of Genetically Modified Foods

[5] John Davison (2010)"GM plants: Science, politics and EC regulations" Plant Science 178(2):94–98

[6] American Association for the Advancement of Science (AAAS), Board of Directors (2012). Legally Mandating GM Food Labels Could Mislead and Falsely Alarm Consumers

[7] Ronald, Pamela (2011). "Plant Genetics, Sustainable Agriculture and Global Food Security". *Genetics* **188** (1): 11–20. doi:10.1534/genetics.111.128553. PMC 3120150. PMID 21546547.

[8] Bett, Charles; Ouma, James Okuro; Groote, Hugo De (August 2010). "Perspectives of gatekeepers in the Kenyan food industry towards genetically modified food". *Food Policy* **35** (4): 332–340. doi:10.1016/j.foodpol.2010.01.003.

[9] American Medical Association (2012). Report 2 of the Council on Science and Public Health: Labeling of Bioengineered Foods

[10] United States Institute of Medicine and National Research Council (2004). Safety of Genetically Engineered Foods: Approaches to Assessing Unintended Health Effects. National Academies Press. Free full-text. National Academies Press. See pp11ff on need for better standards and tools to evaluate GM food.

[11] Key S, Ma JK, Drake PM (June 2008). "Genetically modified plants and human health". *J R Soc Med* **101** (6): 290–8. doi:10.1258/jrsm.2008.070372. PMC 2408621. PMID 18515776.

[12] Berg P, Baltimore D, Boyer HW, Cohen SN, Davis RW, Hogness DS, Nathans D, Roblin R, Watson JD, Weissman S, Zinder ND (1974). "Letter: Potential biohazards of recombinant DNA molecules" (PDF). *Science* **185** (4148): 303. doi:10.1126/science.185.4148.303. PMID 4600381.

[13] Berg, P., Baltimore, D., Brenner, S., Roblin, R. O., and Singer, M. F. (1975). "Summary Statement of the Asilomar Conference on Recombinant DNA Molecules". *Proc. Natl. Acad. Sci. USA* **72** (6): 1981–1984. doi:10.1073/pnas.72.6.1981. PMC 432675. PMID 806076.

[14] Hutt, P.B. (1978). "Research on recombinant DNA molecules: the regulatory issues". *South Calif Law Rev* **51** (6): 1435–50. PMID 11661661.

[15] McHughen A, Smyth S (2008). "US regulatory system for genetically modified [genetically modified organism (GMO), rDNA or transgenic] crop cultivars". *Plant biotechnology journal* **6** (1): 2–12. doi:10.1111/j.1467-7652.2007.00300.x. PMID 17956539.

[16] Bull, A.T., Holt, G. and Lilly, M.D. (1982). *Biotechnology : international trends and perspectives* (PDF). Paris: Organisation for Economic Co-operation and Development.

[17] U.S. Office of Science and Technology Policy (1986). "Coordinated framework for regulation of biotechnology" (PDF). *Fed Regist.* **51** (123): 23302–50. PMID 11655807.

[18] FAO/WHO Report 1991

[19] OECD(1992) Safety Considerations for Biotechnology

[20] OECD(1993) Safety Evaluation of Foods Derived by Modern Biotechnology: Concepts and Principles

[21] OECD (2010) Consensus Document on Molecular Characterisation of Plants Derived from Modern Biotechnology

[22] OECD harmonization webpage

[23] Codex Alimentarius Commission, 2003 Principles and Guidelines on foods derived from biotechnology

[24] Codex Alimentarius Commission, 2004 Food Import & Export

[25] Substantial Equivalence in Food Safety Assessment, Council for Biotechnology Information, March 11, 2001

[26] Joint FAO/WHO Expert Consultation on Biotechnology and Food Safety, Rome, Italy, 30 September to 4 October 1996 p. 4

[27] Organisation for Economic Co-operation and Development. Report of the Task Force for the Safety of Novel Foods and Feeds C(2000)86/ADD1, May 17, 2000 Quote: "Much experience has been gained in the safety assessment of the first generation of foods derived through modern biotechnology, and those countries that have conducted assessments are confident that those GM foods they have approved are as safe as other foods."

[28] Substantial equivalence of antinutrients and inherent plant toxins in genetically modified novel foods, Novak, W. K.; Haslberger, A. G.,Food and Chemical Toxicology Volume 38 (6) p.473-483, 2000

[29] Kok EJ, Kuiper HA (October 2003). "Comparative safety assessment for biotech crops". *Trends Biotechnol.* **21** (10): 439–44. doi:10.1016/j.tibtech.2003.08.003. PMID 14512230.

[30] FAO. Safety aspects of genetically modified foods of plant origin - Consultations 4. Approaches to the Nutritional and Food Safety Evaluation of Genetically Modified Foods[www.fao.org/wairdocs/ae584e/ae584e04.htm]

[31] Transgenic harvest Editorial. *Nature* 467 , pages 633–634, 7 October 2010, doi:10.1038/467633b. Retrieved 9 November 2010

[32] Dunmore, Charlie and Kumwenda, Olivia (6 June 2013) As health fears ebb, Africa looks at easing GM crop bans Reuters, 6 June 2006, Retrieved 9 August 2013

[33] Slides & Tables : Global Status of Commercialized Biotech/GM Crops: 2010 - ISAAA Brief 42-2010 | ISAAA.org

[34] Denge, Mark and Gachenge, Beatrice (4 July 2011) Kenya approves law to allow GM crops Reuters Africa, 4 July 2004, Retrieved 9 November 2011

[35] "Nigeria Passes Law Allowing Genetically Modified Plants". *Biotechnology Law Report* **31** (2): 153–153. 2012. doi:10.1089/blr.2012.9900.

[36] "Ghana to Allow Genetically Modified Crops". *Biotechnology Law Report* **31** (2): 153–154. 2012. doi:10.1089/blr.2012.9901.

[37] Kikulwe, E., J. Wesseler J. Falck-Zepeda (2011): Attitudes, Perceptions, and Trust: Insights from a Consumer Survey Regarding Genetically Modified Banana in Uganda. Appetite 57(2):401-413

[38] Kikulwe, E., E. Birol, J. Wesseler, J. Falck-Zepeda (2011): A Latent Class Approach to Investigating Developing Country Consumers' Demand for Genetically Modified Staple Food Crops: The Case of GM Banana in Uganda. Agricultural Economics. 42:547–560

[39] Maharaj, Davan and Mukwita, Anthony (28 August 2002) Zambia Rejects Gene-Altered U.S. Corn Los Angeles Times. Retrieved 9 November 2011

[40] Zambian Leader Defends Ban On Genetically Altered Foods – New York Times. Nytimes.com (2002-09-04). Retrieved on 2011-02-08.

[41] Zambia Allows Its People To Eat. Consumerfreedom.com. Retrieved on 2011-02-08.

[42] Africans vow to resist any US pressure on GMOs. Reuters. 2/9/2006

[43] World Environment News. Planet Ark. Retrieved on 2011-02-08.

[44] ISAAA Brief 37-2007 - Slides & Tables > ISAAA.org

[45] Inderscience. Inderscience.metapress.com. Retrieved on 2014-01-14.

[46] James, C (2011). "ISAAA Brief 43, Global Status of Commercialized Biotech/GM Crops: 2011". *ISAAA Briefs*. Ithaca, New York: International Service for the Acquisition of Agri-biotech Applications (ISAAA). Retrieved 2012-08-30.

[47] James, Clive (28 January 2015). "Global Status of Commercialized Biotech/GM Crops: 2014". *ISAAA Briefs* (Ithaca, NY, USA: International Service for the Acquisition Of Agri-biotech Applications (ISAAA)) **49**. Retrieved 15 December 2015.

[48] Chen, Mao; Shelton, Anthony; Ye, Gong-yin (2011). "Insect-Resistant Genetically Modified Rice in China: From Research to Commercialization". *Annual Review of Entomology* **56**: 81–101. doi:10.1146/annurev-ento-120709-144810. PMID 20868281.

[49] AgBioForum 5(4): Agricultural Biotechnology Development and Policy in China

[50] TNAU Agritech Portal :: Bio Technology. Agritech.tnau.ac.in (1989-12-05). Retrieved on 2014-01-14.

[51] "India puts on hold first GM food crop on safety grounds". BBC. 9 February 2010. Retrieved 9 February 2010.

[52] "Govt says no to Bt brinjal for now". The Times of India. February 9, 2010. Retrieved February 9, 2010.

[53] Chaudhury, Shoma (March 7, 2010). "The Gene Gun At Your Head". *Tehelka Magazine*. Retrieved 12 August 2013.

[54] "Report of Dr S K Sopory Committee on BNLA106 event" (PDF). *Indian Council of Agricultural Research*. Retrieved 12 August 2013.

[55] "Committee on agriculture - Ministry of Agriculture (Department of agriculture and cooperation) "Cultivation of genetically modified food crops –prospects and effects" Thirty seventh report" (PDF). Retrieved 12 August 2013.

[56] "Monthly Policy Review July 2013" (PDF). *PRSIndia*. p. 9. Retrieved 12 August 2013.

[57] Teiji Takahashi December 2009 Laws and Regulations on Food Safety and Food Quality in Japan Accessed June 15, 2014

[58] Japan's Alien species and LMO Regulation office, Wildlife Division, Nature Conservation Bureau Ministry of the Environment Biosafety Regulations in Japan Accessed June 15, 2014

[59] Australian Government, Department of Agriculture. Last reviewed July 4, 2011 Appendix A - Regulatory Arrangements for GMOs and GM products in Australia and Australia's Major Export Markets for Canola and Cottonseed Accessed June 15, 2014

[60] http://ca.judiciary.gov.ph/cardis/SP00013.pdf

[61] http://edigest.elaw.org/sites/default/files/ph.greenpeacese.pdf

[62] http://edigest.elaw.org/sites/default/files/ph.eggplantsept2014.pdf

[63] http://m.greenpeace.org/international/en/high/press/releases/Philippines-Supreme-Court-bans-development-of-genetically-engineered-products/

[64] Supreme Court of the Philippines. December 12, 2015

[65] Lynch D, Vogel D. (2001). *The Regulation of Gmos in Europe and the United States: A Case-Study of Contemporary European Regulatory Politics.*.

[66] GMO Compass: The European Regulatory System. Retrieved 28 July 2012.

[67] Wesseler J. and N. Kalaitzandonakes (2011): Present and Future EU GMO policy. In Arie Oskam, Gerrit Meesters and Huib Silvis (eds.). EU Policy for Agriculture, Food and Rural Areas. Second Edition, pp. 23-323 – 23-332. Wageningen: Wageningen Academic Publishers

[68] "Health and Consumers: Food and feed safety." (under "What are the National safeguard measures?") Link. Retrieved 28 July 2012.

[69] Beckmann, V., C. Soregaroli, J. Wesseler (2006): Co-Existence Rules and Regulations in the European Union. American Journal of Agricultural Economics 88(5):1193-1199

[70] Cooper, Alice. "Political Indigestion: Germany Confronts Genetically Modified Foods." German Politics 18.4 (2009): 536-558. Academic Search Premier. Web. 6 Dec. 2012.

[71] Beckmann, V., C. Soregaroli, and J. Wesseler (2010). Ex-Ante Regulation and Ex-Post Liability under Uncertainty and Irreversibility: Governing the Coexistence of GM Crops. Economics: The Open-Access, Open-Assessment E-Journal. Vol. 4, 2010-9

[72] Groeneveld, R., J. Wesseler, P. Berentsen (2013): Dominos in the dairy: An analysis of transgenic maize in Dutch dairy farming. Ecological Economics 86(2):107-116

[73] Skevas, T., P. Fevereiro, J. Wesseler (2010): Coexistence Regulations & Agriculture Production: A Case Study of Five Bt Maize Producers in Portugal. Ecological Economics. 69(12):2402-2408.

[74] Staf (19 May 2004) European Union lifts GM food ban BBC. Retrieved 7 September 2012

[75] EU GMO ban was illegal, WTO rules Published 12 May 2006, retrieved 28 July 2012.

[76] *EC – Approval and Marketing of Biotech Products (Disputes DS291, 292, 293)* World Trade Organisation, retrieved 28 July 2012.

[77] Potato Pro Newsletter

[78] "GM Potato Approval 'A Big Step for Germany'" Link. Published 3 March 2010, retrieved 28 July 2012.

[79] Scientific background report AMFLORA potato VIB (Flemish Institute for biotechnology), Belgium. Retrieved 29 October 2010.

[80] Prof. František Sehnal, Prof. Jaroslav Drobnik, Editors, (2009) White book Genetically Modified Crops

[81] "Biotech firms warn EU over pace of GM crop approvals." Link. Published 11 Oct 2011, retrieved 28 July 2012.

[82] Hogan, Michael (5 April 2012) BASF to undertake GMO potato trials in Europe Reuters, Retrieved 7 September 2012

[83] Staff EU register of genetically modified food and feed European Commission. Health and Consumers. EU register of authorised GMOs. Retrieved 15 August 2012

[84] "The Irish Potato: Will Consumers Eat Genetically Modified Spuds?" Published 27 July 2012, retrieved 28 July 2012. Link.

[85] *Nature Biotechnology.* "Europe legitimizes GM crop exclusion zones." Table 1: National bans currently implemented under the 'safeguard clause.' Link. Retrieved 28 July 2012.

[86] Ricroch A, Bergé JB, Kuntz M (February 2010). "Is the German suspension of MON810 maize cultivation scientifically justified?". *Transgenic Res.* **19** (1): 1–12. doi:10.1007/s11248-009-9297-5. PMC 2801845. PMID 19548100.

[87] (AFP) – 8 February 2008 (8 February 2008). "AFP: French GM ban infuriates farmers, delights environmentalists". Afp.google.com. Retrieved 8 March 2010.

[88] French ban on biotech Monsanto corn ruled illegal Agrimony UK. 28 November 2011. Retrieved 30 December 2011

[89] "French ban of Monsanto GM maize rejected by EU." Link. Published 22 May 2012, retrieved 28 July 2012.

[90] Thorsten Severin and Michael Hogan (14 April 2009). "Germany to ban cultivation of GMO maize-Minister". Reuters. Retrieved 8 March 2010.

[91] "Bulgaria Puts Total Ban on GM Crops". Novinite. 18 March 2010. Retrieved 25 June 2012.

[92] "GMO : POLAND PROMISES STRICT CONTROL ON GM CROPS." Europe Agri 25 Mar. 2013: 332575. General OneFile. Web. 11 Oct. 2013.

[93] European Food Safety Authority, 2012. Scientific opinion addressing the safety assessment of plants developed through cisgenesis and intragenesis. EFSA Journal 2012;10(2):2561ff

[94] (27 October 2014) Europe must lift GM food limits to help feed planet, say experts The Daily Telegraph. Retrieved 28 October 2014

[95] "GMO-free zone: Germany tells EU it bans genetically modified crops cultivation". RT. 2015.

[96] Directorate-general for agriculture and rural development. "Economic impact of unapproved gmos on eu feed imports and livestock production" (PDF). European Commission.

[97] *Europa* summaries of EU legislation: Traceability and labelling of GMOs. Link.

[98] GMO Safety. "New coexistence - Guidelines in the EU: Cultivation bans are now permitted".

[99] Coextra

[100] BioTrack Product Database

[101] accessed 21 January 2011

[102] Plucodes.com

[103] "Jeffrey Smith: PLU Codes Do Not Indicate Genetically Modified Produce". Huffingtonpost.com. 25 May 2011. Retrieved 2012-02-07.

[104] Trade barriers seen in EU label for bio-engineered ingredients. (Regulatory and Policy Trends). Business and the Environment 13.11 (Nov 2002): p14(1).

[105] GMO Compass Rapeseed July 27, 2010. Retrieved August 6, 2010.

[106] Canadian Food Inspection Agency - Regulating Agricultural Biotechnology

[107] Genetically Modified Food.

[108] Evans, Brent and Lupescu, Mihai (15 July 2012) Canada - Agricultural Biotechnology Annual – 2012 GAIN (Global Agricultural Information Network) report CA12029. United States Department of Agriculture, Foreifn Agricultural Service. Retrieved 2 November 2012

[109] McHugen, Alan (September 14, 2000). "Chapter 1: Hors-d'oeuvres and entrees/What is genetic modification? What are GMOs?". *Pandora's Picnic Basket.* Oxford University Press. ISBN 978-0198506744.

[110] Staff (28 November 2005) Health Canada - The Regulation of Genetically Modified Food Glossary definition of Genetically Modified: "An organism, such as a plant, animal or bacterium, is considered genetically modified if its genetic material has been altered through any method, including conventional breeding. A 'GMO' is a genetically modified organism.". Retrieved 2 November 2012

[111] Staff (20 July 2012) Voluntary Labelling and Advertising of Foods that are and are not Products of Genetic Engineering Public Works and Government Services Canada. National Standard of Canada. Retrieved 1 November 2012

[112] Making the Market "Safe" for GM Foods: The Case of the Canadian Biotechnology Advisory Committee | Prudham | Studies in Political Economy

[113] Mexico approves planting and sale of GM crops - SciDev.Net

[114] Mexico: controlled cultivation of genetically modified maize

[115] Mike Shanahan Warning issued on GM maize imported to Mexico - SciDev.Net 10 November 2004

[116] Katie Mandell GM maize found 'contaminating' wild strains - SciDev.Net 30 November 2001

[117] David Alire Garcia (12 November 2013). "Past and future collide as Mexico fights over GMO corn". *Reuters.* Retrieved 9 September 2015.

[118] Clive James (2009). "ISAAA Brief 41-2009: Executive Summary: Global Status of Commercialized Biotech/GM Crops The first fourteen years, 1996 to 2009".

[119] United States Regulatory Agencies Unified Biotechnology Website

[120] Emily Marden, Risk and Regulation: U.S. Regulatory Policy on Genetically Modified Food and Agriculture 44 B.C.L. Rev. 733 (2003)

[121] Pollack, Andrew (2015-07-02). "White House Orders Review of Rules for Genetically Modified Crops". *The New York Times*. ISSN 0362-4331. Retrieved 2015-07-03.

[122] FDA page for Q & A on GM Food

[123] FDA page on Regulation of GM Plants in Animal Feed

[124] "Guide to U.S. Regulation of Genetically Modified Food and Agricultural Biotechnology Products" (PDF). *The Pew Initiative on Food and Biotechnology*. Washington, DC: The Pew Charitable Trusts. 2001. Retrieved 2012-06-02.

[125] "Consultation Procedures under FDA's 1992 Statement of Policy - Foods Derived from New Plant Varieties". FDA. October 1997 [June 1996].

[126] Peggy G. Lemaux Genetically Engineered Plants and Foods: A Scientist's Analysis of the Issues (Part I) Annual Review of Plant Biology 59: 771-812 DOI: 10.1146/annurev.arplant.58.032806.103840. Quote: "Although GE foods can be marketed without certain regulatory approvals, to date all products in the marketplace have undergone full review by regulatory agencies regarding safety and content relative to unmodified forms (searchable data on specific events available at 84). Submitting the safety data is in the developer's best interests, however, given the legal liabilities incurred should a problem with the food arise following market introduction"

[127] Marygold Walsh-Dilley (2009) "Localizing control: Mendocino County and the ban on GMOs" Agriculture and Human Values 26(1-2):95-105

[128] http://www.co.jackson.or.us/page.asp?navid= 3967 . http://www.co.josephine.or.us/Files/ 17-58ballottitlewebsite.pdf. http://www.kgw.com/news/ politics/genetically-modified-foods-260011131.html. Archived May 16, 2014 at the Wayback Machine

[129] Reilly, Genevieve (11 December 2013). "Malloy signs state GMO labeling law in Fairfield". *ctpost.com*. Retrieved 7 March 2014.

[130] Herling DJ, Mintz, L, Cohn, Ferris, Glovsky, Popeo, P.C. (12 January 2014). "As Maine Goes, So Goes The Nation? Labeling for Foods Made with Genetically Modified Organisms (GMOs).". *The National Law Review*. Retrieved 8 March 2014.

[131] Terri Hallenbeck, for the Burlington Free Press. April 27, 2014 How GMO labeling came to pass in Vermont

[132] Terri Hallenbeck for The Burlington Free Press. May 8, 2014 Vermont gov signs law to require labels on GMO foods

[133] staff (June 30, 2015). "Federal judge rules Maui GMO ban invalid".

[134] BASF presentation

[135] Argentina poised to update seed law-Argentina, seed, GM, crop, biotech

[136] (2010) Biotech Facts and Trends Honduras 2010 ISAAA. Retrieved 9 August 2013

[137] (2013)GM Crop Events approved in Costa Rica ISAAA GM Approval database. Retrieved 9 August 2013

[138] (2012) Biotech Facts and Trends Columbia 2012 ISAAA. Retrieved 9 August 2013

[139] Herbst, Allyson (6 June 2013) Government Plans to Expand Use of GM Foods Bolivia Weekly. Retrieved 9 August 2013

[140] (2013) Biotech Facts and Trends Paraguay 2013 ISAAA. Retrieved 9 August 2013

[141] (2012) Biotech Facts and Trends Chile 2012 ISAAA. Retrieved 9 August 2013

[142] (2012) Uruguay Annual Biotechnology Report 2012 USDA Foreign Agricultural Service, Global Agricultural Information Network Retrieved 9 August 2013

[143] Venezuela: Chavez Dumps Monsanto. Globalpolicy.org (2004-05-05). Retrieved on 2011-02-08.

[144] Pena, Karla (2008) "Putting Food First in the Constitution of Ecalthough Puador." Food First Institute for Food & Development Policy. https://www.foodfirst.org/en/node/2301. (updated 31 October 2008, accessed 21 May 2008).

[145] (2 October 2012)Transgenic Products Gain Ground Central American Data. Retrieved 9 August 2013

[146] Northwestern.edu Northwestern Journal of Technology and Intellectual Property Paper on: "Consumer Protection" Consumer Strategies and the European Market in Genetically Modified Foods

[147] "Welcome to the Office of the Gene Technology Regulator Website". Office of the Gene Technology Regulator. Retrieved 25 March 2011.

[148] Agriculture - Department of Primary Industries

[149] Rosemary Polya (17 October 2008). "Chronology of genetic engineering regulation in Australia: 1953–2008". Commonwealth of Australia: Science, Technology, Environment and Resources Section. Retrieved 25 March 2011.

[150] GM Crops and Stockfeed

[151] GM Carnations in Australia

[152] Information Paper on Genetically Modified Canola A report by the Ministerial GMO Industry Reference Group Chaired by the Hon Kim Chance MLC. May 2009

[153] "GM canola gets the green light". *The Sydney Morning Herald*. 2003-04-01.

[154] GRDC - Australia's first GM canola crop comes off

[155] Ian Walker for the Global Mail. February 2014. Steve Marsh and the Bad Seeds Accessed July 8, 2014

[156] Crothers, Lindy (29 June 2011) Australia, Agricultural Biotechnology Annual. 2011 USDA Foreign Agricultural Service. Global Agricultural Information Network Report Number AS 1120. Retrieved 29 September 2011

[157] *New Zealand Ministry for the Environment*Genetically modified medicines and food June 2004

[158] "Rotorua GE Tree Trial Remains Environment Threat". Soil and Health Association. 2008-03-17. Retrieved 2009-01-23.

[159] Williams, David (2009-01-20). "GE activists call for trials to be ended". The Press. Retrieved 2009-01-27.

[160] Food Standards Australia New Zealand FSANZ GM food approvals 8 pages, n.d. 2009?

35.6 Text and image sources, contributors, and licenses

35.6.1 Text

- **Genetic engineering** *Source:* https://en.wikipedia.org/wiki/Genetic_engineering?oldid=712491717 *Contributors:* AxelBoldt, Magnus Manske, Sodium, Mav, Bryan Derksen, Zundark, The Anome, Taw, Malcolm Farmer, Vignaux, Rmhermen, William Avery, Anthere, Azhyd, Heron, Bth, Michael Hardy, Tim Starling, EddEdmondson, Booyabazooka, Lexor, Kku, Gabbe, Minesweeper, 168..., Ahoerstemeier, Muriel Gottrop~enwiki, JWSchmidt, BigFatBuddha, Sir Paul, Bogdangiusca, Netsnipe, Evercat, Astudent, Mxn, Kat, Charles Matthews, Guaka, StAkAr Karnak, Zoicon5, Quux, Maximus Rex, Wenteng, Furrykef, Angiotensinogen, SEWilco, Anupamsr, Bjarki S, AnthonyQBachler, PuzzletChung, Robbot, Fredrik, Tomchiuke, Jredmond, R3m0t, Jmabel, Netizen, Lowellian, Samrolken, Thunderbolt16, Babbage, Academic Challenger, Hadal, HaeB, Pengo, Alan Liefting, Giftlite, MPF, Tom harrison, Lupin, Mark Richards, Peruvianllama, Everyking, Frencheigh, DO'Neil, Guanaco, Siroxo, Deus Ex, Edcolins, JRR Trollkien, Kandar, Wmahan, Utcursch, Dullhunk, Nova77, LiDaobing, Antandrus, Beland, Loremaster, PDH, Rd-smith4, Girolamo Savonarola, Jesster79, leairns, Zfr, Sam Hocevar, Neutrality, Joyous!, Jew69, Ukexpat, Syvanen, Clemwang, Deglr6328, Adashiel, Trevor MacInnis, Grunt, Mike Rosoft, Poccil, Discospinster, Solitude, Rich Farmbrough, Vsmith, Bender235, Kbh3rd, Ttguy, Brian0918, Tirdun, Edwinstearns, Hayabusa future, Shanes, RoyBoy, Triona, Bookofjude, Semper discens, Causa sui, Bobo192, Smalljim, Nectarflowed, NightDragon, Jolomo, Arcadian, Vystrix Nexoth, Bawolff, Larryv, Nhandler, Krellis, ADM, Grutness, Alansohn, JYolkowski, Verdlanco, Linmhall, Riana, Nicholas Cimini, Splat, Dougward, Lightdarkness, Seans Potato Business, Mineralogy, Mailer diablo, Katefan0, Snowolf, Brown Shoes22, Atomicthumbs, Super-Magician, ReyBrujo, Yuckfoo, Amorymeltzer, Randy Johnston, TenOfAllTrades, Mnolander, Dzhim, Bsadowski1, Harvestdancer, Xmort~enwiki, Y0u, Abanima, Stemonitis, Pekinensis, Woohookitty, Mindmatrix, JarlaxleArtemis, TigerShark, Swiftblade21, Commander Keane, Duncan.france, Tabletop, Bkwillwm, SCEhardt, Hughcharlesparker, Prashanthns, Gimboid13, DL5MDA, Mandarax, Matilda, Magister Mathematicae, Deadcorpse, FreplySpang, Yurik, Enzo Aquarius, Sjö, Drbogdan, Rjwilmsi, Nightscream, Syndicate, Kinu, Collins.mc, Hulagutten, Vary, Bruce1ee, StephanieM, Mohawkjohn, DoubleBlue, Yamamoto Ichiro, Richard B., Titoxd, Ian Pitchford, Ground Zero, Latka, Nihiltres, Doucher, RexNL, Gurch, Leslie Mateus, Brandon402, Fieryfaith, Brendan Moody, Alphachimp, McDogm, Common Man, King of Hearts, Chobot, Flying Jazz, JesseGarrett, Bgwhite, NSR, WriterHound, Gwernol, Kdehl, The Rambling Man, Yurik-Bot, Wavelength, RobotE, Crotalus horridus, Sceptre, Kafziel, Sarranduin, Postglock, Kazikameuk, Spaully, Lexi Marie, Chris Capoccia, SpuriousQ, Stephenb, Alvinrune, Rsrikanth05, Wimt, Ugur Basak, Big Brother 1984, Anomalocaris, NawlinWiki, Wiki alf, BigCow, Bachrach44, DJ Bungi, Arichnad, Cquan, RazorICE, Dureo, Irishguy, Aaron Brenneman, Shinmawa, Muu-karhu, Misza13, Syrthiss, Dbfirs, Aaron Schulz, BOT-Superzerocool, TastyCakes, DeadEyeArrow, PS2pcGAMER, Haemo, Kkmurray, Saric, FF2010, 21655, Mercury1, Theodolite, Encephalon, Theda, Closedmouth, Arthur Rubin, MrTroy, Brz7, JoanneB, TBadger, Chrishmt0423, Kevin, ArielGold, RunOrDie, Kungfuadam, RG2, Sam Weber, DVD R W, WikiFew, Bibliomaniac15, SaveTheWhales, Veinor, Sarah, Random Tangent, SmackBot, Xephael, Estoy Aquí, KnowledgeOfSelf, Royalguard11, Hydrogen Iodide, Blue520, Clpo13, Anastrophe, Grey Shadow, EncycloPetey, Delldot, Beautiful irony, PJM, Brossow, The Ronin, Edgar181, Apers0n, Cool3, Phorner87, Aksi great, Gilliam, Betacommand, Xylophonic chicken, JRSP, Persian Poet Gal, JDC-MAN, Rmt2m, Pylori, Miquonranger03, Silly rabbit, Sadads, FordPrefect42, HubHikari, Baa, DHN-bot~enwiki, Arg, Darth Panda, Scalene, Can't sleep, clown will eat me, Scott3, Jahiegel, Tim Pierce, MeekSaffron, VMS Mosaic, Computerman45, Addshore, Edivorce, Mr.Z-man, Metroid dragon, GVnayR, Philipwhiuk, Lox, Krich, Mytwocents, Khukri, Nakon, Blake-, Richard001, DoubleAW, Smokefoot, Weregerbil, Drphilharmonic, Aboyall, BinaryTed, Hammer1980, Masterdriverz, Jenaisis, Kukini, Imecs, StN, SevenEightTwo, The Ungovernable Force, Weatherman1126, Victor D, SashatoBot, EMan32x, Nishkid64, ArglebargleIV, Saccerzd, Omgitsasecret, Rklawton, Andy1066, Kuru, John, Gobonobo, Breno, Samwise_90, Tim Q. Wells, 2glack, Joffeloff, Ammardiwan, Ocatecir, NongBot~enwiki, Aleenf1, IronGargoyle, PseudoSudo, WalterWalrus3, PANDA(PersonAmendingNumerousDefectiveApostrophes), JHunterJ, MarkSutton, Slakr, Special-T, Tase, Michael Greiner, ThePI, Tyhopho, DI2000, Mego'brien, Ravewolf, SimonD, Fan-1967, Iridescent, Skapur, Shoeofdeath, Sm18, Pjcotterill6445, J Di, Zukeylukey, CapitalR, Blehfu, Wackodraco, Sabik, Zuljin, Tawkerbot2, Nahteeczur, IronChris, AbsolutDan, Ur mom sux, Fvasconcellos, SkyWalker, JForget, Tarchon, Ale jrb, Toadams, Dycedarg, Silversink, CWY2190, Maximilli, Ckuzyk, Chefperson, ONUnicorn, Mc366509, Phædrus, Whereizben, Abdullahazzam, Yaris678, Vaquero100, Cydebot, Acolyte of Discord, Abeg92, Erik Smith, Vanished user vjhsduheuiui4t5hjri, Michaelas10, Carzmaniac, Llort, ST47, Adolphus79, Dr._jon, Odie5533, Grand Slam 7, Tawkerbot4, DumbBOT, Pdemeez, Adz71, Viridae, SpK, Omicronpersei8, MayaSimFan, Zalgo, Maziotis, Professorgupta, Imanupstart, Casliber, Thijs!bot, Epbr123, Jaxsonjo, O, Ucanlookitup, Pcu123456789, N5iln, Mojo Hand, Newton2, Faijer, John254, A3RO, Mmcknight4, James086, Wallet, Dfrg.msc, Philippe, CharlotteWebb, Chillysnow, Masonian, Mentifisto, Ju66l3r, AntiVandalBot, Luna Santin, Wengero, Quintote, Prolog, TimVickers, Smartse, LibLord, North Shoreman, Wing Nut, Liger14, Pixelface, Altamel, Tempest115, Lfstevens, Manu bcn, AubreyEllenShomo, Myanw, Nimbusjdf, JAnDbot, MER-C, Plantsurfer, Db099221, Xeno, Hut 8.5, PhilKnight, Savant13, Joshua, Acroterion, Magioladitis, Bongwarrior, VoABot II, AuburnPilot, CattleGirl, Mbc362, Swpb, Harelx, Starch~enwiki, SineWave, Singularity, KConWiki, Theroadislong, Giggy, Cerajewski, Animum, Ahecht, Ciaccona, Sehumi555, ANONYMOUS COWARD0xC0DE, DerHexer, Super cool SPACE PIRATE!, TheRanger, MasterRadias, Kornfan71, Atulsnischal, Leaderofearth, Hdt83, MartinBot, G-my, Renski, Tech Nerd, APT, Rettetast, Anaxial, Jay Litman, CalendarWatcher, R'n'B, Kateshortforbob, CommonsDelinker, KTo288, MapleTree, WelshMatt, Lilac Soul, Gate28, Artaxiad, Themanfromsiam, J.delanoy, Trusilver, Rgoodermote, Synapomorphy, Maurice Carbonaro, Bbbbbbbbbb~enwiki, Jlbribeiro~enwiki, Keesiewonder, Lantonov, Blotto adrift, Katalaveno, Willie the Walrein, JayJasper, ECH3LON, AntiSpamBot, (jarbarf), Mrmitt94, Krasniy, Jomaye10, SmilesALot, Rwessel, Bollyx, Suckindiesel, EyeRmonkey, Nanajoth, Cmichael, 2help, Cometstyles, RB972, Tiggerjay, Burzmali, Jamesontai, Remember the dot, Sarregouset, Amandaw91, Inter16, Fritol, D icko24, Useight, Permafrost, Kay4087, Henrik ziegler, Beezhive, Martial75, Idioma-bot, Funandtrvl, Peter1592, Remi0o, Chromancer, Lights, G. Völcker, Zaf191, Broked Foot, Jonwilliamsl, VolkovBot, Morenooso, ABF, Davidvennik, Indubitably, Cookingupastory, Adrian two, Stagyar Zil Doggo, Ryan032, Philip Trueman, TXiKiBoT, Booyah3, Maximillion Pegasus, Malinaccier, Walor, Anonymous Dissident, Ilcastro, Qxz, Olly150, Mynameiswill, OlavN, Russiandrey, Rich Janis, Seraphim, DennyColt, Martin451, Leafyplant, Kiyabyers, Amkered, Shasato, THC Loadee, LeaveSleaves, Andrewrost3241981, David in DC, DBragagnolo, Whatiguana, Mtvdmeent, Dragonwish, FinnWiki, Titchna, Redhourlgass, Psion1369, Shokkman, Screamingman14, WatermelonPotion, The Devil's Advocate, Brianga, Chi-ha tank, Monty845, Bobo The Ninja, MaCRoEco, Symane, GavinTing, Michael Allan, Dry dust, EHonkoop, 2ash star2, SylviaStanley, Davidlg86, Subh83, Slugger18, Calliopejen1, Milnivri, Tiddly Tom, Scarian, WereSpielChequers, Hawk Pidgeon, BotMultichill, Penlib, Jsc83, Phe-bot, Dawn Bard, Caltas, Matthew Yeager, King of Corsairs, Drummer26boy, Jcon123456, Jpfreely, Zsomps6, Xenobiologista, Yintan, Keilana, Flyer22 Reborn, Tiptoety, Radon210, Qst, The Evil Spartan, Oda Mari, Askild, JSpung, Oxymoron83, Steven Crossin, Grogman93, Techman224, Sunrise, Maelgwnbot, Charles Luciano, Mygerardromance, Universeman476, Dan droscher, Krefts, Pinkadelica, Kortaggio, Arcaneangel666, IRKAIN, Forluvoft, Loren.wilton, Martarius, ClueBot, Rumping, The Thing That Should Not Be, Tom blum, Jan1nad, Nnemo, Chessy999, Taroaldo, Gregcaletta, Drmies, Niallm90, Mild Bill Hiccup, Serialace, Gkrajeshrajesh, Yiling1990, CounterVandalismBot, Conscars, Blanchardb, Dharma8, Neverquick, Hackstien77, Mbcudmore, Takeaway, Excirial, Alexbot, Jusdafax, Nudve, PixelBot, Eeekster, Muenda, Muhandes, Vivio Testarossa, Sheeprock777, Doctorrosenrosen, Nikwong, Jakob Theorell, JonatasM, Bubble-icious92, Longgrass123, Stephanecastel, Ottawa4ever, Tommillion, La Pianista, Richard Barrett, Thingg, 96well, Mr.X23, Jadeddissonance, Aitias, Jonverve, ForestDim, Scalhotrod, Versus22, Fledgeaaron, Qwfp, SoxBot III, Rhetzky, Vanished User 1004, RMFan1, Semitransgenic, Skunkboy74, Dylanfromthenorth, XLinkBot, Hoterocodile, Redclock22, Jytdog, Swift as an Eagle, 95cornbread23, Keyoti, Crowbarthe1337h4x0r, AndreNatas, Jasynnash2, Noctibus, Ryan skekler, Bragen, RPBnimrod, HexaChord, W363fuuf6, Brave galahad, Addbot, Proofreader77, Cxz111, Adises, Rini genes, Private Sweety, Wsvlqc, Elvire, Betterusername, Landon1980, Captain-tucker, Atethnekos, Wnrud709, Bottyhunter, Rockliffe, Startstop123, CanadianLinuxUser, Chamal N, DFS454, Debresser, Favonian, Njr45, Traviswilt1, Tide rolls, Lightbot, Krano, Jarble, Rojypala, Sergspergs, Yobot, WikiDan61, Etineskid, 2D, Fraggle81, Jackie, David Tornheim, Dr. 12345678, Punctilius, Azcolvin429, LordChuckles2, Eric-Wester, Synchronism, Backslash Forwardslash, AnomieBOT, Brroga, Ciphers, 1exec1, Killiondude, BlazerKnight, Piano non troppo, Keithbob, AdjustShift, Ambrosiaster, Ulric1313, Flewis, Materialscientist, Limideen, Citation bot, Benhen1997, Ninjasaif, Dazdavo11, Felyza, Maxis ftw, GB fan, Frankenpuppy, Anne O'Nemus, LilHelpa, McGavock, Injust, Sionus, 417.417, Cureden, SHINY TOY GUNS, CONGLOGRANT, Kyran0763, Haljolad, Capricorn42, Hoshin12, Gigemag76, Frisbeedude08, Spotfixer, 4twenty42o, Sban85, DSisyphBot, Gilo1969, Shooblah, Gap9551, Inferno, Lord of Penguins, Michwilley, Thmasd525, Ericherdman, Philoponic, Jba28, Shao Lun, Doulos Christos, Trafford09, WaysToEscape, Erik9, Endothermic, Timehigh, Danisgay123, GliderMaven, Appeltree1, Jcatgrl, FrescoBot, Thebanefulprophet, Deanlurie, Tobby72, Roonaldo10, Jssteinke, Ιωάννης Κιραμαιήτρος, DivineAlpha, Citation bot 1, ScienceGeekling, Pinethicket, DTMGO, Cerevisae, Teratragen, Jschnur, RedBot, Btilm, Serols, KnowledgeRequire, Trappist the monk, Lotje, Dolphin887654, Miracle Pen, A4466, Miller6951, Vanished user aoiowainyr894isdik43, Karl Granda123, SoleFulcheri, HomewrokExpert, RjwilmsiBot, Hyarmendacil, DRAGON BOOSTER, Aircorn, EmausBot, John of Reading, Bernard Teo, La pixy, Tayemesay, Going-Batty, Passionless, K6ka, HiW-Bot, Érico, Minor4th, The Nut, John Mackenzie Burke, Tolly4bolly, Wikiproject1400, Donner60, Rangoon11, Herk1955, TYelliot, Abesam, ClueBot NG, Movyn, Satellizer, Nimajneb74000, Tideflat, Frietjes, Braincricket, Widr, DrChrissy, Oddbodz, Helpful Pixie Bot, HMSSolent, Bibcode Bot, Lowercase sigmabot, BG19bot, NewsAndEventsGuy, Vagobot, Cyberpower678, MusikAnimal, AwamerT, Nammu wiki, TheProfessor, NotWith, Jayadevp13, Shlokrchawla, Kjhvm, Aks23121990, A1candidate, BattyBot, Arr4, Cyberbot II, Timelezz, Дмитрий Дж..., Mogism, Frosty, ComfyKem, Benjaminfowler, Gabelglesia, Me, Myself, and I are Here, Joeinwiki, FallingGravity, CsDix, Sosthenes12, Papoose34328, ProtossPylon, JamesMoose, Jakec, Mew4657, EvergreenFir, ElHef, Rmonerie, Dave Castrillon, Evolution and evolvability, Thevideodrome, Kingofaces43, Puppog, The Herald, My name is not dave, Ginsuloft, Amr94, Jackmebarn, LibraryStudent24, John Hancocks, Collin curtin, Gemeatodorakk, JaconaFrere, BilboBaggined, Ecotfey92, Giancarlobasile, Monkbot, History liveson, Indiamonsoon, Grantstanleywilson, Reawooddna, Silent Singularitarian, Trackteur, Χρυσάνθη Λυκούση, Trystin deablo, Fuckit123, 9696969696969a, AsusparkyEP, Chaudeau, Amtk128, Dragonmagicediter, DigitalOwnage, Frenlex, Govindaharihari, Rubbish computer, Ashurii777, ToonLucas22, Hungryce, KasparBot, Ceannlann gorm, Srednuas Lenoroc, Cmaestaz, Oserosteve, Nialldunne96, Denishya saju, ZAD45, Aleksandersen2002, Kowwe and Anonymous: 1533

- **History of genetic engineering** *Source:* https://en.wikipedia.org/wiki/History_of_genetic_engineering?oldid=710926107 *Contributors:* Edward, Fred Bauder, Stesmo, Rjwilmsi, Wavelength, Hardyplants, Cattus, Zeamays, Headbomb, Smartse, Lfstevens, Oshwah, Duncan.Hull, Macdonald-ross, Kitsunegami, Arjayay, Jytdog, AnomieBOT, Citation bot, Gaba p, Trappist the monk, Aircorn, John of Reading, Deirovic, Antonyevans, Golden.2007, ClueBot NG, Widr, Bibcode Bot, Plantdrew, Mdy66, TheProfessor, Batard0, MeanMotherJr, BattyBot, Cyberbot II, Dexbot, Mogism, Timingbeeverything, Comp.arch, Abubakar n garba, Monkbot, Massivefranklin, BrightonC, Evilqueen 7 and Anonymous: 24

- **Introduction to genetics** *Source:* https://en.wikipedia.org/wiki/Introduction_to_genetics?oldid=700792388 *Contributors:* RodC, Alan Liefting, Nunh-huh, JohnArmagh, Bender235, Reinyday, Arcadian, Gary, Etxrge, Woohookitty, Benhocking, WadeSimMiser, Mandarax, RichardWeiss, Wavelength, Huw Powell, Brec, Shanel, Matriak, Yoninah, RUL3R, Werdna, Zzuuzz, Snalwibma, Federalist51, TestPilot, BiT, Timotheus Canens, Richard001, DMacks, Aaker, 16@r, Dl2000, Leevanjackson, AshLin, Thijs!bot, Horologium, Wmasterj, Widefox, TimVickers, Labongo, Lklundin, Professor marginalia, Kuyabribri, Engineman, Jim.henderson, Vox Rationis, Victor Blacus, J.delanoy, Filll, Nbauman, Uncle Dick, Dr d12, Ryan Postlethwaite, Juliancolton, Tokenhost, Butwhatdoiknow, Z.E.R.O., Locogato, Sintaku, Jackfork, Lova Falk, SylviaStanley, SieBot, Graham Beards, ConfuciusOrnis, Andrewjlockley, Bentogoa, Flyer22 Reborn, Nk.sheridan, Naturespace, Forluvoft, ClueBot, Antarcticadventurer, XmaceX, Artichoker, The Thing That Should Not Be, Sting au, Jaums, Niceguyede, GoEThe, Aua, Deselliers, Exciral, Rhododendrites, Sun Creator, Hthiddler, Jonverve, Johnuniq, Little Mountain 5, Little Stupid, Addbot, Idkmybffjill27, Seipjere, DOI bot, ContiAWB, Ccacsmss, Gail, Ettrig, Luckas-bot, Yobot, Azcolvin429, Backslash Forwardslash, Kristen Eriksen, Starproject, Materialscientist, Citation bot, La comadreja, Xqbot, The Roman Candle, Brandon5485, Sesu Prime, DoctorDNA, Mark Renier, HRoestBot, Tom.Reding, Pedromelcop, Vkil, Lotje, Mtinker86, Eurye, RenamedUser01302013, James.harris.anderson, Ysulaiman, Biosicherheit, ClueBot NG, Harold Web, Wetzelman, Encyclopedant, Luceth, TheProfessor, Minorview, YFdyh-bot, T. Reule, Isarra (HG), Dave Braunschweig, Epicgenius, CsDix, Konijnvalstrik, JaconaFrere, Chaya5260, Monkbot, GinAndChronically, IiKkEe, Slawzkii, DatGuy and Anonymous: 89

- **Genetics** *Source:* https://en.wikipedia.org/wiki/Genetics?oldid=712782838 *Contributors:* Magnus Manske, Mav, Bryan Derksen, The Anome, Ed Poor, Andre Engels, Youssefsan, Christian List, SimonP, AdamRetchless, Graft, Heron, DonDaMon, Olivier, Renata, Stevertigo, Zocky, Lexor, Tango, MichaelJanich, 168..., Ahoerstemeier, Ronz, Den fjättrade ankan~enwiki, LittleDan, Mxn, AhmadH, Deoetzee, Greeard, WhisperToMe, Zoicon5, Steinsky, Peregrine981, Samsara, Pir, Flockmeal, Robbot, Jotomicron, Peak, Kowey, Romanm, Chris Roy, Postdlf, Monk, Fuelbottle, Anthony, Giftlite, Seaeagle04, Inter, Netoholic, Tom harrison, Everyking, Michael Devore, Bensaccount, Niteowlneils, Jfdwolff, Duncharris, Lang rabbie, Dolfin~enwiki, SWAdair, Alan Au, Kandar, Wmahan, Stevietheman, Adenosine, Barneyboo, Alexf, Bact, Antandrus, Onco p53, Loremaster, PDH, ShakataGaNai, APH, Jenks, Karl-Henner, Sayeth, Gseshoyru, Joyous!, Adashiel, Bluemask, Maestrosync, Mormegil, ClockworkTroll, Discospinster, Rich Farmbrough, Guanabot, Dancxjo, Notmasnaid, Cariaso, Bender235, Jaberwocky6669, Kbh3rd, Pt, El C, Lycurgus, Hayabusa future, Art LaPella, RoyBoy, Guettarda, Bobo192, Cretog8, Smalljim, Nectarflowed, John Vandenberg, Fremsley, Elipongo, ParticleMan, MarkHab, KBi, Jojit fb, Haham hanuka, Krellis, Nsaa, Jumbuck, Zachlipton, Siim, Alansohn, Plumbago, Lightdarkness, Mailer diablo, Malo, Snowolf, Laundry, Wtmitchell, ClockworkSoul, Yuckfoo, Suruena, Esparkhu, Amorymeltzer, RainbowOfLight, Sciurinæ, Martian, RyanGerbil10, Ron Ritzman, Bobrayner, Natarajanganesan, Boothy443, Woohookitty, Anilocra, Carcharoth, WadeSimMiser, Sengkang, GregorB, Palica, Turnstep, Dysepsion, Mandarax, RichardWeiss, Graham87, GoldRingChip, Drbogdan, Rjwilmsi, Mayumashu, Hughbl, Tawker, Bubba73, Brighterorange, Bhadani, TBHecht, Orb4peace, Maurog, Yamamoto Ichiro, Ravidreams, Titoxd, FlaBot, SchuminWeb, RobertG, Nihiltres, Nivix, Chanting Fox, Andy85719, RexNL, Gurch, Otets, CoolFox, TeaDrinker, McDogm, DVdm, Gulolopez, WriterHound,

The Rambling Man, YurikBot, Wavelength, Angus Lepper, Sceptre, Phantomsteve, RussBot, Petiatil, Serinde, Loom91, Chris Capoccia, Chaser, Akamad, Stephenb, Grubber, Chaos, Rsrikanth05, Wimt, NawlinWiki, SEWilcoBot, Wiki alf, JohnDenny, Daanschr, Dureo, Raven4x4x, Dead-EyeArrow, Nick123, Nikkimaria, Theda, Jwissick, ASmartKid, BorgQueen, GraemeL, Ybbor, DVD R W, Luk, Sardanaphalus, FieryPhoenix, SmackBot, Unschool, Moeron, Prodego, Hydrogen Iodide, David Shear, Bomac, Momirt, Davewild, Grey Shadow, Frymaster, DLH, One-bravemonkey, Edgar181, HalfShadow, Yamaguchi先生, Gilliam, Jdfoote, Hmains, Andy M. Wang, Amatulic, Bluebot, Kurykh, TimBentley, NCurse, Grimhelm, Raymond arritt, Hichris, Miquonranger03, MalafayaBot, SchfiftyThree, Afasmit, Deli nk, Adamstevenson, Abgeneticist, DHN-bot~enwiki, 7258, Colonies Chris, WikiPedant, Zsinj, Can't sleep, clown will eat me, Onorem, ASwann, JonHarder, TheKMan, Rrburke, JohnJHenderson, TKD, RedHillian, Edivorce, GVnayR, Amazon10x, COMPFUNK2, Flyguy649, Mightyxander, Downtown dan seattle, Bio-hunter, Nakon, Richard001, Bulldog123, Drphilharmonic, LeoNomis, Zeamays, Ohconfucius, Cyberevil, Madeleine Price Ball, SashatoBot, Nishkid64, Kingfish, Shlomke, Linnell, Coredesat, Tim Q. Wells, Goodnightmush, Verdant04, IronGargoyle, Bilby, Werdan7, Bendzh, Op-takeover, SandyGeorgia, Ryulong, Jessezhao, RichardF, Citicat, ShakingSpirit, Roland Deschain, Iridescent, Gholam, Shoeofdeath, Blackprince, Twas Now, RekishiEJ, CapitalR, Woodshed, Harrychiu, Tawkerbot2, Samnuva, JForget, Cogpsych, Ale jrb, Agathman, JohnCD, Leevan-jackson, CWY2190, Dgw, Bernie Wadelheim, Thomas41546, Standonbible, Moreschi, Onemanbandbjm, Nilfanion, Cydebot, Michaelas10, Serkewt, Dancter, Tawkerbot4, Chrislk02, Narayanese, Changmw, Kozuch, Omicronpersei8, Vanished User jdksfajlasd, Heidijane, Gimmetrow, Epbr123, Ultimus, 24fan24, Mojo Hand, Peter Deer, A3RO, Detribe, Zé da Silva, Joymmart, AgentPeppermint, CharlotteWebb, Blathnaid, Thomaswgc, KrakatoaKatie, Cyclonenim, AntiVandalBot, Iat, Luna Santin, Opelio, Doc Tropics, Jj137, TimVickers, Smartse, LibLord, Chill doubt, LéonTheCleaner, JAnDbot, Dan D. Ric, MER-C, Epeefleche, The Transhumanist, Seddon, Infinitenoodles, Andonic, LittleOldMe, Acro-terion, Yahel Guhan, Bencherlite, Efbfweborg, Magioladitis, Bongwarrior, VoABot II, JNW, Jon f, Rivertorch, Steven Walling, SwiftBot, Re-bekahThorn, Bubba hotep, Fabrictramp, Catgut, C.-ting Wu, Animum, BatteryIncluded, Adrian J. Hunter, 28421u2232nfenfcenc, Allstarecho, Emw, DerHexer, GregU, Calltech, DGG, Pvosta, S3000, Atulsnischal, Designquest10, Dave, Hdt83, MartinBot, BetBot~enwiki, MichaelClair, Pwnz0r1377, Wikipedianerd123, Mschel, R'n'B, CommonsDelinker, AlexiusHoratius, Tgeairn, N4nojohn, AlphaEta, J.delanoy, Pharaoh of the Wizards, Filll, DrKay, Bogey97, Colincbn, Cunningham3131, Saeedwiki, Skumarlabot, Paul A. Newman, Acalamari, Katalaveno, Nemvocalist, Vandriel1325, Ignatzmice, Dr d12, Janus Shadowsong, Artgen, Pyrospirit, Joshharsh, Krasniy, NewEnglandYankee, In Transit, I r ud, Oxalate, Ericaschuster, Fjbfour, Gr8white, Entropy, Davcraig75, Cometstyles, Inomyabcs, Burzmali, Xaxx, Useight, CardinalDan, Idioma-bot, Spellcast, Remi0o, Stina335577, Ibnsina786, VolkovBot, IWhisky, DrMicro, Lordmontu, Fbifriday, Macedonian, Bacchus87, Philip Trueman, DoorsA-jar, TXiKiBoT, GimmeBot, Rightfully in First Place, A4bot, Malljaja, GDonato, MarwaQ, Mr10123, Arnon Chaffin, John Carter, Lradrama, Sintaku, Melsaran, JhsBot, Facibene, Lartoven, Mweites, Basketball110, Lant, Lovey73, Fattyjwoods, Dekisugi, Bealegotpinky, Thehelpfu-lone, Ajakajason, Thingg, Acabashi, Aitias, Jonverve, Qwfp, SoxBot III, Egmontaz, C.m.brandenburg, XLinkBot, Hotcrocodile, Rikuansem13, Jytdog, Owl028, Rror, Feinoha, Stickmadicka, Bobcats 23, Borock, Frood, Lngbchderrico, WikiDao, Aunt Entropy, Jr89511, Ejosse1, Lilgoni, Erikamit, HexaChord, CalumH93, Danielleyaya1, Addbot, Some jerk on the Internet, Salwateama2008, DOI bot, Danielkotsias, Shawisland, Blueelectricstorm, Ronhjones, Themdawgs, 39tiro, CanadianLinuxUser, Looie496, Cst17, MrOllie, Glane23, Paris 16, Favonian, LinkFA-Bot, Geniejargon, DubaiTerminator, Norbit22, Tide rolls, Jan eissfeldt, MuZemike, Jarble, Ettrig, Legobot, Luckas-bot, TheSuave, Yobot, Tohd8BohaithuGh1, Mmxx, THEN WHO WAS PHONE?, Jkdelete5, Goodburgernoobhunter, Randombabbleblah, Azcolvin429, AnomieBOT, Inthracis, 1exec1, ThaddeusB, Jim1138, IRP, AdjustShift, Kingpin13, Shogatetus, Flewis, Juan922, Materialscientist, RobertEves92, Citation bot, Lollerskater4life, LilHelpa, Truhiyanu, MauritsBot, Xqbot, ColdDeath~enwiki, Sionus, Capricorn42, Mononomic, Sellyme, Proquence, RadiX, GrouchoBot, Abce2, ProtectionTaggingBot, Omnipaedista, RibotBOT, Mathonius, Amaury, S1066584, Moxy, Methcub, Misortie, A.amitkumar, Griffinofwales, FrescoBot, Pokemon149, Steve Quinn, Commit charge, HamburgerRadio, AstaBOTh15, Pinethicket, Jivee Blau, Voz7, Tinton5, A8UDI, Achaemenes, Reconsider the static, Sturms, Jauhienij, Tim1357, Russot1, Archercondor, RC Howe, Fama Clamosa, Dksnfg, Aequitarum Custos, Kerkukfeneri, Jonkerz, Vrenator, LilyKitty, Clarkej12, Spees112, Suffusion of Yellow, Myplace345, Nascar1996, Tbhotch, Sam'sTheJazzMan, Nomiddlee0824, DARTH SIDIOUS 2, Horneythorney, Onel5969, The Fudginator, RjwilmsiBot, Waters2100, Bhawani Gautam, Aircorn, WildBot, Slon02, EmausBot, HobosTakeMyCash, Pcfreakofbhs, Jerico12, Immunize, Ajraddatz, Dewritech, Go-ingBatty, RA0808, TheManFromYourMom, Zachdragon10, JamesHilt62, Slightsmile, Tommy2010, Wikipelli, Dcirovic, Aidandrummer1, 15turnsm, Fæ, Josve05a, WeijiBaikeBianji, Kiwi128, John Mackenzie Burke, Pacman2580, H3llBot, Mr legumoto, Makecat, Wayne Slam, Aarp65, Ashishkh, Donner60, DesiLady, Aldnonymous, Joannamasel, Jcaraballo, Herk1955, ResidentAnthropologist, Rocketrod1960, Plati-corn, Petrb, Xanchester, ClueBot NG, MickeyVelilla, DHmonkey123, CocuBot, MelbourneStar, This lousy T-shirt, Satellizer, Qazwsxweedede, O.Koslowski, Widr, BeatlesLover, Helpful Pixie Bot, Jujuman5, Christian.peace, Bibcode Bot, Jeraphine Gryphon, BG19bot, TheFallenSol-dier, Rijinatwiki, Tannnnna, Davidiad, J991, FutureTrillionaire, CimanyD, Falkirks, Joydeep, P'tit Pierre, MrBill3, TheProfessor, NotWith, Cehauser1, JZCL, GeneticsSociety, ASHGgenetics, Vanischenu, Shaun, Farter123321, BattyBot, Biosthmors, David.moreno72, Jeffreydavid-speck, 512bits, Biolprof, Popopo8776, JYBot, Mkh388, Harsh 2580, Dexbot, Mogism, Lugia2453, Frosty, SFK2, Lvalance, Ehehcheheh, Jnanni, Reatlas, Joeinwiki, RichardMarioFratini, Probing Mind, Bobisongg.jhh, Vanamonde93, CsDix, Acetotyce, Zeboko14, Seppi333, SarahGWiki, Ginsuloft, JamalHayne, Streverett, Elizabeth sunny, Qwertykeyboard1223, TriXpeediapHd;, Meteor sandwich yum, Publiceditz, Hogwild13, JaconaFrere, Sawtooth72, Grumblegeek, Lmeth, TuxLibNit, Zixus, Monkbot, Fvckmeplease, GinAndChronically, Jasifuentes, Entitymas-terblaster, MusikVarmint, Juloup, Pakornpromkamin, Fionanielsen, Jennifertumer01, ClarDii, Darrenr33, Cynulliad, KeBessy, VicunaBianca, Mrjohnafen, Sarahbratt, Sam4334, KasparBot, STIPMUR, Ch.muhammadasad, Kristasherm, Brennank11, Bellevillea, Rj jabu, SSTflyer, Emily gullu, Greggg230, Belcerv and Anonymous: 1211

- **Genetically modified organism** *Source:* https://en.wikipedia.org/wiki/Genetically_modified_organism?oldid=712929348 *Contributors:* Ax-elBoldt, Magnus Manske, Kpjas, Eloquence, Bryan Derksen, Tarquin, Koyaanis Qatsi, Malcolm Farmer, DanKeshet, Ed Poor, Dachshund, Josh Grosse, Christian List, William Avery, Anthere, David spector, Heron, Quercusrobur, Patrick, RTC, Michael Hardy, Zashaw, Fred

Bauder, Lexor, Kku, Liftarn, Gabbe, Ixfd64, Ahoerstemeier, Ronz, Zannah, JWSchmidt, BigFatBuddha, Glenn, Llull, Mxn, Hashar, Dysprosia, Wik, Katana0182, Big Bob the Finder, Almak, Dogface, Jeeves, Fvw, Owen, Robbot, Chris 73, TimothyPilgrim, Yosri, Lord Bob, Diderot, Hadal, Lupo, Tsavage, Alan Liefting, Mor-enwiki, DocWatson42, Mintleaf-enwiki, Lupin, Fastfission, Maroux, Mboverload, Tweenk, DryGrain, Bobblewik, JRR Trollkien, PeterC, Architeuthis, Chowbok, Dvavasour, Cckkab, Beland, Joeblakesley, OverlordQ, PDH, DragonflySixtyseven, Sidney, Jh51681, Klemen Kocjancic, Canterbury Tail, Lectincircuit, DanielCD, Discospinster, Rich Farmbrough, Guanabot, Vsmith, MuDavid, Martpol, Yersinia-enwiki, ESkog, Hrodrik, Kbh3rd, Ground, Danny B-), Ttguy, Brian0918, Uli, CanisRufus, RoyBoy, Euyyn, CDN99, Bobo192, Viriditas, Arcadian, La goutte de pluie, Rajah, Helix84, Fox1, Nsaa, Emoticon, ADM, Storm Rider, Alansohn, Free Bear, Rd232, Jeltz, Riana, Kurieeto, JanSöderback, Hu, Malo, Bart133, Melaen, Velella, BanyanTree, Knowledge Seeker, Docboat, Amorymeltzer, RainbowOfLight, Mnolander, Deward, SteinbDJ, Ceyockey, Xmort-enwiki, Alkarex, Stemonitis, Bobrayner, Kelly Martin, OwenX, Camw, Pol098, JeremyA, Xaliqen, Knuckles, GregorB, Wayward, Mandarax, Graham87, Ypotier, Jermlai, Sjö, Drbogdan, Rjwilmsi, Brucelee, Kajmal, StephanieM, Brighterorange, MarnetteD, Hungrymouse, Gmo-enwiki, Old Moonraker, Nihiltres, Crazycomputers, Sophrosune, Firehox, RexNL, Wetawran, Gurch, Ayla, Kolbasz, Chobot, Flying Jazz, SirGrant, DVdm, Sasoriza, Digitalme, WriterHound, Vmenkov, UkPaolo, Roboto de Ajvol, Wavelength, Deadlyhead, Jzylstra, Mahahahaneapneap, Mukkakukaku, JarrahTree, Killervogel5, RadioFan, Shaddack, Rsrikanth05, Anomalocaris, Yserarau, Turgonml, Bachrach44, Dialectric, Albedo, FreelanceWizard, Misza13, Bucketsofg, Xompanthy, PS2pcGAMER, LW-enwiki, Hrvoje Simic, Intershark, TheSeer, Mike Serfas, 2over0, Rushyo, Encephalon, Nightryder84, Cyrus Grisham, Arthur Rubin, Empion, Vicarious, CWenger, Shawne, Peter, ArielGold, Whouk, Allens, Mdwyer, RG2, Eitch, Blastwizard, Fritsky-enwiki, SmackBot, Senski, Vermoskitten, Haymaker, KnowledgeOfSelf, McGeddon, Pmaas, Wehwalt, Clpo13, WookieInHeat, Delldot, Jab843, Hardyplants, Brossow, Kintetsubuffalo, Typhoonchaser, Cool3, PeterSymonds, Gilliam, Tyciol, Chris the speller, Kurykh, Persian Poet Gal, RDBrown, Rajanm13, George Church, Mrarfarf, Deli nk, MIB4u, Antonrojo, Primacag, Garble, Can't sleep, clown will eat me, Alexstep uk, Skidude9950, Nixeagle, OSborn, Rrburke, Krich, MrRadioGuy, Ritchie333, Kandarin, Nakon, ALYOSHA, TedE, Jared, Blake-, David cameron, Bronzie, Hex4def6, DenisRS, Hammer1980, DMacks, J.smith, Bidabadi-enwiki, Pilotguy, Kukini, Qmwne235, Evets70, Srikeit, Bloody rox, Bic1313, Soap, Fanx, AmiDaniel, Wtwilson3, Gobonobo, Mrsatanpants, Luizabpr, Shlomke, Minna Sora no Shita, Notme5, IronGargoyle, PseudoSudo, Ben Moore, Ckatz, Stratadrake, Acaryatid, SQGibbon, Mr Stephen, Optakeover, SandyGeorgia, NJA, Ryulong, Lenn0r, H, DI2000, Hu12, BranStark, Iridescent, Shadoman, JoeBot, LadyofShalott, Courcelles, Radiant chains, Tawkerbot2, Ayanoa, Fdot, Sami4-enwiki, Dvdhn, Agathman, Makeemlighter, BeenAroundAWhile, Leevanjackson, Mcswell, GHe, Growerotl, MrFish, Bobnorwal, Trimp, Fnlayson, Nbound, Roman Cheplyaka, MC10, Arrowned, Katherine Tredwell, Epistaxis-enwiki, Tawkerbot4, DumbBOT, Ameliorate!, SpK, Daven200520, Zalgo, Maziotis, Gimmetrow, RickDC, SummonerMarc, Thijs!bot, Epbr123, Mercury-enwiki, CynicalMe, Headbomb, Marek69, Electron9, Horologium, Detribe, James086, Yettie0711, CameoAppearance, Davidhorman, Wallet, AgentPeppermint, Giac83, Raupp, David D., AntiVandalBot, MoogleDan, Majorly, Luna Santin, Seaphoto, Prolog, Telaridge, TimVickers, Smartse, Pizzazzle, Britishbabe94, Darklilac, Joimbob, Lfstevens, Jordan Rothstein, Ghmyrtle, Weaselmaster, Weaselmaster2, Esuzu, JAnDbot, Thechristian, Barek, MER-C, Ermengrabby, Xeno, Vanished user s4irtj34tuvkj12erhskj46thgdg, MaxPont, Magioladitis, VoABot II, Weebiloobil, Carlwev, NoDepositNoReturn, Ishikawa Minoru, Dekimasu, Cyzorb, Singularity, Steven Walling, Midgrid, Catgut, Cerajewski, Emw, Lenticel, TimidGuy, Wook5, Atulsnischal, Yobol, MartinBot, R'n'B, Mlfitzpat, Tgeairn, Cotton-enwiki, J.delanoy, Pharaoh of the Wizards, Trusilver, Euku, Bbbbbbbbbb-enwiki, Mr Rookles, Roman V. Odaisky, Anonywiki, Oceanflynn, BoredTerry, Richard D. LeCour, NewEnglandYankee, Rwessel, DadaNeem, Thegreatestrevenge, Lillialexis, Entropy, Cometstyles, Kdawson66, Remember the dot, Agrofe, Balohmann, Pdcook, Petitepassionz, CardinalDan, G. Völcker, Xorgthezombie, King Lopez, 28bytes, VolkovBot, 1Whisky, Murderbike, Meaningful Username, Armetrek, Dlesjack, Soliloquial, Davidwr, Ryan032, Philip Trueman, Zeuron, TXiKiBoT, Theouhiterabbit, Chris-marsh-usa, Vipinhari, Hqb, Malljaja, Miranda, Someguy1221, Albval, CliffingtonFalls, Clarince63, Phillip2, Martin451, Amkered, THC Loadee, Tpk5010, Raymondwinn, Seb az86556, Extra-introvert-enwiki, BotKung, Wiae, Raucanum, Songrit, Madhero88, Meters, M Laurain, Synthebot, Enviroboy, Turgan, Teraman, Sue Rangell, MaCRoEco, SylviaStanley, Bfpage, Karkian, SieBot, Jwray, Kleshni, Leannet3, Gerakibot, Dawn Bard, Caltas, Matthew Yeager, Yintan, Connor mayes, Keilana, Happysailor, Flyer22 Reborn, Alexbrn, Suwatest, Jojalozzo, Enti342, JSpung, Prestonmag, Stolenname, Android Mouse Bot 3, Hello71, Tombomp, Edog718, Fratrep, Streaks102, Maelgwnbot, Cyfal, Tony Webster, Pikamander2, Struway2, Denisarona, Escape Orbit, Corbettreport, Forest Ash, Forluvoft, Dannmerrill, Atif.t2, Sfan00 IMG, ClueBot, GorillaWarfare, Foxj, Infoeco, The Thing That Should Not Be, Tipdrill, Drmies, Dilla418, PolarYukon, Gkrajeshrajesh, Vikte, Karoline2006, DragonBot, Kitsunegami, Excirial, Erebus Morgaine, Sun Creator, Cenarium, Arjayay, PeaceShot, Domni, Ashdenej, Staygyro, Aitias, Scalhotrod, Versus22, PCHS-NJROTC, Thunderstix, MelonBot, Thompsontough, Phil.Austermann, DumZiBoT, Against the current, Fastily, Roxy the dog, Jytdog, Jovianeye, Rror, Gerhardvalentin, Nepenthes, Avoided, Joeyaa, Jaimetex, Bluenausea, WPjcm, RyanCross, Wyatt915, Addbot, Proofreader77, Willking1979, Freakmighty, Non-dropframe, Otisjimmy1, Blechnic, Neodop, Shirtwaist, NjardarBot, LaaknorBot, DFS454, Glane23, MauriceTrainer, FiriBot, Matter 95, Debresser, Favonian, Ginosbot, SamatBot, Rtz-bot, Immortal Horrors or Everlasting Splendors, F Notebook, Isaaa-knowledge center, Gnr234, Tide rolls, Lightbot, Krano, Jarble, Megaman en m, Swarm, Frehley, Legobot, Luckas-bot, Yobot, Fraggle81, Wikipedian2, CinchBug, David Tornheim, AnomieBOT, DemocraticLuntz, 1exec1, Jim1138, Piano non troppo, Keithbob, Flewis, Materialscientist, ImperatorExercitus, Citation bot, Mechamind90, Maxis ftw, ArthurBot, LilHelpa, Gsmgm, Xqbot, Zad68, Cpichardo, Kyran0763, Capricorn42, 4twenty42o, SGS CTS, DSisyphBot, Gilo1969, NathanielGallion, Gap9551, Abce2, Frankie0607, Erdema, 78.26, Pink Distortion, Edmundosargento, Shadow jams, GliderMaven, FrescoBot, Flixi, Ryryrules100, Tobby72, Krj373, D'ohBot, Cookea2, Endie2009, Mel81, Tylerknight69, Citation bot 2, Ἰωάννης Κυπριμήτρος, DivineAlpha, Wireless Keyboard, Cannolis, Citation bot 1, Javert, Japan Hanno, Fagandslag, WQUirich, Gaba p, Pinethicket, I dream of horses, RS32, Edderso, 10metreh, Bsborden, MastiBot, Serols, Yutsi, Meaghan, Forp, Reconsider the static, IJBall, Tim1357, 19cass20, Trappist the monk, LogAntiLog, Sumone10154, Javierito92, Matt Hubner, Vancouver Outlaw, Defender of torch, Paralympiakos, TheGrimReaper NS, Amkilpatrick, Ivanvector, Informed counsel, Tbhotch, Reach Out to the Truth, Sideways713, Threedogmoon, DARTH SIDIOUS 2, Selam1978, The Utahraptor, Philosmystic, RjwilmsiBot, Hajatvrc, Kelvindann, Krispy805, Aircorn, Autumnalmonk, Deagle AP, EmausBot, Energy Dome, John of Reading, Immunize, ScottyBerg, JamesHilt62, Boleroinferno, Ansleyfones, Tommy2010, HeatherWalker, Wikipelli, Deirovic, K6ka, John Cline, Dsonguyen95, Shuapzv3, Xabier Armendaritz, Kiwi128, John Mackenzie Burke, H3llBot, Mtnsmith, Tolly4bolly, Erianna, Seldnis, Ajchristiano, Ventus55, Ariesgago, Donner60, Ediacara, Charliemydog, VictorianMutant, TYelliot, Biosicherheit, 69wikiwhiz69, Petrb, Helpsome, ClueBot NG, Mechanical digger, Condatyla, Vergilden, IrenaK, This lousy T-shirt, Qarakesek, Satellizer, ForgottenHistory, ToreBKrudtaa, Chester Markel, BrekekekexKoaxKoax, Bryonyj, PoqVaUSA, Cntras, Dreth, Tonehtigah, Brickroseo, Widr, DrChrissy, Codewell, Oddbodz, Helpful Pixie Bot, Tristanisms, Chrisk2000, Bibcode Bot, Plantdrew, Mallock, Tlaree, BG19bot, Orphadeus, Petrarchan47, Vikrumli, PTJoshua, Nt4313, Cyberpower678, Hallows AG, MusikAnimal, Yoda956, Mulengaesk, Dan653, BDQU, Snow Blizzard, TheProfessor, Gggggggggbbbb, Moonman2197, Smettems, Ehi2011!, Cascadiaviolet, Eshetlapidot, WikiLerry, Jakegossman, Kjhvm, Anbu121, Dcm32, Justincheng12345-bot, Berky21, Newprinter, Pratyya Ghosh, Rlruggiero4, Arr4,

Cyberbot II, ChrisGualtieri, GoShow, Jack.Brighty, King-chode, Aa952, IjonTichyIjonTichy, Mojtaba Pezhman, Dexbot, Rushmchotty, Mogism, 331dot, LalahGrace, Axlroseyboy, Lugia2453, Frosty, SFK2, Graphium, Patriotsfan12345, Dmitry Dzhagarov, ComfyKem, HullIntegrity, Goober69, GabeIglesia, Max Stardust, Reatlas, Newburnstown, Epicgenius, Springmoonshine, Howicus, Mattsabe, Riggy C, PepYoung, Eyesnore, Sosthenes12, Tentinator, Everymorning, EvergreenFir, Geraldatyrrell, ElHef, JacobiJonesJr, Valcorwabajak, Babitaarora, Mynameispikajew234432225, Kingclutchpat, Thevideodrome, Kharkiv07, Ashorocetus, Kingofaces43, Ugog Nizdast, Balljuggler9, The Herald, Zenibus, Prokaryotes, Hansmuller, NottNott, Bronat03, FDMS4, Cosmic.krishna, Jackmcbarn, Catverine, MrScorch6200, CFredkin, Sproffit, Anrnusna, Collin curtin, Alexiaarmstrong, Kcheley, Cboudre7, Gemeaiodorakk, Abubakar n garba, Smashbros1234, Arsenal lb, Sree15, Giancarlobasile, Melcous, Monkbot, HowlingAngel, Dziew, Horseless Headman, Vieque, Addisnog, BethNaught, Stmonkeez, Shur0620, Eman235, Sportmedman, Ike1x, Gingergirl90, Mynameis624740, Biiiiiiiigboy, Amortias, Sy9045, Chaudeau, KurodaSho, Jettcrow, LesVegas, D.S. Cordoba-Bahle, The Old Boy In Town, Crystallizedcarbon, Akhi666, Co9man, JoshuaCat, FourViolas, Orduin, Real life swedish fish, Vanessa.cunningham, Provingpoint, Livethetruth, Goldengotatoofdoom, TheGoldenPotato, Dominic Palumbo, Yobrosef, Havalove, Jerodlycett, Tdnuhn, MaxwellBarr, Theawesomethe, Oksanalarysa, Frankiegreg, Agatewood0007, MarcMarangolo, Rindstar04, Ta mere123321123343234, Kkitagor, Gaybo6969, CommonGround2100, 12terminator12, SourceOG12345678910, NerudaPoet, Cole.ciesla, AccountInCompliance, T.Dub090815, Ballsackmcgee13, Soul1313, Jokerboy023, Zardos1984, Gruntsmoker53, Abcdefghijklmnopqrstuvwxyz01, JGD999, Bobdont eat gmo and Anonymous: 1304

- **Genetic engineering techniques** *Source:* https://en.wikipedia.org/wiki/Genetic_engineering_techniques?oldid=705804608 *Contributors:* Ubiquity, Neutrality, Viriditas, BD2412, Sadads, Iridescent, Lfstevens, CommonsDelinker, Deathgecko, Hzh, Harry-enwiki, MadmanBot, Hutcher, Jytdog, Yobot, AnomieBOT, Thehelpfulbot, Trappist the monk, Aircorn, Banyers99, MelbourneStar, Widr, Wbm1058, BG19bot, Dexbot, Mogism, Enock4seth, Evolution and evolvability, The Herald, Noyster, Anrnusna, Monkbot and Anonymous: 15

- **Genetically modified food** *Source:* https://en.wikipedia.org/wiki/Genetically_modified_food?oldid=712570963 *Contributors:* AxelBoldt, Tarquin, Ed Poor, Youssefsan, Arvindn, Rmhermen, Toby Bartels, William Avery, SimonP, Anthere, Graft, R Lowry, Olivier, Frecklefoot, Boud, Michael Hardy, Dante Alighieri, MartinHarper, Ixfd64, Paul Benjamin Austin, Alfio, Ahoerstemeier, Ronz, BigFatBuddha, Bogdangiusca, °¡°, Kat, Aarontay, Clipdude, Daniel Quinlan, Fuzheado, Sanxiyn, Big Bob the Finder, Bevo, Raul654, GPHemsley, Chuunen Baka, Bearcat, Robbot, Paranoid, Friedo, Altenmann, Modulatum, Postdlf, Pingveno, Academic Challenger, Hemanshu, Texture, SoLando, HaeB, Tsavage, Dina, Tobias Bergemann, Alan Liefting, Giftlite, DocWatson42, Fastfission, Zigger, Marcika, Everyking, Bkonrad, Michael Devore, Zoney, Brockert, Jackol, Bobblewik, Edcolins, Stevietheman, Utcursch, Alexf, Antandrus, Beland, Piotrus, PDH, Jossi, DragonflySixtyseven, Popadopolis, Burschik, Orchie, Jh51681, Ratiocinate, Liberlogos, Eisnel, Maestrosync, Mike Rosoft, Dr.frog, Spiffy sperry, Madewokherd, Discospinster, 4pq1injbok, Rich Farmbrough, Rhobite, Wrp103, Florian Blaschke, Prowsej, Zazou, Bender235, ESkog, PT-enwiki, Kbh3rd, Ttguy, Mr. Billion, RoyBoy, Femto, Shoujun, Bobo192, Longhair, Smalljim, Shenme, Cmdrjameson, Adrian-enwiki, I9Q79oL78KiL0QTFHgyc, Dreikin, Giraffedata, La goutte de pluie, TheProject, Fox1, Hagerman, Pearle, Nsaa, Ranveig, Jumbuck, Vanished user lkjsdkf34ij48fjhk4, Zachlipton, Danski14, Alansohn, LtNOWIS, Arthena, Rd232, Lectonar, SlimVirgin, Iris lorain, Kurieeto, Malo, Titanium Dragon, Bart133, Ombudsman, Wtmitchell, Velella, Max Naylor, Mmolander, A.Kurtz, Zenithan, Netkinetic, Bookandcoffee, Yuriviet, Xmort-enwiki, Kenyon, AndyBuckley, Daranz, Stemonitis, Bobrayner, Kelly Martin, OwenX, Woohookitty, Vash The Stampede, Mindmatrix, Killian, Jacobk, Temuler, Camw, Nuggetboy, John Cardinal, Astator, Pol098, Ruud Koot, Tedneeman, Tabletop, Bluemoose, Wtfunkymonkey, GregorB, Prashanthns, Driftwoodzebulin, Mandarax, Elvey, FreplySpang, Jclemens, Sjö, Rjwilmsi, Coemgenus, Nightscream, BruceJee, MZMcBride, Tawker, Mentality, Rebelgecko, StephanieM, The wub, MarnetteD, Sango123, DirkvdM, Yamamoto Ichiro, Leithp, FlaBot, Musical Linguist, Latka, Winhunter, AAMiller, Nihiltres, Nivix, RexNL, KFP, NavarroJ, Alphachimp, ImpalerBugz, King of Hearts, DVdm, Bgwhite, Hall Monitor, NSR, Gwernol, Tone, The Rambling Man, Wavelength, Phantomsteve, Peoplesunionpro, Anonymous editor, Chris Capoccia, Foxxygirltamara, Stephenb, Gaius Cornelius, CambridgeBayWeather, Rsrikanth05, Wimt, Tavilis, NawlinWiki, Wiki alf, Dialectric, Nirvana2013, Grafen, RazorICE, Spikehay, Joelr31, Dureo, Lexicon, Anetode, Brandon, Moe Epsilon, Panscient, Misza13, Zwobot, Syrthiss, Darkfred, Everyguy, JdwNYC, Werdna, The Spith, Slicing, Wknight94, The Halo, Crisco 1492, TheSeer, 2over0, Zzuuzz, PTSE, Tanjir, Closedmouth, Ketsuekigata, KGasso, Josh3580, JuJube, Mokgand, JoanneB, CWenger, Chrishmt0423, Ybbor, Allens, Katieh5584, Junglecat, Mdwyer, RG2, KKL, Hardkorn, One, That Guy, From That Show!, Mtiffany, Reronk, A13ean, Johntiger1, SmackBot, Nick Dillinger, Moeron, Bobet, Robotbeat, Prodego, KnowledgeOfSelf, TestPilot, DCGeist, C.Fred, Jim62sch, Ramdrake, Yuyudevil, Immanuel goldstein, Ttguy2, Timeshifter, Brossow, Cool3, Rhys 100, Gilliam, Skizzik, Anwar saadat, Chris the speller, TheDarkArchon, Dycotiles, Ian13, Calliopejen-enwiki, JDCMAN, Whywhywhy, Billjw, Miquonranger03, MalafayaBot, SchfiftyThree, Deli nk, Uthbrian, CyberSach, TheLeopard, Baa, Wisden17, Antonrojo, Annelid, Cogito-ergo-sum, Chendy, Newmanbe, Dethme0w, Can't sleep, clown will eat me, JoelWhy, Jahiegel, Jefffire, Nixeagle, KateCH, Rrburke, Addshore, Keordina, Jjjsixsix, Laurent666, Flyguy649, Iapetus, Mitar, The-Theb, Nakon, Jared, "alyosha", Pwjb, Lpgeffen, Bmgoau, MBCF, Hgilbert, DenisRS, Chamaeca.cosmica, Wizardman, KeithB, Zeamays, Vina-iwbot-enwiki, Ck lostsword, Bezapt, Arielco, Andersson h-enwiki, Evets70, Byelf2007, Memilygiraffe, DivineBaboon, Lakinekaki, Robofish, This user has left wikipedia, JoshuaZ, NYCJosh, Peterlewis, CredoFromStart, Scetoaux, IronGargoyle, Llamadog903, Ben Moore, A. Parrot, Hvn0413, Timmeh, Stwalkerster, Acaryatid, Beetstra, Noah Salzman, Fangfufu, Waggers, Tuspm, Whomp, Ryulong, KurtRaschke, Dr.K., HarveyWilliams, Galactor213, Caiaffa, Hu12, BranStark, Fredil Yupigo, Woodroar, Iridescent, JoeBot, NativeForeigner, Fuckyourmother, Courcelles, Anger22, Tawkerbot2, Pi, Jh12, AbsolutDan, Switchercat, JForget, CmdrObot, Porterjoh, Eggman64, Stmrlbs, Weezcake, John Riemann Soong, Styler 13, RedRollerskate, Mystylplx, CWY2190, Dgw, NickW557, Denysmonroe81, SEJohnston, FlyingToaster, Growerotl, Science & Spirit, TheTito, Charlie Huggard, MrFish, Cydebot, Yukino91, Samuell, Marqueed, Abeg92, Valentimd, Vanished user 2340rujowicrfj08234irjwfw4, Steel, Grammaticus Repairo, Mato, Michaelas10, Carzmaniac, Snowboarder123, Tawkerbot4, MatthewAJYD, DumbBOT, FastLizard4, Pdemecz, Andheartssemicolon, Karuna8, Omicronpersei8, Vanished User jdksfajlasd, Maziotis, Jfox11, NadirAli, Rymich13, Jadorno, SummonerMarc, Vash1306, Epbr123, Wikid77, Rencheple, Pstanton, Matthegav, Ucanlookitup, PerfectStorm, Adudenamedpuch, Mojo Hand, Headbomb, Marek69, John254, A3RO, TheTruthiness, Detribe, James086, Universe Man, JustAGal, Wallet, Philippe, FreeKresge, Dawnseeker2000, Scottandrewhutchins, Mentifisto, Hmrox, Thadius856, David D., WikiSlasher, Sidasta, AntiVandalBot, Luna Santin, Seaphoto, Peter50, SummerPhD, Prolog, Doc Tropics, Jj137, TimVickers, Smartse, Coyets, Justinmeister, Dylan Lake, Wing Nut, Vanaprasthi, Lfstevens, Fireice, Timtastik, Gökhan, MikeLynch, Markthemac, JAnDbot, Leuko, Husond, Barek, FemCofounder, Depottey, Bhamv, MER-C, Jabam, Wizardboy777, Sophie means wisdom, Andonic, Sitethief, Tergadare, Vanished user s4irtj34tivkj12erhskj46thgdg, Tstrobaugh, PhilKnight, LittleOldMe, Acroterion, DRHagen, MaxPont, Magioladitis, Tprebble, Pedro, Bongwarrior, VoABot II, Professor marginalia, Vintei, Salinecjr, MastCell, JamesBWatson, Steven Walling, WODUP, Cilstr, TDN169, Froid, Animum, Cgingold, Ifpri, Mtd2006, Robotman1974, Computer genius, Allstarecho, Lethaniol, Cooper-42, Vssun, DerHexer, JaGa, Bottre73, .V., TheRanger, Narayan ran, Newlyarrived, Atulsnischal, Yobol, Rustyfence, MartinBot, Gandydancer, EditorBelisarius, Sjjupadhyay–enwiki, Rettetast, Rob Lindsey, Penikett, Keith D, Jay Litman, CommonsDelinker, AlexiusHoratius, Adomnia, Morehoratioalger, PStrait, Wiki Raja, LedgendGamer, Soul-mine, Tgeairn, J.delanoy, Pharaoh of the Wizards, Trusilver, Bogey97, Uncle Dick, Bbbbbbbbbb–enwiki, Vanished user 342562, Hellosparta, Ieseaturtles, Maproom, Pyrosity10, BanGMFoods, McSly, Jeepday, Nigelt, Jayden54, DvirK, Hut 6.5, NewEnglandYankee, Lukenmatt1234, SJP, Dheerav2, Trident lv, Kraftlos, Shoessss, 2help, KylieTastic, Charlie2301, Juliancolton, Entropy, Cometstyles, Kenneth M Burke, Jamesontai, ACBest, Drink Cillit Bang, Balohmann, Krazor, Ceciliachw, Doctoroxenbriery, Useight, WLRoss, Biotechblogger, ThePointblank, CardinalDan, Idioma-bot, Funandtrvl, Spellcast, G. Völcker, Vranak, Meiskam, Bothell130, Tourbillon, CWii, Magmwalsh, Hersfold, Jeff G., Lear's Fool, Benjiwolf, Philip Trueman, Mogley 666, Zdonuke, Sherip23, Vipinhari, A4bot, Sarenne, GDonato, Olinga, Ann Stouter, Karmos, Viduraine, Enwiki 12, JayC, Arnon Chaffin, Aymatth2, Someguy1221, Laveol, Retiono Virginian, Anna Lincoln, Seraphim, Amkered, EWhitted, Jackfork, LeaveSleaves, Noformation, David in DC, Foodgroupie, Ilyushka88, Natg 19, Wingedsubmariner, ACEOREVIVED, Alen ph, Eddieqi, Madhero88, Fortinne, Meters, PnkRckPrncssLM, Falcon8765, Turgan, Burntsauce, Dinggas, Mohamedegal2, Insanity Incarnate, HiDrNick, Bobo The Ninja, Volman33, Nagy, Oliver Cromwell, Logan, ZBrannigan, Runewiki777, Red, SylviaStanley, Thunderwing, Ari21, Gmoafrica, Nargoth, Coffee, Twopenguins, Waldhorn, Calliopejen1, Tresiden, Pizzachicken, Ødipus sic, Qwertythecat, Gerakibot, Burbaum, Dawn Bard, Caltas, Maxwellfazio, Wateva101, Paulerob, Jjeverett, Keilana, Happysailor, Redpandacat, Flyer22 Reborn, Radon210, Alexbrn, Qst, Chhandama, Oda Mari, Doctorfluffy, Arsenal140392, CarolynETaylor, William Henry Harrison, Hello71, AnonGuy, Lightmouse, The posp, Francisco Tevez, Kingdan10, Fratrep, Sunrise, Jimtpat, StaticGull, Capitalismojo, Nimbusania, Pinkadelica, Jons63, Escape Orbit, Thornerag, Explicit, ImageRemovalBot, Faithlessthewonderboy, Twinsday, Loren.wilton, Martarius, Tanvir Ahmmed, ClueBot, ETCGroup, Sennen goroshi, Infoeco, The Thing That Should Not Be, Abhinav, EoGuy, Jan1nad, ImperfectlyInformed, Matsuiny2004, Bullet2811, Randall1190, Ndenison, Drmies, Dosciai2, Boing! said Zebedee, Arystarca, Ryandman768, Trivialist, Puchiko, PMDrive1061, Excirial, Jannisri, Kjramesh, Jusdafax, Darcy7, Erebus Morgaine, SpikeToronto, Vivio Testarossa, Lartoven, Simon D M, Sun Creator, Arjayay, Sflorman, Razorflame, Alan94313, Vally3, MeHateNala, Dekisugi, PeaceShot, The Red, Ottawa4ever, Thehelpfulone, Calor, Peace Makes Plenty, Thingg, Aitias, 7, Scalhotrod, Versus22, SoxBot III, Egmontaz, Goodvac, Shauntp, Vanished user uih38riiw4hjlsd, Chhe, DumZiBoT, Finalnight, Semitransgenic, GM Pink Elephant, UltraCaution, BarretB, XLinkBot, Kurdo777, Spitfire, Jed 20012, Alex naish, Jytdog, Byrne0005, Aivilo123, Stickee, Rror, Nepenthes, Little Mountain 5, Avoided, WikHead, ErkinBatu, Crispdy, Prodiety, Alexius08, JIMfoamy1, Noctibus, Mm40, Vianello, ZooFari, GyRIdashai, Mary 0530, Dieyou12, Laughton.andrew, ZYoay gnoz, Kbdankbot, HexaChord, Thebestofall007, DOI bot, Non-dropframe, AlbinoFerret, Rockliffe, Blueelectricstorm, TutterMouse, Jncraton, Fieldday-sunday, Startstop123, KitchM, Harryboardman, Scientus, CanadianLinuxUser, Redsoxfan1, Ashanda, Cst17, Morning277, Chzz, Deamon138, Podpoet, Nguyendude, Slatedorg, Quercus solaris, 5 albert square, Tyw7, Coolisha 500, Dyuku, Bwrs, Liveidiot, Tide rolls, Verbal, Muiranec, Apteva, Luckas Blade, CountryBot, TruthFaith, Swarm, Ben Ben, आर्षीष भटनागर, Yobot, Worldbruce, Indira Y Reddy, LegalFiction, Nomapsonmytaps, Darx9url, Torquetime, 4fiiikw, Aboalbiss, David Tornheim, SwisterTwister, Azcolvin429, OregonD00d, Out Kindly, DiverDave, AnomieBOT, Nutriveg, DemocraticLuntz, Jim1138, Piano non troppo, AdjustShift, Kingpin13, LMBM2012, Ulric1313, RandomAct, Wilkesn, Reyrod19, Materialscientist, Purple cush, Citation bot, E2eamon, Felyza, Roux-HG, GB fan, Behemoth1795, A.ericksonwayman, LovesMacs, LilHelpa, Micyclesmith, Xqbot, Natandoron9, S h i v a (Visnu), Cureden, Hayley Tales, Addihockey10, Gigemag76, WillHHudson, Adhitthana, MakeBelieveMonster, PraeceptorIP, Stephen1704, Anna Frodesiak, AbigailAbernathy, RadiX, ProtectionTaggingBot, Zefr, Shattered Gnome, Amaury, Adyione, Doulos Christos, Shadowjams, E0steven, Deadquestions, Trelio345, Kahultman, Dougofborg, 728256723nick, Cekli829, Noahjb120, Jameskmonger, Micyclebicycle, Appeltree1, FrescoBot, Djcam, Glitch82, Mario 4052, Krj373, Sky Attacker, Unomi, Azza12321, Roonaldo10, Jtindall100, Michael93555, Recognizance, Galorr, Shil1111, HJ Mitchell, Age Happens, Mel81, Doom Order, Studstud, Drew R. Smith, Venice133, G.Voelcker, GMO lives!!, HamburgerRadio, Atlantia, Citation bot 1, Mosemamenti, Javert, Intelligentsium, 00Ragora00, Paperman12345, Pinethicket, I dream of horses, Kyleleitch, Vicenarian, Ryanandsam, Pepsi334, Haddof, Steppenwolf24, Hamtechperson, BlackHades, Meaghan, Fumitol, Shanmugamp7, RazielZero, White Shadows, ActivExpression, Rob33322, Kelly Fountain, Trappist the monk, Shimjappu, Lotje, Vrenator, TBloemink, MrX, Miracle Pen, Biggs108, Dusty777, Specs112, Suffusion of Yellow, Tbhotch, Reach Out to the Truth, Stephsolis, Minimac, Keegscee, DARTH SIDIOUS 2, Mean as custard, RjwilmsiBot, Donalwilliamkeane, Bento00, Suzeikew, Noommos, Kerollosm, Meer, Aircorn, Eriewag, Bradleystetz, Enauspeaker, Sadalachbya, Orphan Wiki, T3dkjn89q00vl02Cxp1kqs3x7, Immunize, Ajraddatz, Fishvodka, Yt95, RA0808, Philipp Wetzlar, RenamedUser01302013, NotAnonymous0, Kennykiller7, Passionless, Tommy2010, Winner 42, Wikipelli, K6ka, Ginger999333, Professionaleducator, Mz7, Callumrakhit, Ajb555, Evanh2008, Thundermonty, ZéroBot, John Cline, Fæ, Josve05a, Shuipzv3, Точки над Е, Lateg, Thargor Orlando, Newbiepedian, Davidalightfoot, Minor4th, The Nut, Dffgd, Jenks10, Bertman&robin, Gestaltgs, A930913, Rails, SporkBot, Kingofaces42, Tolly4bolly, Sophie Clayton, GMOscience, TyA, Rhettfight, Ventus55, Mayur, Otframp720, Prof. Michael, Jess, Donner60, MysteriousStrangerintheDark, SBaker43, Eduacara, Puffin, Autoerrant, Orange Suede Sofa, Rangoon11, Wakebrdkid, Ace of Raves, Jenny-Tools, TYelliot, DASHBotAV, Bandrow, Prof. Kirsten Sorbie, Xanchester, ClueBot NG, Jack Greenmaven, Hei Tharr, Vergilden, Wictorya, MelbourneStar, Invitrovanitas, 16rsohns, Frietjes, Moneya, O.Koslowski, Rjbz554, Widr, Scottonsocks, Bigsis11, Chillllls, Paddyandstuff, DrChrissy, Spazem, RafikiSykes, Jemuller1985, Helpful Pixie Bot, Lochie Jay, Waterbug42, Titodutta, Calabe1992, Bibcode Bot, Lacrossguy06, DBigXray, Lowercase sigmabot, BG19bot, Virtualerian, Phuong Huy, Petrarchan47, King7505, CinagroErunam, Nt4313, MusikAnimal, PercyWM, Mark Arsten, Canoe1967, Compfreak7, Rm1271, AdventurousSquirrel, Gorthian, Altair, Furthermost, Legendarygottyline, Hurricanefan24, Snow Blizzard, 31baller, Zujua, Glacialfox, Bige97, Vanischenu, Nesbitr, Shisha-Tom, Aisteco, Fylbecatulous, BrianWo, BattyBot, Simeon Dahl, ShepardoftheEarth, Eduardofeld, RichardMills65, Jahmal.council, Ammelemore1987, Cyberbot II, Aikidojohn, BiankaBrown, ChrisGualtieri, TwizteDope, Saedon, Lunabeast27, Sirjustinhe, Thasmae, Dexbot, Kolega2357, Webclient101, Wikijimmy23, Mendez1993, Superspeller6, Riverstogo, Lugia2453, Noelarthur, Jamesx12345, Dmitry Dzhagarov, BigMatow, ComfyKem, Athomeinkobe, GabeIglesia, Newburnstown, Faizan, TeamJayJay, 004598-49ferrero, I am One of Many, Riggy C, PepYoung, Dumpleshumple, Eyesnore, Jodosma, Pueschi, Everymorning, Bbmusicman, Catrunleen, Geraldatyrrell, Greedyyellow, Jjavier2, Spatel2011, Thevideodrome, Buffbills7701, Kingofaces43, Solowriter, Zenibus, Prokaryotes, Saludparatodos, Ginsuloft, Hatagalow, Cheeseface26, Unstarched Juga, Lemonwooed-661, CFredkin, Limstephanas, Mon3oturf, Chris teleki, Bmccoy1111, Collin curtin, Erinbaker21, Catgao, TiaMarie08, Blue AS47, Seashell1, Jocrevans, JaconaFrere, Lakun.patra, Onerealman123, Rocketjones, Walshie901, R14558, Monkbot, Chesnaught555, Shur0620, Ayrıntılı Bilgi, Sportmedman, Harlza123, SageRad, Matrixuniverses, Conny Feij, Co9man, Spacelaserbunnyobama, Sari Cat, Dracolych, KasparBot, LaTonya43, Stefan19799, JJMC89, Sweepy, SuperSwagilagin, Tlake29, Agent of the nine, GrayDuck156, Mahmud Shakheel, Heaviside glow, Jjkajaja, Nleduc8465, Sweetbacteria, Sablelands807, Qzd, Hjung0524, NIGHTMARE230103, Thefiredragon21, Donniedavid and Anonymous: 2131

- **Gene expression** *Source:* https://en.wikipedia.org/wiki/Gene_expression?oldid=712126624 *Contributors:* Mav, Zundark, Aldie, Deb, William Avery, SimonP, Michael Hardy, Lexor, Kku, 168..., CatherineMunro, Julesd, Habj, Quizkajer, Twang, Robbot, Giftlite, Dmb000006, Bensaccount, AlistairMcMillan, Wmahan, Antandrus, PDH, PFHLai, Bender235, PrometheusOne, Plociam, Adambro, Piotr J Kruk, Tmh, Arcadian,

La goutte de pluie, Rajah, Jamyskis, Dan East, Ceyockey, Natarajanganesan, Fenteany, Rjwilmsi, HonoluluMan, Eubot, DannyWilde, YurikBot, Wavelength, RussBot, Gaius Cornelius, Pseudomonas, Bota47, Gzabers, Mbase1235, Modify, Curpsbot-unicodify, Veinor, KnightRider-enwiki, SmackBot, Jforman, Slashme, Lankenau, Edgar181, Apers0n, Zephyris, Yamaguchi先生, Chris the speller, Jethero, Uthbrian, DHN-bot-enwiki, Zven, Yanksox, Wynand.winterbach, Lassefolkersen, Jreedy21, JonHarder, Vidric, Ivkost, TedE, Smokefoot, FrozenMan, Physis, JanBielawski, Harold f, MightyWarrior, Agathman, Leevanjackson, GV wiki, HilJackson, Narayanese, Optimist on the run, הסרפד, Thijs!bot, Giac83, AntiVandalBot, Ais523, Pwhitwor, TimVickers, Fetchcomms, Mr mutwil, Savant13, VoABot II, Wikiality123, Giggy, Emw, StanleyNickarz, Gomm, Squidonius, MartinBot, Nono64, Twong@emmt.ube.ca, Victor Blacus, Trusilver, Boghog, Yonidebot, KDSKDS, SteveChervitzTrutane, NewEnglandYankee, Heero Kirashami, Pdcook, Tmkates, Spellcast, G. Völcker, VolkovBot, TXiKiBoT, Dllahr, Henrytheli, Ndhoang, Ferengi, Martin451, Luuva, ARUNKUMAR P.R, Qlid, Steven Weston, SieBot, Graham Beards, Gerakibot, Radon210, Gordon014, Boppet, NPalmius, Arostron, Literaturegeek, Brunhilda18, Forluvoft, Loren.wilton, ClueBot, Gits (Neo), Snigbrook, ArneLH, Hadrianheugh, GoEThe, DragonBot, Excirial, Iohannes Animosus, Tombadog, Johnuniq, Super-c-sharp, Helixweb, MystBot, Addbot, Jimbothegreek, Fieldday-sunday, Bastion Monk, Thkim75, MrOllie, Download, LaaknorBot, Shakiestone, OlEnglish, Loupeter, Zorrobot, Ettrig, Legobot, Luckas-bot, Yobot, AnomieBOT, Nutriveg, Piano non troppo, Citation bot, ArthurBot, LilHelpa, Sheha, Xqbot, Cpichardo, Kyng, FrescoBot, Mamaberry11, Citation bot 1, Mizhang118, Pinethicket, I dream of horses, Robinhaw, RedBot, Trappist the monk, Bursting74, Rpazoki, Tstormcandy, Omics, Dancojocari, Firefly's luciferase, Billare, Mashin6, EmausBot, Dewritech, GoingBatty, Wikipelli, Deirovic, Tuxedo junction, MansuriUmar, Thecheesykid, Listmeister, ZéroBot, H3llBot, Erianna, Usb10, Robwwilliams, Minnsurfur2, ClueBot NG, Macarenses, Drevie, The Master of Mayhem, Widr, واجب احمد جلالی, Bibcode Bot, Walk&check, Stevetihi, CatPath, Sr.kheradpisheh, Hellod85, Smettems, Toni 001, ChrisGualtieri, SpectraValor, Mohamed 151995, GargantuanDan, Evolution and evolvability, Seppi333, Mangostaniko, Daleslimnsaw, Mon3oturf, Biotechnologyme, Chaya5260, Ethically Yours, Avatar of Horus, TuxLibNit, Monkbot, Alfakini, IiKkEe, Balabmg, Kosaksi12, Carolinegreen26 and Anonymous: 208

- **Genome** *Source:* https://en.wikipedia.org/wiki/Genome?oldid=710505873 *Contributors:* AxelBoldt, Magnus Manske, Marj Tiefert, LC-enwiki, Bryan Derksen, Zundark, Youssefsan, Vanderesch, Marian, Ben-Zin-enwiki, AdamRetchless, Heron, Youandme, Stevertigo, Zashaw, Lexor, Miciah, 168..., Aboerstemeier, Ronz, Julesd, Glenn, Llull, Mxn, Geoff, Wikiborg, Fuzheado, Hgamboa, Steinsky, Tpbradbury, Taxman, ZeWrestler, SEWilco, Samsara, PuzzletChung, Robbot, Schutz, Seglea, Moink, Pifactorial, VanishedUser kfljdfjsg33k, Centrx, Giftlite, HangingCurve, Herbee, No Guru, Brona, Dmb000006, Niteowlneils, AlistairMcMillan, Delta G, Adenosine, Pgan002, Quadell, Onco p53, Savant1984, PDH, Oneiros, Jenks, Zfr, Iwilcox, Thorwald, Mindspillage, Discospinster, Rich Farmbrough, Vsmith, Cyclopia, Loren36, CanisRufus, Summer Song, Mark R Johnson, Robotje, Smalljim, Reinyday, Kbradnam, Arcadian, Giraffedata, Ramujana, Calebe, Alansohn, JongPark, Chino, Karlthegreat, Eric Kvaalen, Arthena, Stephen Turner, ClockworkSoul, Amorymeltzer, Shoefly, Johntex, Ceyockey, Adrian.benko, Dejvid, Firsfron, Woohookitty, Mindmatrix, WadeSimMiser, Nick Thompson, TheAlphaWolf, Turnstep, GSlicer, RichardWeiss, Drbogdan, Rjwilmsi, Matt.whitby, SeanMack, Gsp, Fish and karate, FlaBot, Doucher, Youssefa, Chobot, DVdm, Whosasking, YurikBot, Wavelength, RobotE, Kafziel, Raccoon Fox, RussBot, The Storm Surfer, SpuriousQ, Chaos, Rosieredfield, Sentausa, NawlinWiki, Snek01, A.bit, JHCaulfield, Open2universe, KGasso, Grmagne, Netrapt, For7thGen, Wootini, Katieh5584, Banus, GrinBot-enwiki, Evolver, DVD R W, True Pagan Warrior, SmackBot, Eperotao, Derek Andrews, Paranthaman, Joconnol, Bomac, IstvanWolf, Apers0n, Gilliam, Ohnoitsjamie, Kaiwen1, Rkitko, MartinPoulter, MalafayaBot, George Church, Polyhedron, Sewlong, Frap, Cregox, MrPMonday, Hgilbert, Rich.lewis, DMacks, Cephalodd, FerzenR, PradeepArya1109, Madeleine Price Ball, SashatoBot, Lambiam, John, JH-man, Stwalkerster, ChazYork, Iridescent, Kaarel, Shoeofdeath, Rhetth, Shrimp wong, Patho-enwiki, Lavateraguy, Agathman, Pathh, SocialContext, Moreschi, Pewwer42, Gogo Dodo, Kylie, Thijs!bot, Kfergy, Escarbot, Quintote, TimVickers, Smartse, Minimice, Huttarl, Sluzzelin, Altairisfar, .anacondabot, Acroterion, Sangak, Magioladitis, Equinexus, Allstarecho, Emw, EoD, Sabedon, Genometer, Doctor Faust, Moorelin, Pvosta, Sjjupadhyay-enwiki, R'n'B, Axelv, Nono64, Trusilver, Svetovid, Mikael Häggström, Skier Dude, SteveChervitzTrutane, Lbeaumont, Joshafina, Sabisteb, Ontarioboy, RJASE1, Black Kite, VolkovBot, Larryisgood, TXiKiBoT, Vipinhari, Scilit, Ferengi, Dave Blank, Alexbateman, Synthebot, Lova Falk, Insanity Incarnate, Onceonthisisland, SieBot, ShiftFn, Scarian, Keilana, AnneDELS, ScAvenger lv, ioverka, Lightmouse, Commutator, Adamace99, Alex.muller, OKBot, MarionADelgado, Forluvoft, ClueBot, Tosendo, Randomaperry, Paul Abrahams, Drmies, Jaums, Peteruetz, Ottava Rima, Namazu-tron, Hawkeye356, DragonBot, Gulmammad, Shinkolobwe, Camera-PR, Dnaphd, Jonverve, Two hundred hum, Johnuniq, InternetMeme, PSimeon, Veryhuman, Purnajitphukon, Candelario Hanson, Jbeans, The Rationalist, Hubcap21, Addbot, DOI bot, Betterusername, MartinezMD, Laurinavicius, CanadianLinuxUser, Download, Carptesticle, George Gastin, Kducey, OlEnglish, Xenobot, Luckas-bot, Yobot, Azcolvin429, AnomieBOT, Jim1138, Betawarrior60, Shogatetus, Jo3sampl, Citation bot, Quebec99, Xqbot, TheAMmollusc, Oxwil, Flavonoid, HYanWong, Tomaschwutz, Holycow32989, ProtectionTaggingBot, Abigor, Wet dog fur, CHJL, Realfoxxx, A.amitkumar, FrescoBot, DoctorDNA, Dogposter, Haeinous, Citation bot 1, DrilBot, Tom.Reding, Drinkybird, MondalorBot, Σ, Awesome girl09, TobeBot, Trappist the monk, Fama Clamosa, Lotje, Tomyhoi, Vrenator, Specs112, Stroppolo, Jesse V., Dr. ambitious, Xmteam, RjwilmsiBot, TjBot, Aircorn, EmausBot, WikitanvirBot, Immunize, GoingBatty, Tareqiu, Thecheesykid, Felipe Sobreira Abrahão, DelianDiver, John Mackenzie Burke, Wayne Slam, Sahimrobot, Orange Suede Sofa, ChuispastonBot, Woodsrock, Will Beback Auto, ClueBot NG, Jack Greenmaven, Amigoswiki, Jbkgiants, Zakarps, Cntras, Matt11111111, A.cristianlucian, Helpful Pixie Bot, TrenSur, Bibcode Bot, Gitana127, BG19bot, Waerfeles, Northamerica1000, Californiadreams, MusikAnimal, Silvrous, CitationCleanerBot, Manchester123456789-enwiki, NotWith, Estevezj, BattyBot, Indianpjn, Mkh388, Wenfuli, Dexbot, Mogism, Lugia2453, Sampsonjk, Alglascock, YaguchiA, Haywardie, Genome biologist KS, Bradleysp1, Sumukal, Vasi31, CsDix, Evolution and evolvability, The Herald, Askpat13, Cyborg1981, Quigend, Skr15081997, Monkbot, ShawntheGod, Bobsaw8081, Kaytopi, Χρυσάνθη Λυκούση, ASDFghJKL.2255, M.Jormungand, GenomeEditor, 1wilsonp, MoreTomorrow, 1115crocodileov, KasparBot, Sarush soti and Anonymous: 295

- **Gene** *Source:* https://en.wikipedia.org/wiki/Gene?oldid=713124428 *Contributors:* AxelBoldt, Magnus Manske, Marj Tiefert, Sodium, Lee Daniel Crocker, Eloquence, Mav, Bryan Derksen, The Anome, Taw, Slrubenstein, Alex.tan, Fnielsen, Youssefsan, SimonP, Anthere, AdamRetchless, Graft, Heron, Youandme, JDG, Michael Hardy, Llywrch, Zashaw, Oliver Pereira, Lexor, Shyamal, Kku, TakuyaMurata, CesarB, 168..., Looxix-enwiki, Ellywa, Aboerstemeier, Snoyes, CatherineMunro, JWSchmidt, Susurrus, Rob Hooft, Mxn, Quizkajer, Zarius, Eszett, Wikiborg, Lfh, Dysprosia, Greenrd, Steinsky, Taxman, Dogface, Samsara, Bevo, Shizhao, Fvw, Wilke, Raul654, Jusjih, Qertis, Tsanth, Francs2000, Mrdice, Phil Boswell, Robbot, Ke4roh, Palnu, RedWolf, Romanm, Sverdrup, Rholton, Rhombus, Smb1001, Acegikmo1, Moink, Hadal, UtherSRG, David Gerard, Giftlite, JamesMLane, Gene Ward Smith, Inter, Michael Devore, Jfdwolff, Duncharris, Suspekt-enwiki, Scottveirs, Sundar, Chameleon, Bobblewik, Wmahan, Pgan002, Andycjp, Pcarbonn, Antandrus, Williamb, Onco p53, G3pro, PFHLai, Kaeleto, Sam Hocevar, Sayeth, Neutrality, JohnArmagh, Sonett72, Ehamberg, Ivo, Dpen2000, Frangibility, Archer3, Discospinster, Rich Farmbrough, Qutezuce, Vsmith, Dufresne77, Murtasa, Ponder, MarkS, Bender235, Richard Taylor, Eric Forste, Brian0918, CanisRufus, Mwanner, Susvolans, Saturnight, Guettarda, Gyll, Sole Soul, Bobo192, Longhair, Smalljim, Func, Kbradnam, Shenme, Tmh, Cohesion, Harvestgalaxy, Giraffedata,

Jerryseinfeld, Physicistjedi, Athf1234, Vanished user 19794758563875, Andrewbadr, John Fader, MPerel, Haham hanuka, Kierano, Pearle, Ranveig, Jumbuck, Alansohn, Gary, Etxrge, Craigy144, Echuck215, Snowolf, ClockworkSoul, Amorymeltzer, Sciurinæ, Mikeo, Computerjoe, Ceyockey, Weyes, Nuno Tavares, Jeffrey O. Gustafson, OwenX, Woohookitty, Kzollman, MarcoTolo, Stefanomione, JohnJohn, RichardWeiss, Cuvtixo, FreplySpang, Yurik, Pmj, Sjakkalle, Rjwilmsi, Mayumashu, Fred Hsu, DeadlyAssassin, You wouldnt dare block me., Sigmalmtd, Gjuggler, The wub, Fred Bradstadt, FlaBot, Latka, Jakob Suckale, Nivix, Ayla, McDogm, Silivrenion, King of Hearts, Chobot, Gwernol, Siddhant, YurikBot, Wavelength, RussBot, Serinde, Ytrottier, Splette, RadioFan, Eleassar, Chaos, Wimt, NawlinWiki, Bachrach44, Waldow, RazorICE, Dogcow, Ragesoss, Davemck, Ospalh, Kyle Barbour, Larsobrien, DeadEyeArrow, Bota47, ColinFine, Wknight94, Jwissick, KGasso, Josh3580, Xaxafrad, Dark Tichondrias, Pádraic MacUidhir, Curpsbot-unicodify, Stuhacking, Allens, RG2, DVD R W, Tom Morris, Eog1916, SmackBot, Tigerghost, Saravask, Herostratus, David Shear, Joconnol, Setanta747 (locked), Geno-Supremo, Jab843, IstvanWolf, Cunya, Gilliam, ElspethB, Skizzik, Kurykh, Persian Poet Gal, NCurse, Tito4000, SchfiftyThree, Afasmit, Miguel Andrade, Colonies Chris, Para, Mikker, Trekphiler, Can't sleep, clown will eat me, JonHarder, Yidisheryid, Xiner, Rrburke, Addshore, Jmnbatista, Pwb, Richard001, Drphilharmonic, DMacks, Kukini, Clicketyclack, Cyberevil, Madeleine Price Ball, Cody5, SS2005, Vanished user 9i39j3, ML5, CinnamonITFC, Tim bates, Mgiganteus1, Tlesher, Cyberstrike2000x, Ben Moore, A. Parrot, Jimmy Pitt, Michael Greiner, P199, RMHED, Sasata, Mycophage, Xionbox, Vanished user, Roland Deschain, Iridescent, Courcelles, Dlohcierekim, Yashgaroth, Atomobot, DBooth, Agathman, Pathh, Vh6666, Rabid Lemur, Neelix, Karenjc, Badseed, HalJor, Lightofglory, Luepke, WillowW, Steel, Vanished user vjhsduheuiui4t5hjri, Gogo Dodo, Was a bee, Serkewt, RealThanny, Pascal.Tesson, BDS2006, Tawkerbot4, Major Despard, Psychodolly, Narayanese, Lee, Mikewax, Makwy2, Sweikart, Scarpy, Satori Son, Thijs!bot, Epbr123, MarvintheParanoidAndroid, Opabinia regalis, Xlonlonx, Headbomb, Pjvpjv, Marek69, Basement12, Syimrvm, Inecossa, Joymmart, JustAGal, AgentPeppermint, WhaleyTim, Dawnseeker2000, Natalie Erin, Ju66l3r, AntiVandalBot, Luna Santin, Akradecki, QuiteUnusual, TimVickers, Smartse, Res2216firestar, Sluzzelin, JAnDbot, NBeale, EmersonLowry, MER-C, Andonic, Achero, TransControl, Sgb235, Bongwarrior, VoABot II, Nyq, SHCarter, Think outside the box, Indon, Adrian J. Hunter, Gomm, GetAgrippa, DerHexer, Tickopa, Pax:Vobiscum, Squidonius, Pvosta, Atulsnischal, MartinBot, STBot, Sjjupadhyay~enwiki, Ggrimes, Blumin, AstarothCY, Neoprote, Anaxial, Jay Litman, Mschel, Test100000, Genetics411, Pekaje, LedgendGamer, Ivan T., AlphaEta, J.delanoy, Pharaoh of the Wizards, Filll, Bogey97, Boghog, Uncle Dick, Singularitarian, Tdadamemd, Keesiewonder, Lantonov, Rod57, Katalaveno, Go Me Go, Number 04, Mikael Häggström, Tarotcards, Memestream, SteveChervitzTrutane, Chiswick Chap, Cnsnmrt~enwiki, NewEnglandYankee, MetsFan76, Mohrflies, Cometstyles, GLHamilton, Ja 62, Million Moments, Rpeh, Iamaboy123, Idioma-bot, Tokenhost, Wikieditor06, Tekkaman~enwiki, Gomackay, Jrugordon, VolkovBot, Lear's Fool, Philip Trueman, Hilario w, TXiKiBoT, Antonov86, GDonato, Oxfordwang, Anna Lincoln, Seraphim, Brunton, Cerebellum, Wrt, LeaveSleaves, Wwwwzzzz, Johnraiti, Luuva, Suriel1981, Eubulides, Masterofsuspense, Asha Vaughn, Kineticpro, Wolfrock, WJetChao, Lova Falk, Jaguarlaser, Karlbrezner, Doc James, AlleborgoBot, Thunderbird2, NHRHS2010, Bfpage, SieBot, Coffee, Graham Beards, Jauerback, Caltas, ConfuciusOrnis, Andrewjlockley, Floppy the bunnyx, Happysailor, Flyer22 Reborn, Radon210, The Evil Spartan, JD554, Oda Mari, Rico12121212, Wilson44691, Edwardlai1992, Wombatcat, Hzh, Emolovekills, John11711, Oxymoron83, Harry~enwiki, Blah696969, Scottman07, Bozzo07, Lightmouse, Dr.Kilioth, Sunrise, Sviek, Tezscion, C'est moi, Mike2vil, Mygerardromance, Aminoacid91, Ptr123, Florentino floro, MinaFam, Revilonnud, Arendedwinter, Felizdenovo, Denisarona, Jbray3179, Micheb, Explicit, Chriff, Forluvoft, Jonathanstray, Church, Sfan00 IMG, Elassint, Emilyrader, ClueBot, Tosendo, The Thing That Should Not Be, Matdrodes, Plastikspork, Nnemo, Enthusiast01, Sting au, Blanchardb, LizardJr8, Mustbcrackers, Peteruetz, Namazu-tron, Masterpiece2000, Excirial, Gujulla.rahul, Gulmammad, Cbailey7, Arjayay, Jackrm, Razorflame, Thehelpfulone, JPLeRouzic, Aliciaa90099, 96well, Aitias, Karolno, Versus22, Johnuniq, SoxBot III, Egmontaz, DumZiBoT, Wmarron, XLinkBot, Rikuansem13, Jytdog, Lilied1, Libcub, Avoided, SilvonenBot, Hazelgrace07, PL290, Noctibus, Kswilson, Thatguyflint, HexaChord, Addbot, Ifetobi17, DOI bot, Ronhjones, Fieldday-sunday, Mai-tai-guy, Tide rolls, Slgcat, Ettrig, Legobot, Luckashbot, TheSuave, Yobot, Fraggle81, Denispir, Nallimbot, Tweek49, R500Mom, Szajci, AnomieBOT, Ciphers, Killiondude, IRP, Galoubet, Piano non troppo, AdjustShift, Merube 89, Flewis, Materialscientist, RobertEves92, Helioarnold, Moppatop, The High Fin Sperm Whale, Citation bot, Obersachsebot, Xqbot, Sionus, Cureden, JimVC3, Capricorn42, Acebulf, Larouxp12, Francisco Cardenas, Cjamesh507, SuperG2daMAX, Classcutie7, Proquence, NOrbeck, -), Donkeylove123, GrouchoBot, ProtectionTaggingBot, Nickayy77, Wise Sage150, RibotBOT, Sophus Bie, 2ndAccount, Shadowjams, Goodtoknowbob, Electricsforlife, FrescoBot, DoctorDNA, Tobby72, Mark Renier, Rotideypoc41352, HamburgerRadio, Citation bot 1, Redrose64, AstaBOTh15, Pinethicket, Abductive, Trevor DE, LinDrug, Trevoreno, Abe5red, RedBot, Andallo, Merosenstein, Jauhienij, Lil brooksy, Trappist the monk, DixonDBot, Lam Kin Keung, Lotje, Millmoss, Chihuahuau, Vrenator, Cowlibob, Reaper Eternal, 564dude, Jlc123456789, Xrmach, Suffusion of Yellow, Reach Out to the Truth, Jesse V., Minimac, DARTH SIDIOUS 2, EngineerFromVega, Andrea105, Woogee, Beyond My Ken, Hajatvrc, Somdna, NerdyScienceDude, Slon02, EmausBot, John of Reading, Snow storm in Eastern Asia, Mordgier, Observer6, Jesshay, Dewritech, GoingBatty, Sxoa, Hellogorge10, JamesHilt62, Blin00, Wikipelli, K6ka, Thecheesykid, JSquish, Felipe Sobreira Abrahão, John Cline, Josve05a, Test19791979, Akstttt, John Mackenzie Burke, Access Denied, Makecat, Edmund Gordon Gey, Capim Dourado, TyA, Brandmeister, GrayFullbuster, ClueBot NG, Invitrovanitas, This lousy T-shirt, Piast93, Karland90, Yourmomssmalldick, Curiouscorey, Cntras, O.Koslowski, Historystudent122, Flyleaflaceymosley, Widr, Antiqueight, Bernard10124, Chillllls, FredMcGary, Timflutre, Helpful Pixie Bot, Wannapwray, Plasticman83, Bibcode Bot, Gell0065, Battyofiron03, Juro2351, PhnomPencil, Scratlikesacorns, AvocatoBot, Edward Gordon Gey, Kim Batteau, Mh40, Tharani117, Claviclehorn, Achowat, NicolasAzurrrie, Aamir mrj, Tutelary, Peggy hopper, Giugiu427, Teammm, Cyberbot II, MadGuy7023, Dexbot, Webclient101, CuriousMind01, Lugia2453, Jennes83, Frosty, SFK2, Logan.Phospholipid, Hopefuldonor, Randykitty, Lingzhi, ProtossPylon, Revolution1221, Robert4565, Evolution and evolvability, Chefcouchon, Demi9m7, Seppi333, Djhk12, Druidiliy, Elizabeth sunny, Quigend, EWAN112, Liam225544, JaconaFrere, Saloofbourrow, Lattouf, Melcous, Monkbot, Cegandodge, Urubamba2011, Alexacorpus5, Roshandp1, Kozhimaster12345, HMSLavender, Archiloe, Mycorrhizal.network, Iaritmioawp, Abskrmn, FourViolas, IEditEncyclopedia, Cyrej, Usernamessssssssss, Craftwerker, Catphamileez, JMWSlack, KasparBot, Melissae530, GalaxyGamer2015, CrownTheCat, Glacialfrost and Anonymous: 894

- **Gene therapy** *Source:* https://en.wikipedia.org/wiki/Gene_therapy?oldid=712471275 *Contributors:* AxelBoldt, Derek Ross, Sodium, Bryan Derksen, Edward, Lexor, Dominus, Kku, Ixfd64, Nina, Axlrosen, 168..., Ahoerstemeier, TUF-KAT, Glenn, Marteau, Llull, Tristanb, LordK, Zarius, Wikiborg, Wik, Zoicon5, Steinsky, Furrykef, Saltine, Skeetch, Robbot, Moriori, Fredrik, Kristof vt, RedWolf, Naddy, Chris Roy, Academic Challenger, Acegikmo1, Hadal, GerardM, Dave6, Giftlite, Jfdwolff, Guanaco, Matt Crypto, JRR Tzollkien, Wmahan, Antandrus, Beland, G3pro, PDH, Rdsmith4, Semenko, Sam Hocevar, Gseshoyru, Rgrg, Joyous!, M1ss1ontomars2k4, Esenco~enwiki, Discospinster, Rich Farmbrough, KillerChihuahua, Vsmith, ESkog, Ebrooks, The improbable, Lankiveil, Kanzure, Bobo192, AmosWolfe, Smalljim, Rackham, Tronno, Shenme, :Ajvol:, Vanished user 19794758563875, MPerel, Sam Korn, Jigen III, Alansohn, Gary, Anthony Appleyard, Atlant, Andrew Gray, Wouterstomp, Axl, Seans Potato Business, Mysdaao, Snowolf, Colin Kimbrell, EAi, TenOfAllTrades, Bsadowski1, Vadim Makarov, Johntex, Ceyockey, Tom.k, Woohookitty, 2004-12-29T22:45Z, Mindmatrix, Etacar11, LOL, StradivariusTV, Benbest, Jeff3000, Graham87, FreplySpang, BorisTM, Sjö, Drbogdan, Rjwilmsi, Eyu100, Tawker, Shalin8th, Bhadani, Fish and karate, FlaBot, Ground Zero, Latka, Jakob Suckale,

Margosbot-enwiki, Nivix, RexNL, OrbitOne, BradBeattie, Wjfox2005, Straker, YurikBot, Wavelength, Hairy Dude, Cabiria, Kafziel, Phantomsteve, Sarranduin, Gaius Cornelius, Eleassar, Wimt, NawlinWiki, Semolo75, Erielhonan, Merman, Nick, RUL3R, Samir, Kkmurray, Nlu, ClaesWallin, Ali K. Encephalon, Rrpbgeek, Arthur Rubin, E Wing, Petri Krohn, DGaw, Richardevan, Back ache, Pádraic MacUidhir, Allens, Tom Morris, Choi9999, Minnesota1, SmackBot, Eperotao, Espresso Addict, KnowledgeOfSelf, TestPilot, Hydrogen Iodide, KocjoBot-enwiki, KVDP, Jab843, Apers0n, Skizzik, Squiddy, Tyciol, Chris the speller, Ajh20, RDBrown, NCurse, George Church, HXII, Mark7-2, Wedian, A. B., Yanksox, Poobarb, Can't sleep, clown will eat me, Shalom Yechiel, JonHarder, Rrburke, JimHu, Rashad9607, Pepsidrinka, Wen D House, Iapetus, Jared, Pwjb, Richard001, TheST, Daniel.Cardenas, Kukini, JzG, John, N1person-enwiki, Shlomke, Dumarest, Mgiganteus1, Melody Concerto, Werdan7, Kyoko, Waggers, D12000, Hu12, DabMachine, Llama Man, Walton One, Martious, Igoldste, Freecat, Maelor, Audiosmurf, Tawkerbot2, Dlohcierekim, Yashgaroth, Tannerhelland, Xcentaur, JForget, Ragingbullfrog, CmdrObot, Ale jrb, Sir Vicious, Dycedarg, Maximilli, TomasB, Dgw, Lemmio, ShelfSkewed, Pgr94, Avillia, Cydebot, Peripitus, Striker265, MC10, Steel, MikeLacey, Anonymi, Anthonyhcole, LMAnthony, Studerby, Shekharsuman, Tawkerbot4, Zian, RXPhd, Viridae, Daven200520, Darth Lamchop, Luka Krstulović, QuietApocalypse, Asterphage, Epbr123, Marooned Morlock, Purple Paint, Marek69, Peter Znamenskiy, John254, A3RO, James086, Caravass, S142968, CharlotteWebb, Transhumanist, M0s6p, Ju6613r, David D., AntiVandalBot, Shiaic, JHFTC, Quintote, Smmalarkodi, TimVickers, Smartse, Salgueiro-enwiki, Lfstevens, AubreyEllenShomo, Gökhan, Sluzzelin, Erxnmedia, Blahbleh, MER-C, Robina Fox, Medishome, Owenozier, Andonic, Purlah, Ryan4314, Nancymarion, PhDsharkey, Chronolegion, Bongwarrior, VoABot II, JamesBWatson, Xtothel, Chavoguero, Avicennasis, 1 JethroBT, Dtaciuch, Kimik0123, Kjmoran, DerHexer, Dunnette, Lentivirus, MartinBot, Sjjupadhyay-enwiki, Rettetast, Cmsjustin, Keith D, R'n'B, Axelv, Smokizzy, LedgendGamer, J.delanoy, BigrTex, Chrisbear68, Boghog, Uncle Dick, Ginsengbomb, Aetkin, Thaurisil, Lantonov, Rod57, Ryan Postlethwaite, MarcoLittel, Gurchzilla, Thurinym, Sedital, InspectorTiger, NewEnglandYankee, Burzmali, S.riccardelli, Bonadea, Lcawte, AltuğGüner RC, Speciate, Fr4ntic32, Prfssrbrnss, VolkovBot, CWii, Soliloquial, Philip Trueman, TXiKiBoT, Zabong-enwiki, Oshwah, Tomcoulter, Burpen, Tameeria, Scilit, Thundermaker, Orielglassstudio, GcSwRhIc, Arnon Chaffin, Qxz, Seraphcheir, Brunton, Leafyplant, Chocowulf, Syunne, Guest9999, Wiae, Sam Lievrouw, Lil Devy, Eubulides, Falcon8765, Anna512, Ninjaboi888, K10wnsta, Monty845, Doc James, Logan, AcademicLawyer, SylviaStanley, GreaterWikiholic, Euryalus, Twinkiest, Dawn Bard, Mackenzie kenzy, Yintan, Seushe, Wikilarry, Bendy660, Hzh, Oxymoron83, Waddie1, Steven Crossin, Lightmouse, Sunrise, Paul Gujt, Wuhwuzdat, Mygerardromance, Martarius, Animeronin, ClueBot, NickCT, Binksternet, Patrick.goertz, The Thing That Should Not Be, Shark96z, Drmies, Mild Bill Hiccup, Neverquiek, Peteruetz, Arunsingh16, Swift1357, Excirial, Jens.forsberg, Alexbot, Jusdafax, Muenda, Fantaseedude, Cenarium, Arjayay, World, Aleksd, George moorey, Parister, Dhalexander, Versus22, DumZiBoT, Finalnight, BarretB, XLinkBot, Jcline0, Gwandoya, Jytdog, PervyPirate, Dthomsen8, TenTonParasol, Trabelsiismail, PL290, Mattachu, Noctibus, Ying.b, Vrasralijgra, Addbot, Mlroberts, Sketerpot, Astergan1, DOI bot, Footballerusad, D0762, CanadianLinuxUser, Mac Dreamstate, MrOllie, Download, Anil.gayali, DFS454, West.andrew.g, Arka.s.ghosh, درفش کاویانی, QuadrivialMind, Luckas-bot, Yobot, Fraggle81, ArchonMagnus, Pablocasal380, FriendlyRobotOverlord, Scottyaz, Groaznic, AnomieBOT, AUG, Pyrrhus16, AdjustShift, Kingpin13, Sniperhail, Ulric1313, Flewis, Akmafia, Materialscientist, Human, Citation bot, Tachturbo, JohnnyB256, Hannah.c.l, Xqbot, Sionus, Capricorn42, Spottixer, Gap9551, Makeswell, GrouchoBot, Abce2, Amaury, Cishaurim, Ocolla, Profwatters, MLauba, MeriLinkBot, Shadowjams, Bobiemon, Catpowerzzz, RetiredWikipedian789, Prari, FrescoBot, DoctorDNA, StaticVision, JMS Old Al, Mewulwe, TruthIIPower, Citation bot 1, ScienceGeekling, Differentitles, Machn, Pinethicket, Parvo art, Pepsi334, Bakul 15, Avidmosh, Theduderog, Roboo.jack, Σ, Jandalhandler, Mjs1991, Shashank Reddy.P, Brokshen, Trappist the monk, Pollinosisss, Rbhuvan14, Dmlevy 99, Vrenator, Allen4names, Ballyjoelover, Jesse V., RjwilmsiBot, Altes2009, Phlegat, Mlj9290, Beeritical, Aircorn, Androst9, John of Reading, Orphan Wiki, Immunize, Jesshay, Dewritech, RA0808, Sooday, Klbrain, Akhilan, Sameer Singhal, Potionism, Chachihassard, Michel Awkal, The Nut, Apocryphals, Hazard-SJ, Cobaltcigs, Cheddar3210, Lexusuns, Ready, Donner60, Septuaginta, Jbergste, Orange Suede Sofa, Maymay Gothica, K. the Surveyor, Whoop whoop pull up, ClueBot NG, Rich Smith, Celltherapynews, Mythole, Thatbiotechguy, Snotbot, O.Koslowski, Widr, Incrediblemouse, Thethird33, Pfabio, Bibcode Bot, BG19bot, Virtualerian, Southwestlane, MusikAnimal, Mark Arsten, Lkjhg631, Saiarjun, MrBill3, Biomat77, ConnorGriffin, Zujua, BeccaGutt, Thegreatgrabber, A1candidate, Zzuffuto118, Idenshi, BattyBot, Jamschad, Several Pending, HueSatLum, Teammm, Stigmatella aurantiaca, Cyberbot II, ChrisGualtieri, Dexbot, AlexandrosClericusRuadh, Makecat-bot, Everything Is Numbers, 93, Andyhowlett, Djdoritos, Leahcepko, FallingGravity, Dustin V. S., Evolution and evolvability, Thevideodrome, Bahoganog61799, White adder, Elizabeth sunny, DrIvenkatesh, Prof.Greg.Davidson, Fixuture, Survivorfan1995, Drsoumyadeepb, 22merlin, Giancarlobasile, Monkbot, Boobies5678, Luckyseven77, Rainmaker6, Goblinshark17, Gronk Oz, SusithCM, Yairchaim, Tracy liu China, CV9933, Jafar from disneyland, KasparBot, Primal Zerg, Oserosteve, Ed hernandez, ChrisGuinvlx268, Celtics233, Singular Near, ViktoriaAnselm, Dr.Showclinie, Mschnuckler1, Mrbigknobweeknob, Myrrh128 and Anonymous: 937

- **Heritability** Source: https://en.wikipedia.org/wiki/Heritability?oldid=713189088 Contributors: Mav, Slrubenstein, Michael Hardy, Doom, Quizkajer, Charles Matthews, Steinsky, Samsara, Phil Boswell, Sam Hocevar, Rich Farmbrough, Guettarda, Shenme, Kjkolb, Vanished user 19794758563875, Sam Korn, Officiallyover, Waabu, Kzollman, Lgallindo, Harkenbane, RichardWeiss, BD2412, Rjwilmsi, Klortho, Wobble, Gurch, Pete.Hurd, Avilella, Jadon, Pseudomonas, Maunus, Darrel francis, SmackBot, BahramH, TimBentley, Jprg1966, Marcosantezana, Lassefolkersen, TedE, Infovoria, SashatoBot, BRFrank, Tim bates, Grumpyyoungman01, Beetstra, JoeBot, JRSpriggs, CmdrObot, Agathman, CBM, Memills, Gregor Strasser, Headbomb, Joymmart, WinBot, Danger, Ephery, Sonicsuns, JPG-GR, Genetics411, J.delanoy, Berkeley99, Extransit, Ben Skála, NiallB-enwiki, Guillaume2303, T0mpr1c3, Temporaluser, AlleborgoBot, Dan Polansky, Leeni2000, Erwinloh, Victor Chmara, Tractorboy60, Niceguyedc, GoEThe, Brookjhill, Excirial, Skbkekas, Aprock, AnjaManix, Rror, Kembangraps, Addbot, Tedtoal, Yobot, Amirobot, JackieBot, FangedFaerie, Citation bot, Philngo, FrescoBot, Citation bot 1, Natisto, The.megapode, Brionthorpe, Trappist the monk, Dmoskal, Duoduoduo, Netwalker 88, Tesseract2, Slightsmile, Thecheesykid, WeijiBaikeBianji, Erianna, Tijfo098, Miradre, ClueBot NG, W.andrea, Eoxenford, Vforget, Dexbot, DaybreakGD, Juntilla87, Occurring, Monkbot, TerryAlex, Evol&Glass, KasparBot, Abidemi37 and Anonymous: 99

- **Cell fusion** Source: https://en.wikipedia.org/wiki/Cell_fusion?oldid=694692600 Contributors: Wouterstomp, Stemonitis, GregorB, BD2412, DeadEyeArrow, Xiagu, Lambiam, Rigadoun, Heimstern, Ben Moore, Mr. Vernon, Alaibot, Dancanm, Magioladitis, CommonsDelinker, KylieTastic, Arjun.theone, Escape Orbit, Mr. Granger, DumZiBoT, Addbot, Tcncv, Blethering Scot, Luckas-bot, FoxBot, Amkilpatrick, Microtubules, John of Reading, WeigelaPen, ClueBot NG, BG19bot, Mogism, Dmitry Dzhagarov, FiredanceThroughTheNight, Nickiemckinnon, LibraryStudent24, Blurred Lines, Alexiaarmstrong and Anonymous: 16

- **Hybrid (biology)** Source: https://en.wikipedia.org/wiki/Hybrid_(biology)?oldid=711552642 Contributors: Chexum, The Anome, Rmhermen, SimonP, B4hand, Patrick, Michael Hardy, Lexor, Kku, Matthewmayer, Tannin, Sannse, Skysmith, Ahoerstemeier, Andrewa, Glenn, Evercat, Cherkash, Conti, Andrewman327, Samsara, Wetman, GPHemsley, Dimadick, Robbot, Altenmann, Ashley Y, Academic Challenger, Nilmerg, Sarexpert, Hadal, GerardM, Bene, Rho-enwiki, JamesMLane, MPF, Elf, Lupin, Everyking, Niteowlneils, Chinasaur, Henryhartley, Yekrats, Jason Quinn, Falcon Kirtaran, Christopherlin, John Abbe, Uteursch, Mendel, Sreyan, Yath, Antandrus, JoJan, Lesgles, PDH, Taka, Now3d,

Shotwell, Quill, Heegoop, DanielCD, Cfaikle, Xezbeth, BCwine, Dbachmann, Lachatdelarue, Bender235, Dara, Jpgordon, Dustinasby, Bobo192, Duk, R. S. Shaw, Kjkolb, Nk, The Recycling Troll, Pschemp, Hesperian, Alansohn, Gary, Anthony Appleyard, Plumbago, Kurt Shaped Box, Weft, Malo, RPellessier, Pioneer-12, Peter McGinley, Talkie tim, BerserkerBen, Zntrip, Jeffrey O. Gustafson, Bellhalla, Consequencefree, Samsoncity, SP-KP, Ratzer, Prashanthns, RichardWeiss, BD2412, Galwhaa, OGRastamon, FreplySpang, RxS, Rjwilmsi, Nightscream, Phileas, Jake Wartenberg, Band B, Kalogeropoulos, Rui Silva, Zaurus, Margosbot~enwiki, Nivix, Ewlyahoocom, Pete.Hurd, LeCire~enwiki, BradBeattie, Imnotminkus, Smithbrenon, Chobot, DTOx, Bjwebb, Bgwhite, Dj Capricorn, YurikBot, Wavelength, Rtkat3, RussBot, Sarranduin, Ytrottier, Gaius Cornelius, Bovineone, Curtis Clark, Dysmorodrepanis~enwiki, BigCow, Snek01, Voyevoda, Meekrob, Ladywolf13, Irishguy, Arastep, Zephalis, DRosenbach, Haemo, Nlu, Phgao, 2over0, Aremisasling, Arthur Rubin, E Wing, SMcCandlish, Noodleman, Sarefo, NielsenGW, Johapseudo, Allens, Kungfuadam, CIreland, Crystallina, SmackBot, Victor M, Vicente Selvas, Brya, JohnPomeranz, C.Fred, Vald, KocjoBot~enwiki, Eskimbot, ASarnat, Hardyplants, Edgar181, Commander Keane bot, Skizzik, Rmosler2100, Aaadddaaammm, Improbcat, Kurykh, Rkitko, Persian Poet Gal, Jprg1966, PrimeHunter, Silly rabbit, Octahedron80, Muboshgu, Skinrider, Frap, Alexmcfire, Messybeast, Mr.Z-man, Khoikhoi, Korako, Nakon, Michaelrccurtis, Hgilbert, Abbott75, Lisasmall, Zeamays, Alan G. Archer, Olsdude, SashatoBot, Lambiam, Eliyak, Staalmannen, Titus III, Rigadoun, Jefe619, Pskykosys, Mgiganteus1, Jer13, Ocatecir, Aleenf1, Andypandy.UK, Optakeover, Idon'texist, Peter Horn, SmokeyJoe, Iridescent, NEMT, Newone, Chika11, Tortfeasor, Abyss42, CmdrObot, Lavateraguy, Sashag, Ben 10, Dgw, Rabid Lemur, Iokseng, Montanabw, Funnyfarmofdoom, Sopoforic, Michaelas10, Gogo Dodo, Jayen466, Tkynerd, Tawkerbot4, Christian75, UberScienceNerd, Crowmanyclouds, Smeazel, JamesAM, Thijs!bot, Epbr123, Mojo Hand, Headbomb, MattWatt, SusanLesch, Mentifisto, David D., AntiVandalBot, Seaphoto, KP Botany, Gfroelich, Clamster5, JAnDbot, Natureguy1980, Papa Lima Whiskey, Magioladitis, WolfmanSF, VoABot II, Steven Walling, Animum, Americanhero, Jacobko, Faro0485, Finiteyoda, Glen, JaGa, Tclowers, Salopian, Defenestrating Monday, Barhamd, Scottalter, Atulsnischal, MartinBot, Agricolae, Anaxial, Themania, R'n'B, Levin-bj84, Dr Almost, Lilac Soul, Petter Bøckman, J.delanoy, Adavidb, Phewkin, Colincbn, Octopus-Hands, MrMR78, Melanochromis, EWM27, Dr d12, Mikael Häggström, Skier Dude, TheTyrant, SuzanneKn, Plasticup, Chiswick Chap, Toast222, Milani2, GOTMILK555, Fences and windows, Barneca, TXiKiBoT, Pirxhh, LeaveSleaves, Raymondwinn, BotKung, Maxim, Doug, Roland Kaufmann, @pple, Enviroboy, Fredtheflyingfrog, Fanatix, SieBot, Carlpiercey, Mikememike, VVVBot, Arda Xi, Xenophon777, Flyer22 Reborn, Antonio Lopez, Lightmouse, Fratrep, BfxO, Chrisrus, Sibonakaliso, Bkhaworth, Crioca, Animeronin, ClueBot, Jiří Janoušek, Niceguyedc, Namazu-tron, Richard B. Frost, Pumpmeup, Heathmoor, Vroomfundel, Smiley Boy, KC109, Shinkolobwe, Daedalus Cz, SchreiberBike, Rui Gabriel Correia, ForestDim, Rjbesquire, Dana boomer, Berean Hunter, Marco Liverpool, Chhe, DumZiBoT, Koumz, Jytdog, Koolokamba, Skarebo, Nicolae Coman, Hotbasketdude09, MystBot, Joshnadler, Kajabla, Addbot, Redneck Wizard, Crackaddict221, Fieldday-sunday, Bte99, Download, TimTomTom, Hybrid Dragon, Flakinho, Lightbot, OlEnglish, Jan eissfeldt, Jarble, Drpickem, Yobot, EdwardLane, 2D, Theserialcomma, Anypodetos, Suntag, AnomieBOT, Galoubet, Materialscientist, Citation bot, Deedeedee Manik, Obersachsebot, Xqbot, Capricorn42, DSisyphBot, Hanberke, Josiah117, -), Gtfobabe1993, Nomorelogic, RibotBOT, DDennisM, Jordangillespie, WaysToEscape, Joaquin008, Taka76, FrescoBot, Lothar von Richthofen, Citation bot 1, Pinethicket, I dream of horses, Jonesey95, Kibi78704, Trappist the monk, Jonkerz, Clarkcj12, DragonofFire, CobraBot, Obsidian Soul, Aircorn, Superk1a, EmausBot, WikitanvirBot, Gfoley4, Look2See1, Henrypenn1, Chatpasha, Auró, Thecheesykid, Listmeister, Jargoness, E557, Kelsklan, Medeis, AManWithNoPlan, Tolly4bolly, Justwl, Deutschgirl, Jasonz2z, Cymbelmineer, Shockbolt, Autodidact1, ClueBot NG, Morgankevinj huggle, Jarebika, Goose friend, Frietjes, Old wombat, Breogan2008, Rezabot, K.Darcy, Widr, Helpful Pixie Bot, Wbm1058, Plantdrew, Kailash29792, BG19bot, Dean001, Rm1271, CitationCleanerBot, Glacialfox, Klilidiplomus, Anbu121, BattyBot, Darorcilmir, Malnutrition1234, Jethro B, Dissident93, FoCuSandLeArN, Sminthopsis84, Mogism, Hair, Snarffe, Ueutyi, CsDix, FrigidNinja, Melonkelon, FiredanceThroughTheNight, BanunterX, Ac130hgunr, The Herald, Pewendland, AfadsBad, Ruppia2000, Kzyoung, Golden Bosnian Lily, 17DoTrinhWilliam, Melcous, Fafnir1, Monkbot, Lvpsa1, Crystallizedcarbon, Myth420, Drpetermj, Milap1404, KasparBot, MAX (VIDU!!), Carbon540, Kalum12345678910, Mjohnson55, St-louis14, Duchamp 7, Duggirala shanmukha and Anonymous: 460

- **Recombinant DNA** *Source:* https://en.wikipedia.org/wiki/Recombinant_DNA?oldid=711859459 *Contributors:* Malcolm Farmer, Michael Hardy, Lexor, Mxn, Lfh, Markhurd, Isopropyl, Giftlite, Tweenk, Chowbok, Pgan002, PDH, Thorwald, Discospinster, Zombiejesus, Nina Gerlach, Bender235, El C, John Vandenberg, Viriditas, Adrian~enwiki, Giraffedata, Alansohn, Viridian, Bart133, Zereshk, Richard Arthur Norton (1958-), Woohookitty, Mindmatrix, ApLundell, Graham87, Ansend, Tangotango, Bruce1ee, Davelong, FlaBot, Naraht, TeaDrinker, Chobot, Helios, Gwernol, FrankTobia, YurikBot, Wavelength, Anomalocaris, Moe Epsilon, Kkmurray, Arthur Rubin, Kubra, SmackBot, Froren, Xephael, TestPilot, Pgk, Adrian232, Michaelll, Bryan Nguyen, Gilliam, Oscarthecat, Kurykh, Buermann, DHN-bot~enwiki, Colonies Chris, A. B., Zirconscot, Dreadstar, Drphilharmonic, DMacks, Madeleine Price Ball, Victor D, AmiDaniel, Buchanan-Hermit, Gobonobo, Mgiganteus1, Ben Moore, Jon186, Toastthemost, Jwalte04, Martaous, Ziusudra, Tawkerbot2, BBuchbinder, Ryt, Tifego, CmdrObot, Agathman, Mato, TicketMan, Daniel J. Leivick, Carstensen, Christian75, Pcbene, LachlanA, AntiVandalBot, Luna Santin, 8ohm, Lfstevens, Hotfneier, Res2216firestar, MER-C, RedSharpie, Greensburger, PhilKnight, TransControl, VoABot II, Sstrumello, Chris G, J.delanoy, Maurice Carbonaro, Rod57, Mikael Häggström, Crazypaco, Adawar, Enix150, Juliancolton, RB972, Amkered, Wkpurves, Noformation, MrChupon, Minestrone Soup, SieBot, Oogabooga007, Demantos, Formerly the IP-Address 24.22.227.53, The Evil Spartan, Rachelkroll, Ufinne, Nancy, Maz 255, Mygerardromance, WikiLaurent, Der Rabe Ralf, Martarius, ClueBot, Phoenix-wiki, GorillaWarfare, The Thing That Should Not Be, EoGuy, Arakunem, Gkrajeshrajesh, Michaplot, Ernstblumberg, Excirial, Jusdafax, Vendeka, Jaob, Manderson198, Edwin Okli, 96well, Bbelmont, Maederv15h, Vanished User 1004, DumZiBoT, XLinkBot, Jytdog, Alexius08, Erikamit, Addbot, T.c.w7468, Blechnic, MrOllie, Chzz, Quercus solaris, Tide rolls, Wing, Luckas-bot, Yobot, Vshallrock, AnomieBOT, Apollo1758, Keithbob, Materialscientist, Citation bot, ArthurBot, LilHelpa, Naveensky, Pmlineditor, Peterish, Friedmanfr0, DivineAlpha, Citation bot 1, Pinethicket, I dream of horses, Hallerifreak, Jim Bynum, Amkilpatrick, Telefunkentelefunken, RjwilmsiBot, Aircorn, Salvio giuliano, EmausBot, Tommy2010, Deirovic, JSquish, John Mackenzie Burke, Bosseditor, Tolly4bolly, Donner60, ChuispastonBot, Sharonmil, DASHBotAV, Lv131, ClueBot NG, Widr, Swmmr1928, Helpful Pixie Bot, MusikAnimal, Tyrael123, DaHuzyBru, Klilidiplomus, Kitlit, Hghyux, Stigmatella aurantiaca, Saltwolf, MadGuy7023, Hanauc, Laksh1903, Dexbot, Gabelglesia, Joeinwiki, Epicgenius, JPaestpreornJeolhlna, Evolution and evolvability, The Herald, Vikasna, Ginsuloft, LibraryStudent24, Elizabeth sunny, LauChingYiAngel, Damon4salvatore, Balapasapugazh, Ryan115, Giancarlobasile, Kaseypeesho, Keslingmj, ACB Smith, Saugat Sawin and Anonymous: 307

- **Vector (molecular biology)** *Source:* https://en.wikipedia.org/wiki/Vector_(molecular_biology)?oldid=708581118 *Contributors:* Mike Rosoft, Laklare, Abanima, Woohookitty, BD2412, Jakg, SmackBot, TestPilot, Chris the speller, Ulcha, Stevenrosson, RolandR, Fremte, Freecat, Myasuda, Kanags, RelentlessRecusant, Ebrahim, Anupam, Tstrobaugh, WhatamIdoing, Mikael Häggström, Mercurywoodrose, MajorHazard, Yerpo, Hzh, Sunrise, Mac'ero, Trabelsiismail, Addbot, Mimarob, Vedran12, TaBOT-zerem, Br1answanson, AdmiralHood, Howard McCay, Wireless Keyboard, Jeangabin, Jesse V., EmausBot, John Mackenzie Burke, Bill william compton, Rmashhadi, ClueBot NG, MerllwBot, Helpful Pixie

Bot, Aks23121990, Lugia2453, 路路, Ali Kafroo, Enigmatore and Anonymous: 39

- **Transcription (genetics)** *Source:* https://en.wikipedia.org/wiki/Transcription_(genetics)?oldid=711392670 *Contributors:* Mav, The Anome, Michael Hardy, Zashaw, Lexor, Kku, Menchi, Ronz, JWSchmidt, Александър, Mxn, Ec5618, Fuzheado, Steinsky, Phil Boswell, Kzhr, Timemutt, VanishedUser kfljdfjsg33k, Giftlite, Michael Devore, Bensaccount, Duncharris, Antandrus, G3pro, Oneiros, PFHLai, Grunt, Archer3, DanielCD, Discospinster, Mgtoohey, Pabloes, Perfecto, AKGhetto, Tmh, Arcadian, Sriram sh, Srlasky, Haham hanuka, Alansohn, Terrycojones, SI, Wouterstomp, Riana, Seans Potato Business, Batmanand, Helixblue, Tycho, Amorymeltzer, GabrielF, Ceyockey, RyanGerbil10, Brookie, Woohookitty, Mindmatrix, EnSamulili, Fbv65edel, Mms, Jclemens, Dpv, Sjakkalle, Rjwilmsi, FlaBot, Margosbot~enwiki, Vossman, DVdm, Fenoxielo, Roboto de Ajvol, Wavelength, Postglock, WAvegetarian, Hede2000, Rosieredfield, Shanel, Rmky87, Bucketsofg, WAS 4.250, Joan-neB, Curpsbot-unicodify, Teply, Mengxu, Hughitt1, SmackBot, TestPilot, Martin.Budden, Hydrogen Iodide, Geno-Supremo, Apers0n, Zephyris, Yamaguchi先生, Gilliam, BrotherGeorge, Kazkaskazkasako, Evandrix, Aaadddaaammm, MalafayaBot, Uthbrian, Oxhop, Miguel Andrade, DHN-bot~enwiki, Ribrob, Vidric, Khoikhoi, Totophe64~enwiki, Richard001, Drphilharmonic, Kshieh, Ifan160, Kukini, CoeurDeLion, The under-tow, Nishkid64, ArglebargleIV, AThing, Kuru, Epingchris, Bloodpack, IronGargoyle, Ben Moore, MarkSutton, Slakr, Noah Salzman, Irides-cent, Gentlemaan, Peter M Dodge, Kaarel, Cph3992, JForget, Liam Skoda, CmdrObot, OverlordKain, Bonás, Agathman, Sameerbau, Spot-tydog3, Neelix, Opus118, Yided, Tomjc, Nuplex, Nick.wiebe, RelentlessRecusant, Was a bee, Corpx, Smelissali, Narayanese, Vanished User jdksfajlasd, Phi*n!x, Thijs!bot, Opabinia regalis, Mojo Hand, John254, NERIUM, Kickassso, AntiVandalBot, MoogleDan, BokicaK, Eltanin, NightwolfAA2k5, TimVickers, Smartse, MDG38, Neur0X, Legolost, Kswenson, Dcooper, Jullag, Greensburger, .anacondabot, VoABot II, Lea-Hazel, Antorjal, Squidonius, NunoAgostinho, MartinBot, Cvd5012, Rettetast, Anaxial, R'n'B, Qrex123, Huzzlet the bot, HoergerJ, SU Linguist, Lantonov, Rod57, KDSKDS, Jmajeremy, Mikael Häggström, Martyn Axon, Antony-22, Vanished user 39948282, Useight, Zamftb, Llorenzi, G. Völcker, Hammersoft, Saurabh523, Nburden, AlnoktaBOT, Scresawn, Philip Trueman, Segabud, Z.E.R.O., Ilia Kr., Mishlai, Furfurfur, Gillyweed, Enviroboy, Jedietz03, RaseaC, Pjoef, AlleborgoBot, Kehrbykid, PGWG, ASDZXCQWE, SieBot, Nubiatech, Sophos II, Sakkura, Lucasbfrbot, Flyer22 Reborn, Teethies563, Yerpo, Sunrise, Iknowyourider, DaDrought3, Altzinn, Neta90, Brettbarbaro, Forluvoft, ClueBot, Binksternet, Paulabek, DragonBot, Excirial, PixelBot, Skyuppercutt, Cbailey7, Sydney3803, Frigginacky, SchreiberBike, LeighClesterMolar, Thingg, BVBede, Qwfp, Johnuniq, Local hero, XLinkBot, LostLucidity, Feinoha, Vojtěch Dostál, WikHead, Yvorez1274, Drosilia, Hilwhale, Alohascott, Geir.overland, Addbot, DOI bot, CanadianLinuxUser, R-skin, Lehtv, ChenzwBot, Tassedethe, Tide rolls, Gail, Tedtoal, Luckas-bot, Yobot, Ptbotgourou, Berkay0652, TaBOT-zerem, Amirobot, KamikazeBot, The Flying Spaghetti Monster, Legendre17, AnomieBOT, Ker-fuffler, Jim1138, Kingpin13, Hpswimmer, Materialscientist, Citation bot, Maxis ftw, Quebec99, Marshallsumter, Xqbot, GrouchoBot, Bran-don5485, Vikky2904, E0steven, Thehelpfulbot, Rgocs, FrescoBot, Brianwatson94, S73v3n, Citation bot 1, I dream of horses, Foureyes915, Flecha2, A8UDI, RedBot, Mexican9493, Orenburg1, Bursting74, Περίεργος, Ferrari430man, Amkilpatrick, Jcorry10, Gamingmaster125, Tbhotch, Jesse V., Galneon23, RjwilmsiBot, Pjshort42, J36miles, EmausBot, WikitanvirBot, RA0808, Transitiveinstance, Wikipelli, K6ka, 20Lukianto, Grunny, John Mackenzie Burke, EWikist, Wayne Slam, Tropicalpurplekitty, Donner60, BioPupil, Theislikerice, ChuispastonBot, Jeffpkamp, AnnaJune, ClueBot NG, This lousy T-shirt, Rida97, Liney22, Alex-engraver, Mesoderm, Jogmiers, Brynedal, Widr, Ajdavis5, Yasmeh, Jacobso4, Helpful Pixie Bot, Miguelferig, BG19bot, Bths83Cu87Au06, CatPath, Dhp4, MusikAnimal, Cncmaster, Mdebortoli, Bar-ton1234, Cmeclean22, Lolzpvp, Roleren, Kilidiplomus, SIMONCOHEN, BattyBot, TuringMachine17, Jimw338, Sermadison, Iamozy, Kelvin-song, Dexbot, Saba irshad, Mohamed 151995, Avijitarya64, Guma44, Dreesem, Xxkitsune, Limelightmk, FallingGravity, Lemnaminor, Mel-onkelon, Aadharm, Wally 84, NYBrook098, A Certain Lack of Grandeur, Quenhitran, APsCollegeEditor, Slj758, ChelseaE, Mjt3727, Wiki-cology, Iwilsonp, Rai hamid raza kharal, Cyrej, Zhirzh, Mmm053, KasparBot, Tom29739, Husbro, Anwaaraqeel and Anonymous: 521

- **Transformation (genetics)** *Source:* https://en.wikipedia.org/wiki/Transformation_(genetics)?oldid=713234765 *Contributors:* AxelBoldt, Bryan Derksen, Chuq, Lexor, MichaelJanich, Steinsky, Robbot, Mushroom, Alan Liefting, Dmmaus, Strych, PDH, Discospinster, Nina Gerlach, Yersinia~enwiki, Cmdrjameson, Wisdom89, Arcadian, Alansohn, Arthena, Mariabrenna, Xmort~enwiki, RyanGerbil10, Woohookitty, Dun-can.france, Contele de Grozavesti, Isnow, Tokek, Mandarax, Rjwilmsi, Vary, Bruce1ee, Klortho, BobofBobs, Yamamoto Ichiro, WriterHound, Wavelength, Genomancer, Rosieredfield, Anetode, JHCaufield, IceCreamAntisocial, RunOrDie, Allens, Katieh5584, Zav, Teo64x, SmackBot, Eperotao, TestPilot, Gilliam, Kazkaskazkasako, Chris the speller, RDBrown, Deli nk, Huon, TedE, Jared, Celefin, Gobonobo, Seb951, Bendzh, Hu12, Tawkerbot2, Agathman, Raz1el, Michaelas10, Christian75, Thijs!bot, Gharmon, Headbomb, John254, MER-C, Andrewericoleman, Andonic, NighthawkJ, Adrian J. Hunter, Sabedon, CitizenB, R'n'B, Nono64, Fconaway, Obscurans, Melamed katz, Colincbn, Arup Acharjee, Lantonov, Jeyradan, Dr d12, Wwfboy17, Tjhuang88, AndreasJSbot, G. Völcker, Philip Trueman, Bpavel, Technopat, StrangeTaco, Ychastnik APL, Jackfork, KC Panchal, Dclyde, Mykon~enwiki, Kehrbykid, SieBot, Hzh, Coolbrdr, Louismaddox, Anchor Link Bot, Forluvoft, Clue-Bot, PipepBot, Fyyer, Lutetium, Boing! said Zebedee, Peteruetz, Excirial, -Midorihana-, Noneforall, Khauser32353, Versus22, Wnt, Jytdog, Vojtěch Dostál, Little Mountain 5, Avoided, Jaimetex, Addbot, Freakmighty, DOI bot, Tanhabot, Download, Bernstein0275, Glane23, Chen-zwBot, Lightbot, Wikimono111, Ben Ben, Luckas-bot, Yobot, Ptbotgourou, Denispir, InfoCan, Jim1138, Ulric1313, Materialscientist, Citation bot, JFY, Proquence, GrouchoBot, Ejedias, Citation bot 1, Lorem-ipsum, Pinethicket, I dream of horses, Vgcap, Trappist the monk, Vrena-tor, Miracle Pen, DARTH SIDIOUS 2, RjwilmsiBot, Aircorn, Techhead7890, EmausBot, ScottyBerg, Dcirovic, JSquish, H3llBot, Kusemame, AManWithNoPlan, Siesta2003, Lorem Ip, Biosicherheit, Minnsurfur2, ClueBot NG, CaitlinStewart, Iiii I I I, Rkirkbride, Enfcer, Rezabot, Widr, Helpful Pixie Bot, Werbolo, Dannyparton, CitationCleanerBot, Giftedlyarsenine, BattyBot, Qqnome, Cyberbot II, YFdyh-bot, Illusionof-confusion, Dexbot, Lockerpartnr, Marchino61, DavidLeighEllis, Professor Ramaswamy, Aboperculate, The Herald, Notrium, Sprovenzano15, Annaroe13, Jianhui67, JaconaFrere, Monkbot, RK1030 and Anonymous: 230

- **Gene gun** *Source:* https://en.wikipedia.org/wiki/Gene_gun?oldid=702823462 *Contributors:* SimonP, Frecklefoot, Nealmcb, Lexor, Michael-Janich, Beroe, PDH, TheObtuseAngleOfDoom, DanielCD, Nina Gerlach, LindsayH, Raazer, ZayZayEM, Alansohn, Curious1i, Caesura, RJFJR, Xmort~enwiki, Edsmilde, WriterHound, Chris Capoccia, SmackBot, DLH, Gilliam, Xenoglot, Freecat, Zarex, Thijs!bot, TimVickers, Smartse, Ceastill, Tybo09, Lenticel, Nono64, Blew1500, Mikael Häggström, VolkovBot, Ryan032, Bonze blayk, Kmhkmh, Toyota6291, MaCRoEco, Mikemoral, Xenobiologista, Yerpo, Lightmouse, Iknowyourider, ClueBot, Jytdog, Addbot, AnnaFrance, Zorrobot, Luckas-bot, Killiondude, Citation bot, OneHuman, Sumone10154, Jesse V., Hyarmendacil, Aircorn, WikitanvirBot, ClueBot NG, Arnavchaudhary, Me, Myself, and I are Here, Thevideodrome, Shearflyer, Monkbot, Stacie Croquet, RachelBrooks15, Aspera0 and Anonymous: 61

- **Agrobacterium** *Source:* https://en.wikipedia.org/wiki/Agrobacterium?oldid=712704491 *Contributors:* Josh Grosse, Azhyd, Maximus Rex, Ev-eryking, ChicXulub, Onco p53, PDH, TonyW, DMG413, Ken-ichi~enwiki, Rich Farmbrough, Reeve, CanisRufus, Chino, MarcoTolo, BD2412, Rjwilmsi, Daycd, Gdrbot, Bgwhite, WriterHound, Chris Capoccia, Pseudomonas, Brian Crawford, Spitshine, WAS 4.250, SmackBot, Edgar181, Gilliam, Bookworm89, DocKrin, RDBrown, Zeamays, Seb951, Thatcher, Stmrlbs, Herd of Swine, Cslsweller, Thijs!bot, Liquid-aim-bot,

Smartse, DRHagen, Yoda43, Pvosta, CommonsDelinker, G. Völcker, TXiKiBoT, Albval, Dpbalazs–enwiki, Ninjatacoshell, Hellens, Touchstone42, Fadesga, Niceguyede, Jack-A-Roe, DumZiBoT, Jytdog, Lab-oratory, Kembangraps, Addbot, DOI bot, Jncraton, Yobot, AnomieBOT, Citation bot, Xqbot, GrouchoBot, Kosmoceras, RibotBOT, Citation bot 1, Lanulos, Pinethicket, Adlerbot, Aircorn, WikitanvirBot, Tommy2010, AvicBot, Manubot, Daniel-Brown, Citovsky, Cyberbot II, Gusterix, Lukmiel14, Ajpolino, Monkbot, Crystallizedcarbon, RK1030 and Anonymous: 47

- **Transfection** *Source:* https://en.wikipedia.org/wiki/Transfection?oldid=706195970 *Contributors:* William Avery, Booyabazooka, Theresa knott, Steinsky, Jotomicron, Peak, Delta G, Williamb, PDH, Mike Rosoft, Luxdormiens, Bender235, Aman–enwiki, Raazer, Wisdom89, Arcadian, Alansohn, SemperBlotto, Dbenzhuser, Graham87, Rjwilmsi, Orb4peace, FlaBot, Kvas, Mushin, Chris Capoccia, Rosieredfield, Ngorongoro, Kkmurray, Uartseieu, Maristoddard, Chefyingi, SmackBot, Eperotao, IainP, Tyciol, RDBrown, Fuzzform, Deli nk, Zirconscot, BinaryTed, Leskovsek, Freecat, DewaldNoeth, Mammal4, Cydebot, JFreeman, HilJackson, Calvero JP, Thijs!bot, Headbomb, Ph.eyes, Agreene175, Magioladitis, Hb2019, Xtothel, Adrian J. Hunter, Adavidb, Colincbn, Wstefano, Catherine de Burgh, Larryisgood, Bpavel, Tameeria, Delyde, Bform, Hzh, Vincent.Delauzun, ClueBot, Franamax, XiaoliangWang, Ordinaterr, Jlittle78, Jclne0, Clmay77, Addbot, DOI bot, Martina Steiner, Yobot, Tohd8BohaithuGh1, Doceva, Mardueng, AmericanDaveInScotland, Rockypedia, Citation bot, Obersachsebot, MauritsBot, Cpichardo, Howard McCay, Amssd, FrescoBot, Olga Ovcharenko, Citation bot 1, MATra, Jumabe, Jeangabin, Transfectionista, Jesse V., RjwilmsiBot, Biologist2001, Aircorn, EmausBot, ZéroBot, Matthew Angel, Solde9, Minnsurfur2, Widr, Perryhackett, BG19bot, Makecat-bot, Me, Myself, and I are Here, Joe Jirka, The Herald, Alexiaarmstrong, Monkbot, Ajlopez83, Ankit2070, Ankit.sqz, IJ99 and Anonymous: 93

- **Electroporation** *Source:* https://en.wikipedia.org/wiki/Electroporation?oldid=711918921 *Contributors:* AxelBoldt, Magnus Manske, Bryan Derksen, Enchanter, Lexor, Maximus Rex, Fernkes, Robbot, Orangemike, Christopherlin, Alanl, Fys, CanisRufus, Wouterstomp, Seans Potato Business, Gene Nygaard, Pixie, BD2412, Rjwilmsi, Gurch, Chris Capoccia, Kazulanth, Carl Daniels, JHCaufield, Smoddy on Wheels, 2over0, Isoxyl, SmackBot, Zephyris, Tyciol, RDBrown, Jwy, Jared, MatthewBChambers, Leskovsek, Freecat, Thijs!bot, Edal, TimVickers, Nitricoxide, Dauphiné, Magioladitis, MastCell, DGG, Gwern, Wiki wiki1, Verdatum, Colincbn, Lantonov, Stan J Klimas, Emailpritam, VolkovBot, Malljaja, Matasg, Bangalos, Sabri76, Wprlh, DoctorEric, Lab-oratory, Addbot, Jacopo Werther, DOI bot, Abduallah mohammed, Luckas-bot, Yobot, KDS4444, Citation bot, LilHelpa, MDougM, RibotBOT, B.andinsky, FrescoBot, Citation bot 1, Lmp883, Aircorn, EmausBot, Mattcaffrey, AndrewPapp, ClueBot NG, Marcopolo53, Bibcode Bot, U4ealongan, RelCallie, Ekoms, Anrnusna, Alexiaarmstrong, Morlen44., Marcom instituteBCN, Lakepotter and Anonymous: 46

- **Microinjection** *Source:* https://en.wikipedia.org/wiki/Microinjection?oldid=712354595 *Contributors:* Stone, Tobias Bergemann, Dvavasour, DragonflySixtyseven, Vsmith, RJFJR, Graham87, JonMoulton, Kolbasz, Michele Bini, Bhny, Daniel Mietchen, JHCaufield, Stijndon, Arthur Rubin, SmackBot, Gilliam, Bluebot, Ajh20, Jon513, Igilli, Mr. Random, Mgiganteus1, Beetstra, Freecat, Yashgaroth, CmdrObot, ShelfSkewed, Kupirijo, Alaibot, Karuna8, TimVickers, Jiehanchong, Fabrictramp, Pekaje, J.delanoy, Adavidb, Maxim, Synthebot, ClueBot, The Thing That Should Not Be, EoGuy, Sabri76, DumZiBoT, Alboyle, XLinkBot, Addbot, DougsTech, PlankBot, Jazzvibes, AnomieBOT, KDS4444, Jim1138, Qwertycua456, Sandip90, MauritsBot, DivineAlpha, Pinethicket, Microinjection, Aircorn, EmausBot, Aidothebeast, ClueBot NG, Satellizer, BG19bot, Crh23, Nemui10pm, BattyBot, Calere, Me, Myself, and I are Here, QCRI, The Herald, Racer Omega, Coo coo pigeon, Amitpanchal440 and Anonymous: 28

- **Viral transformation** *Source:* https://en.wikipedia.org/wiki/Viral_transformation?oldid=674803346 *Contributors:* Graeme Bartlett, Arcadian, (aeropagitica), Vissvia, Duncan.france, Rjwilmsi, Koavf, SmackBot, Where, Kaarel, Darkmands, Alaibot, Magioladitis, Sabedon, Pekaje, Lantonov, Tameeria, Yobot, James500, Erik9bot, Jesse V., Aircorn, John of Reading, Peter Karlsen, BG19bot, BattyBot, Jamesmcmahon0, Monkbot, Tatabox8, Richarnj, CarpeDiem90 and Anonymous: 9

- **Lipofection** *Source:* https://en.wikipedia.org/wiki/Lipofection?oldid=678947179 *Contributors:* Lexor, Delta G, Wisdom89, A2Kafir, Alansohn, Rjwilmsi, SmackBot, RDBrown, Freecat, Cydebot, Happypatatoes, Lantonov, Dnaguys, FrescoBot, Jeangabin, Aircorn, EmausBot, Monkbot and Anonymous: 10

- **Transgenesis** *Source:* https://en.wikipedia.org/wiki/Transgenesis?oldid=707171450 *Contributors:* Xanzzibar, Discospinster, Rich Farmbrough, Vsmith, Melaen, Ceyockey, RHaworth, Nightryder84, SmackBot, RDBrown, Kikumbob, Gogo Dodo, Hebrides, JustAGal, Prolog, Smartse, Dougher, R'n'B, Idioma-bot, Birczanin, Enviroboy, MaCRoEco, SimonTrew, Genya Avocado, Forest Ash, ClueBot, Unbuttered Parsnip, Roadfish, BOTarate, MystBot, Addbot, Thymo, LaaknorBot, F Notebook, Luckas-bot, Yobot, Fraggle81, Freikorp, AnomieBOT, GrouchoBot, Omnipaedista, FrescoBot, Steve Quinn, Mparu, Aircorn, ZéroBot, Joonoob, Nicepirate, Puffin, ClueBot NG, Snotbot, Brickrosco, Rhall28, Willnabors, Evanbrem, BG19bot, Bonnie13J, Naapple, Webclient101, Calere, Dmitry Dzhagarov, Me, Myself, and I are Here, Hillbillyholiday, Deceiving Gods, Monkbot, Jim-Siduri, Jordanjlatimer, Abcdefghijklmnopqrstuvwxyz01, Jaimelevinrouge and Anonymous: 31

- **Cisgenesis** *Source:* https://en.wikipedia.org/wiki/Cisgenesis?oldid=693973833 *Contributors:* Gabbe, Julesd, Bearcat, RoyBoy, Gerbil, Closedmouth, RDBrown, Thumperward, Dr. Dan, Smartse, Dougher, Nagy, SylviaStanley, ClueBot, Infoeco, Paul sisco, MystBot, Addbot, Yobot, Eugene-elgato, Aircorn, AManWithNoPlan, Sidescural, Famedog, ClueBot NG, Lipton 7777, UnbelievableError, Dexbot, Mogism, Me, Myself, and I are Here, Mvdileo, Monkbot, Henk Schouten, MrZiggle and Anonymous: 14

- **Gene knockout** *Source:* https://en.wikipedia.org/wiki/Gene_knockout?oldid=686215739 *Contributors:* AxelBoldt, SimonP, Axel Driken, Erik Zachte, Lexor, Shyamal, 168..., JWSchmidt, Llull, Wik, Robbot, Edcolins, Gzuckier, Jokestress, Bender235, Adan, Wisdom89, Alansohn, Seans Potato Business, Pixie, Srborlongan, FreplySpang, Wavelength, Mushin, Sillybilly, Hydrargyrum, SmackBot, WookieInHeat, Auton1, Betacommand, Radagast83, Phuzion, Guillaume777, Stanlekub, ShelfSkewed, Thijs!bot, Alphachimpbot, Alberth2, Adrian J. Hunter, MartinBot, Rupeshsrivastava, TechnoFaye, R'n'B, CommonsDelinker, Fconaway, Tikiwont, Sabisteb, Byers1rm, Kuebi, TXiKiBoT, Macdonald-ross, SieBot, SchreiberBike, BOTarate, Dana boomer, DumZiBoT, Addbot, CarTick, Nodar95, Luckas-bot, Yobot, Ptbotgourou, Darx9url, Rubinbot, Jim1138, Outofthedim, ArthurBot, 4twenty42o, RibotBOT, Dr. Esra, FrescoBot, Redrose64, Dinamik-bot, Jesse V., Doughorner, Aircorn, EmausBot, WikitanvirBot, Nwhand, Abergabe, Onmymindxxo, OhJenniferLyn, Isthereanamenotused, CeraBot, HeavyQuark, Graphium, JPaestpreornJeoihlna and Anonymous: 37

- **Gene knockdown** *Source:* https://en.wikipedia.org/wiki/Gene_knockdown?oldid=713137956 *Contributors:* Nina, 168..., Llull, Rpyle731, Beland, Cmdrjameson, Lectonar, Woohookitty, Rjwilmsi, JonMoulton, Avocado, Kerowyn, YurikBot, Rathfelder, Bluebot, -Marcus-, Twr57, Pgr94, Was a bee, Alaibot, Ph.eyes, Magioladitis, MiltonT, Mike.lifeguard, Mcat2, Squids and Chips, TXiKiBoT, Twooars, SieBot, Malcolmxl5, Flyer22 Reborn, Allmightyduck, PipepBot, Unbuttered Parsnip, DragonBot, Londonsista, Jytdog, Addbot, DOI bot, Touch.and.go, Yobot, TaBOT-zerem, Amirobot, AnomieBOT, Jim1138, Materialscientist, Citation bot, MauritsBot, Xqbot, FrescoBot, Pinethicket, Aircorn, ChuispastonBot, BG19bot, Mogism, RI477Guy, Molecularbiology12, Giancarlobasile, Caftaric, SciDragon and Anonymous: 26

35.6.2 Images

35.6.3 Content license